Emotion Intelligence Based on Smart Sensing

Emotion Intelligence Based on Smart Sensing

Editors

Mincheol Whang
Sung Park

MDPI • Basel • Beijing • Wuhan • Barcelona • Belgrade • Manchester • Tokyo • Cluj • Tianjin

Editors
Mincheol Whang
Sangmyung University
Seoul
Korea

Sung Park
Sangmyung University
Seoul
Korea

Editorial Office
MDPI
St. Alban-Anlage 66
4052 Basel, Switzerland

This is a reprint of articles from the Special Issue published online in the open access journal *Sensors* (ISSN 1424-8220) (available at: https://www.mdpi.com/journal/sensors/special_issues/smart_sens).

For citation purposes, cite each article independently as indicated on the article page online and as indicated below:

LastName, A.A.; LastName, B.B.; LastName, C.C. Article Title. *Journal Name* **Year**, *Volume Number*, Page Range.

ISBN 978-3-0365-6646-7 (Hbk)
ISBN 978-3-0365-6647-4 (PDF)

© 2023 by the authors. Articles in this book are Open Access and distributed under the Creative Commons Attribution (CC BY) license, which allows users to download, copy and build upon published articles, as long as the author and publisher are properly credited, which ensures maximum dissemination and a wider impact of our publications.

The book as a whole is distributed by MDPI under the terms and conditions of the Creative Commons license CC BY-NC-ND.

Contents

About the Editors . . . . **vii**

Sung Park and Mincheol Whang
Special Issue "Emotion Intelligence Based on Smart Sensing"
Reprinted from: *Sensors* **2023**, *23*, 1098, doi:10.3390/s23031098 . . . . **1**

Arturas Kaklauskas, Ajith Abraham, Ieva Ubarte, Romualdas Kliukas, Vaida Luksaite, Arune Binkyte-Veliene, et al.
A Review of AI Cloud and Edge Sensors, Methods, and Applications for the Recognition of Emotional, Affective and Physiological States
Reprinted from: *Sensors* **2022**, *22*, 7824, doi:10.3390/s22207824 . . . . **5**

Pedro Martins, José Silvestre Silva and Alexandre Bernardino
Multispectral Facial Recognition in the Wild
Reprinted from: *Sensors* **2022**, *22*, 4219, doi:10.3390/s22114219 . . . . **85**

Seo-Jeon Park, Byung-Gyu Kim and Naveen Chilamkurti
A Robust Facial Expression Recognition Algorithm Based on Multi-Rate Feature Fusion Scheme
Reprinted from: *Sensors* **2021**, *21*, 6954, doi:10.3390/s21216954 . . . . **103**

Hyunwoong Ko, Kisun Kim, Minju Bae, Myo-Geong Seo, Gieun Nam, Seho Park, et al.
Changes in Computer-Analyzed Facial Expressions with Age
Reprinted from: *Sensors* **2021**, *21*, 4858, doi:10.3390/s21144858 . . . . **129**

Sung Park, Seong Won Lee and Mincheol Whang
The Analysis of Emotion Authenticity Based on Facial Micromovements
Reprinted from: *Sensors* **2021**, *21*, 4616, doi:10.3390/s21134616 . . . . **143**

Kyoung Ju Noh, Chi Yoon Jeong, Jiyoun Lim, Seungeun Chung, Gague Kim, Jeong Mook Lim and Hyuntae Jeong
Multi-Path and Group-Loss-Based Network for Speech Emotion Recognition in Multi-Domain Datasets
Reprinted from: *Sensors* **2021**, *21*, 1579, doi:10.3390/s21051579 . . . . **161**

Jing Zhang, Sung Park, Ayoung Cho and Mincheol Whang
Recognition of Emotion by Brain Connectivity and Eye Movement
Reprinted from: *Sensors* **2022**, *22*, 6736, doi:10.3390/s22186736 . . . . **179**

Anmol Gupta, Gourav Siddhad, Vishal Pandey, Partha Pratim Roy and Byung-Gyu Kim
Subject-Specific Cognitive Workload Classification Using EEG-Based Functional Connectivity and Deep Learning
Reprinted from: *Sensors* **2021**, *21*, 6710, doi:10.3390/s21206710 . . . . **197**

Xiaoliang Zhu, Wenting Rong, Liang Zhao, Zili He, Qiaolai Yang, Junyi Sun and Gendong Liu
EEG Emotion Classification Network Based on Attention Fusion of Multi-Channel Band Features
Reprinted from: *Sensors* **2022**, *22*, 5252, doi:10.3390/s22145252 . . . . **217**

Laura Gutiérrez-Martín, Elena Romero-Perales, Clara Sainz de Baranda Andújar, Manuel F. Canabal-Benito, Gema Esther Rodríguez-Ramos, Rafael Toro-Flores, et al.
Fear Detection in Multimodal Affective Computing: Physiological Signals versus Catecholamine Concentration
Reprinted from: *Sensors* **2022**, *22*, 4023, doi:10.3390/s22114023 . . . . **235**

Ayoung Cho, Sung Park, Hyunwoo Lee and Mincheol Whang
Non-Contact Measurement of Empathy Based on Micro-Movement Synchronization
Reprinted from: *Sensors* **2021**, *21*, 7818, doi:10.3390/s21237818 . . . . **263**

Geesung Oh, Euiseok Jeong, Rak Chul Kim, Ji Hyun Yang, Sungwook Hwang, Sangho Lee and Sejoon Lim
Multimodal Data Collection System for Driver Emotion Recognition Based on Self-Reporting in Real-World Driving
Reprinted from: *Sensors* **2022**, *22*, 4402, doi:10.3390/s22124402 . . . . **277**

Jung-Yong Kim, Hea-Sol Kim, Dong-Joon Kim, Sung-Kyun Im and Mi-Sook Kim
Identification of Video Game Addiction Using Heart-Rate Variability Parameters
Reprinted from: *Sensors* **2021**, *21*, 4683, doi:10.3390/s21144683 . . . . **299**

Sung Park, Si Pyoung Kim and Mincheol Whang
Individual's Social Perception of Virtual Avatars Embodied with Their Habitual Facial Expressions and Facial Appearance
Reprinted from: *Sensors* **2021**, *21*, 5986, doi:10.3390/s21175986 . . . . **313**

About the Editors

Mincheol Whang

Mincheol Whang received the M.S. and Ph.D. degrees in biomedical engineering from the Georgia Institute of Technology, Atlanta, GA, USA, in 1990 and 1994, respectively. Since March 1998, he has been a Professor in the Department of Emotion Engineering, Graduate School, Sangmyung University, Seoul, South Korea. He has published more than 400 academic papers in human–computer interaction, emotion engineering, human factors, and bioengineering and is currently the dean of College of Convergence Engineering, Sangmyung University.

Sung Park

Sung Park received the M.S. degree in HCI from the University of Michigan, Ann Arbor in 2003 and Ph.D. degree in engineering psychology from the Georgia Institute of Technology, Atlanta in 2009. Since 2009, he has been a senior UX designer and UX group leader at Samsung Electronics and SK Telecom. He has led the UX design of voice recognition AI speakers and social robots. He has been a Professor in SCAD (Savannah College of Art and Design) since 2019 before joining Sangmyung University in 2022. His research interest includes human–computer interaction, emotion engineering, human factors, and artificial intelligence.

Editorial

Special Issue "Emotion Intelligence Based on Smart Sensing"

Sung Park [1] and Mincheol Whang [2,*]

[1] Department of Emotion Engineering, Sangmyung University, Seoul 03016, Republic of Korea
[2] Department of Human-Centered Artificial Intelligence, Sangmyung University, Seoul 03016, Republic of Korea
* Correspondence: whang@smu.ac.kr

Emotional intelligence is essential to maintaining human relationships in communities, organizations, and societies. By definition, emotional intelligence refers to how well emotion is recognized and expressed. The level of emotional intelligence of an AI is mainly determined by its ability to accurately and reliably recognize its human counterpart; that is, all next-generation AI devices and services involving VR, AR, and social robots are able to quantitatively track and recognize emotion in real-time during an interaction with a human.

Emotion has been quantified by sensing facial expressions, gestures, and physiological signals such as EEG, ECG, and EDA. In addition, emotion could be more accurately recognized by considering the emotional context, including spatiotemporal variability, the congruency of implicit and explicit responses, the consistency of human action, and human relationships in society. Human emotion includes not only short-term but also long-term responses to patterns and trends in daily life. Lab studies with the aim of sensing emotion should extend to smart sensing, which monitors and tracks emotional variation with a predictable pattern.

This Special Issue explores empirical studies of emotional mechanisms, qualitative and quantitative measurements of emotion, the recognition of emotional contexts, and the application of emotion. Fourteen papers were accepted for publication in this Special Issue entitled "Emotion Intelligence Based on Smart Sensing", which includes papers ranging from lab-based studies aimed at understanding emotional mechanisms to applying emotion recognition in the real world (e.g., in driving, games, education, and virtual avatars). They are summarized below.

The review paper in [1] presents a detailed analysis of over 600 papers related to sensors and methods to understand affective-, emotional-, and physiological-state recognition. Facial action coding and facial expression analysis are long-studied fields, as represented by four articles in our SI. While facial recognition systems in the real world (i.e., in an uncontrolled environment) have evolved with performance improvements, [2] proposed a multi-spectral facial recognition system that overcomes the limitation of a single spectral band in the visible spectrum. The multi-spectral facial recognition system is robust to occlusions (e.g., fog or plastic materials) and low- or no-light environments. The authors of [2] achieved 99.5% (pose variation) and 99.6% (expression variation) Rank-1 scores in the TUFTS multi-spectral database. As AI technology evolves rapidly, so does the facial expression recognition algorithm. The authors of [3] proposed a multi-depth network that classifies facial expressions by being fed reinforced features. A multi-rate-based 3D convolutional neural network (CNN) built on a multi-rate signal process scheme was suggested, and they achieved 96.23% accuracy with the CK+ dataset.

Building an emotionally intelligent system requires a better understanding of human facial expression characteristics. The authors of [4] investigated the differences in the intensity of facial expressions between older (n = 56) and younger adults (n = 113). The participants' facial expressions were elicited using facial expression stimuli. The results indicated that the older adults strongly expressed some negative and neutral emotions. In addition, older adults used more facial muscles than younger adults across emotions.

Citation: Park, S.; Whang, M. Special Issue: "Emotion Intelligence Based on Smart Sensing". *Sensors* 2023, 23, 1098. https://doi.org/10.3390/s23031098

Received: 13 December 2022
Accepted: 12 January 2023
Published: 18 January 2023

Copyright: © 2023 by the authors. Licensee MDPI, Basel, Switzerland. This article is an open access article distributed under the terms and conditions of the Creative Commons Attribution (CC BY) license (https://creativecommons.org/licenses/by/4.0/).

Human facial expressions include facial micromovements, which provide insights into fake expressions. The authors of [5] investigated the characteristics of real and fake facial expressions representing emotions by analyzing participants' facial micromovements. The results indicated significant differences in the micromovement feature variables between the real and fake expression conditions. The differences varied according to facial regions as a function of emotions.

This issue also includes a speech-emotion-recognition study [6] that proposed a multi-path and group-loss-based network (MPGLN) for emotion recognition to support multi-domain adaptation. The authors proposed a model that includes a bidirectional long short-term memory-based temporal feature generator and a transferred feature extractor from the pre-trained VGG-like audio classification model (VGGish). The model learns simultaneously based on multiple losses according to the association of emotion labels in the discrete and dimensional models.

The simultaneous activation of brain regions (i.e., brain connection features) is an essential mechanism of brain activity in emotion recognition, and this issue presents three EEG-based studies that advance such science. The authors of [7] investigated the relationship between brain connectivity (strength and directionality) and eye movement features (left and right pupils, saccades, and fixations) when participants (n = 47) viewed emotion-eliciting content. They found that the connectivity eigenvalues of the long-distance prefrontal lobe, temporal lobe, parietal lobe, and center were related to cognitive activity involving high valance. In addition, saccade movement was correlated with long-distance occipital–frontal connectivity. The authors of [8] investigated model-free functional connectivity metrics along with deep learning to efficiently classify human cognitive workload. They achieved state-of-the-art multi-class classification accuracy of 80.87% using a combination of MI (Mutual Information) and CNN, followed by 75.88% using a combination of PLV (Phase Locking Value) and CNN (at), and 71.87% using MI with LSTM. The authors of [9] constructed a learning emotion EEG dataset (LE-EEG) which captures physiological signals reflecting the emotions of boredom, neutrality, and engagement during learning, and proposed an EEG emotion classification network based on attention fusion (ECN-AF). On the LE-EEG dataset, the proposed model achieved the highest accuracy of 95.87%, demonstrating a 21.49% increase compared to the baseline models.

Biological hormones are relatively less explored, but could provide insights into negative emotions such as fear or panic. The authors of [10] investigated catecholamines, which are hormones released in the body in response to physical or emotional stress. They analyzed physiological signals in reference to catecholamine through an experimental task whereby 21 female volunteers received audiovisual stimuli through an immersive virtual-reality environment.

The essence of emotional intelligence overlaps with empathy, a psychological construct. A system that analyzes whether a human is empathizing is paramount. The authors of [11] suggested a non-contact method for measuring empathy by evaluating the synchronization of facial micro-movements between consumers and people in the media. Their study shows that the non-contact ballistocardiography (BCG) method can be complementary to subjective empathy scales.

Finally, this issue also extends to studies applicable to the real world (e.g., in driving, games, and virtual agents). The authors of [12] proposed a data collection system that collects multimodal emotion datasets during real-world driving. The proposed system includes a self-reportable HMI application into which a driver directly inputs their current emotion state. To demonstrate the collected dataset's validity, the paper provides case studies for statistical analysis, driver face detection, and personalized driver emotion recognition. The authors of [13] used electrocardiograms (ECGs) to investigate heart rate variability (HRV) parameters that can quantitatively characterize game addiction. The participants played the game *League of Legends*, and the experimenter performed ECG measurements during the game at various window sizes and specific events. The correlation and factor analyses were used to find the most effective parameters. The most accurate set of parameters was found to be pNNI20, RMSSD, and LF within 30 s after the "being killed" event. The authors of [14] investigated elements that may affect a the participant's social perceptions (similarity, familiarity, attraction, liking, and involvement) of customized virtual avatars engineered considering the user's facial characteristics. The results indicated that participants felt that the avatar that embodied their habitual expressions was more similar to them than the avatar that did not.

Funding: This research received no external funding.

Conflicts of Interest: The authors declare no conflict of interest.

References

1. Kaklauskas, A.; Abraham, A.; Ubarte, I.; Kliukas, R.; Luksaite, V.; Binkyte-Veliene, A.; Vetloviene, I.; Kaklauskiene, L. A Review of AI Cloud and Edge Sensors, Methods, and Applications for the Recognition of Emotional, Affective and Physiological States. *Sensors* **2022**, *22*, 7824. [CrossRef] [PubMed]
2. Martins, P.; Silva, J.S.; Bernardino, A. Multispectral Facial Recognition in the Wild. *Sensors* **2022**, *22*, 4219. [CrossRef] [PubMed]
3. Park, S.-J.; Kim, B.-G.; Chilamkurti, N. A Robust Facial Expression Recognition Algorithm Based on Multi-Rate Feature Fusion Scheme. *Sensors* **2021**, *21*, 6954. [CrossRef] [PubMed]
4. Ko, H.; Kim, K.; Bae, M.; Seo, M.-G.; Nam, G.; Park, S.; Park, S.; Ihm, J.; Lee, J.-Y. Changes in Computer-Analyzed Facial Expressions with Age. *Sensors* **2021**, *21*, 4858. [CrossRef]
5. Park, S.; Lee, S.W.; Whang, M. The Analysis of Emotion Authenticity Based on Facial Micromovements. *Sensors* **2021**, *21*, 4616. [CrossRef]
6. Noh, K.J.; Jeong, C.Y.; Lim, J.; Chung, S.; Kim, G.; Lim, J.M.; Jeong, H. Multi-Path and Group-Loss-Based Network for Speech Emotion Recognition in Multi-Domain Datasets. *Sensors* **2021**, *21*, 1579. [CrossRef]
7. Zhang, J.; Park, S.; Cho, A.; Whang, M. Recognition of Emotion by Brain Connectivity and Eye Movement. *Sensors* **2022**, *22*, 6736. [CrossRef] [PubMed]
8. Gupta, A.; Siddhad, G.; Pandey, V.; Roy, P.P.; Kim, B.-G. Subject-Specific Cognitive Workload Classification Using EEG-Based Functional Connectivity and Deep Learning. *Sensors* **2021**, *21*, 6710. [CrossRef] [PubMed]
9. Zhu, X.; Rong, W.; Zhao, L.; He, Z.; Yang, Q.; Sun, J.; Liu, G. EEG Emotion Classification Network Based on Attention Fusion of Multi-Channel Band Features. *Sensors* **2022**, *22*, 5252. [CrossRef] [PubMed]
10. Gutiérrez-Martín, L.; Romero-Perales, E.; de Baranda Andújar, C.S.F.; Canabal-Benito, M.; Rodríguez-Ramos, G.E.; Toro-Flores, R.; López-Ongil, S.; López-Ongil, C. Fear Detection in Multimodal Affective Computing: Physiological Signals versus Catecholamine Concentration. *Sensors* **2022**, *22*, 4023. [CrossRef] [PubMed]
11. Cho, A.; Park, S.; Lee, H.; Whang, M. Non-Contact Measurement of Empathy Based on Micro-Movement Synchronization. *Sensors* **2021**, *21*, 7818. [CrossRef] [PubMed]
12. Oh, G.; Jeong, E.; Kim, R.C.; Yang, J.H.; Hwang, S.; Lee, S.; Lim, S. Multimodal Data Collection System for Driver Emotion Recognition Based on Self-Reporting in Real-World Driving. *Sensors* **2022**, *22*, 4402. [CrossRef] [PubMed]
13. Kim, J.-Y.; Kim, H.-S.; Kim, D.-J.; Im, S.-K.; Kim, M.-S. Identification of Video Game Addiction Using Heart-Rate Variability Parameters. *Sensors* **2021**, *21*, 4683. [CrossRef] [PubMed]
14. Park, S.; Kim, S.P.; Whang, M. Individual's Social Perception of Virtual Avatars Embodied with Their Habitual Facial Expressions and Facial Appearance. *Sensors* **2021**, *21*, 5986. [CrossRef] [PubMed]

Disclaimer/Publisher's Note: The statements, opinions and data contained in all publications are solely those of the individual author(s) and contributor(s) and not of MDPI and/or the editor(s). MDPI and/or the editor(s) disclaim responsibility for any injury to people or property resulting from any ideas, methods, instructions or products referred to in the content.

Review

A Review of AI Cloud and Edge Sensors, Methods, and Applications for the Recognition of Emotional, Affective and Physiological States

Arturas Kaklauskas [1,*], Ajith Abraham [2], Ieva Ubarte [3], Romualdas Kliukas [4], Vaida Luksaite [1], Arune Binkyte-Veliene [3], Ingrida Vetloviene [1] and Loreta Kaklauskiene [1]

1. Department of Construction Management and Real Estate, Vilnius Gediminas Technical University, Sauletekio Ave. 11, LT-10223 Vilnius, Lithuania
2. Machine Intelligence Research Labs, Scientific Network for Innovation and Research Excellence, Auburn, WA 98071, USA
3. Institute of Sustainable Construction, Vilnius Gediminas Technical University, Sauletekio Ave. 11, LT-10223 Vilnius, Lithuania
4. Department of Applied Mechanics, Vilnius Gediminas Technical University, Sauletekio Ave. 11, LT-10223 Vilnius, Lithuania
* Correspondence: arturas.kaklauskas@vilniustech.lt

Citation: Kaklauskas, A.; Abraham, A.; Ubarte, I.; Kliukas, R.; Luksaite, V.; Binkyte-Veliene, A.; Vetloviene, I.; Kaklauskiene, L. A Review of AI Cloud and Edge Sensors, Methods, and Applications for the Recognition of Emotional, Affective and Physiological States. *Sensors* **2022**, *22*, 7824. https://doi.org/10.3390/s22207824

Academic Editors: Mario Munoz-Organero, Mincheol Whang and Sung Park

Received: 18 August 2022
Accepted: 12 October 2022
Published: 14 October 2022

Publisher's Note: MDPI stays neutral with regard to jurisdictional claims in published maps and institutional affiliations.

Copyright: © 2022 by the authors. Licensee MDPI, Basel, Switzerland. This article is an open access article distributed under the terms and conditions of the Creative Commons Attribution (CC BY) license (https://creativecommons.org/licenses/by/4.0/).

Abstract: Affective, emotional, and physiological states (AFFECT) detection and recognition by capturing human signals is a fast-growing area, which has been applied across numerous domains. The research aim is to review publications on how techniques that use brain and biometric sensors can be used for AFFECT recognition, consolidate the findings, provide a rationale for the current methods, compare the effectiveness of existing methods, and quantify how likely they are to address the issues/challenges in the field. In efforts to achieve the key goals of Society 5.0, Industry 5.0, and human-centered design better, the recognition of emotional, affective, and physiological states is progressively becoming an important matter and offers tremendous growth of knowledge and progress in these and other related fields. In this research, a review of AFFECT recognition brain and biometric sensors, methods, and applications was performed, based on Plutchik's wheel of emotions. Due to the immense variety of existing sensors and sensing systems, this study aimed to provide an analysis of the available sensors that can be used to define human AFFECT, and to classify them based on the type of sensing area and their efficiency in real implementations. Based on statistical and multiple criteria analysis across 169 nations, our outcomes introduce a connection between a nation's success, its number of Web of Science articles published, and its frequency of citation on AFFECT recognition. The principal conclusions present how this research contributes to the big picture in the field under analysis and explore forthcoming study trends.

Keywords: review; human emotions; affective and physiological states; Plutchik's wheel of emotions; sensors; methods and applications; statistical and multiple criteria analysis; country success and publications maps of the world

1. Introduction

Global research in the field of neuroscience and biometrics is shifting toward the widespread adoption of technology for the detection, processing, recognition, interpretation and imitation of human emotions and affective attitudes. Due to their ability to capture and analyze a wide range of human gestures, affective attitudes, emotions and physiological changes, these innovative research models could play a vital role in areas such as Industry 5.0, Society 5.0, the Internet of Things (IoT), and affective computing, among others.

For hundreds of years, researchers have been interested in human emotions. Reviews on the applications of affective neuroscience include numerous related topics, such as the mirror mechanism and its role in action and emotion [1], the neuroscience of under-standing emotions [2], consumer neuroscience [3], the role of positive emotions in education [4], mapping the brain as the basis of feelings and emotions [5], the neuroscience of positive emotions and affect [6], the cognitive neuroscience of music perception [7], and social cognition in schizophrenia [8]. Applications in neuroscience also include the analysis of cognitive neuroscience [9–11], and brain sensors [12,13], and works in the literature also discuss the recognition of basic emotions using brain sensors [14].

Studies of the applications of affective biometrics can be found in the literature in the fields of brain biometric analysis [15], predictive biometrics [16], keystroke dynamics [17], applications in education [18], consumer neuroscience [19], adaptive biometric systems [20], emotion recognition from gait analyses [21], ECG databases [22], and others. Several works on affective states have integrated multiple biometric and neuroscience methods, but none have included an integrated review of the application of neuroscience and biometrics and an analysis of all of the emotions and affective attitudes in Plutchik's wheel of emotions.

Scientists analyzed various brain and biometric sensors in the reviews [23–26]. Curtin et al. [23], for instance, state that both fNIRS and rTMS sensors have changed significantly over the past decade and have been improved (their hardware, neuronavigated targeting, sensors, and signal processing), thus clinicians and researchers now have more granular control over the stimulation systems they use. Krugliak and Clarke [26], da Silva [24], and Gramann et al. [27] analyzed the use of EEG and MEG sensors to measure functional and effective connectivity in the brain. Khushaba et al. [25] used brain and biometric sensors to integrate EEG and eye tracking for assessing the brain response. Other scientists [28–33] used the following biometric sensors in their studies: heart rate, pulse rate variability, odor, pupil dilation and contraction, skin temperature, face recognition, voice, signature, gestures, and others.

Indeed, the biometrics and neuroscience field has been the focus of studies by many researchers who have achieved significant results. A number of neuroscience studies have analyzed the detection and recognition of human arousal [34], valence [35,36], affective attitudes [36,37], emotional [38–41], and physiological [42] states (AFFECT) by capturing human signals.

Though most neuroimaging approaches disregard context, the hypothesis behind situated models of emotion is that emotions are honed for the current context [43]. According to the theory of constructed emotion, the construction of emotions should be holistic, as a complete phenomenon of brain and body in the context of the moment [44]. Barrett [45] argues that rather than being universal, emotions differ across cultures. Emotions are not triggered—they are created by the person who experiences them. The combination of the body's physical characteristics, the brain (which is flexible enough to adapt to whatever environment it is in), and the culture and upbringing that create that environment, is what causes emotions to surface [45]. Recently, there have been attempts in the academic community to supply contextual (from cultural and other circumstances) analysis [46,47].

Various theories and approaches (positive psychology [48–50], environmental psychology [51–53], ergonomics—human factors science [54–56], environment–behavior studies, environmental design [57–59], ecological psychology [60,61], person–environment–behavior [62], behavioral geography [63], and social ecology research [64] also emphasize emotion context sensitivity.

The objective of this research is to provide an overview of the sensors and methods used in AFFECT (affective, emotional, and physiological states) recognition, in order to outline studies that discuss trends in brain and biometric sensors, and give an integrated review of AFFECT recognition analysis using Plutchik's [65] wheel of emotions as the basis. Furthermore, the research aim is to review publications on how techniques that use brain and biometric sensors can be used for AFFECT recognition. In addition, this is a quantitative study to assess how the success of the 169 countries impacted the number of Web of Science articles on AFFECT recognition techniques that use brain and biometric sensors that were published in 2020 (or the latest figures available).

In this paper, we identify the critical changes in this field over the past 32 years by applying text analytics to 21,397 articles indexed by Web of Science from 1990 to 2022. For this review, we examined 634 publications in detail. We have analyzed the global gap in the area of neuroscience and affective biometric sensors and have aimed to update the current big picture. The aforementioned research findings are the result of this work.

When emotions as well as affective and physiological states are determined by recognition sensors and methods—and, later, when such studies are put to practice—a number of issues arise, and we have addressed these issues in this review. Moreover, our research has filled several research gaps and contributes to the big picture as outlined below:

- A fairly large number of studies around the world apply biometric and neuroscience methods to determine and analyze AFFECT. However, there has been no integrated review of these studies.
- Another missing piece is a review of AFFECT recognition, classification, and analysis based on Plutchik's wheel of emotions theory. We have examined 30 emotions and affective states defined in the theory.
- Information on diversity attitudes, socioeconomic status, demographic and cultural background, and context is missing from many studies. We have therefore identified real-time context data and integrated them with AFFECT data. The correct assessment of AFFECT and predictions of imminent behavior are becoming very important in a highly competitive market.
- To demonstrate a few of the aforementioned new research areas in practice, we have developed our own metric, the Real-time Vilnius Happiness Index (Section 4), among other tools. These studies have used integrated methods of biometrics and neuroscience, which are widely applied in various fields of human activity.
- In this research, we therefore examine a more complex problem than any prior studies.

The following sections present the results of this study, a discussion, the conclusions we can draw, and avenues for future research. The method is presented in Section 2. Section 3 summarizes the emotion models. In Section 4, we discuss about brain and biometrics AFFECT sensors, classifications of biometric and neuroscience methods and technologies, emotions and explores the use of traditional, non-invasive neuroscience methods (Section 4) and widely used and advanced physiological and behavioral biometrics (Section 4). Section 4 also summarizes prior research and studies techniques for the recognition of arousal, valence, affective attitudes, and emotion-al and physiological states (AFFECT) in more detail. We summarize existing research on users' demographic and cultural backgrounds, socioeconomic status, diversity attitudes, and the context in Section 5. We present our research results in Section 6, evaluation of biometric systems in Section 7, and finally, a discussion and our conclusions in Section 8.

2. Method

The research method we used can be broken down as follows: (1) formulating the research problem; (2) examining the most popular emotion models, identifying the best option among them for our research (Section 3), and creating the Big Picture for the model; (3) carrying out a review of publications in the field (Section 4); (4) raising and confirming two hypotheses; (5) collecting data; (6) using the INVAR method for multiple criteria analysis of 169 countries; (7) determining correlations; (8) developing three maps to illustrate the way the success of the 169 countries impacts the number of Web of Science articles on AFFECT (emotional, affective, and physiological states) recognition and their citation rates; (9) developing three regression models; and (10) consolidating the findings, providing a rationale for the current methods, comparing the effectiveness of existing methods, and quantifying how likely they are to address the issues and challenges in the field. The following ten steps of the method describe the proposed algorithm and its experimental evaluation in detail.

Furthermore, the research aim is to review publications on how techniques that use brain and biometric sensors can be used for AFFECT recognition, consolidate the findings, provide a rationale for the current methods, compare the effectiveness of existing methods, and quantify how likely they are to address the issues/challenges in the field (Step 1). We have analyzed the global gap in the area of neuroscience and affective biometric sensors and have set the goal of updating the current big picture. The findings of the research above framed the problem.

Step 2 of the research was to examine the most popular emotion models (Section 3) and identify the best option among them for our research. We have chosen the Plutchik's wheel of emotions and one of the main reasons is that the model enables integrated analysis of human emotional, affective, and physiological states.

Step 3 was to review sensors, methods, and applications that can be used in the recognition of emotional, affective, and physiological states (Section 4). We have identified the major changes in the field over the past 32 years through a text analysis of 21,397 articles indexed by Web of Science from 1990 to 2022. We searched for keywords in three databases (Web of Science, ScienceDirect, Google Scholar) to identify studies investigating the use of both neuroscience and affective biometric sensors. A total of 634 studies that used both neuroscience and affective biometric sensor techniques in the study methodology were included, and no restrictions were placed on the date of publication. Studies which investigated any population group were at any age or gender were considered in this work.

A set of keywords related to biometric and neuroscience sensors were used for the above search of three databases. Two main sets of keywords "sensors + biometrics + emotions" and "sensors + neuroscience/brain + emotions" were used in our main search. More specific search terms related to biometrics (i.e., eye tracking, blinking, iris, odor, heart rate), neuroscience/brain techniques (i.e., EEG, MEG, TMS, NIRS, SST) and their components (i.e., algorithms, functionality, performance) were also used to refine the search. For each candidate article, the full text was accessed and reviewed to determine its eligibility. The primary results and article conclusions were identified, and discrepancies were resolved by way of discussion. The studies differed significantly in terms of protocol design, signal processing, stimulation methods, the equipment used, the study population, and statistical methods.

In Step 4, two central hypotheses were raised and confirmed:

Hypothesis 1. *There is an interconnection between a country's success, its number of Web of Science articles published, and its citation frequency on AFFECT recognition. When there are changes in the country's success, its number of Web of Science articles published, and its citation times on AFFECT recognition, the countries' 7 cluster boundaries remain roughly the same (Section 6).*

Hypothesis 2. *Increases in a country's success usually go hand in hand with a jump in its number of Web of Science articles published and its citation times on AF-FECT recognition.*

Next, in Step 5, we collected data. The determination of the success of 169 countries and the results obtained are described in detail in a study by Kaklauskas et al. [66]. This study used data [66] from the framework of variables taken from a number of databases and websites, such as the World Bank, Eurostat-OECD, the World Health Organization, Global Data, Global Finance, Transparency International, Freedom House, Knoema, Socioeconomic Data and Applications Center, Heritage, the Global Footprint Network, Climate Change Knowledge Portal (World Bank Group, Washington, DC, USA), the Institute for Economics and Peace, and Our World in Data; global and national statistics and publications were also used. We based our research calculations on publicly available data from 2020 (or the latest available).

We used the INVAR method [67] to conduct a multi-criteria examination of the 169 nations—the outcomes can be found in Section 6 (Step 6). This method determines a combined indicator for whole nation success. This combined indicator is in direct proportion to the corresponding impact of the values and significances of the specified indicators on a nation's success. The INVAR method was used to conduct multiple criteria analyzes of different groups of countries, such as the former Soviet Union [68], Asian countries [69], and the global analysis of 169 [66] and 173 [70] countries.

The study's 7th step presents the median values of the correlations for 169 countries, its publications, and citations (Section 6). It was found that the median correlation of the dependent variable of the Publications—Country Success model with the independent variables (0.6626) is higher than in the Times Cited—Country Success model (0.5331). Therefore, it can be concluded that the independent variables in the Publications—Country Success model are more closely related to the dependent variable than in the Times Cited—Country Success model.

In Step 8, we developed three maps that illustrate the way the success of the 169 countries impacts the number of Web of Science articles on AFFECT (emotional, affective, and physiological states) recognition and their citation rates. The Country's Success and AFFECT Recognition Publications (CSP) Maps of the World are a convenient way to illustrate how the three predominant CSP dimensions (a country's success, the numbers of publications, and the frequency of articles being cited) are interconnected for the 169 countries, while the CSP models allow for these connections to be statistically analyzed from various perspectives. It also allows for CSP dimensions to be forecast based on the country's success criteria. In other words, the CSP models give us a more detailed analysis of the CSP dimensions through statistical and multi-criteria analysis, while the CSP maps (Section 6) are more of a way to present the results in a visual manner. The amount of data available is gradually increasing, as is the knowledge gained from research conducted around the world. As a result, the CSP models are becoming better and better, and providing a clearer reflection of the actual picture. This means that they can effectively facilitate research and innovation policy decisions.

In Step 9, we created two regression models (Section 6). For the multiple linear regressions, we used IBM SPSS V.26 to build two regression models on 15 indicators of country success [66] and the three predominant CSP dimensions (Section 6). Step 9 entailed the construction of regression models for the number of publications and their citation rates, and the calculation of the effect size indicators describing them. Two dependent variables and 15 independent variables were analyzed to construct these regression models. The process was as follows:

- Construction of regression models for the numbers of publications and their citations.
- Calculation of statistical (Pearson correlation coefficient (r), standardized beta coefficient (β), coefficient of determination (R^2), standard deviation, p-values) and non-statistical (research context, practical benefit, indicators with low values) effect size indicators describing these regression models.

It was found that changes in the values of the Country Success variable explain the variance of the Publications variable by 89.5%, and the variance of the Times Cited variable by 54.0%. Additionally, when the value of the Country Success variable increases by 1%, the value of Publications increases by 1.962% and Times Cited—by 2.101%. As the success of a country increased by 1%, the numbers of Web of Science articles published and their citations grew by 1.962% and 2.101%, respectively. A reliability analysis of the compiled regression models allows us to conclude that the models are suitable for analysis ($p < 0.05$). The 15 country success indicators explained 69.4% and 51.18% of the number of Web of Science articles published and their citations, respectively.

Step 10 was to assess the biometric systems under analysis: the rationale behind the available biometric and brain approaches was outlined, the efficacy of existing methods compared, and their ability to address issues and challenges present in the field determined (Section 7).

3. Emotion Models

First, this chapter will discuss emotion models in more detail. Then, we will choose the best option for our research and look at the Big Picture, i.e., the links between the selected emotion model and biometric and brain sensors, and the trends.

Emotional responses are natural to humans, and evidence shows they influence thoughts, behavior, and actions. Emotions fall into different groups related to various affects, corresponding to the current situation that is being experienced [71]. People encounter complex interactions in real life, and respond to them with complex emotions that often can be blends [72]. Emotional responses are the way for our brain and body to deal with our environment, and that is why they are fluid and depend on the context around us [73].

Two fundamental viewpoints form the basis in approaches to the classification of emotions: (a) emotions are discrete constructs and they have fundamental differences, and (b) emotions can be grouped and characterized on a dimensional basis [74]. These classifications (emotions as discrete categories and dimensional models of emotion) are briefly analyzed next.

In word recognition, alternative models have so far received little interest, and one example is the discrete emotion theory [75]. This theory posits that there is a limited set of universal basic emotions hardwired through evolution, and that each of the wide variety of affective experiences can essentially be categorized into this limited set [76,77]. The discrete emotion theory states that many emotions can be distinguished on the basis of expressive, behavioral, physiological, and neural features [78]. The definition of emotions provided by Fox [79] states they are consistent and discrete responding processes that can include verbal, physiological, behavioral, and neural mechanisms. They are triggered and changed by external or internal stimuli or events and respond to the environment. Russell and Barrett [80] argue that, unlike the discrete emotion theory, their alternative models can account for the rich context-sensitivity and diversity of emotions. Emotion blends could be of three kinds: (a) Positive-blended emotions were blends of only positive emotions; (b) negative-blended emotions were blends of only negative emotions; and (c) mixed emotions were blends of both positive and negative emotions, as well as neutral ones. The way teachers have described blended emotions reflects that mathematics teaching involves many and complex tasks, where the teacher has to continuously keep gauging the level of progress [81].

Emotional dimensions represent the classes of emotion. Categorized emotions can be characterized in a dimensional form, with each emotion located in a different location in space, for example in 2D (the circumplex model, "consensual" model of emotion, and vector model) or 3D (the Lövheim cube, the pleasure–arousal–dominance (PAD) emotional state model, and Plutchik's model) [82].

The circumplex model [83] proposes that two independent neurophysiological systems: One of the systems is related to arousal (activated/deactivated) and to valence (a pleasure–displeasure continuum), and the other to valence (a continuum from pleasure to displeasure) and to arousal (activation–deactivation) [84]. Each emotion can be understood as having varying valence and arousal, and is a linear combination of these two dimensions, or as varying valence and arousal [83,85]. We already applied the Russel's circumplex model of emotions to perform a review of the human emotion recognition of sensors and methods [85].

The vector model comprises two vectors. The model holds that there is an underlying dimension of arousal with a binary choice of valence that determines direction, and an underlying dimension of arousal. This results in there being two vectors that, both starting at zero arousal and neutral valence and zero arousal, proceed as straight lines, one in a positive, and one in the direction of negative valence and the other in the direction of positive valence. Typically, the vector model uses direct scaling of the dimensions of each individual stimulus individually in this model [86,87].

The positive activation–negative activation (PANA) or "consensual" model of emotion, also known as positive activation/negative activation (PANA), assumes that there are two separate systems—positive affect and negative affect. In the PANA model, the vertical axis represents low to high positive affect, and the horizontal axis of this model represents low to high negative affect (low to high). The vertical axis represents positive affect (low to high) [88]. There are two uncorrelated and independent dimensions: Positive Affect (PA), represents the extent (from low to high) to which a person shows enthusiasm for life. The second factor is Negative Affect (NA), and NA represents the extent to which a person is feeling upset or unpleasantly aroused. Positive Affect and Negative Affect are independent and uncorrelated dimensions [89].

The Pleasure–Arousal–Dominance (PAD) Emotional-State Model, offers a general three-dimensional approach to measuring emotions [90]. This 3D model captures emotional response, and includes the three dimensions of pleasure–displeasure (P), arousal–nonarousal (A), and dominance–submissiveness (D) as basic factors of emotional response [91]. The initials PAD stand for pleasure, arousal, and dominance, which span different emotions. For instance, pleasure can be happy/unhappy, hopeful/despairing, satisfied/unsatisfied, pleased/annoyed, content/melancholic, and relaxed/bored. Arousal can be excited/calm, stimulated/relaxed, wide-awake/sleepy, jittery/dull, frenzied/sluggish, and aroused/unaroused. Dominance can be important/awed, dominant/submissive, influential/influenced, controlling/controlled, in control/cared-for, and autonomous/guided [92]. The neuro-decision and neuro-correlation tables, the inverted U-curve theory, the PAD emotional state model, neuro-decision making, and neuro-correlation tables are used to evaluate the impact of digital twin smart spaces (such as indoor air quality, a level of the lighting intensity and colors, learning materials, images, smells, music, pollution, and others) on users, and track their response dynamics in real time, and to then react to this response [93].

The PAD is composed of three different subscales, reflecting pleasure, arousal, and dominance. These can represent different emotions; for example, the pleasure states include happy (unhappy), pleased (annoyed), satisfied (unsatisfied), contented (melancholic), hopeful (despairing) and relaxed (bored), while the arousal states include stimulated (relaxed), excited (calm), frenzied (sluggish), jittery (dull), wide awake (sleepy) and aroused (unaroused), and the dominance states include controlling (controlled), influential (influenced), in control (cared for), important (awed), dominant (submissive), and autonomous (guided) [92]. The affective space model makes it possible to visualize the distribution of emotions along the two axes of valance (V) and arousal (A). Using this model, different emotions can be identified, such as happiness, calmness, fear, and sadness [94].

Swedish neurophysiologist Lövheim proposed that a cube of emotion is the direct relation between certain specific combinations of the levels of the three signal substances (serotonin, noradrenaline, and dopamine) and eight basic emotions [95]. A three-dimensional model, the Lövheim cube of emotion, was presented where there is a model with each of the signal substances of form represented as the axes of a coordinated system, and each corner of this 3D space holding one of the eight basic emotions is placed in the eight corners. In this model, anger is produced by the combination of high noradrenaline, high dopamine, and low serotonin [96].

The eight main categories of emotions defined by Robert Plutchik in 1980s include two equal groups opposite to each other: half are positive emotions and the other half are negative ones [97]. To visualize eight primary emotion dimensions, which are fear, trust, surprise, anticipation, anger, joy, disgust and sadness, eight sectors have been isolated [98]. The Emotion Wheel shows each of the eight basic emotions highlighted with a recognizable color [99]. When we add another dimension, the Wheel of Emotions becomes a cone with its vertical dimension representing intensity. Moving from the outside towards the wheel's center emotions intensify and this fact is highlighted by the indicator color. The intensity of emotions is decreasing towards the outer edge and the color, correspondingly, becomes less intense [98,99]. When feelings intensify one feeling can turn into another: annoyance into rage, serenity into ecstasy, interest into vigilance, apprehension into terror, acceptance into admiration, pensiveness into grief, distraction into amazement, and, if left unchecked, boredom can become loathing [98]. Some emotions have no color marking. They are a mix of two primary emotions [98,99]. Joy and anticipation, for instance, combine to become optimism. When anticipation combines with anger it becomes aggressiveness. The combination of trust and fear is submission, joy and trust combine to become love, surprise and fear become awe, the pair of disgust and anger becomes contempt, sadness and disgust combine to become remorse, and surprise and sadness become disapproval [100].

After the analysis of the said emotion models, we have made the decision to choose Plutchik's wheel of emotions for our research. The ability to analyze human emotional, affective, and physiological states in an integrated manner offered by this model is one of the main reasons of our choice. The wheel is briefly discussed below.

Several ways to classify emotions have been proposed in the field of psychology. For that purpose, the basic emotions are first identified and then they allow clustering with any other more complex emotion [101]. Plutchik [65] proposed a classification scheme based on eight basic emotions arranged in a wheel of emotions, similar to a color wheel. Just like complementary colors, this setup allows the conceptualization of primary emotions by placing similar emotions next to each other and opposites 180 degree apart. Plutchik's wheel of emotions classifies these eight basic emotions grounded on the physiological aim [102]. Emotions are coordinated with the body's physiological responses. For example, when you are scared, your heart rate typically increases and your palms become sweaty. There is ample empirical evidence that suggests that physiological responses accompany emotion [103]. Another parallel with colors is the fact that some emotions are primary emotions and other emotions are derived by combining these primary emotions. The two models share important similarities, and such modelling can also serve as an analytical tool to understand personality. In this case, a third dimension has been added to the circumplex model to represent the intensity of emotions. The structural model of emotions is, therefore, shaped like a cone [104]. Figure 1 demonstrates Plutchik's wheel of emotions, biometrics and brain sensors, and trends and interdependence in this Big Picture stage. At the center of the circles is Plutchik's wheel of emotions. Plutchik's wheel of emotions also includes affective attitudes (interest, boredom). Plutchik [65] notes that the same instinctual source of energy is discharged as part of the emotion felt and the underlying peripheral physiological process. Emotions can be of various levels of arousal or degrees of intensity [105]. Looking at the intensity of Plutchik's eight basic emotions, Kušen et al. [106] identified variations in emotional valence. The first circle, therefore, analyses, directly or indirectly, human arousal, valence, affective attitudes, and emotional and physiological states (AFFECT). Human AFFECT can be measured by means of neuroscience and biometric techniques. The market and global trends are a constant force affecting neuroscience and biometric technologies and their improvement. Based on the analysis of global sources [107–110] and our experience, Figure 1 presents brain and biometric sensors, as well as technique trends. Sensors will be able to integrate more and more new technologies and collect a greater variety of data, as they will become more accurate, more flexible, cheaper, smaller, greener, and more energy-efficient [108–110]. Network neuroscience, a new explicitly integrative approach towards brain structure and function, seeks new ways to record, map, model, and analyze what constitutes neurobiological systems and what interactions happen inside them. The computational tools and theoretical framework of modern network science, as well as the availability of new empirical tools to map extensively and record the way shifting patterns link molecules, neurons, brain areas and social systems, are two trends enabling and driving this approach [107].

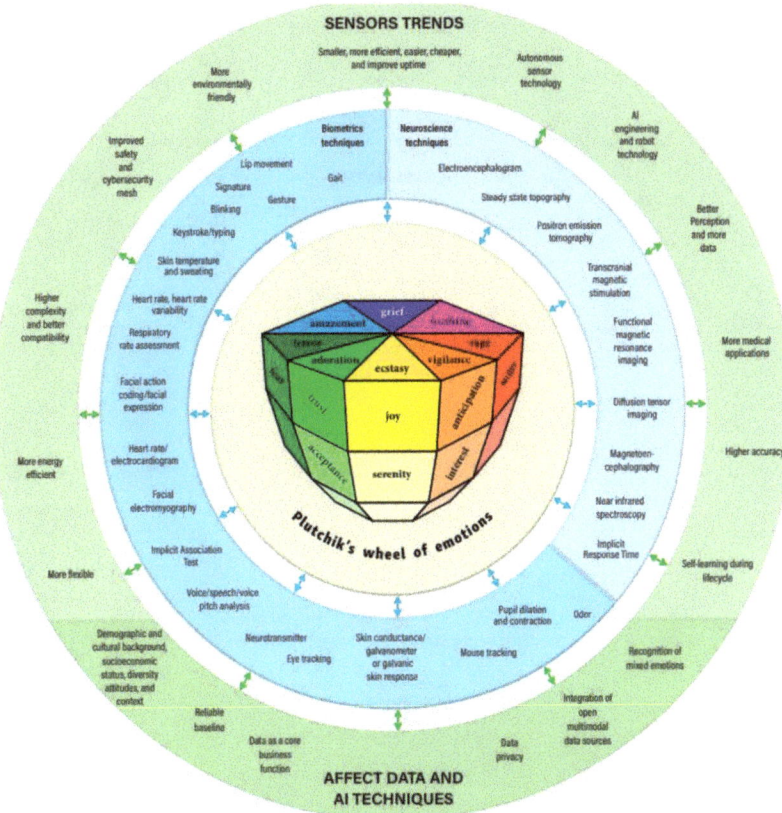

Figure 1. Plutchik's wheel of emotions, biometrics and neuroscience sensors, and trends.

Figure 2 shows numerous sciences and areas in which neuroscience and biometrics analyze the AFFECT. According to Sebastian [111], neuroeconomics is the study of the effect of anticipating money decisions on our brain. It has solidified as an entirely academic and unifying field that ventures to describe the techniques of the decision-making process; and reiterates economic behavior and decision-making process with economic disposition. The procedure of neuroeconomics involves the integration of behavioral experiments and brain imaging in order to more clearly appreciate the workings behind individual and collective decision-making [112]. Serra [113] reported that neuroeconomics researchers utilize neuroimaging devices such as functional magnetic resonance imaging (fMRI), magnetic resonance imaging (MRI), transcranial magnetic stimulation (rTMS), and transcranial direct-current stimulation (tDCS), positron emission tomography (PET) and electroencephalography (EEG). The majority of challenges probed by neuroeconomics researchers are basically similar to the problems a marketing researcher would acknowledge as aspects of their functional domain [114]. Kenning and Plassmann [115] has also defined neuroeconomics as the implementation of neuroscientific methods in the evaluation and appreciation of economically significant behavior.

Figure 2. Neuroscience and biometric branches analyzing AFFECT in various sciences and fields.

According to Wirdayanti and Ghoni [116], neuromanagement entails psychology, the biological aspect of humans for decision-making in management sciences. As stated Teacu Parincu et al. [117], neuromanagement is targeted at investigating the acts of the human brain and mental performances whenever people are confronted with management challenges, using cognitive neuroscience, in addition to other scientific disciplines and technology, to evaluate economic and managerial problems. Its focal point is on neurological activities that are related to decision-making and develops personal as well as organizational intelligence (team intelligence). It also centers on the planning and management of people (for example, selection, training, group interaction and leadership) [118].

Neuro-Information Science can be defined as the science that observes neurophysiological reactions that are connected with the peripheral nervous system; that is then connected to conventional cognitive activities. Michalczyk et al. [119] stated that neuro-information-systems research has developed into a conventional approach in the information systems (IS) discipline for evaluating and appreciating user behavior. Riedl et al. [120] and Michalczyk et al. [119] concluded that Neuro-information-systems comprise studies that are centered on all types of neurophysiological techniques, such as functional magnetic resonance imaging (fMRI), electroencephalograhy (EEG), fNIRS (functional near-infrared spectroscopy), electromyography (EMG), hormone studies, or skin conductance and heart rate evaluations, as well as magnetoencephalography (MEG) and eye-tracking (ET).

Neuro-Industrial Engineering brought about by the synergy between neuroscience and industrial engineering has afforded resolutions centered on the physiological status of people. Ma et al. [121] reported that NeuroIE secures its objective and real data by analyzing human brain and physiological indexes with advanced brain AFFECT devices and biofeedback technology, evaluating the data, adding neural activities as well as physiological status in the process of evaluation; as new constituents of operations management, and finally understanding better human–machine integration by modifying work environment and production system in line with people's reaction to the system, preventing mishaps and enhancing efficiency and quality. According to Ma et al. [121], Neuro-Industrial Engineering is centered on humans and lays hold of human physiological status data (e.g., EEG, EMG, GSR and Temp). Zev Rymer [122] also stated that the application of Neuro-Industrial Engineering is multidisciplinary in that it cuts across the neurological sciences (particularly neurology and neurobiology) in addition to different fields of engineering disciplines such as simulation, systems modeling, robotics, signal processing, material sciences, and computer sciences. The area encompasses a range of topics and applications; for example, neurorobotics, neuroinformatics, neuroimaging, neural tissue engineering, and brain–computer interfaces.

As soon as a user contacts an insurer, a bank or any other call center, a version of Cogito's software known as Dialog could be active in the background, assisting the client service agent to deal with the client. Should the user become upset or angry, the client service agent can ensure that necessary actions are taken to satisfy the client. According to Cogito, this service is known as "digital intuition". Its usefulness in call centers cannot be overemphasized as it can give feedback about real-time communications. The speed at which speeches are made by the callers as well as the dynamic range of their voices can also be analyzed by the software. For example, significant variations in pitch and stresses in caller's tones could signify excitement or anger. Less significant dynamism, a monotonous flat tone, could imply a lack of interest or unconcern. Some companies make use of the software to assist their employees engage new patients for healthcare projects that help control health challenges such as obesity or asthma. Cogito is among recent profit-based research companies whose focus are on the evaluation of signals subconsciously given off by people which exposes their mindset. The evaluation of these kinds of social-signals is beneficial beyond call centers and meeting rooms. According to Hodson [123], keeping track of conversations during surgeries or plane cockpits could assist surgeons and pilots to be aware of whether their colleagues are really attentive to their directives, possibly preserving lives.

Several areas where we can apply the technology of recognizing emotions from speech include human–computer interactions and call centers [124].

4. Brain and Biometric AFFECT Sensors

4.1. Classifications

Globally, several classifications of biometric and neuroscience methods and technologies are used. Our research focuses on neuroscience methods that are non-invasive. The use of non-invasive brain stimulation is widespread in studies of neuroscience [125]. The non-invasive neuroscience methods are: transcranial magnetic stimulation (TMS), electroencephalography (EEG), magnetoencephalography (MEG), positron emission tomography (PET), functional magnetic resonance imaging (fMRI), near infrared spectroscopy (NIRS), diffusion tensor imaging (DTI), steady-state topography (SST), and others [126–134]. These non-invasive neuroscience methods are described in detail in Section 3. In the future, the authors of this article plan to analyze invasive neuroscience methods, too.

Biometrics can be physical or behavioral. In the first case, emotions can be identified by their physical features, including face, and in the second case by their behavioral characteristics, including gait, voice, signature, and typing patterns [135]. Various sensors can measure physiological signals, known as biometrics, capturing the response of bodily systems to things that are experienced through our senses, but also things imagined, by tracking sleep architecture, heart rate variability (HRV), respiratory rate (RR), and heart rate (RHR) [136].

Scientific literature classifies biometrics into certain types. Stephen and Reddy [137] and Banirostam et al. [138], for instance, classify biometrics into three categories: physiological, behavioral, and chemical/biological. Yang et al. [139] distinguish physiological and behavior traits. Kodituwakku [140] believes biometric technology can be classified into two general categories: physiological biometric techniques and behavioral biometric techniques. Jain et al. [141] and Choudhary and Naik [142] also classify biometrics into two categories: physiological and behavioral. In the literature, not only signature, voice, and gait are considered behavioral biometric features, but also ECG, EMG, and EEG [143], while other authors distinguish cognitive biometrics [144,145], including electroencephalography (EEG), electrocardiography (ECG), electrodermal response (EDR), blood pulse volume (BVP), near-infrared spectroscopy (NIR), electromyography (EMG), eye trackers (pupillometry), hemoencephalography (HEG), and related technologies [145]. Some scientific sources claim that eye tracking is a behavioral biometric [146], while others claim that it is a measurement in physiological computing [147]. Physiological biometrics measures the physiological signals to determine identity as well as authenticating and analyzing users emotions. Respiration, perspiration, heartbeat, eye-reactions to light, brain activity, emotions, and even body odor can be measured for numerous purposes, including physical and logical access control, payments, health monitoring, liveness detection, and neuromarketing among them [136].

Scientists identify the following AFFECT biometric types [139–142,148–150]:

- Physiological features: facial patterns, odor, pupil dilation and contraction, skin conductance, heart rate, respiratory rate, temperature, blood volume pulse, and others.
- Behavioral features: gait, keystroke, mouse tracking, signature, handwriting, speech/voice, and others.
- The authors of this article have used the classification of biometrics proposed by the abovementioned authors (physiological and behavioral features).

Biometric technologies are usually divided into those of first and second generation [151]. First-generation biometrics can confirm a person's identity in a quick and reliable way, or authenticate them in different contexts, and law enforcement is one of the areas where such solutions are employed in practice [152]. The primary purpose of first-generation biometrics is identity verification, such as facial recognition, and the technology is built around simple sensors that capture physical features and store them for later use [153]. Second-generation biometrics can also be used to detect emotions, with electro-physiologic and behavioral biometrics (e.g., based on ECG, EEG, and EMG) as examples of such technologies [154]. Second-generation biometrics measure individual patterns of learned behavior or physiological processes, rather than physical traits, and are also known as behavioral biometrics [155]. Second-generation biometrics usage has the ability to analyze/evaluate emotions and detect intentions [156]. The use of second-generation biometrics enables wireless data collection regarding the body. The data can then be used to infer an individual's intent and emotions, as well as emotion tracking across spaces [151,157]. We examine only physiological effects affected by emotional reactions (i.e., second-generation biometrics), and the use of biometric patterns for the identification of individuals is not discussed in this study.

A diverse range of AI algorithms have been applied for AFFECT recognition, for example machine learning, artificial neural networks, search algorithms, expert systems, evolutionary computing, natural language processing, metaheuristics, fuzzy logic, genetic algorithms, and others. Some of the most important supervised (classification, regression), unsupervised (clustering), and reinforcement learning algorithms of machine learning are common as tools in biometrics or neuroscience research to detect emotions and affective attitudes, and are listed below:

- Among classification algorithms the most common choices are: naïve Bayes [158–160], Decision Tree [161–163], Random Forest [164–166], Support Vector Machines [167–169], and K Nearest Neighbors [170–172].
- Among regression algorithms the usual choices are: linear regression [173–175], Lasso Regression [176,177], Logistic Regression [178–180], Multivariate Regression [181,182], and Multiple Regression Algorithm [183,184].
- Among clustering algorithms the most common choices in biometrics or neuroscience research are: K-Means Clustering [185–187], Fuzzy C-means Algorithm [188,189], Expectation-Maximization (EM) Algorithm [190], and Hierarchical Clustering Algorithm [188,191,192].
- Among reinforcement learning algorithms the most common choices are: deep reinforcement learning [193–195] and inverse reinforcement learning [196].

4.2. Brain AFFECT Devices and Sensors

Neuroscience is associated with multiple fields of science, for example chemistry, computation, psychology, philosophy, and linguistics. Various research areas of neuroscience include behavioral, molecular, operative, evolutionary, cellular, and therapeutic features of the neurotic system. The neuroscience market encompasses technology (electrophysiology, neuro-microscopy, whole-brain imaging, neuroproteomics analysis, animal behavior analysis, neuro-functional study, etc.), components (services, instrument, and software) and end-users (healthcare centers, research institutions and academic, diagnostic laboratories, etc.) [197]. Global Industry Analysts Inc. (San Jose, CA, USA) [197] has previously grouped the global neuroscience market into instrument, software, and services based on components.

Neuroscience provides valuable perceptions concerning the structural design of the brain and neurological, physical, and psychological activities. It helps neurologists to appreciate the various components of the brain that can assist in the development of medications and techniques to handle and avoid many neurological anomalies. The rising death rate as a result of several neurological disorders, such as Parkinson's disease, Alzheimer's, schizophrenia, and other brain-related health challenges, represents the basic factor controlling the neuroscience market growth [198]. According to Neuroscience Market [198], the increasing request for neuroimaging devices and the progressive brain mapping research and evaluation projects are other crucial growth-inducing factors.

Neuroscience covers a whole range of branches, such as, neuroevolution, neuroanatomy, developmental neuroscience, neuroimmunology, cellular neuroscience, neuropharmacology, clinical neuroscience, cognitive neuroscience, nanoneuroscience, molecular neuroscience, neurogenetics, neuroethology, neurochemistry, neurophysics, paleoneurobiology, neurology, and neuro-ophthalmology.

Other branches of neuroscience analyze AFFECT in various related sciences and fields, such as affective neuroscience [199,200], neuroinformatics [201,202], neuroimaging [203,204], systems neuroscience [205,206], computational neuroscience [207,208], neurophysiology [51,209], behavioral neuroscience [210,211], neural engineering [212,213], neuroeconomics [214,215], neurolinguistics [216,217], neuropsychology [218–220], neurophilosophy [221–223], neuroaesthetics [224–226], neurotheology [227–229], neuropolitics [230–232], neurolaw [233–235], social neuroscience [236,237], cultural neuroscience [238,239], neuroliterature [240–242], neurocinema [243–245], neuromusicology [246–248], and neurogastronomy [249,250].

For example, Lim [251] identifies the following neuroscientific techniques for neuromarketing:

- Electromagnetic methods, including magnetoencephalography (MEG), electroencephalography (EEG), and steady-state topography (SST). MEG involves the magnetic fields produced by the brain (its natural electrical currents) and is used to track the changes that occur when participants see or interact with various presentation outputs. EEG is related to the ways in which brainwaves change and is used to detect changes

when participant see or interact with various promoting outputs (an electrode band or helmet is used for this purpose). SST measures a steady-state visually evoked potential, and is used to determine how brain activities change depending on the task;
- Metabolic methods, including positron emission tomography (PET) and functional magnetic resonance imaging (fMRI). PET is used to examine the metabolism of glucose within the brain with great accuracy by tracing radiation pulses, while fMRI is used to measure blood flow in the brain to determine changes in brain activity;
- Electrocardiography (ECG), which uses external skin electrodes to measure electrical changes related to cardiac cycles;
- Facial electromyography (fEMG), which amplifies tiny electrical impulses to record the physiological properties of the facial muscles;
- Transcranial Magnetic Stimulation (TMS), which is used to observe the effects of promoting output on behavior by temporarily disrupting specific brain activities. TMS is a non-invasive, safe brain stimulation method. By means of a strong electromagnet, this technique momentarily generates a short-lived virtual lesion, i.e., disrupts information processing in one of brain regions. If stimulation interferes with performing a certain task, the affected brain region is, then, necessary for normal performance of the task [252].

Table 1 demonstrates traditional non-invasive neuroscience methods.

Table 1. Traditional non-invasive neuroscience methods.

Methods	Author(s)	Description
Electroencephalography (EEG)	[111,253–266]	EEGs capture brainwave variations, using recorded amplitudes to monitor mental states that include alpha waves (relaxation), beta waves (wakefulness), delta waves (sleep), and theta waves (calmness) [255]. An EEG signal comprises five brain waves and measuring the activity of certain brain areas can reveal the state of the subject's cortical activation. Each wave is characterized by different amplitudes and frequencies, and corresponds to distinct cognitive states [265].
Magnetoencephalography (MEG)	[111,253–256,259,260,267]	Using magnetic potentials, an MEG records brain activity at the scalp level. A helmet with sensitive detectors is placed on the subject's head to track the signal [255], and the MEG detects the magnetic fields produced by electromagnetic fields [111].
Transcranial Magnetic Stimulation (TMS) (Figure 3)	[111,251,253,255,258,260, 267]	TMS modulates the activity of certain brain areas located 1–2 cm below the skull, without reaching the neocortex, using magnetic induction [255]. When TMS is used, short electromagnetic impulses are applied at the scalp level. This instrument can stimulate or inhibit a particular cortical area [111].
Near Infrared Spectroscopy (NIRS)	[267–269]	NIRS measures hemodynamic alterations accompanying brain activation and is a simple bedside technique [269]. NIRS makes use of the near-infrared region of the electromagnetic spectrum (about 700–2500 nm). Measurements are taken of light scattered from the surface of and through a sample, and NIR reflectance spectra can give rapid insight into the properties of a material without altering the sample [268].
Steady-State Topography (SST)	[251,253,255,256,260]	SST can be applied to track high-speed changes and measure the activity of the human brain. This tool is very commonly used in neuromarketing research and cognitive neuroscience [255].

Table 1. Cont.

Methods	Author(s)	Description
Functional Magnetic Resonance Imaging (fMRI) (Figure 4)	[111,251,253–256,258–261,263,264,266,267]	fMRI is suitable for use within neuromarketing studies, as brain activity can be measured in subjects performing certain tasks or experiencing marketing stimuli. It allows for the observation of deep brain structures, and hence can reveal patterns [255]. fMRI can also measure increases in oxygen levels in the blood flow to the brain and can detect the active cortical regions [111].
Positron Emission Tomography (PET) (Figure 4)	[111,251,253,254,256,259–261,267]	The subject is injected with a radioactive substance, and the flow of the substance is then measured. Significant increases in the flow are seen in activated areas [111].
Diffusion Tensor Imaging (DTI) (Figure 5)	[267,270,271]	This is an MRI-based neuroimaging technique that allows the user to estimate the location, anisotropy and orientation of the brain's white matter tracts [271]. DTI makes it possible to visualize and characterize white matter fasciculi in two and three dimensions [270].

For clarity, several descriptions of traditional neuroscience methods are presented below.

Wearable healthcare devices store a lot of sensitive personal information which makes the security of these devices very essential. Sun et al. [272] proposed an acceleration-based gait recognition method to improve gait-based elderly recognition. Gait is also a good indicator in health assessment, Majumder et al. [273] created a simple wearable gait analyzer for the elderly to support healthcare needs.

Lim [251] states that neuroscientific methods and tools include those that track, chart, and record the activity of a person's neural system and brain in relation to a certain behavior, and neurological representations of this activity can then be generated to shed light on how an individual's brain and nervous system respond when the person is exposed to a stimulus. In this way, neuroscientists can observe the neural processes as they happen in real time. There are three main types of neuroscientific method: those that track what is happening inside the brain (metabolic and electromagnetic activity); those that track what is happening at the neural level outside the brain; and those that can influence neural activity (Table 1, Figure 1).

Non-invasive neuroscience technical information is provided in detail in various research literature about the origin of the measured signal and the engineering/physical principle of the sensors for EEG [274–276], MEG [277–279], TMS [280–282], etc.

Gannouni et al. [283] have proposed a new approach with EEG signals used in emotion recognition. To achieve better emotion recognition using brain signals, Gannouni et al. [283] applied a novel adaptive channel selection method. The basis of this method is the acknowledgment that different persons have unique brain activity that also differs from one emotional state to another. Gannouni et al. [283] argue that emotion recognition using EEG signals needs a multi-disciplinary approach, encompassing areas such as psychology, engineering, neuroscience, and computer science. With the aim of improving the reproducibility of emotion measurement based on EEG, Apicella et al. [35] have proposed an emotional valence detection method for a system based on EEG, and their experiments proved an accuracy of 80.2% in cross-subject analysis and 96.1% in within-subject analysis. Dixson et al. [284] have pointed out that facial hair may interfere with detection of emotional expressions in a visual search. However, facial hair may also interfere with the detection of happy expressions within the face in the crowd paradigm, rather than facilitating an effect of anger superiority as a potential system for threat detection.

Wang et al. [285] introduced an EEG-based emotion recognition system to classify four emotion states (joy, sadness, fear, and relaxed). Their experiments used movie elicitation to acquire EEG signals from their subjects [285]. The way in which meditation influences emotional response was investigated via EEG functional connectivity of selected brain regions as the subjects experienced happiness, anger, sadness or were relaxed, before and after meditation.

Neurometrics is a quantitative EEG method. Looking at individual records, this method provides a reproducible, precise estimate of deviations from normal. Only sufficient amount of good quality raw data transformed for Gaussian distributions, correlated with age, and corrected taking into account intercorrelations among measures ensure meaningful and reliable results [286]. Businesses, government agencies, and individuals use neurometric information when they need timely and profitable decisions. Techniques based on neurometric information are applied to make profitable business decisions. These techniques are based on biometric information, eye tracking, facial action coding and implicit response testing, and are used to understand and record human sentiments and other related feedback [161].

The fronto-striatal network is involved in a range of cognitive, emotional, and motor processes, such as decision-making, working memory, emotion regulation, and spatial attention. Practice shows that intermittent theta burst transcranial magnetic stimulation (iTBS) modulates the functional connectivity of brain networks. Treatments of mood disorders usually involve high stimulation intensities and long stimulation intervals in transcranial magnetic stimulation (TMS) (Figure 3) therapy [287].

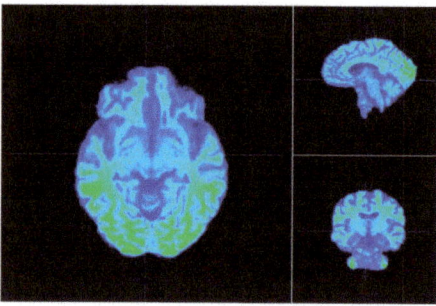

Figure 3. Resting state TMS brain scan image [287].

One of imaging techniques is FDG-PET/fMRI (simultaneous [18F]-fluorodeoxyglucose positron emission tomography and functional magnetic resonance imaging). This technique makes it possible to image the cerebrovascular hemodynamic response and cerebral glucose uptake. These two sources of energy dynamics in the brain can provide useful information. Another greatly useful technique for characterizing interactions between distributed brain regions in humans has been resting-state fMRI connectivity, while metabolic connectivity can be a complementary measure to investigate the dynamics of the brain network. Functional PET (fPET), a new approach with high temporal resolution, can be used to measure fluoro-D-glucose (FDG) uptake and looks like a promising method to assess the dynamics of neural metabolism [288]. Figure 4 shows raw images of signal intensity variation across the brain for one individual subject.

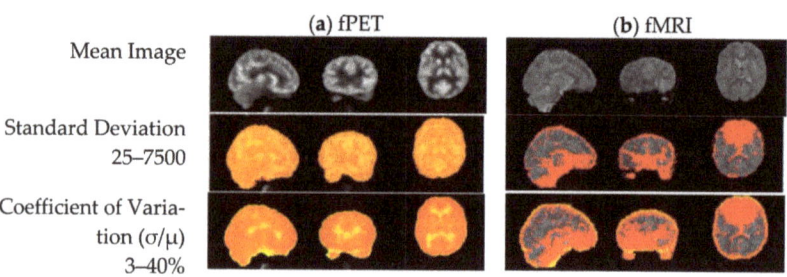

Figure 4. Raw images of fPET and fMRI scans [288].

Many biological tissues comprised of fibers, which are groups of cells aligned in a uniform direction, have anisotropic properties. In the human brain, for instance, within its white matte regions, axons usually form complex fiber tracts that enable anatomical communication and connectivity. Non-invasive tools can show the groups of axonal fibers visually. One of them is diffusion tensor magnetic resonance medical imaging (DTI), which is one particular method or application of the broader Diffusion-Weighted Imaging (DWI). The basic principle behind this technique is that water diffuses more slowly as it moves perpendicular to the preferred direction, whereas in the direction aligned with the internal structure the diffusion is more rapid. The DTI outputs can be further used to compute diffusion anisotropy measures such as the fractional anisotropy (FA). The principal direction of the diffusion tensor can also be used to obtain estimates related to the white matter connectivity in the brain. Figure 5 shows an example of DTI tractography, or visualization of the white matter connectivity [289].

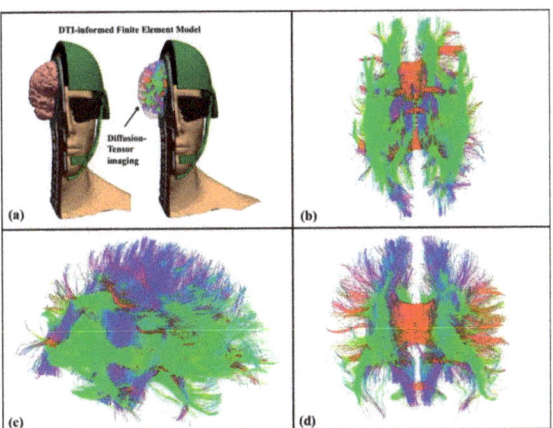

Figure 5. DTI can be used to construct a transversely isotropic model by overlaying axonal fiber tractography on a finite element mesh: (**a**) DTI-informed Finite Element Model; tractography shows complex fibers from (**b**) the dorsal view, (**c**) the right lateral side view, and (**d**) the posterior view. Cartography of the tracts' position, direction by color: red for right-left, blue for foot-head, green for anterior-posterior [289].

4.3. Physiological and Behavioral Biometrics

Physiological biometrics (as opposed to behavioral biometrics) is a category of approaches that refers to physical measurements of the human body, including face, pupil constriction and dilation [290]. When a recognition system is based on physiological characteristics it can ensure a comparatively high accuracy [291]. The ubiquity of electronics such as cell phones and computers, and evolving sensor technology offer human beings new possibilities to track their behavioral and physiological features and evaluate the associated biometric results. Advances in mobile devices mean they now have many efficient and complex sensors. Biometric technology often contributes to mobile application growth, including online transaction efficiency, mobile banking, and voting. The global market for biometric systems is wide and comprises many different segments such as healthcare, transportation and logistics, security, military and defense, government, consumer electronics, and banking and finance [292].

Table 2 presents widely used physiological and behavioral biometrics.

Table 2. Physiological and behavioral biometrics.

Technique	Author(s)	Description
Physical/Physiological Features		
Eye Tracking (ET) (Figure 6)	[111,251,253–261,264–267]	ET determines the areas at which the subject is looking and for how long, and also tracks the movement of the subject's eyes and changes in pupil dilation while the subject looks at stimuli. With this technique, behavior and cognition can be studied without measuring brain activity [255]. By measuring eye movements and visual attention, an eye tracker determines the point of regard [265].
Blinking	[261,264,293]	Eye blinking forms the basis of the new biometric emotions identifier proposed by Abo-Zahhad et al. [293]. These authors outline where eye blinking signals come from and give an overview of the features of the EOG signals from which the eye blinking waveform is extracted.
Iris characteristics		User-oriented examinations were applied to find the relationships between personality and three common iris characteristics: pigment dots, crypts, and contraction furrows [294]. Dark-eyed individuals typically have higher scores for neuroticism and extraversion [295], sociability [296], and ease of emotional arousal [297].
Facial Action Coding (FC)/Facial Expression Analysis Surveys (Figure 7)	[253–258,260,261,263–265,298]	FC uses a video camera to track micro-expressions that correspond to certain subconscious reactions. The activity of the facial muscles is tracked [255]. Scientists and practitioners have developed various open data datasets (KaoKore Dataset, CelebFaces At-tributes Dataset, etc.) and applied elicitation techniques (gamification, virtual reality) in practice.
Facial Electromyography (fEMG) (Figure 8)	[251,253–256,259–263,298,299]	fEMG is used in measuring and evaluating the physiological properties of facial muscles [255].
Odor	[300]	This a method of emotion recognition based on an individual's odor [300]. An emotional mood, for example a period of depression, may affect body odor [301].
Keystroke dynamics and mouse movements (Figure 9)	[302]	AFFECT states can be determined by how a person moves a computer mouse while sitting at a computer.
Skin Conductance (SC)/Galvanometer or Galvanic Skin Response (GSR)	[111,251,253,255,256,258,260–262,264,265,267]	SC is highly correlated with the rate of perspiration, and is often linked to stress as well as to the processes happening in the nervous system [261]. SC methods measure arousal based on tiny changes in conductance that occur when something activates the autonomic nervous system [255]. The sympathetic branch of the autonomic nervous system controls the skin's sweat glands, and the activity of the glands determines the galvanic skin response [265].
Heart rate (HR)/Electrocardiogram (ECG)	[19,111,251,256,261,303]	An ECG is used to measure the electrical activity of the heart [261]. An ECG relies on cardiac electrical activity and measures the electrical impulses that travel through the heart with each beat, causing the heart muscle to pump blood. In ECGs of a normal heartbeat, the timing of the lower and top chambers of the heart is charted [303].
Respiratory Rate Assessment (RRA)	[111,261,304]	Respiratory rate, one of fundamental vital signs, is sensitive to various pathological situations (clinical deterioration, pneumonia, adverse cardiac events, etc.), as well as stressors [304].
Skin temperature (SKT)	[305]	SKT data can be used to measure the thermal responses of human skin. SKT depends on the complex relationship between blood perfusion in the skin layers, heat exchange with the environment, and the central warmer regions of the skin [305]

Table 2. Cont.

Technique	Author(s)	Description
Photoplethysmography (PPG) or Blood volume pulse (BVP)	[305]	Changes in the amplitudes of PPG signals are related to the level of tension in a human being. PPG is a simple, non-invasive method of taking measurements of the cardiac synchronous changes in the blood volume [305].
Trapezium electromyogram	[306]	EMG is a technique that can be used to evaluate and record the electrical activity generated by skeletal muscle [306], for example the trapezius muscle [307].
Neurotransmitter (NT)	[251,308]	Brain neurotransmitters are particular chemical substances that act as messengers in chemical synaptic transmissions and can transmit emotive information. They have excitability and inhibitive abilities [308].
Voice/Speech/Voice Pitch Analysis (VPA)	[263,267,300,309,310]	This is a method of emotion recognition that relies on the person's voice.
Implicit Association Test (IAT)	[255,264,311]	IAT measures individual behavior and experience by assessing the reaction times of subjects to determine their inner attitudes. The subjects are given two cognitive tasks, and measurements are taken of the speed at which they associate two distinct concepts (brands, advertisements, etc.) with two distinct assessed features. IATs can be used to identify hierarchies of products by means of comparisons [255].
Mouse Tracking (MT)	[257,312]	Recognition of a user's emotions is possible based on their mouse movements. Users can be classified by extracting features from raw data on mouse movements and employing complex machine learning techniques (e.g., a support vector machine (SVM)) and basic machine learning techniques (e.g., k-nearest neighbor) [312].
Signature (Figure 9)	[298–300,309]	Emotions can be identified by their handwriting style, and in particular their signature.
Gait (Figure 9)	[298–300,309]	This method allows for emotions recognition based on a person's walking style or gait [300].
Lip Movement	[299]	Lip movement measurements are a recently developed form of biometric emotions recognition that is very similar to the way a deaf person determines what is being said by tracking lip movements [299].
Gesture	[298,309]	Gesture recognition is used to identify emotions rather than a person, and gestures are grouped into certain categories [298].
Keystroke/Typing Recognition (Figure 9)	[169,300]	In this method, the unique characteristics of a person's typing style are used for emotions identification purposes [300].

Most of today's eye tracking systems are video-based, with an eye video camera and infrared illumination. Eye tracking systems can be categorized as tower-mounted, mobile, or remote based on how they interface with the environment and the user (Figure 6) and different video-based eye tracking systems are required depending on the experiment, the environment, and the type of activity to be studied [313]. Researchers have used eye-tracking for behavioral research.

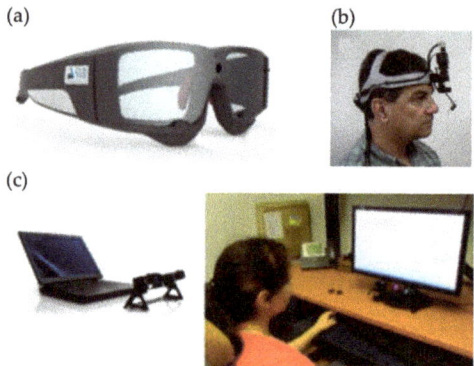

Figure 6. Sample of various kinds of eye-tracking tools: (**a**) eye-tracking glasses [314]; (**b**) helmet-mounted [315]; (**c**) remote or table [316].

The left image in Figure 7 shows the last frame of an expression showing surprise on a sample face from Cohn–Kanade database and highlights the trajectories (the bright lines that change color from darker to brighter from their start to end) followed by each tracked feature point. Figure 7. The application of the dense flow method (right) and the result of applying the feature optical flow on the subset of 15 points (left) [317].

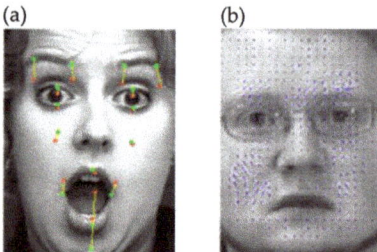

Figure 7. Facial expression recognition: (**a**) feature point tracking; (**b**) dense flow tracking [317].

A group of participants were tested to record the facial EMG (fEMG) activity. Following the guidelines for fEMG placement recommended by Fridlund and Cacioppo, two 4-mm bipolar miniature silver/silver chloride (Ag/AgCl) skin electrodes were placed on their left corrugator supercilii and zygomaticus major muscle regions (Figure 7) [318]. To avoid bad signals or other unwanted influences, the BioTrace software (on NeXus-32) was used to visualize and, if necessary, correct the biosignals before each recording. Figure 8 shows the arrangement of fEMG electrodes on the M. zygomaticus major and M. corrugator supercilii. An example of a filtered electromyography (EMG) signal is shown on the right side [319].

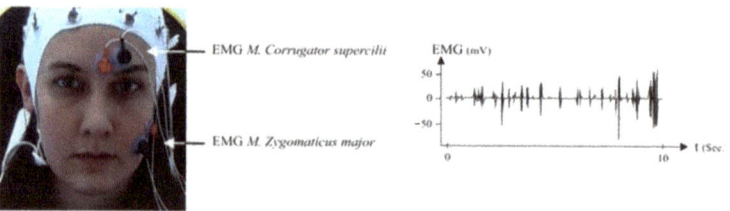

Figure 8. Placement of fEMG electrodes and a sample of a filtered EMG signal [319].

Humans have a range of biometric traits that can be a basis for various biometric recognition systems (Figure 9). The other biometrics traits are iris, face thermogram, gait, keystroke pattern, voice, face, and signature. They can have different significance. For example, iris scan has high accuracy, medium long term stability and medium security level, while voice recognition has low accuracy, low long term stability and low security level [320]. The choice of the biometric traits, however, invariably depends on the availability of the dataset's samples, the application, the value of tolerance accepted, and the level of complexities [150].

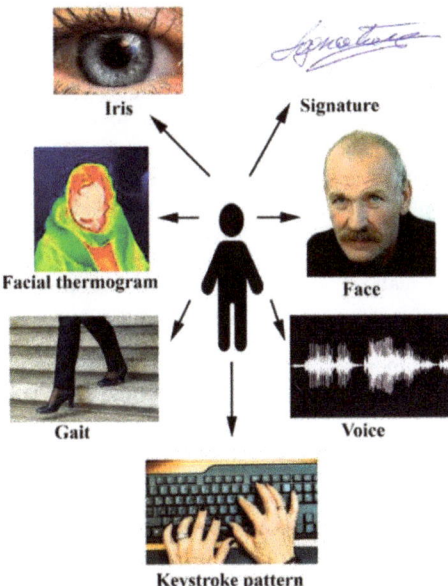

Figure 9. Other examples of biometric traits.

Biometric sensors are transducers that change the biometric traits of a person, such as face, voice, and other characteristics, into an electrical signal. These sensors read or measure speed, temperature, electrical capacity, light, and other types of energy. Different technologies are available with digital cameras, sensor networks, and complex combinations. One type of sensor is required in every biometric device, and biometric sensors are a key feature of emotions recognition technology. Biometrics can be used in a microphone for voice capture or in a high-definition camera for facial recognition [321].

Jain et al. [141] state that enrolment and emotions recognition are two main phases in biometric emotions recognition systems. The enrolment phase means acquiring an individual's biometric data to be stored in the database along with the emotions recognition details. The recognition phase uses the stored data to compare the data with the re-acquired biometric data of the same individual, to determine emotions. A biometric system is, therefore, a pattern recognition system consisting of a database, sensors, a feature extractor, and a matcher.

Loaiza [322] states that overall physiological effects related to emotional reactions depend on three types of autonomic variables: (1) the cardiac system, including blood pressure, cardiac cycles, and heart rate variability; (2) respiration, including amplitude, respiration period, and respiratory cycles; and (3) electrodermal activity, including resistance, responses, and skin conductance levels. Ekman [77] report that different emotions can have very different autonomic variables. For instance, in contrast to someone in a happy state, an angry person had a higher heart rate and temperature. Furthermore, the feeling of fear was also accompanied by higher heart rate. Pace-Schott et al. [323] argue that the ability to regulate physiological state and regulation of emotion are two inseparable features. Physiological feelings contribute to emotion regulation, reproduction, and survival.

Many works have focused on emotion detection using different techniques [35,283,284, 324–327]. Specific tasks (e.g., WASSA-2017, SemEval) have also included emotion detection tasks that cover four categories of emotions (anger, fear, sadness, and joy) [320]. According to Saganowski et al. [326], the most common approach to the use of physiological signals in emotion recognition is to (1) collect and clean data; (2) to preprocess, synchronize, and integrate signal; (3) to extract and select features; and (4) to train and validate machine learning models.

Signals are a natural expression of the human body; they can be used with great success in the classification of emotional states. EEGs, temperature measurements, or electrocardiograms (ECGs) are examples of such physiological signals. They can help us to classify emotional states such as anger, sadness, or happiness, and can be captured by different sensors to identify individual differences. The goal of all of these physiological methods is to evaluate consumer attention and to obtain a particular message noticed, and their performance in this area is commendable. The advantages of these techniques include their creative and versatile placement, the stimulation of interest through novel means that capture attention, the ability to directly target and personalize messages, and lower implementation costs [328]. To study marketing trends, Singh et al. [328] recommend avoiding costly research methods such as fMRI and EEG, and instead using smaller and cheaper galvanic readings and eye tracking (ET) to investigate brain responses. These authors also propose a fuzzy rule-based algorithm to anticipate consumer behavior by detecting six facial expressions from still images.

Various organizations are contributing to the progress of biometric standards, such as international standards organizations (International Electrotechnical Commission, ISO-JTC1/SC37, London, UK), national standards bodies (American National Standards Institute, New York, NY, USA), standards-developing organizations (International Committee for Information Technology Standards, American National Institute of Standards and Technology, Information Technology Laboratory), and other related organizations (International Biometrics and Identification Association, International Biometric Group, Biometric Consortium, Biometric Center of Excellence) [329]. De Angel et al. [330] give rise to numerous recommendations to begin improving the generalizability of the research and generating a more standardized approach to sensing in depression.

- Sample recommendations include reporting on recruitment strategies, sampling frames and participation rates; increasing the diversity of the study population by enrolling participants of different ages and ethnicities; reporting basic demographic data such as age, gender, ethnicity, and comorbidities; and measuring and reporting participant engagement and acceptability in terms of attrition rates, missing data, and/or qualitative data.
- Furthermore, in machine learning models—describing the model selection strategy, performance metrics and parameter estimates in the model with confidence intervals or nonparametric equivalents.
- Recommendations for data collection and analysis include using established and validated scales for depression assessment; presenting any available evidence on the validity and reliability of the sensor or device used; describing in sufficient detail so as to enable replication, data processing and feature construction; and providing a definition and description of how missing data is handled.
- Recommendations for data sharing include making the code used for feature extraction available within an open science framework and sharing anonymized datasets in data repositories.
- The key recommendation is recognizing the need for consistent reporting in this area. The fact that many studies—especially in the field of computer science—fail to report basic demographic information. A common framework should be developed that has

standardized assessment and analysis tools and reliable feature extraction and missing data descriptions, and has been tested in more representative populations.

Neuromarketing, neuroeconomics, neuromanagement, neuro-information systems, neuro-industrial engineering, products, services, call centers studies use various instruments and techniques to measure user psychological states. Some of these tools are more complex than others, and the results that are produced can vary widely [331]. They fall into three major categories: the first two contain tools used for neuroimaging (medical devices offering in vivo information on the nervous system) and use techniques that measure brain electrical activity and neuronal metabolism, while the third contains tools used to evaluate neurophysiological indicators of the mental states of an individual. Leading neuroimaging tools such as fMRI and PET fall into the first category, while EEG, MEG, and other less invasive and cheaper neuroimaging devices that measure electrical activity in the brain [332] fall into the second category, and tools that track and record individual signals of broader physiological reaction and response measurements (e.g., electro-dermal activity, ET, etc.) fall into the third category.

Next, we overview the literature and examine the various types of arousal, valence, affective attitudes, and emotional and physiological states (AFFECT) recognition methods in more detail. A summary of the outcomes is provided in Table 3.

The combination of several different approaches to the recognition and classification of emotional state (also known as multimodal emotion recognition) is currently a research area of great interest, especially since the use of different physiological signals can provide huge amounts of data. Since each physiological can make a significant impact on the ability to classify emotions [333]. Table 3 presents an overview of studies related to the recognition of valence, arousal, emotional states, physiological states, and affective attitudes (affect). A brief overview of some of these studies follows.

Table 3. An overview of studies on arousal, valence, affective attitudes, and emotional and physiological states (AFFECT) recognition.

Stimulus	AFFECT	Methods	Reference
Recording of dances, video	Anger, fear, grief, and joy	GSR, eye movement (Figure 6)	[334]
Neurophysiological research from 2009 to 2016	Overview of the existing works in emotion	EEG	[335]
Affective stimuli	Surprise, disgust, anger, fear, happiness, and sadness	EEG	[336]
The visual stimuli, black and white photographs of 10 different models	Happy, sad	MEG	[337]
20 face actors, each displaying happy, neutral, and fearful facial expressions	Happy, neutral, fearful	MEG	[338]
Task-irrelevant emotional and neutral pictures	Pleasant, unpleasant	TMS	[339]
A subset of music videos from the Dataset for Emotions Analysis using Physiological signals (DEAP) dataset	Valence, arousal	fNIRS, EEG	[340]
Emotional faces for the emotion perception test	Pleasant, unpleasant, neutral	fMRI	[341]
-	Stress	PET	[342]
Video	Happiness, sadness, disgust, anxiety, pleasant, unpleasant, neutral	PET	[343]

Table 3. *Cont.*

Stimulus	AFFECT	Methods	Reference
Facial Emotion Selection Test (FEST)	Positive, negative	DTI	[344]
Real time biometric-emotional data collection from depersonalized passersby	Neutral, happiness, sadness, surprised, anger, scared, valence, arousal, disgust, interest, confusion, boredom	Emotional, Affective and Biometrical States Analytics of the Built Environment Method	[345]
Real time data collection	Happy, sad, angry, surprised, scared, disgusted, valence, arousal	Method of an Affective Analytics of Demonstration Sites	[346]
Scanning a human-centered built environment, real time data collection	Sadness, disgust Happiness, anger, fear surprise, boredom, neutral, arousal, valence, confusion, and interest	Affect-Based Built Environment Video Analytics	[347]
Remote real time data	Happiness, arousal, valence	Video Neuro-advertising Method	[93,348]
Smelling strips	Happy, radiant, well-being, soothed, energized, romantic, sophisticated, sensual, adventurous, comforted, amused, interested, nostalgic, revitalized, self-confident, surprised, free, desirable, daring, excited	IRT	[349]
Text	Positive and negative valence	Eye tracking (ET)	[350]
21 video fragments	High/low arousal, high/moderate/low valence	Eye tracking (ET)	[351]
Crypts	Feelings, tendermindedness, warmth, trust and positive emotions	Iris	[294]
The simulation environment	Wellness/malaise, relaxation/tension, fatigue/excitement	Retina	[352]
Colors	Surprise, Happiness, Disgust, Anger, Sadness and Fear	Blinking, heart rate	[353]
HSV color space	Fear, disgust, surprise, joy, anticipation, sadness, anger, trust	Blinking	[354]
Review of existing novel facial expression recognition systems	Anger, disgust, fear, happiness, sadness, surprise and neutral	Facial expression recognition	[355]
Destination promotional videos	Pleasure, arousal	Skin conductance, facial electromyography	[355]
Games scenario between a human user and a 3D humanoid agent	Arousal, valence, fear, frustrated, relaxed, joyful, excited	Electromyography, skin conductance	[356]
Dramatic film	Real-time emotion estimation	EEG, Heart Rate, Galvanic Skin Response	[357]
Emotional state of a driver while in an automobile	Happy, anger	Electrocardiogram (ECG)	[358]
Music	Pleasure, unpleasure	Heart and respiratory rates	[359]
Trier Social Stress Test	Stress, relax	Respiratory rate and heart rate	[360]
Voice- and speech-pattern analysis	Normal, angry, panic	Voice, speech	[361]
Implicit anxiety-related self-concept	Shame, guilt proneness, anxiety, anger-hostility	Implicit Association Test	[362]

Table 3. Cont.

Stimulus	AFFECT	Methods	Reference
Case studies	Self-control, happiness, anger, fear, sadness, surprise, and anxiety	Mouse Tracking	[302]
Academic study website	Neutral, positive, negative	Mouse Tracking	[363]
Motor improvisation task	Joy, sadness, and a neutral control emotion	Signature	[364]
-	Neutral, joy, anger, sadness	Gait	[365]
Text	Neutral, joy, surprise, fear, anger, disgust, sadness	Lip Movement	[366]
Dataset	Anger, disgust, fear, happiness, sadness, and surprise	Keystroke dynamics	[367]
Recall of past emotional life episodes	Valence, arousal	EEG	[368]
Physiological emotional database for real participants	Valence, arousal	Peripheral signals, EEG	[369]
Data from wearable sensors on subject's skin	High/neutral/low arousal and valence	ECG, EEG, electromyography (EMG)	[370]
Real time heartbeat rate and skin conductance	High/low arousal and valence	GSR, temperature, breathing rate, blood pressure, EEG	[371]
Multimedia contents based on IPTV, mobile social network service, and blog service	Pleasant, unpleasant	GSR, skin temperature, heart rate	[372]
Stress stimuli	High/low valence, high/low arousal	GSR, heart rate, ECG	[373]
CCD-capture human face, measure user's physiological data	Pleasant, unpleasant	GSR, photoplethysmogram (PPG), skin temperature	[374]
Music videos	High/low arousal, high/low valence	EEG	[375]
Detect the current mood of subjects	High/low arousal, high/low valence	EEG	[376]
DEAP database	Joy, fear, sadness, relaxation	EEG, back-propagation neural network	[377]
Hjorth features, statistics features, high order crossing features	Happy, calm, sad, scared	EEG, CNN, LSTM recurrent neural networks	[378]
Thirty film clips	Serenity, hope, joy, awe, love, gratitude, amusement, interest, pride, inspiration	EEG	[379]
Transcendental meditation	Ecstasy	EEG	[380]
Ultimatum game	Acceptance	EEG	[381]
Driving a car equipped	Trust	EEG, GSR	[382]
12 prototypes that were designed based on the framework of diachronic opposite emotions	Amazement, happiness	EEG, SD tests	[383]
Audio-visual emotion database	Pleasure, irritation, sorrow, amazement, disgust, and panic	-	[384]
Sleep measures	Grief	EEG	[385]
Real episodes from subjects' lives	Grief, anger	EEG	[386]

Table 3. Cont.

Stimulus	AFFECT	Methods	Reference
Virtual environment consisting of three types of cues	Pensiveness relaxation, non-arousal, stress	EEG	[387]
Patient with dramatic, episodic, seizure-related rage and violence	Rage and aggression	Video-EEG recording	[388]
DEAP database	Rage	EEG, multiclass-common spatial patterns	[389]
Brainstem auditory evoked potentials	Rage and self-injurious behavior	EEG, brainstem evoked potentials (BAEPs)	[390]
Acoustic annoyance	Annoyance	EEG	[391]
70 dBA white noise and pure tones at 160 Hz, 500 Hz and 4000 Hz	Annoyance	EEG	[392]
30 pictures from International Affective Picture System	Neutral, joy, sadness anger, surprise, valence (positive and negative), contempt, fear, disgust	EEG	[393]
Movie clips	Anger, fear, anxiety, disgust, contempt, joy, happiness	EEG	[394]
Emotional factor	Aggressiveness	EEG	[395]
Buss–Durkee questionnaire	Aggressiveness	EEG	[396,397]
Reward anticipation	Anticipation	EEG	[398]
Structured Clinical Interview for DSM-IV	Anticipation	EEG, fMRI	[399]
DEAP database	High/low valence and arousal	EEG	[400–405]
Reading and reflection task about Muslims	Disapproval	EEG, ANOVA	[406]
Simulated train driving	Fatigue and distraction	EEG, Multi-type feature extraction, CatB-FS algorithm	[407]
Faces (the participant's own face, the face of a stranger, and a celebrity's face)	Admiration	EEG, 18-Items Narcissistic Admiration and Rivalry Questionnaire	[408]
Presentation of 12 virtual agents	Acceptance	EEG and the virtual agent's acceptance questionnaire (VAAQ)	[409]
English prosocial and opposite antisocial words in a sentence	Approval and disapproval	EEG, ANOVA	[410]
Data from Facebook comments	Enjoyment (peace and ecstasy), sadness (disappointment and despair), fear (anxiety and terror), anger (annoyance and fury), disgust (dislike and loathing) surprise, other (neutral)	Natural language processing (NLP); convolutional neural network (CNN) and long short-term memory (LSTM); Random Forest and support vector machine (SVM), standard Vietnamese social media emotion corpus (UIT-VSMEC)	[411]
Video clips	Pride, love, amusement, joy, inspiration, gratitude, awe, serenity, interest, hope	fNIRS	[412]
User's interaction with a web page	Arousal/valence anxiety and aggressiveness	Facial expressions, Facial Action Coding System, specialized questionnaires	[413]
An investment game that uses artificial agents	Trust	EEG	[285]

Table 3. Cont.

Stimulus	AFFECT	Methods	Reference
Simulated autonomous system	Trust	EEG and GSR	[382]
The iCV-MEFED dataset. For each subject in the iCV-MEFED dataset, five sample images were captured.	Neutral, angry, contempt, happy, happily surprised, surprisingly fearful, surprised	Facial emotion recognition (Figure 7), CNN; Inception-V3 network	[414]
Dynamic emotional facial expressions were generated by using FACSGen	Contempt, disgust, sadness, neutral	ANOVA, Participants completed emotion scales	[415]
Film clips	Pride, love, amusement, joy, inspiration, gratitude, awe, serenity, interest, hope	EEG, multidimensional scaling (MDS), intra-class correlation coefficients (ICCs)	[379]
Simulated driving system	Vigilance	EEG and forehead electrooculogram (EOG), eye tracking (Figure 6)	[416]
DEAP dataset	Optimism, pessimism, calm	EEG, CNN	[166]
Music	Relaxing-calm, sad-lonely, amazed-surprised, quiet-still, angry-fearful, happy-pleased	Binary relevance (BR), label powerset (LP), random k-label sets (RAKEL), SVM	[417]
Music	Happiness, love, anger and sadness	EEG, SVM, Multi-Layer Perceptron (MLP), and K-nearest Neighbor (K-NN)	[418]
Three sets of pictures	Anticipation	Facial emotions (Figure 7), action observation network (AON), two-alternative forced-choice procedure, Reaction times (RT), ANOVA	[419]
Individuals enacted aggressive actions, angry facial expressions and other non-aggressive emotional gestures	Aggressive actions and anger	Kinect infrared sensor camera: hand movement, body posture, head gesture, face (Figure 9), and speech. SVM and the rule-based features	[420]
Images of faces from the Ekman and Friesen series of Pictures of Facial Affect	Grief	Facial Expression of Emotion Test (Figure 7)	[421]
Music	Soothing, engaging, annoying and boring	FBS fusion of three-channel forehead biosignals, ECG	[422]
Films	Amusement, anger, grief, and fear	Fingertip blood oxygen saturation (OXY), GSR, HR	[423]
Polish emotional database, database consists of 12 emotional states	Rage, anger, annoyance, grief, sadness, pensiveness, ecstasy, joy, serenity, terror, fear, apprehension	Speech, KNN Algorithm	[424]
Video	Nonverbal behaviors signaling dominance and submissiveness	Implicit association test, body language, MANOVA	[425]
Music	High/low valence, high/low arousal	EMG, EEG, HRV, GSR	[426]
The external auditory canal is warmed or cooled with water or air	High and low arousal	Electrodermal activity (EDA), HRV, activity tracker, EMG, SKT	[427]
After-image experiments, direct visual observation, photography of the eyes, recording of the corneal reflex	High/low valence, high/low arousal	GSR, EMG	[428]

Table 3. Cont.

Stimulus	AFFECT	Methods	Reference
Assessment of emotional states experienced by racing drivers	Sadness, fear, anger, surprise, happiness, and disgust	ECG, EMG, respiratory rate, GSR	[429]
Dataset of standardized facial expressions	Happiness, sadness, anger, disgust, fear, and surprise	Facial Action Coding (FC)	[430]
Neighbor sounds	Arousal, valence	fEMG, heart rate (HR), electrodermal activity (EDA)	[431]
Audio visual stimuli	Joy, sadness, anger, fear	ECG	[432]
Playing with the infant to elicit laughter	Joy	Skin temperature (SKT)	[433]
Two different kinds of video inducing happiness and sadness	Happiness, sadness	Photoplethysmography (PPG), skin temperature (SKT)	[434]
International Affecting Picture System (IAPS) pictures	Joy, sadness, fear, disgust, neutrality, amusement	Electromyogram signal (EMG), respiratory volume (RV), skin temperature (SKT), skin conductance (SKC), blood volume pulse (BVP), heart rate (HR)	[435]
Movie and music video clips	Arousal, valence	Electrooculogram (EOG), electrocardiogram (EEG) trapezium electromyogram (EMG)	[436]
Audio/visual	Anger, happiness, sadness, pleasure	GSR, EMG, respiratory rate, ECG	[437]

Many scientists and practitioners have earned acclaim and honor for their research in areas such as diagnostics, large-scale screening, analysis, monitoring, and categorizations of people by COVID-19 symptoms. Their work relied on early warning systems, wearable technologies, the Internet of Medical Things, IoT based systems, biometric monitoring technologies, and other tools that can assist in the COVID-19 pandemic. Javaid et al. [438] review how different industry 4.0 technologies (e.g., AI, IoT, Big data, Virtual Reality, etc.) can help reduce the spread of disease. Kalhori et al. [439] and Rahman et al. [440] discuss the digital health tools to fight COVID-19. Various sensors and mobile devices to detect the disease, reduce its spread, and measure different symptoms are also widely discussed. Rajeesh Kumar et al. [441] propose a system to identify asymptotic patients using IoT-based sensors, measuring blood oxygen level, body temperature, blood pressure, and heartbeat. Stojanović et al. [442] propose a phone headset to collect information about respiratory rate and cough, Xian et al. [443] present a portable biosensor to test saliva. Chamberlain et al. [444] presented distributed networks of Smart thermometers track COVID-19 transmission epicenters in real-time.

Neurotransmitters (NT) are billions of molecules constantly needed to keep human brains functioning. They are chemical messengers that carry, balance, and boost signals travelling between nerve cells (neurons) and other cells in the body. Many different psychological and physical functions can be affected by these chemical messengers, including fear, appetite, mood, sleep, heart rate, breathing rate, concentration and learning [445]. Lim [251] has also outlined new ways of exploiting neuromarketing research to achieve a better understanding of the brain and neural activity and hence advance marketing science. Lim [251] highlighted three main aspects: (i) antecedents (such as the product, physical evidence, the price of the product, the place where everything is happening, promotion, the process involved, people); (ii) the process; and (iii) the consequences for the target market (behavioral outcomes before, during and after the act of buying) and the marketing organization (visits, sales, awareness, equity). Agarwal and Xavier [253] described the most popular neuromarketing tools, including event-related potential (ERP) (P300), EEG, and fMRI, and explained how these tools could be applied in marketing. A business and marketing article [256] lists the three categories of neuroscientific techniques that are applied in business and advertising research (Tables 1 and 2, Figure 1) as follows:

1. Methods that monitor what is happening in the brain (i.e., the physiological activity of the CNS);
2. Methods that record what is happening elsewhere in the body (i.e., the physiological activity of the PNS);
3. Other techniques for tracking behavior and conduct.

Ganapathy [260] groups neuromarketing tools into three categories (Tables 1 and 2). Farnsworth [258] gives information that can be essential when deciding on the best neuromarketing method or technique to help stakeholders understand research methods relating to human behavior at a glance, while Saltini [264] gives a short list of neuromarketing tools (Tables 1 and 2). A system developed by CoolTool [257] allows several neuromarketing tools to be used separately or combined.

Although individual neuroscientific tools for neuromarketing, neuroeconomics, neuromanagement, neuro-information systems, neuro-industrial engineering, products, services, call centers have been developed by many researchers (for example [111,251,253–270,293,298–300,303,309,311,312,328,446–448], a review and analysis of the complete range of tools used in neuromarketing, neuroeconomics, neuromanagement, neuro-information systems, neuro-industrial engineering, products, services, call centers research has not yet been carried out. Thorough examinations of the range of research tool alternatives that are available for neuroscience are also often missing from research in this area. We have therefore compiled a complete list of neuroscience techniques for neuromarketing, neuroeconomics, neuromanagement, neuro-information systems, neuro-industrial engineering, products, services, call centers. Humans experience emotions and their associated feelings (e.g., gratitude, curiosity, fear, sadness, disgust, happiness, and pride) on a daily basis. Yet, in case of affective disorders such as depression and anxiety, emotions can become destructive. Thus the focus on understanding emotional responsiveness is not surprising in neuroscience and psychological science [449]. So neuroscience techniques analyze emotional, affective and physiological states tracking neural/electrical activity [335–340,450,451] or neural/metabolic activity [341–344,349,447,452,453] within the brain. This is also presented in Table 3.

For example, neuromarketing techniques can complement business decisions and make them more profitable, using the automated mining of opinions, attitudes, emotions and expressions from speech, text, emotions, neuron activity and other database-fed sources. Advertisements that are adjusted based on such information can engage the target audience more effectively and make a better impact on the audience, and this may translate into better sales and higher margins. In an attempt to enhance corporate branding and advertising routines, various factors have been studied, such as emotional appeal and sensory branding, to ensure that companies deliver the right message and that customers perceive the right message [171].

Affect recognition is widely used in gaming to create affect-aware video games and other software. Alhargan et al. [454] present affect recognition in an interactive gaming environment using eye-tracking. Szwoch and Szwoch [455] give a review of automatic multimodal affect recognition of facial expressions and emotions. Krol et al. [456] combined eye-tracking and brain–computer interface (BCI) and created a completely hands-free game Tetris clone where traditional actions (i.e., block manipulation) are performed using gaze control. Elor et al. [457] measure heart rate and galvanic skin response (GSR) with Immersive Virtual Reality (iVR) Head-Mounted Display (HMD) systems paired with exercise games to show how exercise games can positively affect physical rehabilitation.

Stress is a relevant health problem among students, so Tiwari, Agarwal [458] present a stress analysis system to detect stressful conditions of the student, including measurement of GSR and electrocardiogram (ECG) data. Nakayama et al. [459] suggest measuring heart rate variability as a method to evaluate nursing students stress during simulation to provide a better way to learn.

A literature review can reveal the most popular types of traditional and non-traditional neuromarketing methods. According to Sebastian [111], focus groups are one of the more traditional marketing methods, while various neuroscience techniques have also been applied to record the metabolic activity of the body and the electrical activity of the brain (transcranial magnetic stimulation (TMS), electroencephalography (EEG), functional magnetic resonance imaging, magnetoencephalography (MEG), and positron-emission tomography (PET)).

Electronic platforms are not the only possibility for non-traditional marketing, and Tautchin and Dussome [460] believe that traditional media can also be reimagined in new forms, such as guerrilla marketing, local displays, vehicle wraps, scaffolding, and even bubble cloud ads or aerial banners. In addition to giving high-quality feedback data, non-traditional techniques can also help in the evaluation of business decisions and conclusions [328].

Based on factors such as skin texture, gender, and SC, wearable biometric GSR sensors could be used to identify whether a person is in a sad, neutral, or happy emotional state. To understand marketing strategies better and to improve ads, other biometric sensors such as pulse oximeters and health bands could be used in the future to make automated predictions of emotions [461]. The galvanic skin response (GSR) method has an important limitation—it does not provide information on valence. The usual way to address this issue is to use other emotion recognition methods. They provide additional details and thus enable detailed analysis. Table 3 lists studies where GSR is used to measure emotions.

Eye tracking (ET) is used to record the frequencies of choices; sensor features are extracted and matched with certain preference labels to determine mutual dependences and to discover which brain regions are active when a certain choice task is performed. High values for alpha, beta and theta waves have been reported in the occipital and frontal brain regions, with a high degree of synchronization. A hidden Markov model is a popular tool for time-series data modeling, and researchers have successfully used this approach to build brain–computer-interface tools with EEG signals, counting mental task classification, medical applications and eye movement tracking [462].

A classification model based on SVM architecture, developed by Lakhan et al. [463], can predict the level of arousal and valence in recorded EEG data. Its core is a feature extraction algorithm based on power spectral density (PSD).

Multimodal frameworks that combine several modalities to improve results have recently become popular in the domain of human–computer interaction. A combination of modalities can give a more efficient user experience since the strengths of one modality can offset the weaknesses of another and the usability can be increased. These systems recognize and combine different inputs, taking into account certain contextual and temporal constraints and thus facilitating interpretation. Kong et al. [464] created a way of using two different sensors and calibrating them to achieve simultaneous gesture recording. Hidden Markov Model (HMM) was used for all single- and double-handed gesture recognition. Multimodality means that several unimodal solutions are combined into a system, meaning that multiple solutions can be combined into a single best solution using optimization algorithms [464].

The automatic emotion recognition system proposed by El-Amir et al. [465] uses a combination of four fractal dimensions and detrended fluctuation analysis, and is based on three bio-signals, GSR, EMG, and EEG. Using two emotional dimensions, the signals were passed to three supervised classifiers and assigned to three different emotional groups, with a maximum accuracy for the valence dimension of 94.3% and a maximum accuracy for the arousal dimension of 94%. This approach is based on external signals such as facial expressions and speech recognition, which means that it is simple and that no special equipment is required. The limitations of this approach are that emotions can be faked, and that these types of recognition methods fail with disabled people and people with certain diseases. Other approaches are based on electromyography, ECGs, SC, EEGs, and other physiological signals that are spontaneous and cannot be consciously controlled [465].

Plassmann et al. [466] as well as Perrachione and Perrachhione [467] carried out exciting studies in an attempt to determine how marketing stimuli lead to buying decisions. They applied neurosciences to marketing in order to create better models and to understand of how a buyer's brain and emotions operate. Gruter [468] states that a wide range of techniques and tools are used to measure consumer responses and behavior. Three approaches that are used in neuromarketing can give access to the brain: input and output models, internal reflexes, and external reflexes.

Leon et al. [469] present a real-time recognition and classification method based on physiological signals to track and detect changes in emotions from a neutral state to either a positive or negative (i.e., non-neutral) state. They used the residual values of auto-associative neural networks and the statistical probability ratio test in their approach. When the proposed methodology was implemented to process a recognition level of 71.4% was achieved [469]. Monajati et al. [470] also investigated the recognition of negative emotional states, using the three physiological signals of galvanic skin response, respiratory rate and heart rate. Fuzzy-ART was applied to analyze the physiological responses and to recognize negative emotions. An overall accuracy of 94% was achieved in determining which emotions were negative as opposed to neutral [470].

Andrew et al. [471] described investigations of brain responses to modern outdoor advertising, focusing on memorability, visual attention, desirability, and emotional intensity. They also described ways in which the latest imaging tools and methods could be applied to monitor subconscious emotional responses to outdoor media in many forms, from multisensory advertising screens to simple paper posters. Andrew et al. [471] explained the cognitive processes behind their success, not solely in the context of the advertising to which people are typically exposed outside their homes, but also in the broader digital world. Andrew et al. findings have fundamental implications for media campaign planning, design, and development, identifying the possible role of outdoor advertising compared to other media, and possible ways of combining different media platforms and making them work for the benefit of advertisers.

Kaklauskas et al. [472] integrated Damasio's somatic marker hypothesis with biometric systems, multi-criteria analysis techniques, statistical investigation, a neuro-questionnaire, and intelligent systems to produce the INVAR neuromarketing system and method. INVAR can measure the efficiency of both a complete video advertisement and its separate frames. This system can also determine which frames make viewers interested, confused, disgusted, happy, scared, surprised, angry, sad, bored, or confused; can identify the utmost positive or negative video advertisement; measure the consequence of a video advertisement on long-term and short-term memory; and perform other functions.

Lajante and Ladhari [473] applied peripheral psychophysiology measures in their research, based on the assumption that measures of emotion and cognition such as SC responses and facial EMGs could make a significant contribution to new ideas about consumer decision making, judgments and behaviors. These authors believe that their approach can help in applying affective neuroscience to the field of consumer services and retailing.

Michael et al. [474] aimed to understand the ways in which unconscious and direct cognitive and emotional responses underlie preferences for particular travel destinations. A 3×5 factorial design was run in order to better understand the unconscious responses of consumers to possible travel destinations. The factors considered in this study were the type of stimulus (videos, printed names, and images) and the travel destination (New York, London, Hong Kong, Abu Dhabi, and Dubai). ET can provide reliable tracking of cognitive and emotional responses over time. The authors suggested that decisions on travel destinations have both a direct and an unconscious component, which may affect or drive overt preferences and actual choices.

Harris et al. [448] investigated ways of measuring the effectiveness of social ads of the emotion/action type, and then of making these ads more effective using consumer neuroscience. Their research offers insights into changes in behavioral intent brought about by effective ads and gives an improved understanding of ways of making good use of social messages regarding a certain action, challenge or emotion that may be needed to help save lives. It can also reduce spending on social marketing campaigns that end up being ineffectual.

Libert and Van Hulle [475] argue that the development of economically practicable solutions involving human–machine interactions (HMI) and mental state monitoring, and neuromarketing that can benefit severely disabled patients has put brain–computer interfacing (BCI) in the spotlight. The monitoring of a customer's mental state in response to watching an ad is interesting, at least from the perspective of neuromarketing managers. The authors propose a method of monitoring EEGs and predicting whether a viewer will show interest in watching a video trailer or will show no interest, skipping it prematurely. They also trained a k-nearest neighbor (kNN), a support vector machine (SVM), and a random forest (RF) classifier to carry out the prediction task. The average single-subject classification accuracy of the model was as follows: 73.3% for viewer interest and 75.803% for skipping using SVM; 78.333% for viewer interest and 82.223% for skipping using kNN; and 75.555% for interest and 80.003% for skipping using RF.

Jiménez-Marín et al. [476] showed that sensory marketing tends to accumulate user experiences and then exploit them to bring the users closer to the product they are evaluating, thus motivating the final purchase. However, several issues need to be considered when these techniques are applied to reach the desired outcomes, and it is important to be aware of recent advances in neuroscience. The authors explore the concept of sensory marketing, pointing out its possibilities for application and its various typologies.

Cherubino et al. [477] highlighted the new technological advances that have been achieved over the last decade, which mean that research settings are now not the only scenarios in which neurophysiological measures can be employed and that it is possible to study human behavior in everyday situations. Their review aimed to discover effective ways to employ neuroscience technologies to gain better insights into human behavior related to decision making in real-life situations, and to determine whether such applications are possible.

Monica et al. [478] explored the cognitive understanding and usability of banking web pages. They reviewed the theoretical literature on user experience in online banking services research, with a focus on ET as a research tool, and then selected two Romanian banking websites to study consumer attention, while consumers were navigating the sites, and memory, after their visits. The research findings showed that the layout and information display can make web pages more or less usable and can have an effect on cognitive understanding.

Singh et al. [328] discussed various methods of feature extraction for facial emotion detection. The algorithm they proposed could detect a total of six facial emotions, using a fuzzy rule-based system. During their experiment, neurometrics were recorded using a system comprising MegaMatcher software, Grove-GSR Sensor V1.2, and a 12-megapixel Hikvision IP camera. The participants were asked to watch a set of video ads for a range of well-known cosmetic products and wore SC sensors and sat in front of a camera that monitored their responses. Singh et al. [328] analyzed the cognitive processes of university students in relation to advertising and compliance with the code of self-regulation. A quantitative and qualitative methodology based on facial expressions, ET techniques and focus groups was used for this purpose. The results suggested that online game operators could be clearly identified. A high interaction of the public within the exhibition of supposed skills of the successful player and welcome bonuses also exists, and there was shown to be a lack of knowledge of the visual elements of awareness, a trivialization of compulsive gambling, and sexist attitudes towards women attracting public attention. A positive public attitude towards gaming was also observed by Singh et al. [328]; it was seen as a healthy form of leisure that was compatible with family and social relationships.

Goyal and Singh [461] proposed the use of research-based approaches for the automatic recognition of human affective facial expressions. These authors created an intelligent neural network-based system for the classification of expressions from extracted facial images. Several basic and specialized neural networks for the detection of facial expressions were used for image extraction.

Electromyography measures and assesses electric potentials in muscle cells. In medical settings, this method is used to identify nerve and muscle lesions, while in emotion recognition this method is used to look for correlations between emotions and physiological responses. Most EMG-based studies examine facial expressions drawing on the hypothesis that facial expressions take part in emotional responses to various stimuli. The hypothesis was first proposed by Ekman and Friesen in 1978; they described the relationships between basic emotions, facial muscles, and the actions they trigger. Morillo et al. [479] used low-cost EEG headsets and applied discrete classification techniques to analyze scores given by subjects to individual TV ads, using artificial neural networks, the C4.5 algorithm and the Ameva discretization algorithm. A sample of 1400 effective advertising campaigns was studied by Pringle et al. [480], who determined that promotions with exclusively emotional content achieved around double (31% vs. 16%) success as those with only rational content, while compared to campaigns with mixed emotional and rational content, the exclusively emotional campaigns performed only slightly better (31% vs. 26%).

According to Takahashi [481] some of the available emotion recognition systems in facial expressions or speech look at several emotional states such as fear, teasing, sadness, joy, surprise, anger, disgust, and neutral. Takahashi [481] investigated emotion recognition based on five emotional states (fear, anger, sadness, joy, and relaxed).

The authors [353,355–357,359,360,371–374] carried out an in-depth analysis of how blood pressure, SC, heart rate and body temperature depend on stress and emotions. Figures suggest that work-related stress costs the EU countries at least EUR 20 billion annually. Stress experienced at work can cause anxiety, depression, heart disease and increased chronic fatigue which can have a considerable negative impact on creativity, competitiveness and work productivity.

Research worldwide shows that people exposed to stress can experience higher blood pressure and heart rate. Light et al. [482] analyzed cases of daily elevated stress levels and looked at the effects on fluctuations in systolic and diastolic blood pressure. Gray et al. [483] investigated how systolic and diastolic blood pressure can be affected by psychological stress, while Adrogué and Madias [484] described the effects of chronic, emotional and psychological stress on blood pressure. The unanimous conclusion of research in this area is that diastolic and systolic blood pressure and heart rate depend on stress and can increase depending on the level of stress.

Blair et al. [485] analyzed the effect of stress on heart rate and concluded that heart rate rises sharply within three minutes of the onset of stress and starts to fall only after another five to six minutes. Gasperin et al. [486] concluded that high blood pressure was affected by chronic stress. A number of studies have shown that patients with heart rates higher than 70 beats per minute are more likely to develop cardiovascular diseases and to die from them; tests show that a rapid heartbeat increases the risk of heart attack by 46%, heart insufficiency by 56% and death by 34%.

Sun et al. [487] proposed an activity-aware detection scheme for mental stress. Twenty participants took part in their experiment, and galvanic skin response, ECG, and accelerometer data were recorded while they were sitting, standing, and walking. Baseline physiological measurements were first taken for each activity, and then for participants exposed to mental stressors. The accelerometer was used to track activity, and the data gave a classification accuracy between subjects of 80.9%, while the 10-fold cross-validation accuracy for the classification of mental stress reached 92.4%. This study focused on physiological signals for example photoplethysmography and galvanic skin response. The neural network configurations (both recurrent and feed forward) were examined and a comprehensive performance analysis showed that the best option for stress level detection was layer recurrent neural networks. For a sample of 19 automotive drivers, this evaluation achieved an average sensitivity of 88.83%, a precision of 89.23% and a specificity of 94.92% [488].

Palacios et al. [489] applied a new process involving two databases containing utterances under stress by men and women. Four classification methods were used to identify these utterances and to organize them into groups. The methods were then compared in terms of their final scores and quality performance.

Fever occurs when the body's thermoregulatory set point increases, and many findings suggest that the rise in core temperature induced by psychological stress can be seen as fever. A fever of psychological origin in humans might then be a result of this mechanism [490].

Wu and Liang [491] presented a training and testing procedure for emotion recognition based on semantic labels, acoustic prosodic information and personality traits. A recognition process based on semantic labels was applied, using a speech recognizer to identify word sequences, and HowNet, a Chinese knowledge base, was used as the source for deriving the semantic word sequence labels. The emotion association rules (EARs) of the word sequences were then mined by applying a text-based mining method, and the relationships between the EARs and emotional states were characterized using the MaxEnt model. In a second approach based on acoustic prosodic information, emotional salient segments (ESSs) were detected in utterances and their prosodic and acoustic features were extracted, including pitch-related, formant, and spectrum attributes. The next step was the construction of base-level classifiers using SVM, gaussian mixture models (GMM) and MLP, which were then combined (using MDT) by selecting the most promising option for emotion recognition based on acoustic prosodic information. The process ended when the final emotional state was determined. A weighted product fusion method was applied to combine the outputs produced by the two types of recognizers. The personality traits of the specific speaker, as determined from the Eysenck personality questionnaire, were then taken into consideration to examine their impact and personalize the emotion recognition scheme [491].

A hybrid analysis method for online reviews proposed by Nilashi et al. [492] allows for the ranking of factors affecting the decisions of travelers in their choice of green hotels with spa services. This method combined text mining, predictive learning techniques and multiple criteria decision-making methods, and was proposed for the first time in the context of hospitality and tourism, with an emphasis on green hotel customer grouping based on online customer feedback. Nilashi et al. [492] used the latent Dirichlet analysis method to analyze textual reviews, a self-organizing map for cluster analysis, the neuro-fuzzy method to measure customer satisfaction, and the TOPSIS method to rank the features of hotels. The proposed method was tested by analyzing travelers' reviews of 152 Malaysian hotels. The findings of this research offer an important method of hotel selection by travelers, by means of user-generated content (UGC), while hotel managers can use this approach to improve their marketing strategies and service quality.

A neuromarketing method for green, energy-efficient and multisensory homes, proposed by Kaklauskas et al. [493], can be used to determine the conditions that are required. The multisensory dataset (physiological and emotional states) collected as part of this research contained about 200 million data points, and the analysis also included noise pollution and outdoor air pollution (volatile organic compounds, CO, NO_2, and PM10). This article discussed specific case studies of energy-efficient and green buildings as a demonstration of the proposed method. The results matched findings from both current and previous studies, showing that the correlation between age and environmental responsiveness has an inverse U shape and that age is an important factor affecting interest in eco-friendly, energy-efficient homes.

The VINERS method and biometric techniques developed by Kaklauskas et al. [494] for the analysis of emotional states, physiological reactions and affective attitudes were used to determine which locations are the best choice and then to show neuro ads of available homes offered for sale. Homebuyers were grouped into rational segments, taking into account consumer psychographics and behavior (happy, angry or sad, and valence and heart rate) and their demographic profiles (age, gender, marital status, children or no children, education, main source of income). A rational video ad for the respective rational segment was then selected. This study aimed to combine the somatic marker hypothesis, neuromarketing, biometrics and the COPRAS method, and to develop the VINERS method for use with multi-criteria analysis and the neuromarketing of the best places to live. The case study presented in the article demonstrated the VINERS method in practice.

Etzold et al. [495] examined the case of users booking appointments online, and the ways in which they interacted with the webpage interface and visualizations. The main point was to determine whether a new interface for online booking was easy to navigate and successful in attracting user attention. In this study, the authors particularly wanted to determine whether a new, more expensive customer website was seen as more user-friendly and supportive than the older, cheaper alternative. An empirical study was carried out by tracking users eye movements as they were navigating the existing website of Mercedes-Benz, a car manufacturer, and then a new, updated version of the same company's website. A total of 20 people were observed, and evaluations of their ET data suggested that the new service appointment booking interface could be further improved. Scan-paths and heatmaps demonstrated that the old website was superior [495].

In recent years, many different emotional values, such as the net emotional value (NEV), the service encounter emotional value (SEEVal), and others, have been analyzed. Attempts have been also made to put them into practice [496–503]. These studies are overviewed below. To calculate NEV, the average score for negative emotions (stressed, dissatisfied, frustrated, unhappy, irritated, hurried, disappointed, neglected) is subtracted from the average score for positive emotions (cared for, stimulated, happy, pleased, trusting, valued, focused, safe, interested, indulgent, energetic, exploratory). The average score obtained this way can be used to characterize a client's feelings about a service or a product [499]. A higher value of NEV indicates that the relationships forged by a business are more reliable. One advantage of the NEV is that it characterizes the total balance of a consumer's feelings related to products or services, and thus reveals the value drivers. The relationship between NEV and client satisfaction is linear [500].

The NEV can be used to highlight both aspects that need to be improved, and those that are positive. Since the NEV is calculated based on a subtraction, the result may be either a negative or a positive number. The overall score can indicate what is happening with the client at an emotional level, and suggest ways to use this to gain competitive advantage [501].

The SEEVal is another measure proposed by Bailey et al. [504], and is the sum of the NEV experienced by the client and the NEV experienced by the product or service provider's employee. The client's end results linked to SEEVal are typically loyalty, satisfaction, pleasure, and voluntary benevolence [504]. The IGI Global Dictionary defines an emotional value as a set of positive moods (feeling good or being happy) resulting from products or services and contained in the value gain from the customers' emotional states or feelings when using the products or services (IGI Global Dictionary). Emotional value acts as a moderator, and has significant effects on the roles of social, functional, epistemic, conditional and environmental values [497].

Zavadskas et al. [505] examined data on potential buyers to analyze the hedonic value in one-to-one marketing situations. They used the neutrosophic PROMETHEE technique to examine arousal, valence, affective attitudes, emotional and physiological states (AFFECT), and argued that hedonic value is tied to several factors including customers' social and psychological data, client satisfaction, criteria of attractiveness, aesthetics, and economy, the sales site rental price, emotional factors, and indicators of the purchasing process. Their research showed that an analysis of the aforementioned data on potential buyers can make an important contribution to more effective one-to-one marketing. The case study cited in this work concerned two sites in Vilnius and intended to calculate the hedonic value of these sites during the Kaziukas Fair.

The ROCK Video Neuroanalytics and associated e-infrastructure were established as part of the H2020 ROCK project. This project tracked passers-by at ten locations across

Vilnius. One of our outputs is the real-time Vilnius Happiness Index (Figure 10 and https://api.vilnius.lt/happiness-index, accessed on 5 September 2022). The project also involved a number of additional actions (https://Vilnius.lt/en/category/rock-project/, accessed on 5 September 2022).

The intensity of the most intense negative emotion (scared, disgusted, sad, angry) subtracted from the intensity of "happiness" equals valence [430]. This way the single score of valence combines both positive and negative emotions. Our pool of data comprised 208 million data points analyzed using SPSS Statistics, a statistical software suite. Figure 10b presents the average values of valence per hour on weekdays. Every hour, the changes of average valence among Vilnius passers-by were recorded. Valence was measured every second and these values were accumulated by weekdays (marked in the chart with specific colors) at 95% confidence intervals. The y-axis shows the average values of valence (which fluctuates between −1 to 1) for each full day, for seven days, and the x-axis shows the hour starting at midnight [348].

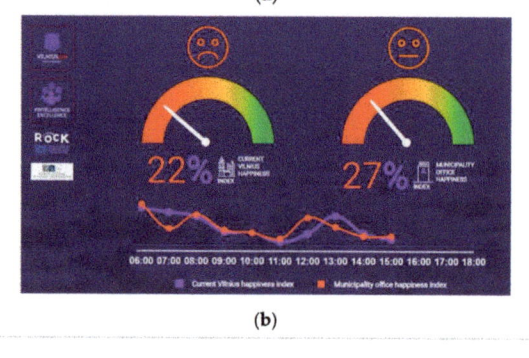

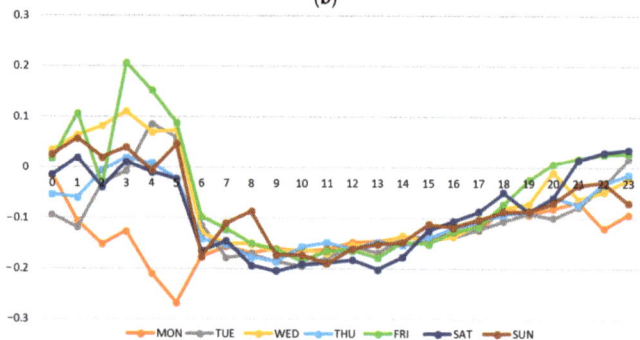

Figure 10. Real-time Vilnius Happiness Index (**a**) and the mean magnitudes of valence, by the hour, on weekdays (**b**).

5. Users' Demographic and Cultural Background, Socioeconomic Status, Diversity Attitudes, and Context

Emotions are a means to engage in a relationship with others: Anger means that the person refuses to accept a specific treatment from others and expresses that they feel entitled to something more. Anger is expressed with the aim of influencing, controlling, and fixing the behavior of others [506].

Through emotions, people can adaptively respond to opportunities and demands they face around them [507–509]. When people face everyday stressors, stressful transitions, ongoing challenges, and acute crises, the adaptive function of emotions is evident in all of these situations. Emotions also depend on context [510]. This means that emotions are most effective when people express them in the situational contexts for which the emotions most likely evolved. In addition, they are specifically most likely to promote adaptation in such scenarios. The experience of anger, for instance, is adaptive because it motivates the focus of energies and the mobilization of resources toward an effective response. When a person expresses anger, adaptive mechanisms are also at work because it shows the person's willingness, and perhaps even ability, to defend themselves. Emotional responses are sensitive to contexts, and are therefore, an integral part of our ways to adapt to daily life and the environment [511].

The ability to modify emotion responses according to changing context may be an important element of psychological adjustment [510]. An individual's capacity to modify emotion responses taking into account the demands of changing contexts (i.e., environmental or interpersonal) is particularly relevant. This mechanism is known as emotion context sensitivity [511].

Cultural and gender differences in emotional experiences have been identified in previous research [512]. For instance, these authors used the Granger causality test to establish how a person's cultural background and situation affect emotion. The conclusions drawn by [513] propose a top-down mechanism where gender and age can impact the brain mechanisms behind emotive imagery, either directly or by interacting with bottom-up stimuli.

Cultural neuroscientists are studying how cultural traits such as values, beliefs, and practices shape human affective, emotional, and physiological states (AFFECT) and behavior. Hampton and Varnum [514] have reviewed theoretical accounts on how culture impacts internal experiences and outward expressions of emotion, as well as how people opt to regulate them. They also analyze cultural neuroscience research that investigates how emotion regulation varies in different cultural groups.

Thus far, differences between nations have largely been the focus in studies of culture in social neuroscience. Culture impacts more than just our behavior—it also plays a role in how we see and interpret the world [515]. For instance, socioeconomic factors such as education, occupation, and income have a significant impact on how a person thinks. In one study, working-class Americans were shown to exhibit a more context-dependent thought process, similar to the collectivist patterns seen in other countries. Individuals of a lower social class in terms of their socio-economic status agreed with contextual explanations of economic trends, broad social outcomes, and emotions [516].

Gallo and Matthews [517] looked at the indirect evidence that socioeconomic status is associated with negative emotions and cognition, and that negative emotions and cognition are associated with target health status. They also proposed a general framework for understanding the roles of cognitive–emotional factors, arguing that low socioeconomic status causes stress, and impairs a person's reserve capacity for managing it, thus heightening emotional and cognitive vulnerability.

Choudhury et al. [518] explore critical neuroscience, a field of inquiry that probes the social, cultural, political, and economic contexts and assumptions that form the basis for behavioral and brain science research.

Numerous studies have illustrated that depending on the specific demographic background, there are major differences between users' emotions, behavior, and perceived usability. According to Goldfarb and Brown [519], scientific research is characterized by racial, cultural, and socioeconomic prejudices, which lead to demographic homogeneity in participation. This in turn spurs inaccurate representations of neurological normalcy and leads to poor replication and generalization.

According to Freud, the unconscious is a depository for socially unacceptable ideas, wishes or desires, traumatic memories, and painful emotions that psychological repression had pushed out of consciousness [520]. HireVue, which is a global front-runner in AI technologies, is one of the top emotional AI companies that is now turning to biosensors that read non-conscious data in lieu of facial coding methods to measure emotions [521].

The ideas of what it means to have good relationships and to be a good person differ in different cultural contexts [522]. People's emotional lives are closely related to these different ideas of how people see themselves and their relationships: Emotions usually match the cultural model [523,524]. Therefore, rather than being random, cultural variation in emotions matches the cultural ideals of ways to be a good person and to maintain good relationships with other people [506].

Aside from being biologically driven, emotion is also influenced by environment, as well as cultural or social situations. Culture can constrain or enhance the way emotions are felt and are expressed in different cultural contexts, and it can influence emotions in other ways. Studies have consistently shown cross-cultural differences in the levels of emotional arousal. Eastern culture, for instance, is related to low arousal emotions, whereas Western culture is related to high arousal emotions [525]. Many findings in cross-cultural research suggest that decoding rules and cultural norms influence the perception of anger [526]. Scollon et al. [527] look at five cultures (Asian American, European American, Hispanic, Indian, and Japanese) to assesses the way emotions are experienced in these cultures. Pride shows the greatest cultural differences [527]. As emotions are fundamentally genetically determined, different ones are perceived in similar ways throughout most nations or cultures [528].

6. Results

The present article aims to bridge the affective biometrics and neuroscience gap in existing knowledge, in order to contribute to the overall knowledge in this area. We also aim to provide information on the knowledge gaps in this area and to chart directions for future research.

We conclude this review by discussing unanswered questions related to the next generation of AFFECT detection techniques that use brain and biometric sensors.

By performing text analytics of 21,397 articles that were indexed by Web of Science from 1990 to 2022, we examined the key changes in this area within the last 32 years. Scientific output relating to AFFECT detection techniques using brain and biometric sensors is steadily increasing. As this trend suggests, there has been continuous growth in the number of papers published in the field, with the total number of articles appearing between 2015 and 2021 nearing the total number of articles published over the previous 25 years (1990 to 2014). In light of the increasing commercial and political interest in brain and biometric sensor applications, this trend is likely to continue.

With ground-breaking emerging technologies and the growing spread of Industry 5.0 and Society 5.0, AFFECT should be analyzed by taking into account demographic and cultural background, socioeconomic status, diversity attitudes, and context. Advanced computational models will be needed for this approach.

Quite a few biometric and neuroscience studies have been performed in the world, where AFFECT detection takes into account demographic and cultural background (age, gender, ethnicity, race, major diagnoses, and major medical history); socioeconomic status (education, income, and occupation); diversity attitudes; and context. Yet, to the best of our knowledge, none of the technologies available in the world offer AFFECT detection that incorporates political views, personality traits, gender, race, diversity attitudes, and cross-cultural differences in emotion.

Sometimes confusion exists in the spirit of some research about physiological effects due to emotional reactions and biometric patterns with regard to individual identification. To resolve this confusion, we analyze only physiological effects caused by emotional reactions (i.e., second generation biometrics; Section 3) in the part of the review discussing biometrics. Biometric patterns for individual identification are not analyzed in this research.

Human emotions can be determined by physiological signals, facial expressions, speech, and physical clues, such as posture and gestures. However, social masking—when people either consciously or unconsciously hide their true emotions—often renders the latter three ineffective. Physiological signals are therefore often a more accurate and objective gauge of emotions [529]. For instance, researchers [530,531] performed many studies to analyze physiological signals and unconscious emotion recognition. Nonetheless, our years of research experience have proven that in public spaces, facial expressions, speech, and physical clues, such as posture and gestures, are much more convenient and effective.

Emotion recognition can be more accurate when human expressions are analyzed looking at multimodal sources such as texts, physiological signals, videos, or audio content [532]. Integrated information from signals such as gestures, body movements, speech, and facial expressions helps detect various emotion types [533]. Statistical methods, knowledge-based techniques, and hybrid approaches are three main emotion classification approaches in emotion recognition [534].

The emotional dimensions follow the approach of representing the emotion classes. Categorized emotions can be represented in a dimensional form with each emotion placed in a distinct position in space: either 2D (Circumplex model, "Consensual" Model of Emotion, Vector Model,) or 3D (Lövheim Cube, Pleasure-Arousal-Dominance [PAD] Emotional-State Model, Plutchik's model, PAD Emotional-State Model), with each emotion occupying a distinct position in space. Most dimensional models have dimensions of valence and arousal or intensity or arousal dimensions: Valence dimension indicates how much and to what degree an emotion is pleasant or unpleasant, whereas arousal dimension differentiates between showing its state, either that of activation or deactivation [82]. The objectives of our study were most in line with Plutchik's 'wheel of emotions' model, which we used in this research.

The use of artificial intelligence to recognize emotions and affective attitudes is a comparatively promising field of investigation. To make the most of artificial intelligence, multiple modalities in context should be generally used. Artificial intelligence has enabled biometric recognition and the efficient unpacking of human emotions and affective and physiological responses and has contributed considerably to advances in the field of pattern recognition in biometrics, emotions, and affective attitudes. Many different AI algorithms are used in the world, such as machine learning, artificial neural networks [535–537], search algorithms [166,538,539], expert systems [540,541], evolutionary computing [542,543], natural language processing [544,545], metaheuristics, fuzzy logic [546–548], genetic algorithm [549–551], and others.

Based on our review, presented in Sections 1–5, we find that investigators should develop procedures to guarantee that AI models are appropriately used and that their specifications and results are reported consistently. There is a need to create innovative AI and machine learning techniques.

Based on the review (Sections 1–5), investigators should develop procedures to guarantee that AI models are appropriately used and that their specifications and results are reported consistently. There is a necessity to create innovative AI and machine learning techniques.

The existing emotion recognition approaches all need data, but the training of machine learning algorithms requires annotated data, and obtaining such data is usually a challenge [552]. The use of AI models may become less complex, and AI algorithms faster when certain database techniques are applied. These techniques can also provide AI capability inside databases. Supporting AI training inside databases is a challenging task. One of the challenges is to store a model in databases, so that its parallel training is possible with multiple tenants involved in its training and use, at the same that security and privacy issues are taken care of. Another challenge is to update a model, especially in case of dynamic data updates [553]. The following datasets can help with the task of classifying different emotion types from multimodal sources such as physiological signals, audio content, or videos: BED [554], MuSe [555], MELD [544,556], UIT-VSMEC [411] HUMAINE [557], IEMOCAP [558], Belfast database [559], SEMAINE [560], DEAP [561], eNTERFACE [384], and DREAMER [562]. Github [563], for instance, provides a list of all public EEG-datasets such as High-Gamma Dataset (128-electrode dataset from 14 healthy subjects with about 1000 four-second trials of executed movements, 13 runs per subject), Motor Movement/Imagery Dataset (2 baseline tasks, 64 electrodes, 109 volunteers), and Left/Right Hand MI (52 subjects).

The findings also suggest that the development of more powerful algorithms cannot address the perception, reading, and evaluation of the complexity of human emotions, by making an integrated analysis of users' demographic and cultural background (age, gender, ethnicity, race, major diagnoses, and major medical history); socioeconomic status (education, income, and occupation); diversity attitudes; and context. We can only hope that the future will bring further research to address this issue and help to develop more advanced AFFECT technologies that can better cope with issues such as demographic and cultural background (age, gender, ethnicity, race, major diagnoses and major medical history); socioeconomic status (education, income and occupation); diversity attitudes; and context (weather conditions, pollution, etc.).

Worldwide research has yet to resolve several problems, and additional research areas have arisen, such as missing data analysis, potential bias reduction, a lack of stringent data collection and privacy laws, application of elicitation techniques in practice, open data and other data-related issues. Olivas et al. [564] for instance, analyze various methods for handling missing data:

- Missing data imputation techniques: analysis of the variable containing missing data (Mean, Regression, Hot Deck, Multiply Imputation) and analysis of relationships between variables for a case containing missing data (Imputation based on Machine Learning: Neural Network, Self-organizing map, K-NN, Multilayer perceptron);
- Case deletion (Listwise Deletion (Complete-case), Pairwise Deletion);
- Approaches that take into account data distributions (Bayesian methods, Model-based likelihood, Maximum Likelihood with EM).

It was found that the median correlation of the dependent variable of the Publications—Country Success model with the independent variables (0.6626) is higher than in the Times Cited—Country Success model (0.5331). Therefore, it can be concluded that the independent variables in the Publications—Country Success model are more closely related to the dependent variable than in the Times Cited—Country Success model (Figure 11).

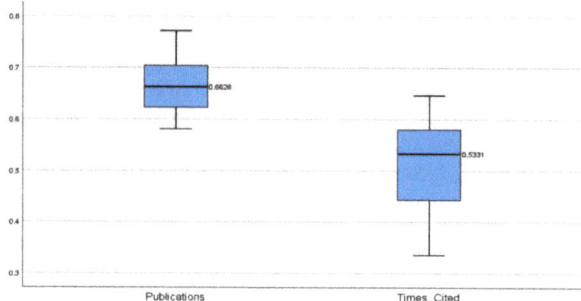

Figure 11. Distribution of correlations based on 15 criteria applied to 169 countries, their publications, and citations, as a CSP map.

The CSP maps of the world that have been compiled for this research provide a visualization of two aspects. A country's success (x-axis) is one of the aspects, while the publications dimensions (CSPN and CSPC; y-axis) are the other (Figures 12 and 13). The publications (x-axis) are one of the aspects, while the publications times cited dimensions (y-axis) are the other in Figure 14. The CSP maps group the countries into the same eight clusters as the Inglehart–Welzel 2020 Cultural Map of the World (English-speaking, Catholic Europe, Protestant Europe, Orthodox Europe, West and South Asia, African-Islamic, Confucian, and Latin America) [565]. Two clusters—English-speaking and Protestant Europe—have been merged into one because of their shared history, religion, cultures, and degree of economic development. The parallels between the two aforementioned clusters have been confirmed by numerous studies [566]. The Inglehart–Welzel 2020 Cultural Map of the World includes many institutional, technological, psychological, and economic variables that demonstrate strong perceptible correlations [567]. The country success indicators in the CSP maps can be characterized as a large set of variables within the criteria system, such as politics, human development and well-being, the environment, macroeconomics, quality of life, and values based.

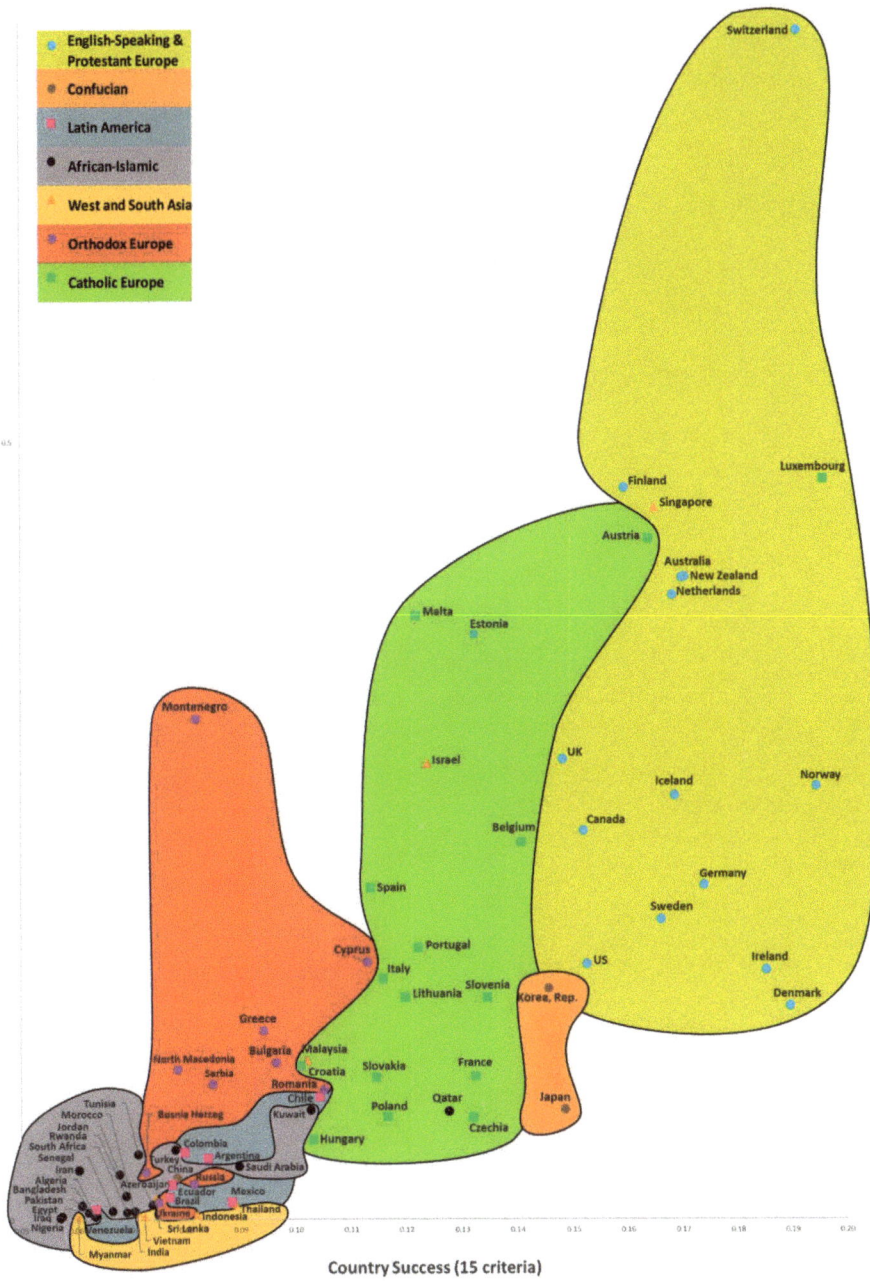

Figure 12. CSP map showing the success of countries in terms of the numbers of publications on AFFECT recognition (CSPN) in Web of Science journals with impact factor.

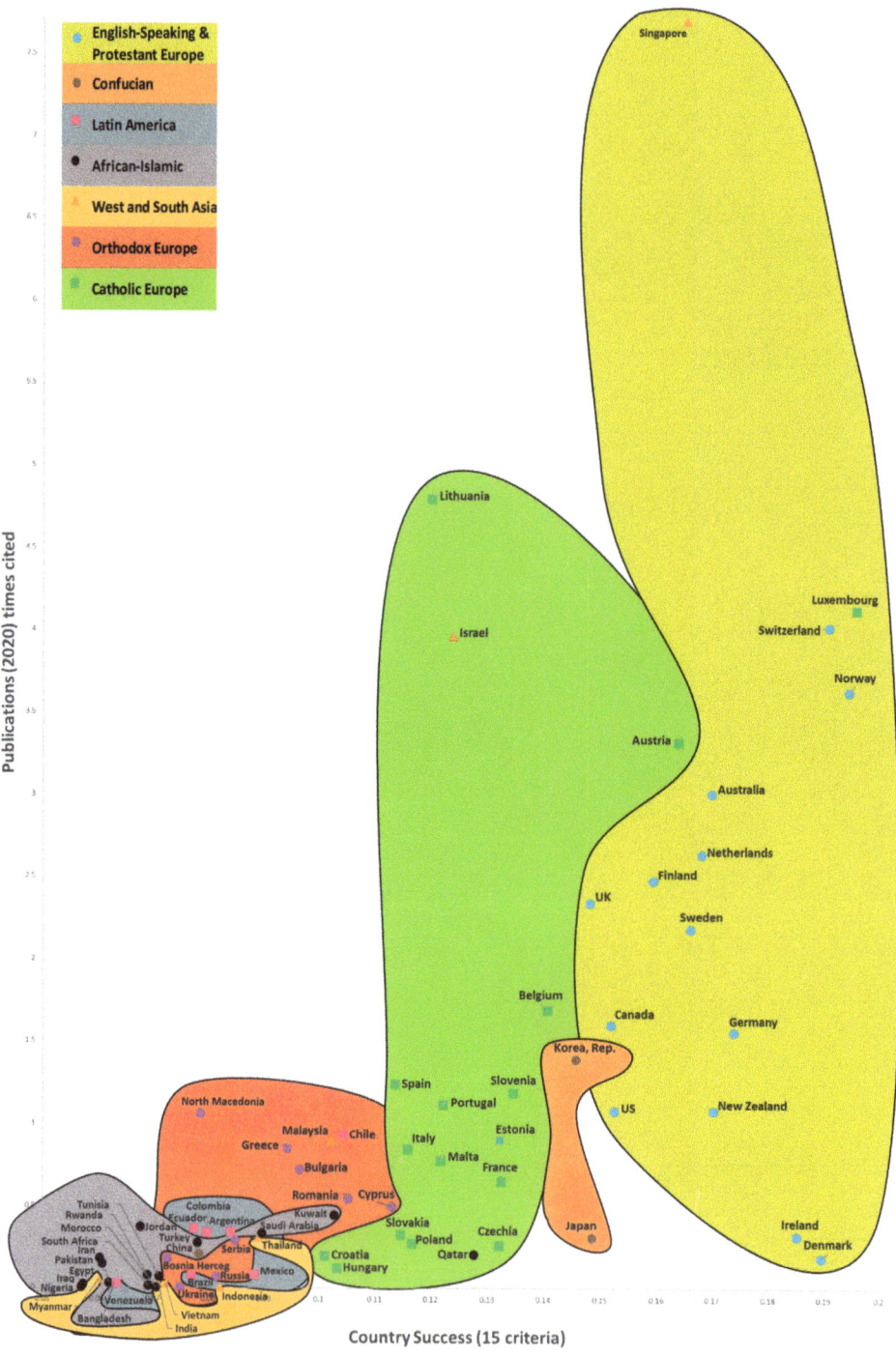

Figure 13. CSP map showing the success of countries in terms of the number of citations of their publications on AFFECT recognition (CSPC) in Web of Science journals with impact factor.

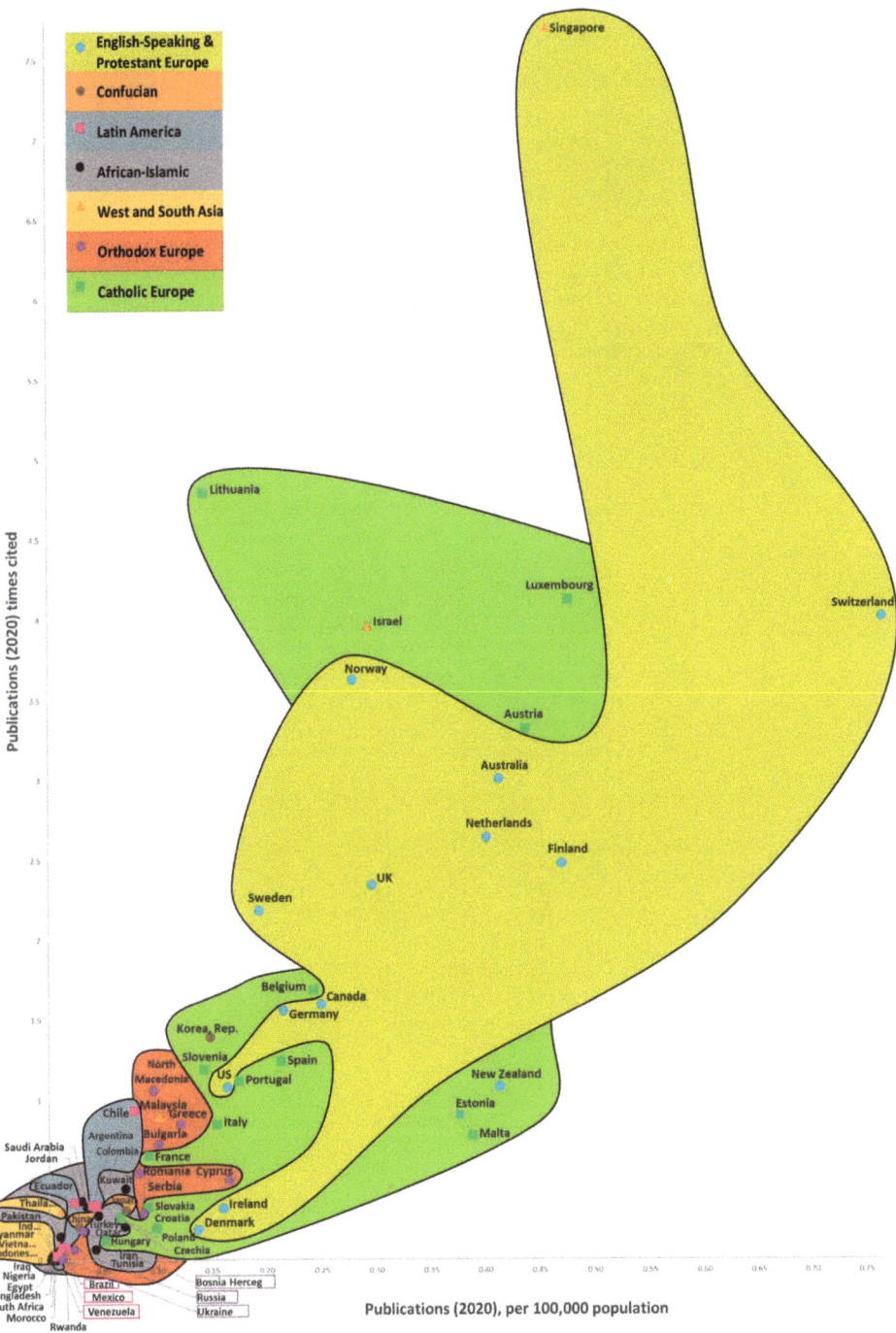

Figure 14. CSP map showing the number of articles on AFFECT recognition and the numbers of citations in Web of Science journals with impact factor.

In addition, this is a quantitative study to assess how the success of the 169 countries impacted the number of Web of Science articles published in 2020 on AFFECT recognition techniques that use brain and biometric sensors (or the latest figures available).

For the multiple linear regressions, we used IBM SPSS V.26 to build two regression models on 15 indicators of country success and the two predominant CSP dimensions. Two CSP regression models were developed based on an analysis of 15 independent variables and two dependent variables. The 15 independent variables and the two regression models are summarized in Tables 4–8. Table 4 contains descriptive statistics for two of the CSP models. The minimum and maximum values indicate the value range for each variable in the set of values that the variable in question can take. The average value of the full range that each variable can take is the mean and is usually equal to the arithmetical average. The standard deviation is a measure of the dispersion in the values of the variable in relation to the mean. Kurtosis is a measure of whether the values are heavy-tailed or light-tailed relative to the center of the distribution, whereas skewness is a measure of the symmetry of the distribution of the values. Acceptable values are considered to be between -3 and $+3$ for skewness, and between -10 and $+10$ for kurtosis. When the skewness is close to zero and kurtosis is close to three, the distribution of the values of the variable within the specified value range is in line with a normal distribution.

Step 9 entailed the construction of regression models for the number of publications and their citation rates, and the calculation of the ES indicators describing them. Two dependent variables and 15 independent variables were analyzed to construct these regression models. The process was as follows:

- Construction of regression models for the numbers of publications and their citations.
- Calculation of statistical effect size (ES) indicators describing these regression models. ES is a value used in statistics to measure the strength of the relationship between two variables, or to calculate a sample-size estimate of that amount [568]. An ES may reflect the regression coefficient in a regression, the correlation between two variables, the mean difference, or the risk of a specific event occurring [569]. Guidelines developed by Durlak [570] provide advice on the ESs to use in research, and how to calculate and interpret them. We used these guidelines, and applied the following five measures of ES, as these indicators are crucial for meta-analysis and could be computed from our measurements:
    - Pearson correlation coefficient (r): Beta weights and structure coefficients r are the two sets of coefficients that can provide a more perceptive stereoscopic view of the dynamics of the data [571]. Interpretation may be also improved through the use of other results (e.g., [572]).
    - Standardized beta coefficient (β): Theoretically, the highest-ranking variable is the one with the largest total effect, since β is a measure of the total effect of the predictor variables [573].
    - Coefficient of determination (R^2): This is a measurement of the accuracy of a CSP model. The outcome is represented by the dependent variables of the model. The closer the coefficient of determination to one, the more variability the model explains. R^2 can therefore be used to determine the proportion of the variation in the dependent variable that can be predicted by examining the independent variables [573].
    - Standard deviation: If this is too high, it will render the measurement virtually meaningless [574].
    - p-values. There is no direct relationship between the p-value and the size, and a small p-value may be associated with a small, medium, or large effect. There is also no direct relationship between the ES and its practical or clinical significance: a lower ES for one outcome may be more important than a higher ES for another outcome, depending on the circumstances [570].
- Calculation of non-statistical ES measures, which may better indicate the significance of the relationships between pairs of variables in our two models:

○ Research context: Durlak [570] argues that ESs must be interpreted in the context of other research.
○ Practical benefit: As this is an intuitive measure, practical benefit can allow stakeholders to make more accurate assessments of whether the research findings published can significantly improve their ongoing projects [575].
○ Indicators with low values: These are usually easier to improve than indicators with high values.

Table 4. Descriptive statistics for the dependent variables of two models.

Descriptive Statistics	Descriptive Statistics of 2 Models Dependent Variables	
	Publications—Country Success	Times Cited—Country Success
	Model 1 (CSPN)	Model 2 (CSPC)
Mean	0.1354	0.9279
Median	0.0785	0.3297
Maximum	0.7642	7.7034
Minimum	0.0015	0.0000
Standard Deviation	0.1557	1.3893
Skewness	1.5533	2.4316
Kurtosis	5.3614	9.8641
Observations	166	165

Based on the results of descriptive statistics, it can be concluded that the values of the dependent variables of the models used in the study demonstrate normal distribution (skewness < 10 and kurtosis < 10), which allows for the use of parametric analysis methods in the analysis.

Table 5. Goodness-of-fit testing for two models.

Independent Variables	Dependent Variables	
	Publications—Country Success	Times Cited—Country Success
	Model 1 (CSPN)	Model 2 (CSPC)
GDP per capita	0.7725 *** (1.2062)	0.6368 *** (7.1524)
GDP per capita in PPP	0.6975 *** (8.4298)	0.6467 *** (7.3418)
Ease of doing business ranking	−0.4821 *** (−4.7652)	−0.4390 *** (−4.2317)
Corruption perceptions index	0.7624 *** (1.5319)	0.6341 *** (7.1014)
Human development index	0.6717 *** (7.8530)	0.5347 *** (5.4799)
Global gender gap	0.4797 *** (4.7348)	0.3354 *** (3.0834)
Happiness index	0.7037 *** (8.5774)	0.5315 *** (5.4340)
Environmental performance index	0.6939 *** (8.3444)	0.5166 *** (5.2256)
Freedom and control	−0.5808 *** (−6.1782)	−0.3832 *** (−3.5932)
Economic freedom	0.6535 *** (7.4765)	0.5801 *** (6.1681)
Democracy Index	0.6227 *** (6.8912)	0.4429 *** (4.2777)
Unemployment rate	−0.1860 (−1.6398)	−0.1642 (−1.4412)
Healthy life expectancy	0.6312 *** (7.0471)	0.5194 *** (5.2635)
Fragile state index	−0.7229 *** (−9.0606)	−0.5405 *** (−5.5634)
Economic decline index	−0.6358 *** (−7.1339)	−0.5597 *** (−5.8487)

Standardized beta coefficients: *** significant at $\alpha = p < 0.001$.

A correlation analysis found that the strongest relationship in the Publications—Country Success model is between the dependent variable Publications and the independent variable GDP per Capita. Meanwhile, in the Times Cited—Country Success model, the strongest relationship is between the variables of Times Cited and GDP per Capita in PPP. It was also found that in both models, the relationships between the dependent variables and the independent variables are statistically significant ($p < 0.001$), except for the relationships between the dependent variables and the Unemployment Rate variable.

Table 6. Descriptive statistics for two models.

Descriptive Statistics	Descriptive Statistics of 2 Models	
	Publications—Country Success	Times Cited—Country Success
	Model 1 (CSPN)	Model 2 (CSPC)
Pearson's correlation coefficient ($|r|$)	0.6272	0.5142
Coefficient of determination (R^2)	0.6943	0.5114
Adjusted R^2	0.6191	0.3912
Standard deviation	0.1557	1.3693
p values (probability level)	0.0000	0.0000
F	9.2356	4.2570

A reliability analysis of the compiled regression models allows us to conclude that the models are suitable for analysis ($p < 0.05$). It was also found that the changes in the values of the independent variables used in the models explain the variance of the Publications variable by 69.4%, and the variance of the Times Cited variable by 51.1%.

Table 7. Standardized beta coefficient values of the dependent variables.

	Independent Variables	Standardized Beta Coefficient Values of the Dependent Variables	
		Publications—Country Success	Times Cited—Country Success
		Model 1 (CSPN)	Model 2 (CSPC)
1	GDP per capita	0.7735 **	−0.0853
2	GDP per capita in PPP	−0.5123 *	0.5304 *
3	Ease of doing business ranking	0.2535	0.1599
4	Corruption perceptions index	0.2392	0.3633
5	Human development index	0.1697	−0.1836
6	Global gender gap	−0.0228	0.0703
7	Happiness index	0.0800	−0.0916
8	Environmental performance index	−0.0601 **/	0.1819
9	Freedom and control	−0.0299	0.0846
10	Economic freedom	0.4558	0.3239
11	Democracy Index	−0.1524	0.0577
12	Unemployment rate	0.0353	0.0552
13	Healthy life expectancy	0.0047	0.0696
14	Fragile state index	−0.0008	0.0246
15	Economic decline index	0.0147	−0.0301

Standardized beta coefficients: * significant at—$p < 0.1$, ** significant at $p < 0.01$.

An analysis of the standardized coefficients of the model allows us to conclude that changes in the GDP per Capita variable have the biggest impact on changes in the Publications variable. The GDP per Capita in PPP variable also have a significant impact. Meanwhile, the Times Cited variable is most affected by the GDP per Capita in PPP variable, which has a statistically significant effect on the dependent variable.

Table 8. How country success and its factors influence the two indicators.

Publications—Country Success	Times Cited—Country Success
Model 1 (CSPN)	Model 2 (CSPC)
When a country's success increases by 1%, the indicator improves by	
1.962%	2.101%
The 17 independent variables explain the dependent variable under analysis by	
89.5%	54.0%

To confirm Hypothesis 1, we built two CSP models, which are formal representations of the CSP maps. These models demonstrate that on average, an increase of 1% in a country's success leads to an average improvement by 0.203% in the country's two CSPN and CSPC dimensions. As the success of a country increased by 1%, the numbers of Web of Science articles published and their citations grew by 1.962% and 2.101%, respectively. Figures 12 and 13 also illustrate that an increase in a country's success goes hand in hand with a jump in its CSPN and CSPC dimensions, thus confirming Hypothesis 1.

Hypothesis 2 was based on the results of the analysis pertinent to the CSP models, as well as on the correlations found between the 169 countries and the 15 indicators [66]. A clear visual confirmation of Hypotheses 1 and 2 are also provided by Figures 12 and 13, which show the specific groupings of countries in the seven clusters examined in this study. These models may be of major significance for policy makers, R&D legislators, businesses, and communities.

7. Evaluation of Biometric Systems

In this chapter, we outline the rationale behind the current biometrics and brain approaches, compare the efficacy of existing methods, and determine whether or not they are capable of addressing the kinds of issues and challenges associated with the field (with figures). Biometric systems have several drawbacks in terms of their precision, acceptability, quality, and security. They are generally evaluated based on aspects such as (1) data quality; (2) usability; (3) security; (4) efficiency; (5) effectiveness; (6) user acceptance and satisfaction; (7) privacy; and (8) performance.

Data quality measures the quality of biometric raw data [576,577]. This type of assessment is generally used to quantify biometric sensors and can also be used to enhance the system performance. According to the International Organization for Standardization ISO 13407:1999 [578], usability is defined as "[t]he extent to which a product can be used by specified users to achieve specified goals with effectiveness, efficiency, and satisfaction in a specified context of use" [579]:

- In this context, efficiency means that users must be able to accomplish the tasks easily and in a timely manner. It is generally measured as task time;
- Here, effectiveness means that users are able to complete the desired tasks without excessive effort. This is generally measured by common metrics such as the completion rate and number of errors, for example the failure-to-enroll rate (FTE) [580];
- User satisfaction measures the user's acceptance of and satisfaction with the system. It is generally measured by looking at a number of characteristics, such as ease of use and trust in the system. Even if the performance of one biometric system exceeds that of another in terms of performance, this will not necessarily mean that it will be more operational or acceptable.

Security measures the robustness of a biometric system (including algorithms, architectures, and devices) against attack. The International Organization for Standardization ISO/IEC FCD 19792 [581] specifically addresses processes for evaluating the security of such systems [579].

Unlike traditional methods, biometric systems do not provide a 100% reliable answer, and it is almost impossible to obtain such a response. In a secure biometric system, there is a trade-off between recognition performance and protection performance (security and privacy). The reason behind this trade-off arises from the unclear concept of security, which requires a more standardized framework for evaluation purposes. If this gap can be closed, an algorithm could be developed that would jointly reduce both of them. ISO 19795 contained standards for performance metrics and evaluation methodologies for traditional biometric systems. In addition to performance testing, it provided metrics related to the storage and processing of biometric information [582]. ISO/IEC 24745 specifies that, unlike privacy, security is delivered at the system level. In general, the ability of a system to maintain the confidentiality of information with the use of the provided countermeasures (such as access control, integrity of biometric references, renewability, and revocability) is referred as its security factor. When seeking to bypass the security of a biometric system, an invader may impersonate a genuine user to gain access to and control over various services and sensitive data. Privacy refers to secrecy at the information level. The following criteria were proposed in ISO/IEC 24745 for the purpose of evaluating the privacy offered by biometric protection algorithms: irreversibility, unlinkability, and confidentiality [583].

The discriminating powers of all biometric technologies rely on the extent of entropy, with the following used as performance indicators for biometric systems [584–587]: False match rate (FMR); False non-match rate (FNMR); Relative operating characteristic or receiver operating characteristic (ROC); Crossover error rate or equal error rate (CER or EER); Failure to enroll rate (FER or FTE), and Failure to capture rate (FTC).

Specific advantages and disadvantages are characteristic to each biometric technology. Table 9 shows these comparisons.

Table 9. Benefits and limitations of biometric technologies.

Tool	Benefits	Limitations
Electroencephalography (EEG)	Can be used to measure rapid changes in neural activity by the millisecond [588] Minimally invasive and/or commercial research packages are available [588] Participants can move around and benefit from enriched/social environments [588] Uses portable instruments and natural environments; there is long tradition of well-controlled experiments; measurement processes requiring several hours are possible in practice [589]	It is difficult to pinpoint neural signals from particular brain areas (poor spatial resolution) [588] Measurements from structures deep within the brain (e.g., nucleus accumbens) are not possible [588] Published studies on biometrics based on this signal have used high-cost medical equipment [590] Subjects have reported discomfort since it is necessary to apply scalp neck gel to improve conduction between electrodes [590]
Functional magnetic resonance imaging (fMRI)	Has the ability to observe activity in small structures [588] Differentiates signal from neighboring areas [588] Measurements of the whole brain are possible [588]	Physically restrictive; participants lie on their back in the scanner and cannot move around [588] Expensive, and equipment is in high demand [588] Equipment cannot be removed from the laboratory; the sequence of the activities is difficult to monitor [589]
MEG (magnetoencephalography)	Some MEG study protocols are quite well suited for design studies; there is a long tradition of well-controlled experiments based on EEG; optimal space-time-resolution [589]	Equipment cannot be removed from the laboratory; the location of existing brain activity is relatively difficult to determine [589]
Electrocardiogram (ECG)	Highly reliable source providing precise features of the electrical and physiological activity taking place with an individual; high performance has been noted in prior research on this signal [591]; it can easily be fused with other signals [592]	One of the great difficulties listed in the literature is a lack of user acceptance, as its implementation at the physical level makes it fairly uncomfortable [593]; body posture can also affect cardiac signals [594]
MRI (magnetic resonance imaging) [589]	Good for studies comparing groups of people	Equipment cannot be removed from the laboratory
PET (positron emission tomography) [589]	Good for comparing groups of people or natural tasks	Radioactive tracer is injected into participants; equipment cannot be removed from the laboratory
Eye tracking [588]	Offers strong nuanced data on visual attention and gaze pathways, and can be integrated with pupillometry	Does not measure inferences, the valence of the response, thoughts, or emotions
Iris [595]	Unique data; input is stable throughout lifetime; non-intrusive	Large data template; images are frequently improperly focused; single-source; high cost

Table 9. *Cont.*

Tool	Benefits	Limitations
NIRS (near-infrared spectroscopy) [589]	Uses portable instruments and natural environments; some NIRS study protocols are well suited for design studies; measurement processes requiring several hours are possible in practice	Difficulties in determining the location of brain activity; few groups are using NIRS for cognitive studies as yet
Transcranial magnetic stimulation (TMS/tDCS) [588]	Can be used to show causality	Limited to investigating the function of brain surface areas. Can generally only lessen (TMS/tDCS) or increase (tDCS) neural activity in a general sense; cannot test for specific levels of activity or influence specific circuits
Forehead electrooculogram (EOG)	These signals are low cost, and are not invasive [596]	Electrodes used for the acquisition of the signals can present instability to eye flicker [597]; signals are highly affected by noises in the immediate vicinity [596]
Skin conductance response (SCR), heart rate, pupil dilation [588]	Simple; well validated. Unobtrusive equipment; allows for more natural interactions with the environment	Cannot distinguish between positive and negative arousal
Lips [598]	Easy acquisition and lip characteristics; it is possible to extract the outline even if the person has a beard or a moustache	An image of the lips cannot be acquired when they are moving
Facial electromyography (fEMG), facial affective coding [599]	This is a precise and sensitive method for measuring emotional expression. Unlike self-reports, fEMG does not depend on language and does not require cognitive effort or memory. Yields large amounts of data and is continuous and scalable (hence more credible). Dynamic tracking of emotional (potentially unconscious) responses to ongoing stimuli/information. Can measure facial muscle activities for the sake of balancing weakly evocative emotional stimuli. Less intrusive than other physiological measures such as fMRI and EEG. Automatic facial encoding software/algorithms are available	The technique is intrusive and may alter natural expression. The number of muscles that can be triggered is limited by how many electrodes can be attached to the face. Requires electrodes to be directly attached to the face (in a lab). Certain medicines that act on the nervous system, such as muscle relaxants and anticholinergics, can impact the final electromyography (EMG) result
Gait	Convenient and non-intrusive (2D); subjects can be evaluated covertly, without their knowledge [595]	During the assessment stage, light affects the results; clothing may affect detection [46]. Data may alter throughout a lifetime (injuries, training, footwear); specialist personnel required for data processing; large data template [595]
Body motion [595]	Unique and various sources of data, small template size	Time consuming; subject must cooperate with reader; specialist personnel required for data processing

Upon completing the literature analysis, we then compared biometric technologies looking at the following seven parameters: universality, distinctiveness/uniqueness, permanence, collectability, performance, acceptability, and circumvention (Table 10). Another set of comparisons was the strengths and weaknesses characteristic to biometric technologies and related to their ease of use, error incidence, accuracy, user acceptance, long term stability, cost, template sizes, security, social acceptability, popularity, speed, and whether or not they have been socially introduced (Table 11). The working characteristics of various biometrics differ, as does their accuracy, and depend on the design of their operation. The level of security and the kinds of possible errors are also different in each biometric approach; the denial of access to the biometric sample holders is possible caused by various factors such as aging, cold, weather conditions, physical damages, and so on [600,601]. Other researchers also look at FAR, FRR, CER, and FTE in their comparisons of biometric technologies (Table 12).

Table 10. Comparison of biometric technologies by seven characteristics (traits).

	Universality	Uniqueness or Distinctiveness	Permanence	Collectability	Performance	Acceptability	Circumvention
Iris/pupil	High [141,602–606]	High [141,602–606]	High [141,602–606]	Medium [141,602–606]	High [141,602–606]	Low [141,602,604–606] Medium [603]	High [141,602] Low [603–606]
Face	High [141,602–606]	Low [141,602,604–606] Medium [603]	Medium [141,602–606]	High [141,602–606]	Low [141,602–606]	High [141,602–606]	Low [141,602] High [603–606]
Odor	High [602–606]	High [602–606]	High [602–606]	Low [602–606]	High [602,603] Low [604–606]	Low [602] Medium [603–606]	Low [602–606]
Keystroke dynamics and mouse movements, Mouse Tracking	Low [141,602,604–606]	Low [141,602,604–606]	Low [141,602,604–606]	Medium [141,602,604–606]	Low [141,602,604–606]	Medium [141,602,604–606]	Medium [141,602,604–606]
Skin temperature -thermogram	High [141,604–606]	High [141,604–606]	Low [141,604–606]	High [141,604–606]	Medium [141,604–606]	High [141,604–606]	High [141] Low [604–606]
Voice/Speech/Voice Pitch Analysis (VPA)	Medium [141,602,604–606]	Low [141,602,604–606]	Low [141,602,604–606]	Medium [141,602,604–606]	Low [141,602,604–606]	High [141,602,604–606]	Low [141,602] High [604–606]
Signature	Low [141,602–606]	Low [141,602–606]	Low [141,602–606]	High [141,602–606]	Low [141,602,604–606] Medium [606]	High [141,602–606]	Low [141,602] High [603–606]
Gait	Medium [602,604–606] High [603]	Low [141,604–606] Medium [603]	Low [141,604–606] Medium [603]	High [602–606]	Low [602–606]	High [602,604–606] Medium [603]	Medium [602–606]

Table 11. Comparison of biometric technologies by various attributes.

	Easy of Use	Error Incidence	Accuracy	User Acceptance	Long Term Stability	Cost	Size of Template	Security	Socially Introduced	Social Acceptability	Popularity	Speed
Eye Tracking (ET)			0.5°–1° [607]			Low–High [608]						
Iris/pupil	Medium [602,603,609]	Lighting [602,609], Lighting, glasses [603]	Very High [602,609], High [320,603,610,611]	Medium [602,609]	High [602,609,610], Medium [320,603]	High [320,603,611]	Small [320]	Medium [320], High [603], Very high [609]	1995 [603]	Medium-Low [610,611]	Medium [603]	Medium [603]
Face	Medium [602,609], High [603]	Lighting, age glasses, hair [602,603,609]	High [602,609], Low [320,602], Medium-Low [610,611]	Medium [602,609]	Medium [602,609], Low [320,602]	High [320], Medium [602,610,611]	Large [320]	Low [320], Medium [602,610,611]	2000 [603]	High [610,611]	High [603]	Medium [603]
Keystroke dynamics and mouse movements, Mouse Tracking	Low [602]	Device, weather [602]	Low [602]		Low [602]	Medium [602]		Low [602]	2005 [603]		Low [603]	Medium [603]
Voice/Speech/Voice Pitch Analysis (VPA)	High [602,603,609]	Noise, colds [602,603,609]	High [602,609], Low [320,603], Medium [610,611]	High [602,609]	Medium [602,603,609], Low [320]	Medium [320,610,611], Low [603]	Small [320]	Low [320], High [603], Medium [609]	1998 [603]	High [610,611]	High [603]	High [603]
Signature	High [602,603,609]	Changing signature [602,603,609]	High [602,609], Medium [320,603], Low [610,611]	High [602], Very High [609]	Medium [602,609], Low [320,603]	Low [320], Medium [603,610,611]	Medium [320]	Low [320], High [603], Medium [609]	1970 [603]	High [610,611]	High [603]	High [603]
Gait			Medium [610]			Medium [610]				Low [610]		
Lip Movement			Medium [603]		Medium [603]	Medium [603]	Small [603]	High [603]				
Gesture			Low [612]									

55

Table 12. Comparison of performance metrics for biometric technologies by various authors.

	FAR	FRR	CER	FTE
Iris/pupil	0.94% [603] 0.0001–0.94 [613] 2.4649% [614]	0.99% [603] 0.99–0.91 [613] 2.4614% [614]	0.01% [603]	0.50% [603]
Face	1% [603] 16% [614]	10% [603] 16% [614]		3.1% [615]
Keystroke dynamics and mouse movements, Mouse Tracking	7% [603] 0.01% [614]	0.10% [603] 4% [614]	1.80% [603]	
Voice/Speech/Voice Pitch Analysis (VPA)	2% [603,613] 7% [614]	10% [603,613] 7% [614]	6% [603]	0.5% [615]

Multimodal biometric systems take advantage of multiple sensors or biometrics to remove the restrictions of unimodal biometric systems [616]. While unimodal biometric systems are restricted by the integrity of their identifier, the change of several unimodal systems having the same restrictions is low [617]. Multimodal biometric systems can fuse these unimodal systems sequentially, simultaneously, both ways, or in series, meaning sequential, parallel, hierarchical, and serial integration modes, respectively. For instance, final results of decision level fusion of multiple classifiers are joined using methods such as majority voting [616]. This multimodal analysis will assist in identifying the actual reasons of such issues with the current biometrics and brain approaches, as well as the restrictions of the existing state-of-the-art approaches and technologies.

An efficient way to combine multiple classifiers Is needed when an array of classifiers outputs is developed. Various architectures and schemes have been proposed for joining multiple classifiers. The most popular methods are majority vote and weighted majority vote. In majority vote, the right class is the one most selected by various classifiers. If all the classifiers show different classes or in the event of a tie, then the one with the highest overall output is chosen to be the right class. Vote averaging method averages the separate classifier outputs confidence for every class over the entire ensemble. The class output with the highest average value is selected to be the right class [618]. The vote averaging method has been used to measure the efficacy of existing biometrics methods (Tables 10 and 11). In our case, High (Very High) was assigned 3 points, Medium was assigned 2, and Low was assigned 1. The calculations did not evaluate some qualitative indicators, such as error incidence and socially introduced. Additionally, not all biometrics technologies had data on the analyzed indicators. As a result, eye tracking we not evaluated in this case due to a lack of data. The highest average number of points was collected by Skin temperature-thermogram (2.57), Iris/pupil (2.43), Face (2.30), and Signature (2.09). Many of the metrics for biometric technologies in Tables 9–12 are analyzed in detail throughout the article.

8. Discussion and Conclusions

Nevertheless, there are still unanswered questions that need to be addressed. We evaluated the evidence available to find a relationship between brain and biometric sensor data and AFFECT in order to determine the primary digital signals for AFFECT. The multidisciplinary literature used was from the disciplines of engineering, computer science, neuroscience, physiology, psychology, mathematical modeling, and cognitive science. The distinct conventions of these disciplines resulted in certain variegations, depending on the features and characteristics of the research results being focused on. The literature under analysis has small sample sizes, short follow-up times, and significant differences in the quality of the reports, which limits the interpretability of the pooled results. On average, the current AFFECT detection techniques that use brain and biometric sensors achieved a classification accuracy greater than 70%, which seems sufficient for practical applications. As part of this review, several issues that need to be addressed were identified, as well as numerous recommendations and directions for future AFFECT detection and recognition research being suggested. They are listed below:

- Many studies fail to report information on demographic and cultural background, socioeconomic status, diversity attitudes, and context, and AFFECT papers often have limited descriptions of feature extraction and analysis. This has a significant impact on the interpretation of their findings. Sample recommendations include reporting on participant enrolment and selection approaches and analysis of demographic and cultural background (age, gender, ethnicity, race, major diagnoses, and major medical history); socioeconomic status (education, income and occupation), diversity attitudes, and context. In order to improve the ability of researchers to assess the strength of evidence, one of the first steps should be the development of this kind of consistent reporting.
- Behavioral traits (e.g., gesture, keystroke, voice) change over time, and therefore are less stable. Multiple interactions are typically required to set a reliable baseline. Injury, illness, age, and stress can also cause changes in behavioral traits. Many of the studies on AFFECT recognition examined brain and biometric data under different AFFECT while overlooking the baseline (spontaneous) brain and biometric data.
- The literature did not contain brain and biometric sensor-based AFFECT recognition of mixed emotions (parallel involvement of negative and positive emotions). We study the 30 primary, secondary, and tertiary dyads of Plutchik's wheel of emotions, creating mixed emotions.
- Researchers need a set of guidelines to ensure AI models (artificial neural networks, evolutionary computing, natural language processing; metaheuristics, fuzzy logic, genetic algorithm) are correctly applied, and that their specifications and results are consistently reported (the model selection strategy, parameter estimates in the model with confidence intervals, performance metrics, etc.). There is also a need to further develop advanced AI and machine learning techniques (multi-modal learning, neuroscience-based deep learning, automated machine learning, self-supervised deep learning, Quantum ML, Tiny ML, System 2 deep learning).
- More results are also needed to identify which of the elicitation techniques applied in practice are effective, and in which cases they work best, taking into account the type of information obtained, the stakeholders' (developers, end-users, etc.) characteristics, the context, and other factors. More data sets need to be created that use active elicitation techniques, such as various games, as these are better at mimicking real-life experiences and bringing about emotions. Gamification is a current trend that uses game methods for real-life AFFECT elicitation.
- Recommendations also state that the two sources of potential bias (AFFECT interpretation algorithmic biases, data sources and input) in multi-feature studies should be reduced, and a wider variety of multimodal samples should be used.
- Missing data analysis has some gaps, for example missing data descriptions and how missing data is handled, and most appropriate methods should be applied in AFFECT recognition. As far as missing data goes, the literature had major shortcomings.
- As algorithms improve, accuracy is growing, but this significantly depends on the data sets used. Some gaps and a lack of discussion have also been noted concerning the question of whether the integrated brain and biometric sensors used in this research are reliable and appropriate for AFFECT detection.
- A trend related to emotional AI businesses (Realeyes, Affectiva, etc.) that expand their global operations in regions with less stringent data collection and privacy laws has not been sufficiently examined globally.
- The recommendations for open science include the proposal to share and reuse open multimodal AFFECT data, information, knowledge, and science practices (publications and software) by preparing a Data Management Plan that would address any

important aspects of making data findable, accessible, interoperable, and reusable, or FAIR. Open data analysis should also include recognized and validated scales for AFFECT evaluation; any accessible confirmation on the reliability and validity of the AFFECT device and sensor applied should be presented. The open datasets have usually sought to obtain higher accuracy by using different sets of stimuli and groups of participants.

Emotional acculturation, happens when people, on contact with a different culture, learn new ways to express their emotions [619], incorporate new cultural values in their existing set, and then adjust their emotions to suit these new values [620–623]. This may be a research area in affective computing that needs more studies and focus. With growing global integration, emotional acculturation will become increasingly important, and advanced computational models will be needed to simulate the related processes. M.-T. Ho et al. [624] believe that this may be a key thematic change in the decades to come. The findings also suggest that developing more powerful algorithms cannot solve the perception, reading and evaluation of the complexity of human emotions. Instead, the complex modulators that affective and emotional states stem from need to be better understood by the scientific community. We can only hope that the future will bring further research that will remedy this and help develop more advanced technologies that can better cope with issues such as gender, race, diversity attitudes, and cross-cultural differences in emotion [624].

The substantial improvements in the development of affordable and simple to utilize sensors for recognizing AFFECT have resulted in numerous studies being conducted. For this review, we studied in detail 634 articles. We focused on recent state-of-the-art AFFECT detection techniques. We also took existing data sets into account. As this review illustrates, exploring the relationship between brain and biometric signals and AFFECT is a formidable undertaking, and novel approaches and implementations are continually being expanded.

The evaluation of the intensity of human AFFECT is a complex process which requires the use of a multidirectional approach. The main difficulties of this process include variations in the nature of human beings, social aspects, etc., due to these methods, which fits for average evaluation of customers majority, but shows poor results in personalized cases and vice versa. Moreover, the reliability of evaluations of human emotions strongly depends on the number of biometric parameters used, and the measurement methods and sensors applied. It is well known that a higher reliability of recognition can be achieved by increasing the number of parameters, but this will also increase the need for certain equipment and will slow down the evaluation process. The selection of measurement methods and sensors is no less important in the successful recognition of emotions. Contact measurement methods give the most reliable results, but their implementation is relatively complicated and may even be frightening for potential customers. The best solution in this case is non-contact measurement methods, that is, contact methods which do not require special preparation and allow measurements to be taken without the knowledge of the customer.

Future research possibly could focus on areas of reaction to emotion development stage, while sensing and evaluation became faster than emotion recognition by person itself.

This research has addressed the various issues that emerge when affective and physiological states, as well as emotions, are determined by recognition methods and sensors and when such studies are later applied in practice. The manuscript presents the key results on the contribution of this research to the big picture. These results are summarized below:

- Many studies around the world apply neuroscience and biometric methods to identify and analyze human valence, arousal, emotional and physiological states, and affective attitudes (AFFECT). An integrated review of these studies is, however, yet missing.
- In view of the fact that no reviews of AFFECT recognition, classification and analysis based on Plutchik's wheel of emotions theory are available, our study has examined the full spectrum of thirty affective states and emotions defined in the theory.

- We have demonstrated the identification and integration of contextual (pollution, weather conditions, economic, social, environmental, and cultural heritage) [342] and macro-environmental [568] data with data on AFFECT states.
- The authors of the article have presented their own Real-time Vilnius Happiness Index (Figure 10a) and other systems and outputs to demonstrate several of the aforementioned new research areas in practice.

Information on diversity attitudes, socioeconomic status, demographic and cultural background, and context is missing in many studies. In this study, we have identified real-time context [347] data and have integrated them with AFFECT data. For example, the ROCK Video Neuroanalytics system and associated e-infrastructure were established as part of the H2020 ROCK project, in which passers-by were tracked at 10 locations across Vilnius [348]. One of the outputs was the real-time Vilnius Happiness Index (Figure 10 and https://api.vilnius.lt/happiness-index, accessed on 5 September 2022), and the project also involved a number of additional activities (https://Vilnius.lt/en/category/rock-project/, accessed on 5 September 2022) [625,626].

The analysis of the global gap In the area of affective biometric and brain sensors presented in this study and our aim of contributing to the current state of research in this area have led to the aforementioned research results.

Based on the evaluation of biometric systems performed in Section 7 and the conclusions presented in Chapter 8, future AFFECT biometrics and neuroscience development directions and guidelines are visible. We performed the above analysis by extensively discussing biometric and neuroscience methods and domains in the article.

Additionally, Sections 2 and 6 present statistical and multiple criteria analysis across 169 nations, our outcomes demonstrate a connection between a nation's success, its number of Web of Science articles published, and its frequency of citation on AFFECT recognition. This analysis demonstrates which country's success metrics significantly influence future AFFECT biometrics and neuroscience development.

Advancements in the development of biometric and neuroscience sensors and their applications are summarized in this review. Regardless of the encouraging progress and new applications, the lack of replicated work and the widely divergent methodological approaches suggest the need for further research. The interpretation of current research directions, the technical challenges of integrated neuroscience and affective biometric sensors, and recommendations for future works are discussed. The reviewed literature revealed a host of traditional and recent challenges in the field, which were examined in this article and are presented below.

Biometric research aims to provide computers with advanced intelligence so that they can automatically detect, capture, process, analyze, and identify digital biometric signals—in other words, so they can "see and hear". In addition to being one of the basic functions of machine intelligence, this is also one of the most significant challenges that we face in theoretical and applied research [627].

There are still many challenging issues in terms of improving the accuracy, efficiency, and usability of EEG-based biometric systems. There are also problems concerning the design, development and deployment of new security-related BCI applications, such as personal authentication for mobile devices, augmented and virtual reality, headsets and the Internet [628]. Albuquerque et al. [628] have presented the recent advances of EEG-based biometrics and addressed the challenges in developing EEG-based biometry systems for various practical applications. They have also put forth new ideas and directions for future development, such as signal processing and machine learning techniques; data multimodal (EEG, EMG, ECG, and other biosignals) biometrics; pattern recognition techniques; preprocessing, feature extraction, recognition and matching; protocols, standards and interfaces; cancellable EEG biometrics; security and privacy; and information fusion for biometrics involving EEG data, virtual environment applications, stimuli sets and passive BCI technology.

Some of these challenges (accuracy, efficiency, usability, etc.) are analyzed in the article. Each of these features can be examined in more detail. For example, Fierrez et al. [629] analyzed five challenges in multiple classifiers in biometrics: design of robust algorithms from uncooperative users in unconstrained and varying scenarios; better understanding about the nature of biometrics; understanding and improving the security; integration with end applications; understanding and improving the usability. "Design of robust algorithms from uncooperative users in unconstrained and varying scenarios" is a challenge that has been a major focus of biometrics research for the past 50 years [2], but the performance level for many biometric applications in realistic scenarios is still not adequate [629].

Recently, new challenges in the field have been appearing; some of which are presented below as an example. Sivaraman [630] argues that in the age of AI and machine learning, cyberattacks are more powerful and are sometimes able to crack biometric systems. Additionally, these attacks will become more frequent. Multimodal biometrics are increasingly important, where a combination of biometrics is used for greater security. The pandemic has resulted in changes to the biometric algorithm of various modalities. Facial recognition algorithms have been improved to recognize people wearing masks and cosmetics. Updates like these may improve the accuracy of biometrics systems. Biometric devices will take web and cloud-based applications to the next level, as many organizations will continue to operate remotely [630].

Furthermore, a few problems have not been solved, and additional research fields have emerged, namely: biometric and neuroscience technologies lack privacy, are invasive and persons do not like to share their personal data and be identified; lack of protection from hacking; lack of accuracy; a quite expensive life cycle (brief, design, development, set up, running, operation, etc.); lack of capability to read some human features; customer satisfaction is not always guaranteed; human figure form recognition and examination of figure fragments, examination of head vibrations, and human electrical fields are inefficient.

Author Contributions: Conceptualization and methodology, A.K.; investigation, A.K., A.A., I.U., R.K., V.L., A.B.-V., I.V. and L.K.; resources, writing—review, editing and visualization, A.K., A.A., I.U., R.K., V.L., A.B.-V., I.V. and L.K.; supervision, A.K. All authors have read and agreed to the published version of the manuscript.

Funding: This work was supported as part of the 'Building information modeling-based tools and technologies toward fast and efficient RENovation of residential buildings—BIM4REN' project, which received funding from the European Union's Horizon 2020 research and innovation program under grant agreement No. 820773. This research was also supported via Project No. 2020-1-LT01-KA203-078100 "Minimizing the influence of coronavirus in a built environment" (MICROBE) from the European Union's Erasmus+ program.

Institutional Review Board Statement: Not applicable.

Informed Consent Statement: Not applicable.

Data Availability Statement: All extracted data are included in the manuscript.

Conflicts of Interest: The authors declare no conflict of interest.

Abbreviations

AFFECT	arousal, valence, affective attitudes, emotional and physiological states
AI	Artificial Intelligence
AISs	artificial intelligence subsystems
AON	action observation network
BAEPs	brainstem evoked potentials
BCI	brain–computer interface
BIM4Ren	Building Information Modelling based tools and technologies toward fast and efficient RENovation of residential buildings
BNCI	brain/neuronal computer interaction
BR	binary relevance

BTL	below the line
CNCP	model Collective Neuromarketing Consumer Persuasion Model
CNN	convolutional neural network
DEAP	Dataset for Emotions Analysis using Physiological signals
DTI	diffusion tensor imaging
DWI	diffusion-weighted imaging
EARs	emotion association rules
ECG	electrocardiography
EDA	electrodermal activity
EEG	electroencephalography
EMG	electromyography
EMSs	engagement marketing subsystems
EOG	electrooculogram
ERP	event-related potential
ESSs	emotional salient segments
ET	eye tracking
FA	fractional anisotropy
FC	facial action coding
FDG	fluoro-D-glucose
FDG-PET/fMRI	simultaneous [18 F]-fluorodeoxyglucose positron emission tomography and functional magnetic resonance imaging
fEMG	facial electromyography
fMRI	functional magnetic resonance imaging
fNIRS	functional near-infrared spectroscopy
fPET	functional positron emission tomography
GMM	gaussian mixture models
GSR	galvanometer or galvanic skin response
HMI	human–machine interactions
HMM	hidden Markov model
HR	heart rate
HVAC	heating, ventilation, and air conditioning
ICCs	intra-class correlation coefficients
IoT	Internet of Things
IRT	implicit reaction time
IS	information systems
iTBS	intermittent theta burst transcranial magnetic stimulation
K-NN	K-nearest neighbor
LP	label powerset
LSTM	long short-term memory
MDS	multidimensional scaling
MEG	magnetoencephalography
MLP	multi-layer perceptron
MRI	magnetic resonance imaging
MT	mouse tracking
N5PSC	neuromarketing, neuroeconomics, neuromanagement, neuro-information systems, neuro-industrial engineering, products, services, call centers
NEV	net emotional value
NIRS	near infrared spectroscopy
NLP	natural language processing
NT	neurotransmitter
PET	positron emission tomography
PPG	photoplethysmogram
PSD	power spectral density
RAKEL	random k-label sets
RF	random forest

ROCK	Regeneration and Optimization of Cultural heritage in creative and Knowledge cities
RRA	respiratory rate assessment
RT	reaction times
rTMS	transcranial magnetic stimulation
SC	skin conductance
SD	tests SDS denaturation test
SEEVal	the service encounter emotional value
SST	steady-state topography
SVM	support vector machine
tDCS	transcranial direct-current stimulation
TMS	transcranial magnetic stimulation
UIT-VSMEC	standard Vietnamese social media emotion corpus
VAAQ	virtual agent's acceptance questionnaire
VPA	voice pitch analysis
VR	virtual reality

References

1. Rizzolatti, G.; Sinigaglia, C. The Mirror Mechanism: A Basic Principle of Brain Function. *Nat. Rev. Neurosci.* **2016**, *17*, 757–765. [CrossRef] [PubMed]
2. Spunt, R.P.; Adolphs, R. The Neuroscience of Understanding the Emotions of Others. *Neurosci. Lett.* **2019**, *693*, 44–48. [CrossRef] [PubMed]
3. Berčík, J.; Neomániová, K.; Mravcová, A.; Gálová, J. Review of the Potential of Consumer Neuroscience for Aroma Marketing and Its Importance in Various Segments of Services. *Appl. Sci.* **2021**, *11*, 7636. [CrossRef]
4. Li, L.; Gow, A.D.I.; Zhou, J. The Role of Positive Emotions in Education: A Neuroscience Perspective. *Mind Brain Educ.* **2020**, *14*, 220–234. [CrossRef]
5. Cromwell, H.C.; Papadelis, C. Mapping the Brain Basis of Feelings, Emotions and Much More: A Special Issue Focused on 'The Human Affectome'. *Neurosci. Biobehav. Rev.* **2022**, *137*, 104672. [CrossRef]
6. Alexander, R.; Aragón, O.R.; Bookwala, J.; Cherbuin, N.; Gatt, J.M.; Kahrilas, I.J.; Kästner, N.; Lawrence, A.; Lowe, L.; Morrison, R.G.; et al. The Neuroscience of Positive Emotions and Affect: Implications for Cultivating Happiness and Wellbeing. *Neurosci. Biobehav. Rev.* **2021**, *121*, 220–249. [CrossRef]
7. Vuust, P.; Heggli, O.A.; Friston, K.J.; Kringelbach, M.L. Music in the Brain. *Nat. Rev. Neurosci.* **2022**, *23*, 287–305. [CrossRef]
8. Green, M.F.; Horan, W.P.; Lee, J. Social Cognition in Schizophrenia. *Nat. Rev. Neurosci.* **2015**, *16*, 620–631. [CrossRef]
9. Bunge, S.A. How We Use Rules to Select Actions: A Review of Evidence from Cognitive Neuroscience. *Cogn. Affect. Behav. Neurosci.* **2004**, *4*, 564–579. [CrossRef]
10. Lieberman, M.D. Social Cognitive Neuroscience: A Review of Core Processes. *Annu. Rev. Psychol.* **2007**, *58*, 259–289. [CrossRef]
11. Sawyer, K. The Cognitive Neuroscience of Creativity: A Critical Review. *Creat. Res. J.* **2011**, *23*, 137–154. [CrossRef]
12. Byrom, B.; McCarthy, M.; Schueler, P.; Muehlhausen, W. Brain Monitoring Devices in Neuroscience Clinical Research: The Potential of Remote Monitoring Using Sensors, Wearables, and Mobile Devices. *Clin. Pharmacol. Ther.* **2018**, *104*, 59–71. [CrossRef]
13. Johnson, K.T.; Picard, R.W. Advancing Neuroscience through Wearable Devices. *Neuron* **2020**, *108*, 8–12. [CrossRef]
14. Soroush, M.Z.; Maghooli, K.; Setarehdan, S.K.; Motie Nasrabadi, A. A Review on EEG Signals Based Emotion Recognition. *Int. Clin. Neurosci. J.* **2017**, *4*, 118–129. [CrossRef]
15. Gui, Q.; Ruiz-Blondet, M.V.; Laszlo, S.; Jin, Z. A Survey on Brain Biometrics. *ACM Comput. Surv.* **2019**, *51*, 1–38. [CrossRef]
16. Fairhurst, M.; Li, C.; Da Costa-Abreu, M. Predictive Biometrics: A Review and Analysis of Predicting Personal Characteristics from Biometric Data. *IET Biom.* **2017**, *6*, 369–378. [CrossRef]
17. Zhong, Y.; Deng, Y. A Survey on Keystroke Dynamics Biometrics: Approaches, Advances, and Evaluations. In *Gate to Computer Science and Research*; Zhong, Y., Deng, Y., Eds.; Science Gate Publishing P.C.: Thrace, Greece, 2015; Volume 2, pp. 1–22. [CrossRef]
18. Hernandez-de-Menendez, M.; Morales-Menendez, R.; Escobar, C.A.; Arinez, J. Biometric Applications in Education. *Int. J. Interact. Des. Manuf.* **2021**, *15*, 365–380. [CrossRef]
19. Berčík, J.; Horská, E.; Gálová, J.; Margianti, E.S. Consumer neuroscience in practice: The impact of store atmosphere on consumer behavior. *Period. Polytech. Soc. Manag. Sci.* **2016**, *24*, 96–101. [CrossRef]
20. Pisani, P.H.; Mhenni, A.; Giot, R.; Cherrier, E.; Poh, N.; Ferreira de Carvalho, A.C.P.d.L.; Rosenberger, C.; Amara, N.E.B. Adaptive Biometric Systems: Review and Perspectives. *ACM Comput. Surv.* **2020**, *52*, 1–38. [CrossRef]
21. Xu, S.; Fang, J.; Hu, X.; Ngai, E.; Guo, Y.; Leung, V.C.M.; Cheng, J.; Hu, B. Emotion Recognition from Gait Analyses: Current Research and Future Directions. *arXiv* **2020**, arXiv:2003.11461. [CrossRef]
22. Merone, M.; Soda, P.; Sansone, M.; Sansone, C. ECG Databases for Biometric Systems: A Systematic Review. *Expert Syst. Appl.* **2017**, *67*, 189–202. [CrossRef]
23. Curtin, A.; Tong, S.; Sun, J.; Wang, J.; Onaral, B.; Ayaz, H. A Systematic Review of Integrated Functional Near-Infrared Spectroscopy (FNIRS) and Transcranial Magnetic Stimulation (TMS) Studies. *Front. Neurosci.* **2019**, *13*, 84. [CrossRef]

24. da Silva, F.L. EEG and MEG: Relevance to Neuroscience. *Neuron* **2013**, *80*, 1112–1128. [CrossRef]
25. Khushaba, R.N.; Wise, C.; Kodagoda, S.; Louviere, J.; Kahn, B.E.; Townsend, C. Consumer Neuroscience: Assessing the Brain Response to Marketing Stimuli Using Electroencephalogram (EEG) and Eye Tracking. *Expert Syst. Appl.* **2013**, *40*, 3803–3812. [CrossRef]
26. Krugliak, A.; Clarke, A. Towards Real-World Neuroscience Using Mobile EEG and Augmented Reality. *Sci. Rep.* **2022**, *12*, 2291. [CrossRef]
27. Gramann, K.; Jung, T.-P.; Ferris, D.P.; Lin, C.-T.; Makeig, S. Toward a New Cognitive Neuroscience: Modeling Natural Brain Dynamics. *Front. Hum. Neurosci.* **2014**, *8*, 444. [CrossRef]
28. An, B.W.; Heo, S.; Ji, S.; Bien, F.; Park, J.-U. Transparent and Flexible Fingerprint Sensor Array with Multiplexed Detection of Tactile Pressure and Skin Temperature. *Nat. Commun.* **2018**, *9*, 2458. [CrossRef]
29. Gadaleta, M.; Radin, J.M.; Baca-Motes, K.; Ramos, E.; Kheterpal, V.; Topol, E.J.; Steinhubl, S.R.; Quer, G. Passive Detection of COVID-19 with Wearable Sensors and Explainable Machine Learning Algorithms. *NPJ Digit. Med.* **2021**, *4*, 166. [CrossRef]
30. Hayano, J.; Tanabiki, T.; Iwata, S.; Abe, K.; Yuda, E. Estimation of Emotions by Wearable Biometric Sensors Under Daily Activities. In *2018 IEEE 7th Global Conference on Consumer Electronics (GCCE), Osaka, Tokyo, 18–21 October 2022*; IEEE: Nara, Japan, 2018; pp. 240–241. [CrossRef]
31. Oostdijk, M.; van Velzen, A.; van Dijk, J.; Terpstra, A. State-of-the-Art in Biometrics for Multi-Factor Authentication in a Federative Context. *Identity* **2016**, *14*, 15.
32. Salman, A.S.; Salman, A.S.; Salman, O.S. Using Behavioral Biometrics of Fingerprint Authentication to Investigate Physical and Emotional User States. In Proceedings of the Future Technologies Conference (FTC) 2021, Volume 2; Arai, K., Ed.; Lecture Notes in Networks and Systems. Springer International Publishing: Cham, Switzerland, 2022; Volume 359, pp. 240–256. [CrossRef]
33. Zhang, Y.-J. Biometric Recognition. In *Handbook of Image Engineering*; Springer: Singapore, 2021; pp. 1231–1256.
34. Maffei, A.; Angrilli, A. E-MOVIE—Experimental MOVies for Induction of Emotions in Neuroscience: An Innovative Film Database with Normative Data and Sex Differences. *PLoS ONE* **2019**, *14*, e0223124. [CrossRef]
35. Apicella, A.; Arpaia, P.; Mastrati, G.; Moccaldi, N. EEG-Based Detection of Emotional Valence towards a Reproducible Measurement of Emotions. *Sci. Rep.* **2021**, *11*, 21615. [CrossRef] [PubMed]
36. Tost, H.; Reichert, M.; Braun, U.; Reinhard, I.; Peters, R.; Lautenbach, S.; Hoell, A.; Schwarz, E.; Ebner-Priemer, U.; Zipf, A.; et al. Neural Correlates of Individual Differences in Affective Benefit of Real-Life Urban Green Space Exposure. *Nat. Neurosci.* **2019**, *22*, 1389–1393. [CrossRef] [PubMed]
37. Mashrur, F.R.; Rahman, K.M.; Miya, M.T.I.; Vaidyanathan, R.; Anwar, S.F.; Sarker, F.; Mamun, K.A. An Intelligent Neuromarketing System for Predicting Consumers' Future Choice from Electroencephalography Signals. *Physiol. Behav.* **2022**, *253*, 113847. [CrossRef] [PubMed]
38. Asadzadeh, S.; Yousefi Rezaii, T.; Beheshti, S.; Meshgini, S. Accurate Emotion Recognition Using Bayesian Model Based EEG Sources as Dynamic Graph Convolutional Neural Network Nodes. *Sci. Rep.* **2022**, *12*, 10282. [CrossRef] [PubMed]
39. Čeko, M.; Kragel, P.A.; Woo, C.-W.; López-Solà, M.; Wager, T.D. Common and Stimulus-Type-Specific Brain Representations of Negative Affect. *Nat. Neurosci.* **2022**, *25*, 760–770. [CrossRef]
40. Prete, G.; Croce, P.; Zappasodi, F.; Tommasi, L.; Capotosto, P. Exploring Brain Activity for Positive and Negative Emotions by Means of EEG Microstates. *Sci. Rep.* **2022**, *12*, 3404. [CrossRef]
41. Sitaram, R.; Ros, T.; Stoeckel, L.; Haller, S.; Scharnowski, F.; Lewis-Peacock, J.; Weiskopf, N.; Blefari, M.L.; Rana, M.; Oblak, E.; et al. Closed-Loop Brain Training: The Science of Neurofeedback. *Nat. Rev. Neurosci.* **2017**, *18*, 86–100. [CrossRef]
42. Del Negro, C.A.; Funk, G.D.; Feldman, J.L. Breathing Matters. *Nat. Rev. Neurosci.* **2018**, *19*, 351–367. [CrossRef]
43. Pugh, Z.H.; Choo, S.; Leshin, J.C.; Lindquist, K.A.; Nam, C.S. Emotion Depends on Context, Culture and Their Interaction: Evidence from Effective Connectivity. *Soc. Cogn. Affect. Neurosci.* **2022**, *17*, 206–217. [CrossRef]
44. Barrett, L.F. *How Emotions Are Made: The Secret Life of the Brain*; Houghton Mifflin Harcourt: Boston, MA, USA, 2017.
45. Barrett, L.F. The Theory of Constructed Emotion: An Active Inference Account of Interoception and Categorization. *Soc. Cogn. Affect. Neurosci.* **2017**, *12*, 1–23. [CrossRef]
46. Basiri, M.E.; Nemati, S.; Abdar, M.; Cambria, E.; Acharya, U.R. ABCDM: An Attention-Based Bidirectional CNN-RNN Deep Model for Sentiment Analysis. *Future Gener. Comput. Syst.* **2021**, *115*, 279–294. [CrossRef]
47. Parry, G.; Vuong, Q. Deep Affect: Using Objects, Scenes and Facial Expressions in a Deep Neural Network to Predict Arousal and Valence Values of Images. *arXiv preprint* **2021**. [CrossRef]
48. Gendron, B.; Kouremenou, E.-S.; Rusu, C. Emotional Capital Development, Positive Psychology and Mindful Teaching: Which Links? *Int. J. Emot. Educ.* **2016**, *8*, 63–74.
49. Houge Mackenzie, S.; Brymer, E. Conceptualizing Adventurous Nature Sport: A Positive Psychology Perspective. *Ann. Leis. Res.* **2020**, *23*, 79–91. [CrossRef]
50. Li, C. A Positive Psychology Perspective on Chinese EFL Students' Trait Emotional Intelligence, Foreign Language Enjoyment and EFL Learning Achievement. *J. Multiling. Multicult. Dev.* **2020**, *41*, 246–263. [CrossRef]
51. Bower, I.; Tucker, R.; Enticott, P.G. Impact of Built Environment Design on Emotion Measured via Neurophysiological Correlates and Subjective Indicators: A Systematic Review. *J. Environ. Psychol.* **2019**, *66*, 101344. [CrossRef]
52. Cassidy, T. *Environmental Psychology: Behaviour and Experience in Context*; Contemporary Psychology Series; Psychology Press: Hove, UK, 1997.

53. Cho, H.; Li, C.; Wu, Y. Understanding Sport Event Volunteers' Continuance Intention: An Environmental Psychology Approach. *Sport Manag. Rev.* **2020**, *23*, 615–625. [CrossRef]
54. Lin, S.; Döngül, E.S.; Uygun, S.V.; Öztürk, M.B.; Huy, D.T.N.; Tuan, P.V. Exploring the Relationship between Abusive Management, Self-Efficacy and Organizational Performance in the Context of Human–Machine Interaction Technology and Artificial Intelligence with the Effect of Ergonomics. *Sustainability* **2022**, *14*, 1949. [CrossRef]
55. Privitera, M.; Ferrari, K.D.; von Ziegler, L.M.; Sturman, O.; Duss, S.N.; Floriou-Servou, A.; Germain, P.-L.; Vermeiren, Y.; Wyss, M.T.; de Deyn, P.P.; et al. A Complete Pupillometry Toolbox for Real-Time Monitoring of Locus Coeruleus Activity in Rodents. *Nat. Protoc.* **2020**, *15*, 2301–2320. [CrossRef]
56. Rebelo, F.; Noriega, P.; Vilar, E.; Filgueiras, E. Ergonomics and Human Factors Research Challenges: The ErgoUX Lab Case Study. In *Advances in Ergonomics in Design*; Rebelo, F., Ed.; Lecture Notes in Networks and Systems; Springer International Publishing: Cham, Switzerland, 2021; Volume 261, pp. 912–922. [CrossRef]
57. Khan, F. Making Savings Count. *Nat. Energy* **2018**, *3*, 354. [CrossRef]
58. Zhang, B.; Kang, J. Effect of Environmental Contexts Pertaining to Different Sound Sources on the Mood States. *Build. Environ.* **2022**, *207*, 108456. [CrossRef]
59. Zhu, B.-W.; Xiao, Y.H.; Zheng, W.-Q.; Xiong, L.; He, X.Y.; Zheng, J.-Y.; Chuang, Y.-C. A Hybrid Multiple-Attribute Decision-Making Model for Evaluating the Esthetic Expression of Environmental Design Schemes. *SAGE Open* **2022**, *12*, 215824402210872. [CrossRef]
60. Silva, P.L.; Kiefer, A.; Riley, M.A.; Chemero, A. Trading Perception and Action for Complex Cognition: Application of Theoretical Principles from Ecological Psychology to the Design of Interventions for Skill Learning. In *Handbook of Embodied Cognition and Sport Psychology*; MIT Press: Boston, MA, USA, 2019; pp. 47–74.
61. Szokolszky, A. Perceiving Metaphors: An Approach from Developmental Ecological Psychology. *Metaphor Symb.* **2019**, *34*, 17–32. [CrossRef]
62. Van den Berg, P.; Larosi, H.; Maussen, S.; Arentze, T. Sense of Place, Shopping Area Evaluation, and Shopping Behaviour. *Geogr. Res.* **2021**, *59*, 584–598. [CrossRef]
63. Argent, N. Behavioral Geography. In *International Encyclopedia of Geography: People, the Earth, Environment and Technology*; Richardson, D., Castree, N., Goodchild, M.F., Kobayashi, A., Liu, W., Marston, R.A., Eds.; John Wiley & Sons, Ltd: Oxford, UK, 2017; pp. 1–11. [CrossRef]
64. Schwarz, N.; Dressler, G.; Frank, K.; Jager, W.; Janssen, M.; Müller, B.; Schlüter, M.; Wijermans, N.; Groeneveld, J. Formalising Theories of Human Decision-Making for Agent-Based Modelling of Social-Ecological Systems: Practical Lessons Learned and Ways Forward. *SESMO* **2020**, *2*, 16340. [CrossRef]
65. Plutchik, R. *The Emotions*, Rev. ed.; University Press of America: Lanham, MD, USA, 1991.
66. Kaklauskas, A.; Milevicius, V.; Kaklauskiene, L. Effects of Country Success on COVID-19 Cumulative Cases and Excess Deaths in 169 Countries. *Ecol. Indic.* **2022**, *137*, 108703. [CrossRef]
67. Kaklauskas, A. Degree of project utility and investment value assessments. *Int. J. Comput. Commun. Control.* **2016**, *11*, 666–683. [CrossRef]
68. Kaklauskas, A.; Herrera-Viedma, E.; Echenique, V.; Zavadskas, E.K.; Ubarte, I.; Mostert, A.; Podvezko, V.; Binkyte, A.; Podviezko, A. Multiple Criteria Analysis of Environmental Sustainability and Quality of Life in Post-Soviet States. *Ecol. Indic.* **2018**, *89*, 781–807. [CrossRef]
69. Kaklauskas, A.; Dias, W.P.S.; Binkyte-Veliene, A.; Abraham, A.; Ubarte, I.; Randil, O.P.C.; Siriwardana, C.S.A.; Lill, I.; Milevicius, V.; Podviezko, A.; et al. Are Environmental Sustainability and Happiness the Keys to Prosperity in Asian Nations? *Ecol. Indic.* **2020**, *119*, 106562. [CrossRef]
70. Kaklauskas, A.; Kaklauskiene, L. Analysis of the impact of success on three dimensions of sustainability in 173 countries. *Sci. Rep.* **2022**, *12*, 14719. [CrossRef]
71. Barrett, L.F. Solving the Emotion Paradox: Categorization and the Experience of Emotion. *Pers. Soc. Psychol. Rev.* **2006**, *10*, 20–46. [CrossRef] [PubMed]
72. Puce, A.; Latinus, M.; Rossi, A.; da Silva, E.; Parada, F.; Love, S.; Ashourvan, A.; Jayaraman, S. Neural Bases for Social Attention in Healthy Humans. In *The Many Faces of Social Attention*; Puce, A., Bertenthal, B.I., Eds.; Springer International Publishing: Cham, Switzerland, 2015; pp. 93–127. [CrossRef]
73. Shablack, H.; Becker, M.; Lindquist, K.A. How Do Children Learn Novel Emotion Words? A Study of Emotion Concept Acquisition in Preschoolers. *J. Exp. Psychol. Gen.* **2020**, *149*, 1537–1553. [CrossRef] [PubMed]
74. Izard, C.E. Basic Emotions, Natural Kinds, Emotion Schemas, and a New Paradigm. *Perspect. Psychol. Sci.* **2007**, *2*, 260–280. [CrossRef] [PubMed]
75. Briesemeister, B.B.; Kuchinke, L.; Jacobs, A.M. Discrete Emotion Effects on Lexical Decision Response Times. *PLoS ONE* **2011**, *6*, e23743. [CrossRef] [PubMed]
76. Ekman, P. An Argument for Basic Emotions. *Cogn. Emot.* **1992**, *6*, 169–200. [CrossRef]
77. Ekman, P. Facial Expressions. In *Handbook of Cognition and Emotion*; Dalgleish, T., Power, M.J., Eds.; John Wiley & Sons, Ltd: Chichester, UK, 1999; pp. 301–320. [CrossRef]
78. Colombetti, G. From Affect Programs to Dynamical Discrete Emotions. *Philos. Psychol.* **2009**, *22*, 407–425. [CrossRef]

79. Fox, E. *Emotion Science: Cognitive and Neuroscientific Approaches to Understanding Human Emotions*; Palgrave Macmillan: Basingstoke, UK; New York, NY, USA, 2008.
80. Russell, J.A.; Barrett, L.F. Core Affect, Prototypical Emotional Episodes, and Other Things Called Emotion: Dissecting the Elephant. *J. Personal. Soc. Psychol.* **1999**, *76*, 805–819. [CrossRef]
81. Cross Francis, D.I.; Hong, J.; Liu, J.; Eker, A.; Lloyd, K.; Bharaj, P.K.; Jeon, M. The Dominance of Blended Emotions: A Qualitative Study of Elementary Teachers' Emotions Related to Mathematics Teaching. *Front. Psychol.* **2020**, *11*, 1865. [CrossRef]
82. Hakak, N.M.; Mohd, M.; Kirmani, M.; Mohd, M. Emotion Analysis: A Survey. In *2017 International Conference on Computer, Communications and Electronics (Comptelix)*, Jaipur, India, 1–2 July 2017; IEEE: Jaipur, India, 2017; pp. 397–402. [CrossRef]
83. Posner, J.; Russell, J.A.; Peterson, B.S. The Circumplex Model of Affect: An Integrative Approach to Affective Neuroscience, Cognitive Development, and Psychopathology. *Develop. Psychopathol.* **2005**, *17*, 715–734. [CrossRef]
84. Eerola, T.; Vuoskoski, J.K. A Comparison of the Discrete and Dimensional Models of Emotion in Music. *Psychol. Music* **2011**, *39*, 18–49. [CrossRef]
85. Dzedzickis, A.; Kaklauskas, A.; Bucinskas, V. Human Emotion Recognition: Review of Sensors and Methods. *Sensors* **2020**, *20*, 592. [CrossRef]
86. Bradley, M.M.; Greenwald, M.K.; Petry, M.C.; Lang, P.J. Remembering Pictures: Pleasure and Arousal in Memory. *J. Exp. Psychol. Learn. Mem. Cogn.* **1992**, *18*, 379–390. [CrossRef]
87. Rubin, D.C.; Talarico, J.M. A Comparison of Dimensional Models of Emotion: Evidence from Emotions, Prototypical Events, Autobiographical Memories, and Words. *Memory* **2009**, *17*, 802–808. [CrossRef]
88. Watson, D.; Tellegen, A. Toward a Consensual Structure of Mood. *Psychol. Bull.* **1985**, *98*, 219–235. [CrossRef]
89. Karbauskaitė, R.; Sakalauskas, L.; Dzemyda, G. Kriging Predictor for Facial Emotion Recognition Using Numerical Proximities of Human Emotions. *Informatica* **2020**, *31*, 249–275. [CrossRef]
90. Mehrabian, A. Framework for a Comprehensive Description and Measurement of Emotional States. *Genet. Soc. Gen. Psychol. Monogr.* **1995**, *121*, 339–361.
91. Mehrabian, A. Correlations of the PAD Emotion Scales with Self-Reported Satisfaction in Marriage and Work. *Genet. Soc. Gen. Psychol. Monogr.* **1998**, *124*, 311–334.
92. Detandt, S.; Leys, C.; Bazan, A. A French Translation of the Pleasure Arousal Dominance (PAD) Semantic Differential Scale for the Measure of Affect and Drive. *Psychol. Belg.* **2017**, *57*, 17. [CrossRef]
93. Kaklauskas, A.; Bucinskas, V.; Dzedzickis, A.; Ubarte, I. Method for Controlling a Customized Microclimate in a Building and Realization System Thereof. European Patent Application. EP 4 020 134 A1, 7 February 2021.
94. Nor, N.M.; Wahab, A.; Majid, H.; Kamaruddin, N. Pre-Post Accident Analysis Relates to Pre-Cursor Emotion for Driver Behavior Understanding. In Proceedings of the 11th WSEAS International Conference on Applied Computer Science, Rovaniemi, Finland, 18–20 April 2012; World Scientific and Engineering Academy and Society (WSEAS): Stevens Point, WI, USA; pp. 152–157.
95. Kolmogorova, A.; Kalinin, A.; Malikova, A. Non-Discrete Sentiment Dataset Annotation: Case Study for Lövheim Cube Emotional Model. In *Digital Transformation and Global Society*; Alexandrov, D.A., Boukhanovsky, A.V., Chugunov, A.V., Kabanov, Y., Koltsova, O., Musabirov, I., Eds.; Communications in Computer and Information Science; Springer International Publishing: Cham, Switzerland, 2020; Volume 1242, pp. 154–164. [CrossRef]
96. Lövheim, H. A New Three-Dimensional Model for Emotions and Monoamine Neurotransmitters. *Med. Hypotheses* **2012**, *78*, 341–348. [CrossRef]
97. Mohsin, M.A.; Beltiukov, A. Summarizing Emotions from Text Using Plutchik's Wheel of Emotions. In *Proceedings of the 7th Scientific Conference on Information Technologies for Intelligent Decision Making Support (ITIDS 2019)*; Atlantis Press: Ufa, Russia, 2019. [CrossRef]
98. Donaldson, M. A Plutchik's Wheel of Emotions—2017 Update. 2018. Available online: https://www.uvm.edu/~mjk/013%20Intro%20to%20Wildlife%20Tracking/Plutchik's%20Wheel%20of%20Emotions%20-%202017%20Update%20_%20Six%20Seconds.pdf (accessed on 5 September 2022).
99. Mulder, P. Robert Plutchik's Wheel of Emotions. 2018. Available online: https://www.toolshero.com/psychology/wheel-of-emotions-plutchik/ (accessed on 5 September 2022).
100. Kołakowska, A.; Landowska, A.; Szwoch, M.; Szwoch, W.; Wróbel, M.R. Modeling Emotions for Affectaware Applications. In *Information Systems Development and Applications*; Faculty of Management, University of Gdańsk: Gdańsk, Poland, 2015; pp. 55–69.
101. Suttles, J.; Ide, N. Distant Supervision for Emotion Classification with Discrete Binary Values. In *Computational Linguistics and Intelligent Text Processing*; Gelbukh, A., Hutchison, D., Kanade, T., Kittler, J., Kleinberg, J.M., Mattern, F., Mitchell, J.C., Naor, M., Eds.; Lecture Notes in Computer Science; Springer: Berlin/Heidelberg, Germany, 2013; Volume 7817, pp. 121–136. [CrossRef]
102. Six seconds The Emotional Intelligence Network. Plutchik's Wheel of Emotions: Exploring the Emotion Wheel. Available online: https://www.6seconds.org/2022/03/13/plutchik-wheel-emotions/ (accessed on 5 September 2022).
103. Karnilowicz, H.R. The Emotion Wheel: Purpose, Definition, and Uses. Available online: https://www.berkeleywellbeing.com/emotion-wheel.html (accessed on 17 August 2022).
104. Cambria, E.; Livingstone, A.; Hussain, A. The Hourglass of Emotions. In *Cognitive Behavioural Systems*; Esposito, A., Esposito, A.M., Vinciarelli, A., Hoffmann, R., Müller, V.C., Hutchison, D., Kanade, T., Eds.; Lecture Notes in Computer Science; Springer: Berlin/Heidelberg, Germany, 2012; Volume 7403, pp. 144–157. [CrossRef]
105. Plutchik, R.; Kellerman, H. *Theories of Emotion*; Academic Press: Cambridge, MA, USA, 2013.

106. Kušen, E.; Strembeck, M.; Cascavilla, G.; Conti, M. On the Influence of Emotional Valence Shifts on the Spread of Information in Social Networks. In Proceedings of the 2017 IEEE/ACM International Conference on Advances in Social Networks Analysis and Mining 2017, Sydney, Australia, 31 July–3 August 2017; ACM: Sydney, Australia, 2017; pp. 321–324. [CrossRef]
107. Bassett, D.S.; Sporns, O. Network Neuroscience. *Nat. Neurosci.* **2017**, *20*, 353–364. [CrossRef]
108. Deion, A. 8 Top Trends of Future Sensors. 2021. Available online: https://community.hackernoon.com/t/8-top-trends-of-future-sensors/57483 (accessed on 5 September 2022).
109. Gartner; Panetta, K. Gartner Top Strategic Technology Trends for 2021. 2020. Available online: https://www.gartner.com/smarterwithgartner/gartner-top-strategic-technology-trends-for-2021 (accessed on 17 August 2022).
110. Kobus, H. Future Sensor Technology: 21 Expected Trends. Available online: https://www.sentech.nl/en/rd-engineer/21-sensor-technology-future-trends/ (accessed on 5 September 2022).
111. Sebastian, V. Neuromarketing and Evaluation of Cognitive and Emotional Responses of Consumers to Marketing Stimuli. *Procedia-Soc. Behav. Sci.* **2014**, *127*, 753–757. [CrossRef]
112. Sawe, N.; Chawla, K. Environmental Neuroeconomics: How Neuroscience Can Inform Our Understanding of Human Responses to Climate Change. *Curr. Opin. Behav. Sci.* **2021**, *42*, 147–154. [CrossRef]
113. Serra, D. Neuroeconomics: Reliable, Scientifically Legitimate and Useful Knowledge for Economists? 2020. Available online: https://hal.inrae.fr/hal-02956441 (accessed on 5 September 2022).
114. Braeutigam, S. Neuroeconomics—From Neural Systems to Economic Behaviour. *Brain Res. Bull.* **2005**, *67*, 355–360. [CrossRef]
115. Kenning, P.; Plassmann, H. NeuroEconomics: An Overview from an Economic Perspective. *Brain Res. Bull.* **2005**, *67*, 343–354. [CrossRef]
116. Wirdayanti, Y.N.; Ghoni, M.A. Neuromanagement Under the Light of Maqasid Sharia. *Al Tijarah* **2020**, *5*, 63–71. [CrossRef]
117. Teacu Parincu, A.M.; Capatina, A.; Varon, D.J.; Bennet, P.F.; Recuerda, A.M. Neuromanagement: The Scientific Approach to Contemporary Management. *Proc. Int. Conf. Bus. Excell.* **2020**, *14*, 1046–1056. [CrossRef]
118. Arce, A.L.; Cordero, J.M.B.; Mejía, E.T.; González, B.P. Tools of Neuromanagement, to Strengthen the Leadership Competencies of Executives in the Logistics Areas of the Auto Parts Industry. *StrategyTechnol. Soc.* **2020**, *10*, 36–63.
119. Michalczyk, S.; Jung, D.; Nadj, M.; Knierim, M.T.; Rissler, R. BrownieR: The R-Package for Neuro Information Systems Research. In *Information Systems and Neuroscience*; Davis, F.D., Riedl, R., vom Brocke, J., Léger, P.-M., Randolph, A.B., Eds.; Lecture Notes in Information Systems and Organisation; Springer International Publishing: Cham, Switzerland, 2019; Volume 29, pp. 101–109. [CrossRef]
120. Riedl, R.; Léger, P. Neuro-Information-Systems (NeuroIS). In *Association for Information Systems*; Springer: Berlin/Heidelberg, Germany, 2016. [CrossRef]
121. Ma, Q.; Ji, W.; Fu, H.; Bian, J. Neuro-Industrial Engineering: The New Stage of Modern IE—From the Human-Oriented Perspective. *Int. J. Serv. Oper. Inform.* **2012**, *7*, 150–166. [CrossRef]
122. Rymer, W.Z. Neural Engineering. Encyclopedia Britannica. 2018. Available online: https://www.britannica.com/science/neural-engineering (accessed on 5 September 2022).
123. Hodson, H. Hang on Your Every Word. *New Sci.* **2014**, *222*, 20.
124. Tzirakis, P.; Zhang, J.; Schuller, B.W. End-to-End Speech Emotion Recognition Using Deep Neural Networks. In *2018 IEEE International Conference on Acoustics, Speech and Signal Processing (ICASSP), Calgary, AB, Canada, 15–20 April 2018*; IEEE: Calgary, AB, Canada, 2018; pp. 5089–5093. [CrossRef]
125. Parkin, B.L.; Ekhtiari, H.; Walsh, V.F. Non-Invasive Human Brain Stimulation in Cognitive Neuroscience: A Primer. *Neuron* **2015**, *87*, 932–945. [CrossRef]
126. Annavarapu, R.N.; Kathi, S.; Vadla, V.K. Non-Invasive Imaging Modalities to Study Neurodegenerative Diseases of Aging Brain. *J. Chem. Neuroanat.* **2019**, *95*, 54–69. [CrossRef] [PubMed]
127. Bergmann, T.O.; Karabanov, A.; Hartwigsen, G.; Thielscher, A.; Siebner, H.R. Combining Non-Invasive Transcranial Brain Stimulation with Neuroimaging and Electrophysiology: Current Approaches and Future Perspectives. *NeuroImage* **2016**, *140*, 4–19. [CrossRef] [PubMed]
128. Cao, M.; Galvis, D.; Vogrin, S.J.; Woods, W.P.; Vogrin, S.; Wang, F.; Woldman, W.; Terry, J.R.; Peterson, A.; Plummer, C.; et al. Virtual Intracranial EEG Signals Reconstructed from MEG with Potential for Epilepsy Surgery. *Nat. Commun.* **2022**, *13*, 994. [CrossRef] [PubMed]
129. Currà, A.; Gasbarrone, R.; Cardillo, A.; Trompetto, C.; Fattapposta, F.; Pierelli, F.; Missori, P.; Bonifazi, G.; Serranti, S. Near-Infrared Spectroscopy as a Tool for in Vivo Analysis of Human Muscles. *Sci. Rep.* **2019**, *9*, 8623. [CrossRef]
130. De Camp, N.V.; Kalinka, G.; Bergeler, J. Light-Cured Polymer Electrodes for Non-Invasive EEG Recordings. *Sci. Rep.* **2018**, *8*, 14041. [CrossRef]
131. Etchell, A.C.; Civier, O.; Ballard, K.J.; Sowman, P.F. A Systematic Literature Review of Neuroimaging Research on Developmental Stuttering between 1995 and 2016. *J. Fluen. Disord.* **2018**, *55*, 6–45. [CrossRef]
132. Peters, J.C.; Reithler, J.; de Graaf, T.A.; Schuhmann, T.; Goebel, R.; Sack, A.T. Concurrent Human TMS-EEG-FMRI Enables Monitoring of Oscillatory Brain State-Dependent Gating of Cortico-Subcortical Network Activity. *Commun. Biol.* **2020**, *3*, 40. [CrossRef]
133. Shibasaki, H. Human Brain Mapping: Hemodynamic Response and Electrophysiology. *Clin. Neurophysiol.* **2008**, *119*, 731–743. [CrossRef]

134. Silberstein, R.B.; Nield, G.E. Brain Activity Correlates of Consumer Brand Choice Shift Associated with Television Advertising. *Int. J. Advert.* **2008**, *27*, 359–380. [CrossRef]
135. Uludag, U.; Pankanti, S.; Prabhakar, S.; Jain, A.K. Biometric Cryptosystems: Issues and Challenges. *Proc. IEEE* **2004**, *92*, 948–960. [CrossRef]
136. Presby, D.M.; Capodilupo, E.R. Biometrics from a Wearable Device Reveal Temporary Effects of COVID-19 Vaccines on Cardiovascular, Respiratory, and Sleep Physiology. *J. Appl. Physiol.* **2022**, *132*, 448–458. [CrossRef]
137. Stephen, M.J.; Reddy, P. Implementation of Easy Fingerprint Image Authentication with Traditional Euclidean and Singular Value Decomposition Algorithms. *Int. J. Adv. Soft Comput. Its Appl.* **2011**, *3*, 1–19.
138. Banirostam, H.; Shamsinezhad, E.; Banirostam, T. Functional Control of Users by Biometric Behavior Features in Cloud Computing. In *2013 4th International Conference on Intelligent Systems, Modelling and Simulation, Bangkok, Thailand, 29–30 January 2013*; IEEE: Bangkok, Thailand, 2013; pp. 94–98. [CrossRef]
139. Yang, W.; Wang, S.; Hu, J.; Zheng, G.; Chaudhry, J.; Adi, E.; Valli, C. Securing Mobile Healthcare Data: A Smart Card Based Cancelable Finger-Vein Bio-Cryptosystem. *IEEE Access* **2018**, *6*, 36939–36947. [CrossRef]
140. Kodituwakku, S.R. Biometric Authentication: A Review. *Int. J. Trend Res. Dev.* **2015**, *2*, 113–123.
141. Jain, A.; Hong, L.; Pankanti, S. Biometric Identification. *Commun. ACM* **2000**, *43*, 90–98. [CrossRef]
142. Choudhary, S.K.; Naik, A.K. Multimodal Biometric Authentication with Secured Templates—A Review. In *2019 3rd International Conference on Trends in Electronics and Informatics (ICOEI), Tirunelveli, India, 23–25 April 2019*; IEEE: Tirunelveli, India, 2019; pp. 1062–1069. [CrossRef]
143. Kim, J.S.; Pan, S.B. A Study on EMG-Based Biometrics. *Internet Serv. Inf. Secur. (JISIS)* **2017**, *7*, 19–31.
144. Maiorana, E. Deep Learning for EEG-Based Biometric Recognition. *Neurocomputing* **2020**, *410*, 374–386. [CrossRef]
145. Revett, K. Cognitive Biometrics: A Novel Approach to Person Authentication. *IJCB* **2012**, *1*, 1–9. [CrossRef]
146. Prasse, P.; Jäger, L.A.; Makowski, S.; Feuerpfeil, M.; Scheffer, T. On the Relationship between Eye Tracking Resolution and Performance of Oculomotoric Biometric Identification. *Procedia Comput. Sci.* **2020**, *176*, 2088–2097. [CrossRef]
147. Cho, Y. Rethinking Eye-Blink: Assessing Task Difficulty through Physiological Representation of Spontaneous Blinking. In Proceedings of the 2021 CHI Conference on Human Factors in Computing Systems, Yokohama, Japan, 8–13 May 2021; ACM: Yokohama, Japan, 2021; pp. 1–12. [CrossRef]
148. Abdulrahman, S.A.; Alhayani, B. A Comprehensive Survey on the Biometric Systems Based on Physiological and Behavioural Characteristics. *Mater. Today Proc.* **2021**, In Press, Corrected Proof. S2214785321048513. [CrossRef]
149. Allado, E.; Poussel, M.; Moussu, A.; Saunier, V.; Bernard, Y.; Albuisson, E.; Chenuel, B. Innovative Measurement of Routine Physiological Variables (Heart Rate, Respiratory Rate and Oxygen Saturation) Using a Remote Photoplethysmography Imaging System: A Prospective Comparative Trial Protocol. *BMJ Open* **2021**, *11*, e047896. [CrossRef] [PubMed]
150. Dargan, S.; Kumar, M. A Comprehensive Survey on the Biometric Recognition Systems Based on Physiological and Behavioral Modalities. *Expert Syst. Appl.* **2020**, *143*, 113114. [CrossRef]
151. Mordini, E.; Tzovaras, D.; Ashton, H. Introduction. In *Second Generation Biometrics: The Ethical, Legal and Social Context*; Mordini, E., Tzovaras, D., Eds.; The International Library of Ethics, Law and Technology; Springer: Dordrecht, The Netherlands, 2012; Volume 11, pp. 1–19. [CrossRef]
152. Fuster, G.G. Artificial Intelligence and Law Enforcement: Impact on Fundamental Rights (European Parliament 2020). 2020. Available online: http://www.europarl.europa.eu/supporting-analyses (accessed on 5 September 2022).
153. Ghilardi, G.; Keller, F. Epistemological Foundation of Biometrics. In *Second Generation Biometrics: The Ethical, Legal and Social Context*; Mordini, E., Tzovaras, D., Eds.; The International Library of Ethics, Law and Technology; Springer: Dordrecht, The Netherlands, 2012; Volume 11, pp. 23–47. [CrossRef]
154. Riera, A.; Dunne, S.; Cester, I.; Ruffini, G. Electrophysiological biometrics: Opportunities and risks. In *Second Generation Biometrics: The Ethical, Legal and Social Context*; Mordini, E., Tzovaras, D., Eds.; The International Library of Ethics, Law and Technology; Springer: Dordrecht, The Netherlands, 2012; Volume 11, pp. 149–176. [CrossRef]
155. Smith, M.; Mann, M.; Urbas, G. *Biometrics, Crime and Security*; Law, science and society; Routledge: New York, NY, USA, 2018.
156. Simó, F.Z. Then and Now. *Profuturo* **2019**, *9*, 78–90. [CrossRef]
157. U.S Department of Homeland Security. Future Attribute Screening Technology. 2014. Available online: https://www.dhs.gov/sites/default/files/publications/Future%20Attribute%20Screening%20Technology-FAST.pdf (accessed on 5 September 2022).
158. Alhalaseh, R.; Alasasfeh, S. Machine-Learning-Based Emotion Recognition System Using EEG Signals. *Computers* **2020**, *9*, 95. [CrossRef]
159. Ma, X.; Jiang, X.; Jiang, Y. Increased Spontaneous Fronto-Central Oscillatory Power during Eye Closing in Patients with Multiple Somatic Symptoms. *Psychiatry Res. Neuroimaging* **2022**, *324*. [CrossRef]
160. Ramesh, S.; Gomathi, S.; Sasikala, S.; Saravanan, T.R. Automatic Speech Emotion Detection Using Hybrid of Gray Wolf Optimizer and Naïve Bayes. *Int. J. Speech Technol.* **2021**, 1–8. [CrossRef]
161. Moses, E.; Clark, K.R.; Jacknis, N.J. The Future of Advertising: Influencing and Predicting Response Through Artificial Intelligence, Machine Learning, and Neuroscience. In *Advances in Business Information Systems and Analytics*; Chkoniya, V., Ed.; IGI Global: Hershey, PA, USA, 2021; pp. 151–166. [CrossRef]
162. Sun, L.; Fu, S.; Wang, F. Decision Tree SVM Model with Fisher Feature Selection for Speech Emotion Recognition. *J. Audio Speech Music Proc.* **2019**, *2019*, 2. [CrossRef]

163. Sun, L.; Zou, B.; Fu, S.; Chen, J.; Wang, F. Speech Emotion Recognition Based on DNN-Decision Tree SVM Model. *Speech Commun.* **2019**, *115*, 29–37. [CrossRef]
164. Chen, L.; Su, W.; Feng, Y.; Wu, M.; She, J.; Hirota, K. Two-Layer Fuzzy Multiple Random Forest for Speech Emotion Recognition in Human-Robot Interaction. *Inf. Sci.* **2020**, *509*, 150–163. [CrossRef]
165. Rai, M.; Husain, A.A.; Sharma, R.; Maity, T.; Yadav, R. Facial Feature-Based Human Emotion Detection Using Machine Learning: An Overview. In *Artificial Intelligence and Cybersecurity*; CRC Press: Boca Raton, FL, USA, 2022; pp. 107–120.
166. Zhang, J.; Yin, Z.; Chen, P.; Nichele, S. Emotion Recognition Using Multi-Modal Data and Machine Learning Techniques: A Tutorial and Review. *Inf. Fusion* **2020**, *59*, 103–126. [CrossRef]
167. Aouani, H.; Ben Ayed, Y. Deep Support Vector Machines for Speech Emotion Recognition. In *Intelligent Systems Design and Applications*; Abraham, A., Siarry, P., Ma, K., Kaklauskas, A., Eds.; Advances in Intelligent Systems and Computing; Springer International Publishing: Cham, Switzerland, 2021; Volume 1181, pp. 406–415. [CrossRef]
168. Bhavan, A.; Chauhan, P.; Hitkul; Shah, R.R. Bagged Support Vector Machines for Emotion Recognition from Speech. *Knowl.-Based Syst.* **2019**, *184*, 104886. [CrossRef]
169. Miller, C.H.; Sacchet, M.D.; Gotlib, I.H. Support Vector Machines and Affective Science. *Emot. Rev.* **2020**, *12*, 297–308. [CrossRef]
170. Abo, M.E.M.; Idris, N.; Mahmud, R.; Qazi, A.; Hashem, I.A.T.; Maitama, J.Z.; Naseem, U.; Khan, S.K.; Yang, S. A Multi-Criteria Approach for Arabic Dialect Sentiment Analysis for Online Reviews: Exploiting Optimal Machine Learning Algorithm Selection. *Sustainability* **2021**, *13*, 10018. [CrossRef]
171. Singh, B.K.; Khare, A.; Soni, A.K.; Kumar, A. Electroencephalography-Based Classification of Human Emotion: A Hybrid Strategy in Machine Learning Paradigm. *Int. J. Comput. Vis. Robot.* **2019**, *9*, 583–598. [CrossRef]
172. Yudhana, A.; Muslim, A.; Wati, D.E.; Puspitasari, I.; Azhari, A.; Mardhia, M.M. Human Emotion Recognition Based on EEG Signal Using Fast Fourier Transform and K-Nearest Neighbor. *Adv. Sci. Technol. Eng. Syst. J.* **2020**, *5*, 1082–1088. [CrossRef]
173. Assielou, K.A.; Haba, C.T.; Gooré, B.T.; Kadjo, T.L.; Yao, K.D. Emotional Impact for Predicting Student Performance in Intelligent Tutoring Systems (ITS). *Int. J. Adv. Comput. Sci. Appl.* **2020**, *11*, 219–225. [CrossRef]
174. Lenzoni, S.; Bozzoni, V.; Burgio, F.; de Gelder, B.; Wennberg, A.; Botta, A.; Pegoraro, E.; Semenza, C. Recognition of Emotions Conveyed by Facial Expression and Body Postures in Myotonic Dystrophy (DM). *Cortex* **2020**, *127*, 58–66. [CrossRef]
175. Li, Y.; Zheng, W.; Cui, Z.; Zong, Y.; Ge, S. EEG Emotion Recognition Based on Graph Regularized Sparse Linear Regression. *Neural Process Lett.* **2019**, *49*, 555–571. [CrossRef]
176. Loos, E.; Egli, T.; Coynel, D.; Fastenrath, M.; Freytag, V.; Papassotiropoulos, A.; de Quervain, D.J.-F.; Milnik, A. Predicting Emotional Arousal and Emotional Memory Performance from an Identical Brain Network. *NeuroImage* **2019**, *189*, 459–467. [CrossRef]
177. Tottenham, N.; Weissman, M.M.; Wang, Z.; Warner, V.; Gameroff, M.J.; Semanek, D.P.; Hao, X.; Gingrich, J.A.; Peterson, B.S.; Posner, J.; et al. Depression Risk Is Associated with Weakened Synchrony Between the Amygdala and Experienced Emotion. *Biol. Psychiatry Cogn. Neurosci. Neuroimaging* **2021**, *6*, 343–351. [CrossRef]
178. Doma, V.; Pirouz, M. A Comparative Analysis of Machine Learning Methods for Emotion Recognition Using EEG and Peripheral Physiological Signals. *J. Big Data* **2020**, *7*, 18. [CrossRef]
179. Pan, C.; Shi, C.; Mu, H.; Li, J.; Gao, X. EEG-Based Emotion Recognition Using Logistic Regression with Gaussian Kernel and Laplacian Prior and Investigation of Critical Frequency Bands. *Appl. Sci.* **2020**, *10*, 1619. [CrossRef]
180. Rafi, T.H.; Farhan, F.; Hoque, M.Z.; Quayyum FMRafi, T.H.; Farhan, F.; Hoque, M.Z.; Quayyum, F.M. Electroencephalogram (EEG) Brainwave Signal-Based Emotion Recognition Using Extreme Gradient Boosting Algorithm. *Ann. Eng.* **2020**, *1*, 1–19.
181. Jackson-Koku, G.; Grime, P. Emotion Regulation and Burnout in Doctors: A Systematic Review. *Occup. Med.* **2019**, *69*, 9–21. [CrossRef]
182. Shams, S. Predicting Coronavirus Anxiety Based on Cognitive Emotion Regulation Strategies, Anxiety Sensitivity, and Psychological Hardiness in Nurses. *Q. J. Nurs. Manag.* **2021**, *10*, 25–36.
183. Scribner, D.R. Predictors of Shoot–Don't Shoot Decision-Making Performance: An Examination of Cognitive and Emotional Factors. *J. Cogn. Eng. Decis. Mak.* **2016**, *10*, 3–13. [CrossRef]
184. Smith, G. Be Wary of Black-Box Trading Algorithms. *JOI* **2019**, *28*, 7–15. [CrossRef]
185. Hajarolasvadi, N.; Demirel, H. 3D CNN-Based Speech Emotion Recognition Using K-Means Clustering and Spectrograms. *Entropy* **2019**, *21*, 479. [CrossRef] [PubMed]
186. Morawetz, C.; Riedel, M.C.; Salo, T.; Berboth, S.; Eickhoff, S.B.; Laird, A.R.; Kohn, N. Multiple Large-Scale Neural Networks Underlying Emotion Regulation. *Neurosci. Biobehav. Rev.* **2020**, *116*, 382–395. [CrossRef] [PubMed]
187. Zou, L.; Guo, Q.; Xu, Y.; Yang, B.; Jiao, Z.; Xiang, J. Functional Connectivity Analysis of the Neural Bases of Emotion Regulation: A Comparison of Independent Component Method with Density-Based k-Means Clustering Method. *Technol. Health Care* **2016**, *24*, S817–S825. [CrossRef] [PubMed]
188. Mohammed, N.S.; Abdul Hassan, K.A. The Effect of the Number of Key-Frames on the Facial Emotion Recognition Accuracy. *Eng. Technol. J.* **2021**, *39*, 89–100. [CrossRef]
189. Shi, F.; Dey, N.; Ashour, A.S.; Sifaki-Pistolla, D.; Sherratt, R.S. Meta-KANSEI Modeling with Valence-Arousal FMRI Dataset of Brain. *Cogn. Comput.* **2019**, *11*, 227–240. [CrossRef]
190. Kaunhoven, R.J.; Dorjee, D. Mindfulness Versus Cognitive Reappraisal: The Impact of Mindfulness-Based Stress Reduction (MBSR) on the Early and Late Brain Potential Markers of Emotion Regulation. *Mindfulness* **2021**, *12*, 2266–2280. [CrossRef]

191. Li, G.; Zhang, W.; Hu, Y.; Wang, J.; Li, J.; Jia, Z.; Zhang, L.; Sun, L.; von Deneen, K.M.; Duan, S.; et al. Distinct Basal Brain Functional Activity and Connectivity in the Emotional-Arousal Network and Thalamus in Patients With Functional Constipation Associated With Anxiety and/or Depressive Disorders. *Psychosom. Med.* **2021**, *83*, 707–714. [CrossRef]
192. Xiao, G.; Ma, Y.; Liu, C.; Jiang, D. A Machine Emotion Transfer Model for Intelligent Human-Machine Interaction Based on Group Division. *Mech. Syst. Signal Processing* **2020**, *142*, 106736. [CrossRef]
193. Li, H.; Xu, H. Deep Reinforcement Learning for Robust Emotional Classification in Facial Expression Recognition. *Knowl.-Based Syst.* **2020**, *204*, 106172. [CrossRef]
194. Li, Y.; Chen, Y. Research on Chorus Emotion Recognition and Intelligent Medical Application Based on Health Big Data. *J. Healthc. Eng.* **2022**, *2022*, 1363690. [CrossRef]
195. Yakovyna, V.; Khavalko, V.; Sherega, V.; Boichuk, A.; Barna, A. Biosignal and Image Processing System for Emotion Recognition Applications. In Proceedings of the IT&AS'2021: Symposium on Information Technologies & Applied Sciences, Bratislava, Slovakia, 5 March 2021; pp. 181–191.
196. Chan, J.C.P.; Ho, E.S.L. Emotion Transfer for 3D Hand and Full Body Motion Using StarGAN. *Computers* **2021**, *10*, 38. [CrossRef]
197. Global Industry Analysts Inc. Neuroscience—Global Market Trajectory & Analytics. 2021. Available online: https://www.prnewswire.com/news-releases/new-analysis-from-global-industry-analysts-reveals-steady-growth-for-neuroscience-with-the-market-to-reach-36-2-billion-worldwide-by-2026--301404252.html (accessed on 5 September 2022).
198. Neuroscience Market. Global Industry Analysis, Size, Share, Growth, Trends, and Forecast, 2021–2031. Available online: https://www.transparencymarketresearch.com/neuroscience-market.html (accessed on 17 August 2022).
199. Celeghin, A.; Diano, M.; Bagnis, A.; Viola, M.; Tamietto, M. Basic Emotions in Human Neuroscience: Neuroimaging and Beyond. *Front. Psychol.* **2017**, *8*, 1432. [CrossRef]
200. Sander, D.; Nummenmaa, L. Reward and Emotion: An Affective Neuroscience Approach. *Curr. Opin. Behav. Sci.* **2021**, *39*, 161–167. [CrossRef]
201. Podladchikova, L.N.; Shaposhnikov, D.G.; Kozubenko, E.A. Towards Neuroinformatic Approach for Second-Person Neuroscience. In *Advances in Neural Computation, Machine Learning, and Cognitive Research IV*; Kryzhanovsky, B., Dunin-Barkowski, W., Redko, V., Tiumentsev, Y., Eds.; Studies in Computational Intelligence; Springer International Publishing: Cham, Switzerland, 2021; Volume 925, pp. 143–148. [CrossRef]
202. Tan, C.; Liu, X.; Zhang, G. Inferring Brain State Dynamics Underlying Naturalistic Stimuli Evoked Emotion Changes with DHA-HMM. *Neuroinform* **2022**, *20*, 737–753. [CrossRef]
203. Blair, R.J.R.; Meffert, H.; White, S.F. Psychopathy and Brain Function: Insights from Neuroimaging Research. In *Handbook of Psychopathy*; The Guilford Press: New York, NY, USA, 2018; pp. 401–421.
204. Blair, R.J.R.; Mathur, A.; Haines, N.; Bajaj, S. Future Directions for Cognitive Neuroscience in Psychiatry: Recommendations for Biomarker Design Based on Recent Test Re-Test Reliability Work. *Curr. Opin. Behav. Sci.* **2022**, *44*, 101102. [CrossRef]
205. Hamann, S. Integrating Perspectives on Affective Neuroscience: Introduction to the Special Section on the Brain and Emotion. *Emot. Rev.* **2018**, *10*, 187–190. [CrossRef]
206. Shaffer, C.; Westlin, C.; Quigley, K.S.; Whitfield-Gabrieli, S.; Barrett, L.F. Allostasis, Action, and Affect in Depression: Insights from the Theory of Constructed Emotion. *Annu. Rev. Clin. Psychol.* **2022**, *18*, 553–580. [CrossRef]
207. Hackel, L.M.; Amodio, D.M. Computational Neuroscience Approaches to Social Cognition. *Curr. Opin. Psychol.* **2018**, *24*, 92–97. [CrossRef]
208. Smith, R.; Lane, R.D.; Nadel, L.; Moutoussis, M. A Computational Neuroscience Perspective on the Change Process in Psychotherapy. In *Neuroscience of Enduring Change*; Oxford University Press: New York, NY, USA, 2020; pp. 395–432. [CrossRef]
209. Hill, K.E.; South, S.C.; Egan, R.P.; Foti, D. Abnormal Emotional Reactivity in Depression: Contrasting Theoretical Models Using Neurophysiological Data. *Biol. Psychol.* **2019**, *141*, 35–43. [CrossRef]
210. Kontaris, I.; East, B.S.; Wilson, D.A. Behavioral and Neurobiological Convergence of Odor, Mood and Emotion: A Review. *Front. Behav. Neurosci.* **2020**, *14*, 35. [CrossRef]
211. Kyrios, M.; Trotzke, P.; Lawrence, L.; Fassnacht, D.B.; Ali, K.; Laskowski, N.M.; Müller, A. Behavioral Neuroscience of Buying-Shopping Disorder: A Review. *Curr. Behav. Neurosci. Rep.* **2018**, *5*, 263–270. [CrossRef]
212. Wang, J.; Cheng, R.; Liao, P.-C. Trends of Multimodal Neural Engineering Study: A Bibliometric Review. *Arch. Comput. Methods Eng.* **2021**, *28*, 4487–4501. [CrossRef]
213. Wu, X.; Zheng, W.-L.; Li, Z.; Lu, B.-L. Investigating EEG-Based Functional Connectivity Patterns for Multimodal Emotion Recognition. *J. Neural Eng.* **2022**, *19*, 016012. [CrossRef]
214. Balconi, M.; Sansone, M. Neuroscience and Consumer Behavior: Where to Now? *Front. Psychol.* **2021**, *12*, 705850. [CrossRef] [PubMed]
215. Serra, D. Decision-Making: From Neuroscience to Neuroeconomics—An Overview. *Theory Decis.* **2021**, *91*, 1–80. [CrossRef]
216. Hinojosa, J.A.; Moreno, E.M.; Ferré, P. Affective Neurolinguistics: Towards a Framework for Reconciling Language and Emotion. *Lang. Cogn. Neurosci.* **2020**, *35*, 813–839. [CrossRef]
217. Wu, C.; Zhang, J. Emotion Word Type Should Be Incorporated in Affective Neurolinguistics: A Commentary on Hinojosa, Moreno and Ferré (2019). *Lang. Cogn. Neurosci.* **2020**, *35*, 840–843. [CrossRef]
218. Burkitt, I. Emotions, Social Activity and Neuroscience: The Cultural-Historical Formation of Emotion. *New Ideas Psychol.* **2019**, *54*, 1–7. [CrossRef]

219. Gluck, M.A.; Mercado, E.; Myers, C.E. *Learning and Memory: From Brain to Behavior*; Worth Publishers: New York, NY, USA, 2008.
220. Shaw, S.D.; Bagozzi, R.P. The Neuropsychology of Consumer Behavior and Marketing. *Soc. Consum. Psychol.* **2018**, *1*, 22–40. [CrossRef]
221. Al-Rodhan, N.R.F. *Emotional Amoral Egoism: A Neurophilosophy of Human Nature and Motivations*, 1st ed.; The Lutterworth Press: Cambridge, UK, 2021. [CrossRef]
222. Carrozzo, C. Scientific Practice and the Moral Task of Neurophilosophy. *AJOB Neurosci.* **2019**, *10*, 115–117. [CrossRef]
223. Northoff, G. Neurophilosophy and Neuroethics: Template for Neuropsychoanalysis? In *Neuropsychodynamic Psychiatry*; Boeker, H., Hartwich, P., Northoff, G., Eds.; Springer International Publishing: Cham, Switzerland, 2018; pp. 599–615. [CrossRef]
224. Chatterjee, A.; Coburn, A.; Weinberger, A. The Neuroaesthetics of Architectural Spaces. *Cogn. Process.* **2021**, *22*, 115–120. [CrossRef]
225. Li, R.; Zhang, J. Review of Computational Neuroaesthetics: Bridging the Gap between Neuroaesthetics and Computer Science. *Brain Inf.* **2020**, *7*, 16. [CrossRef]
226. Nadal, M.; Chatterjee, A. Neuroaesthetics and Art's Diversity and Universality. *WIREs Cogn. Sci.* **2019**, *10*, e1487. [CrossRef]
227. Klemm, W. Expanding the Vision of Neurotheology: Make Neuroscience Religion's Ally. *J. Spiritual. Ment. Health* **2020**, *24*, 1–16. [CrossRef]
228. Klemm, W.R. Whither Neurotheology? *Religions* **2019**, *10*, 634. [CrossRef]
229. Newberg, A. Chapter Three. Neuroscience and Neurotheology. In *Neurotheology*; Columbia University Press: New York, NY, USA, 2018; pp. 46–66. [CrossRef]
230. Haas, I.J.; Warren, C.; Lauf, S.J. Political Neuroscience: Understanding How the Brain Makes Political Decisions. In *Oxford Research Encyclopedia of Politics*; Redlawsk, D., Ed.; Oxford University Press: Oxford, UK, 2020. [CrossRef]
231. Murphy, E. Anarchism and Science. In *The Palgrave Handbook of Anarchism*; Levy, C., Adams, M.S., Eds.; Springer International Publishing: Cham, Switzerland, 2019; pp. 193–209. [CrossRef]
232. Yun, J.H.; Kim, Y.; Lee, E.-J. ERP Study of Liberals' and Conservatives' Moral Reasoning Processes: Evidence from South Korea. *J. Bus. Ethics* **2022**, *176*, 723–739. [CrossRef]
233. Bush, S.S.; Tussey, C.M. Neuroscience and Neurolaw: Special Issue of Psychological Injury and Law. *Psychol. Inj. Law* **2013**, *6*, 1–2. [CrossRef]
234. Schleim, S. Real Neurolaw in the Netherlands: The Role of the Developing Brain in the New Adolescent Criminal Law. *Front. Psychol.* **2020**, *11*, 1762. [CrossRef]
235. Shen, F.X. The Law and Neuroscience Bibliography: Navigating the Emerging Field of Neurolaw. *Int. J. Leg. Inf.* **2010**, *38*, 352–399. [CrossRef]
236. Long, M.; Verbeke, W.; Ein-Dor, T.; Vrtička, P. A Functional Neuro-Anatomical Model of Human Attachment (NAMA): Insights from First- and Second-Person Social Neuroscience. *Cortex* **2020**, *126*, 281–321. [CrossRef]
237. Weisz, E.; Zaki, J. Motivated Empathy: A Social Neuroscience Perspective. *Curr. Opin. Psychol.* **2018**, *24*, 67–71. [CrossRef]
238. Chiao, J.Y. Developmental Aspects in Cultural Neuroscience. *Dev. Rev.* **2018**, *50*, 77–89. [CrossRef]
239. Chiao, J.Y. Cultural neuroscience: A once and future discipline. *Progress in brain research* **2009**, *178*, 287–304. [CrossRef]
240. Antolin, P. "I Am a Freak of Nature": Tourette's and the Grotesque in Jonathan Lethem's Motherless Brooklyn. *Transatlantica* **2019**, *1*, 1–20. [CrossRef]
241. Burn, S.J. The Gender of the Neuronovel: Joyce Carol Oates and the Double Brain. *Eur. J. Am. Stud.* **2021**, *16*, 1–17. [CrossRef]
242. Rahaman, V.; Sharma, S. Reading an Extremist Mind through Literary Language: Approaching Cognitive Literary Hermeneutics to R.N. Tagore's Play the Post Office for Neuro-Computational Predictions. In *Cognitive Informatics, Computer Modelling, and Cognitive Science*; Elsevier: Amsterdam, The Netherlands, 2020; pp. 197–210. [CrossRef]
243. Ceciu, R.L. Neurocinematics, the (Brain) Child of Film and Neuroscience. *J. Commun. Behav. Sci.* **2020**, *1*, 46–62.
244. Moghadasi, A.N. Evaluation of Neurocinema as An Introduction to an Interdisciplinary Science. *CINEJ* **2020**, *8*, 307–323. [CrossRef]
245. Olenina, A.H. Sergei Eisenstein, Neurocinematics, and Embodied Cognition: A Reassessment. *Discourse* **2021**, *43*, 351–382. [CrossRef]
246. Bearman, H. Music & The Brain–How Music Affects Mood, Cognition, and Mental Health. 2018. Available online: https://www.naturalnootropic.com/music-and-the-brain/ (accessed on 15 August 2022).
247. Garg, A.; Chaturvedi, V.; Kaur, A.B.; Varshney, V.; Parashar, A. Machine Learning Model for Mapping of Music Mood and Human Emotion Based on Physiological Signals. *Multimed. Tools Appl.* **2022**, *81*, 5137–5177. [CrossRef]
248. Liu, Y. Research on the Characteristics and Functions of Brain Activity in Musical Performance. *Acad. J. Humanit. Soc. Sci.* **2020**, *3*, 71–79.
249. Berčík, J.; Paluchová, J.; Neomániová, K. Neurogastronomy as a Tool for Evaluating Emotions and Visual Preferences of Selected Food Served in Different Ways. *Foods* **2021**, *10*, 354. [CrossRef]
250. Girona-Ruíz, D.; Cano-Lamadrid, M.; Carbonell-Barrachina, Á.A.; López-Lluch, D.; Esther, S. Aromachology Related to Foods, Scientific Lines of Evidence: A Review. *Appl. Sci.* **2021**, *11*, 6095. [CrossRef]
251. Lim, W.M. Demystifying Neuromarketing. *J. Bus. Res.* **2018**, *91*, 205–220. [CrossRef]
252. Sliwinska, M.W.; Vitello, S.; Devlin, J.T. Transcranial Magnetic Stimulation for Investigating Causal Brain-Behavioral Relationships and Their Time Course. *J. Vis. Exp.* **2014**, *89*, e51735. [CrossRef] [PubMed]

253. Agarwal, S.; Xavier, M.J. Innovations in Consumer Science: Applications of Neuro-Scientific Research Tools. In *Adoption of Innovation*; Brem, A., Viardot, É., Eds.; Springer International Publishing: Cham, Switzerland, 2015; pp. 25–42. [CrossRef]
254. Bakardjieva, E.; Kimmel, A.J. Neuromarketing Research Practices: Attitudes, Ethics, and Behavioral Intentions. *Ethics Behav.* **2017**, *27*, 179–200. [CrossRef]
255. Bercea, M.D. Anatomy of Methodologies for Measuring Consumer Behavior in Neuromarketing Research. In Proceedings of the Lupcon Center for Business Research (LCBR) European Marketing Conference, Ebermannstadt, Germany, 9 August 2012.
256. Bitbrain. Business & Marketing. The 7 Most Common Neuromarketing Research Techniques and Tools. 2019. Available online: https://www.bitbrain.com/blog/neuromarketing-research-techniques-tools (accessed on 15 August 2022).
257. CoolTool. How To Choose the Most Suitable NeuroLab Technology. Available online: https://cooltool.com/blog/-infographics-how-to-choose-the-most-suitable-neurolab-technology (accessed on 17 August 2022).
258. Farnsworth, B. Neuromarketing Methods [Cheat Sheet]. 2020. Available online: https://imotions.com/blog/neuromarketing-methods/ (accessed on 17 August 2022).
259. Fortunato, V.C.R.; Giraldi, J.D.M.E.; de Oliveira, J.H.C. A Review of Studies on Neuromarketing: Practical Results, Techniques, Contributions and Limitations. *J. Manag. Res.* **2014**, *6*, 201–220. [CrossRef]
260. Ganapathy, K. Neuromarketing: An Overview. *Asian Hosp. Healthc. Manag.* 2019. Available online: https://www.asianhhm.com/healthcare-management/current-concepts-on-neuromarketing (accessed on 15 August 2022).
261. Gill, G. Innerscope Research Inc. *JITE DC* **2012**, *1*, 5. [CrossRef]
262. Ohme, R.; Matukin, M.; Pacula-Lesniak, B. Biometric Measures for Interactive Advertising Research. *J. Interact. Advert.* **2011**, *11*, 60–72. [CrossRef]
263. Nazarova, R.; Lazizovich, T.K. Neuromarketing—A Tool for Influencing Consumer Behavior. *Int. J. Innov. Technol. Econ.* **2019**, *5*, 11–14. [CrossRef]
264. Saltini, T. Some Neuromarketing Tools. 2015. Available online: https://tiphainesaltini.wordpress.com/2015/03/10/some-neuromarketing-tools/ (accessed on 15 August 2022).
265. Stasi, A.; Songa, G.; Mauri, M.; Ciceri, A.; Diotallevi, F.; Nardone, G.; Russo, V. Neuromarketing Empirical Approaches and Food Choice: A Systematic Review. *Food Res. Int.* **2018**, *108*, 650–664. [CrossRef]
266. Yağci, M.I.; Kuhzady, S.; Balik, Z.S.; Öztürk, L. In Search of Consumer's Black Box: A Bibliometric Analysis of Neuromarketing Research. *J. Consum. Consum. Res.* **2018**, *10*, 101–134.
267. Klinčeková, S. Neuromarketing—Research and Prediction of the Future. *Int. J. Manag. Sci. Bus. Adm.* **2016**, *2*, 54–58. [CrossRef]
268. Malvern Panalytical. Near-Infrared (NIR) Spectroscopy. Available online: https://www.malvernpanalytical.com/en/products/technology/spectroscopy/near-infrared-spectroscopy/ (accessed on 15 August 2022).
269. Villringer, A.; Planck, J.; Hock, C.; Schleinkofer, L.; Dirnagl, U. Near Infrared Spectroscopy (NIRS): A New Tool to Study Hemodynamic Changes during Activation of Brain Function in Human Adults. *Neurosci. Lett.* **1993**, *154*, 101–104. [CrossRef]
270. Assaf, Y.; Pasternak, O. Diffusion Tensor Imaging (DTI)-Based White Matter Mapping in Brain Research: A Review. *J. Mol. Neurosci.* **2008**, *34*, 51–61. [CrossRef]
271. Imagilys. Diffusion Tensor Imaging. Available online: https://www.imagilys.com/diffusion-tensor-imaging-dti/ (accessed on 15 August 2022).
272. Sun, F.; Zang, W.; Gravina, R.; Fortino, G.; Li, Y. Gait-Based Identification for Elderly Users in Wearable Healthcare Systems. *Inf. Fusion* **2020**, *53*, 134–144. [CrossRef]
273. Majumder, S.; Mondal, T.; Deen, M.J. A Simple, Low-Cost and Efficient Gait Analyzer for Wearable Healthcare Applications. *IEEE Sens. J.* **2019**, *19*, 2320–2329. [CrossRef]
274. Arvaneh, M.; Tanaka, T. Brain–Computer Interfaces and Electroencephalogram: Basics and Practical Issues. In *Signal Processing and Machine Learning for Brain—Machine Interfaces*; 2018; Available online: http://dl.konkur.in/post/Book/Bargh/Signal-Processing-and-Machine-Learning-for-Brain-Machine-Interfaces-%5Bkonkur.in%5D.pdf#page=16 (accessed on 15 August 2022).
275. Hantus, S. Continuous EEG Monitoring: Principles and Practice. *J. Clin. Neurophysiol.* **2019**, *37*, 1. [CrossRef]
276. Tyagi, A.; Semwal, S.; Shah, G. A Review of Eeg Sensors Used for Data Acquisition. *Int. J. Comput. Appl.* **2012**, *1*, 13–18.
277. Burgess, R.C. MEG Reporting. *J. Clin. Neurophysiol.* **2020**, *37*, 545–553. [CrossRef]
278. Harmsen, I.E.; Rowland, N.C.; Wennberg, R.A.; Lozano, A.M. Characterizing the Effects of Deep Brain Stimulation with Magnetoencephalography: A Review. *Brain Stimul.* **2018**, *11*, 481–491. [CrossRef]
279. Seymour, R.A.; Alexander, N.; Mellor, S.; O'Neill, G.C.; Tierney, T.M.; Barnes, G.R.; Maguire, E.A. Interference Suppression Techniques for OPM-Based MEG: Opportunities and Challenges. *NeuroImage* **2021**, *247*, 118834. [CrossRef]
280. Shirinpour, S. Tools for Improving and Understanding Transcranial Magnetic Stimulation. 2020. Available online: https://hdl.handle.net/11299/217801 (accessed on 15 August 2022).
281. Shirinpour, S.; Hananeia, N.; Rosado, J.; Tran, H.; Galanis, C.; Vlachos, A.; Jedlicka, P.; Queisser, G.; Opitz, A. Multi-Scale Modeling Toolbox for Single Neuron and Subcellular Activity under Transcranial Magnetic Stimulation. *Brain Stimul.* **2021**, *14*, 1470–1482. [CrossRef]
282. Widhalm, M.L.; Rose, N.S. How Can Transcranial Magnetic Stimulation Be Used to Causally Manipulate Memory Representations in the Human Brain? *WIREs Cogn. Sci.* **2019**, *10*, e1469. [CrossRef]
283. Gannouni, S.; Aledaily, A.; Belwafi, K.; Aboalsamh, H. Emotion Detection Using Electroencephalography Signals and a Zero-Time Windowing-Based Epoch Estimation and Relevant Electrode Identification. *Sci. Rep.* **2021**, *11*, 7071. [CrossRef]

284. Dixson, B.J.W.; Spiers, T.; Miller, P.A.; Sidari, M.J.; Nelson, N.L.; Craig, B.M. Facial Hair May Slow Detection of Happy Facial Expressions in the Face in the Crowd Paradigm. *Sci. Rep.* **2022**, *12*, 5911. [CrossRef]
285. Wang, X.-W.; Nie, D.; Lu, B.-L. EEG-Based Emotion Recognition Using Frequency Domain Features and Support Vector Machines. In *Neural Information Processing*; Lu, B.-L., Zhang, L., Kwok, J., Eds.; Lecture Notes in Computer Science; Springer Berlin Heidelberg: Berlin/Heidelberg, Germany, 2011; Volume 7062, pp. 734–743. [CrossRef]
286. John, E.R. Principles of Neurometries. *Am. J. EEG Technol.* **1990**, *30*, 251–266. [CrossRef]
287. Alkhasli, I.; Sakreida, K.; Mottaghy, F.M.; Binkofski, F. Modulation of Fronto-Striatal Functional Connectivity Using Transcranial Magnetic Stimulation. *Front. Hum. Neurosci.* **2019**, *13*, 190. [CrossRef]
288. Jamadar, S.D.; Ward, P.G.D.; Close, T.G.; Fornito, A.; Premaratne, M.; O'Brien, K.; Stäb, D.; Chen, Z.; Shah, N.J.; Egan, G.F. Simultaneous BOLD-FMRI and Constant Infusion FDG-PET Data of the Resting Human Brain. *Sci. Data* **2020**, *7*, 363. [CrossRef]
289. Kraft, R.H.; Dagro, A.M. Design and Implementation of a Numerical Technique to Inform Anisotropic Hyperelastic Finite Element Models Using Diffusion-Weighted Imaging. 2011. Available online: https://apps.dtic.mil/sti/pdfs/ADA565877.pdf (accessed on 15 August 2022).
290. Koong, C.-S.; Yang, T.-I.; Tseng, C.-C. A User Authentication Scheme Using Physiological and Behavioral Biometrics for Multitouch Devices. *Sci. World J.* **2014**, *2014*, 781234. [CrossRef]
291. Heydarzadegan, A.; Moradi, M.; Toorani, A. Biometric Recognition Systems: A Survey. *Int. Res. J. Appl. Basic Sci.* **2013**, *6*, 1609–1618.
292. Shingetsu. Global Biometric Systems Market. 2021. Available online: https://www.shingetsuresearch.com/biometric-systems-market/?gclid=Cj0KCQiAybaRBhDtARIsAIEG3kkQZsv-1LwHknyBvnAfURBeXvBbB-uk9YGdpwf22Uw6waMmssmt1ycaAr9hEALw_wcB (accessed on 29 July 2022).
293. Abo-Zahhad, M.; Ahmed, S.M.; Abbas, S.N. A Novel Biometric Approach for Human Identification and Verification Using Eye Blinking Signal. *IEEE Signal Process. Lett.* **2015**, *22*, 876–880. [CrossRef]
294. Larsson, M.; Pedersen, N.L.; Stattin, H. Associations between Iris Characteristics and Personality in Adulthood. *Biol. Psychol.* **2007**, *75*, 165–175. [CrossRef]
295. Gentry, T.A.; Polzine, K.M.; Wakefield, J.A. Human Genetic Markers Associated with Variation in Intellectual Abilities and Personality. *Personal. Individ. Differ.* **1985**, *6*, 111–113. [CrossRef]
296. Gary, A.L.; Glover, J.A. *Eye Color, Sex, and Children's Behavior*; Nelson-Hall Publishers: Chicago, IL, USA, 1976.
297. Markle, A. Eye Color and Responsiveness to Arousing Stimuli. *Percept. Mot. Ski.* **1976**, *43*, 127–133. [CrossRef] [PubMed]
298. Bailador, G.; Sanchez-Avila, C.; Guerra-Casanova, J.; de Santos Sierra, A. Analysis of Pattern Recognition Techniques for In-Air Signature Biometrics. *Pattern Recognit.* **2011**, *44*, 2468–2478. [CrossRef]
299. Miller, W. Different Types of Biometrics. 2019. Available online: https://www.ibeta.com/different-types-of-biometrics/ (accessed on 29 July 2022).
300. Biometrics Institute. Types of Biometrics. Available online: https://www.biometricsinstitute.org/what-is-biometrics/types-of-biometrics/ (accessed on 29 July 2022).
301. Chen, D.; Haviland-Jones, J. Human Olfactory Communication of Emotion. *Percept. Mot. Ski.* **2000**, *91*, 771–781. [CrossRef] [PubMed]
302. Kaklauskas, A.; Zavadskas, E.K.; Seniut, M.; Dzemyda, G.; Stankevic, V.; Simkevičius, C.; Stankevic, T.; Paliskiene, R.; Matuliauskaite, A.; Kildiene, S.; et al. Web-Based Biometric Computer Mouse Advisory System to Analyze a User's Emotions and Work Productivity. *Eng. Appl. Artif. Intell.* **2011**, *24*, 928–945. [CrossRef]
303. American Heart Association. Electrocardiogram (ECG or EKG). 2015. Available online: https://www.heart.org/en/health-topics/heart-attack/diagnosing-a-heart-attack/electrocardiogram-ecg-or-ekg (accessed on 29 July 2022).
304. Nicolò, A.; Massaroni, C.; Schena, E.; Sacchetti, M. The Importance of Respiratory Rate Monitoring: From Healthcare to Sport and Exercise. *Sensors* **2020**, *20*, 6396. [CrossRef]
305. Wang, B.; Zhou, H.; Yang, G.; Li, Y.; Yang, H. Human Digital Twin (HDT) Driven Human-Cyber-Physical Systems: Key Technologies and Applications. *Chin. J. Mech. Eng.* **2022**, *35*, 11. [CrossRef]
306. Nahavandi, S. Industry 5.0—A Human-Centric Solution. *Sustainability* **2019**, *11*, 4371. [CrossRef]
307. Lugovic, S.; Dunder, I.; Horvat, M. Techniques and Applications of Emotion Recognition in Speech. In *2016 39th International Convention on Information and Communication Technology, Electronics and Microelectronics (MIPRO), Opatija, Croatia, 30 May–3 June 2016*; IEEE: Opatija, Croatia, 2016; pp. 1278–1283. [CrossRef]
308. Ge, Y.; Liu, J. Psychometric Analysis on Neurotransmitter Deficiency of Internet Addicted Urban Left-behind Children. *J. Alcohol Drug Depend.* **2015**, *3*, 1–6. [CrossRef]
309. Lafta, H.A.; Abbas, S.S. Effectiveness of Extended Invariant Moments in Fingerprint Analysis. *Asian J. Comput. Inf. Syst.* **2013**, *01*, 78–89.
310. Singh, J.; Goyal, G.; Gill, R. Use of Neurometrics to Choose Optimal Advertisement Method for Omnichannel Business. *Enterp. Inf. Syst.* **2020**, *14*, 243–265. [CrossRef]
311. Fiedler, K.; Bluemke, M. Faking the IAT: Aided and Unaided Response Control on the Implicit Association Tests. *Basic Appl. Soc. Psychol.* **2005**, *27*, 307–316. [CrossRef]
312. Simons, S.; Zhou, J.; Liao, Y.; Bradway, L.; Aguilar, M.; Connolly, P.M. Cognitive Biometrics Using Mouse Perturbation. US Patent Application US14/011,351, 20 March 2014.

313. Martinez-Marquez, D.; Pingali, S.; Panuwatwanich, K.; Stewart, R.A.; Mohamed, S. Application of Eye Tracking Technology in Aviation, Maritime, and Construction Industries: A Systematic Review. *Sensors* **2021**, *21*, 4289. [CrossRef]
314. Skvarekova, I.; Skultety, F. Objective Measurement of Pilot's Attention Using Eye Track Technology during IFR Flights. *Transp. Res. Procedia* **2019**, *40*, 1555–1562. [CrossRef]
315. Eachus, P. *The Use of Eye Tracking Technology in the Evaluation of E-Learning: A Feasibility Study*; University of Salford: Manchester, UK, 2008; pp. 12–14.
316. Sharafi, Z.; Soh, Z.; Guéhéneuc, Y.-G. A Systematic Literature Review on the Usage of Eye-Tracking in Software Engineering. *Inf. Softw. Technol.* **2015**, *67*, 79–107. [CrossRef]
317. Gonzalez-Sanchez, J.; Chavez-Echeagaray, M.E.; Atkinson, R.; Burleson, W. ABE: An Agent-Based Software Architecture for a Multimodal Emotion Recognition Framework. In *2011 Ninth Working IEEE/IFIP Conference on Software Architecture, Washington, United States, 20–24 June 2011*; IEEE: Boulder, CO, USA, 2011; pp. 187–193. [CrossRef]
318. Borkhataria, C. The Algorithm That Could End Office Thermostat Wars: Researchers Claim New Software Can Find the Best Temperature for Everyone. 2017. Available online: https://www.dailymail.co.uk/sciencetech/article-4979148/The-algorithm-end-office-thermostat-war.html (accessed on 29 July 2022).
319. Rukavina, S.; Gruss, S.; Hoffmann, H.; Tan, J.-W.; Walter, S.; Traue, H.C. Affective Computing and the Impact of Gender and Age. *PLoS ONE* **2016**, *11*, e0150584. [CrossRef]
320. Saini, R.; Rana, N. Comparison of Various Biometric Methods. *Int. J. Adv. Sci. Technol.* **2014**, *2*, 24–30.
321. Elprocus. Biometric Sensors—Types and Its Working. 2022. Available online: https://www.elprocus.com/different-types-biometric-sensors/ (accessed on 29 July 2022).
322. Loaiza, J.R. Emotions and the Problem of Variability. *Rev. Phil. Psych.* **2021**, *12*, 329–351. [CrossRef]
323. Pace-Schott, E.F.; Amole, M.C.; Aue, T.; Balconi, M.; Bylsma, L.M.; Critchley, H.; Demaree, H.A.; Friedman, B.H.; Gooding, A.E.K.; Gosseries, O.; et al. Physiological Feelings. *Neurosci. Biobehav. Rev.* **2019**, *103*, 267–304. [CrossRef]
324. Dolensek, N.; Gehrlach, D.A.; Klein, A.S.; Gogolla, N. Facial Expressions of Emotion States and Their Neuronal Correlates in Mice. *Science* **2020**, *368*, 89–94. [CrossRef]
325. Kamila, S.; Hasanuzzaman, M.; Ekbal, A.; Bhattacharyya, P. Investigating the Impact of Emotion on Temporal Orientation in a Deep Multitask Setting. *Sci. Rep.* **2022**, *12*, 493. [CrossRef]
326. Saganowski, S.; Komoszyńska, J.; Behnke, M.; Perz, B.; Kunc, D.; Klich, B.; Kaczmarek, Ł.D.; Kazienko, P. Emognition Dataset: Emotion Recognition with Self-Reports, Facial Expressions, and Physiology Using Wearables. *Sci. Data* **2022**, *9*, 158. [CrossRef]
327. Swanborough, H.; Staib, M.; Frühholz, S. Neurocognitive Dynamics of Near-Threshold Voice Signal Detection and Affective Voice Evaluation. *Sci. Adv.* **2020**, *6*, eabb3884. [CrossRef]
328. Singh, R.; Baby, B.; Suri, A. A Virtual Repository of Neurosurgical Instrumentation for Neuroengineering Research and Collaboration. *World Neurosurg.* **2019**, *126*, e84–e93. [CrossRef]
329. Alonso-Fernandez, F.; Fierrez, J.; Ortega-Garcia, J. Quality measures in biometric systems. *IEEE Secur. Priv.* **2011**, *10*, 52–62. [CrossRef]
330. De Angel, V.; Lewis, S.; White, K.; Oetzmann, C.; Leightley, D.; Oprea, E.; Lavelle, G.; Matcham, F.; Pace, A.; Mohr, D.C.; et al. Digital health tools for the passive monitoring of depression: A systematic review of methods. *NPJ Digit. Med.* **2022**, *5*, 3. [CrossRef]
331. Kable, J.W. The Cognitive Neuroscience Toolkit for the Neuroeconomist: A Functional Overview. *J. Neurosci. Psychol. Econ.* **2011**, *4*, 63–84. [CrossRef]
332. Zurawicki, L. *Neuromarketing: Exploring the Brain of the Consumer*; Springer: Berlin/Heidelberg, Germany, 2010. [CrossRef]
333. Magdin, M.; Prikler, F. Are Instructed Emotional States Suitable for Classification? Demonstration of How They Can Significantly Influence the Classification Result in An Automated Recognition System. *IJIMAI* **2019**, *5*, 141–147. [CrossRef]
334. Camurri, A.; Lagerlöf, I.; Volpe, G. Recognizing Emotion from Dance Movement: Comparison of Spectator Recognition and Automated Techniques. *Int. J. Hum.-Comput. Stud.* **2003**, *59*, 213–225. [CrossRef]
335. Alarcao, S.M.; Fonseca, M.J. Emotions Recognition Using EEG Signals: A Survey. *IEEE Trans. Affect. Comput.* **2017**, *10*, 374–393. [CrossRef]
336. Kim, M.-K.; Kim, M.; Oh, E.; Kim, S.-P. A Review on the Computational Methods for Emotional State Estimation from the Human EEG. *Comput. Math. Methods Med.* **2013**, *2013*, 573734. [CrossRef]
337. Xu, Q.; Ruohonen, E.M.; Ye, C.; Li, X.; Kreegipuu, K.; Stefanics, G.; Luo, W.; Astikainen, P. Automatic Processing of Changes in Facial Emotions in Dysphoria: A Magnetoencephalography Study. *Front. Hum. Neurosci.* **2018**, *12*, 186. [CrossRef] [PubMed]
338. Bublatzky, F.; Kavcıoğlu, F.; Guerra, P.; Doll, S.; Junghöfer, M. Contextual Information Resolves Uncertainty about Ambiguous Facial Emotions: Behavioral and Magnetoencephalographic Correlates. *NeuroImage* **2020**, *215*, 116814. [CrossRef] [PubMed]
339. Van Loon, A.M.; van den Wildenberg, W.P.M.; van Stegeren, A.H.; Ridderinkhof, K.R.; Hajcak, G. Emotional Stimuli Modulate Readiness for Action: A Transcranial Magnetic Stimulation Study. *Cogn. Affect. Behav. Neurosci.* **2010**, *10*, 174–181. [CrossRef] [PubMed]
340. Bandara, D.; Velipasalar, S.; Bratt, S.; Hirshfield, L. Building Predictive Models of Emotion with Functional Near-Infrared Spectroscopy. *Int. J. Hum.-Comput. Stud.* **2018**, *110*, 75–85. [CrossRef]
341. Bae, S.; Kang, K.D.; Kim, S.W.; Shin, Y.J.; Nam, J.J.; Han, D.H. Investigation of an Emotion Perception Test Using Functional Magnetic Resonance Imaging. *Comput. Methods Programs Biomed.* **2019**, *179*, 104994. [CrossRef]

342. Dweck, M.R. Multisystem Positron Emission Tomography: Interrogating Vascular Inflammation, Emotional Stress, and Bone Marrow Activity in a Single Scan. *Eur. Heart J.* **2021**, *42*, 1896–1897. [CrossRef]
343. Reiman, E.M. The Application of Positron Emission Tomography to the Study of Normal and Pathologic Emotions. *J. Clin. Psychiatry* **1997**, *58* (Suppl. S16), 4–12.
344. Takahashi, M.; Kitamura, S.; Matsuoka, K.; Yoshikawa, H.; Yasuno, F.; Makinodan, M.; Kimoto, S.; Miyasaka, T.; Kichikawa, K.; Kishimoto, T. Uncinate Fasciculus Disruption Relates to Poor Recognition of Negative Facial Emotions in Alzheimer's Disease: A Cross-sectional Diffusion Tensor Imaging Study. *Psychogeriatrics* **2020**, *20*, 296–303. [CrossRef]
345. Kaklauskas, A.; Abraham, A.; Dzemyda, G.; Raslanas, S.; Seniut, M.; Ubarte, I.; Kurasova, O.; Binkyte-Veliene, A.; Cerkauskas, J. Emotional, Affective and Biometrical States Analytics of a Built Environment. *Eng. Appl. Artif. Intell.* **2020**, *91*, 103621. [CrossRef]
346. Kaklauskas, A.; Jokubauskas, D.; Cerkauskas, J.; Dzemyda, G.; Ubarte, I.; Skirmantas, D.; Podviezko, A.; Simkute, I. Affective Analytics of Demonstration Sites. *Eng. Appl. Artif. Intell.* **2019**, *81*, 346–372. [CrossRef]
347. Kaklauskas, A.; Zavadskas, E.K.; Bardauskiene, D.; Cerkauskas, J.; Ubarte, I.; Seniut, M.; Dzemyda, G.; Kaklauskaite, M.; Vinogradova, I.; Velykorusova, A. An Affect-Based Built Environment Video Analytics. *Autom. Constr.* **2019**, *106*, 102888. [CrossRef]
348. Kaklauskas, A.; Bardauskiene, D.; Cerkauskiene, R.; Ubarte, I.; Raslanas, S.; Radvile, E.; Kaklauskaite, U.; Kaklauskiene, L. Emotions Analysis in Public Spaces for Urban Planning. *Land Use Policy* **2021**, *107*, 105458. [CrossRef]
349. Porcherot, C.; Raviot-Derrien, S.; Beague, M.-P.; Henneberg, S.; Niedziela, M.; Ambroze, K.; McEwan, J.A. Effect of Context on Fine Fragrance-Elicited Emotions: Comparison of Three Experimental Methodologies. *Food Qual. Prefer.* **2022**, *95*, 104342. [CrossRef]
350. Child, S.; Oakhill, J.; Garnham, A. Tracking Your Emotions: An Eye-Tracking Study on Reader's Engagement with Perspective during Text Comprehension. *Q. J. Exp. Psychol.* **2020**, *73*, 929–940. [CrossRef]
351. Tarnowski, P.; Kołodziej, M.; Majkowski, A.; Rak, R.J. Eye-Tracking Analysis for Emotion Recognition. *Comput. Intell. Neurosci.* **2020**, *2020*, 2909267. [CrossRef]
352. Coutinho, E.; Miranda, E.R.; Cangelosi, A. Towards a Model for Embodied Emotions. In *2005 Portuguese Conference on Artificial Intelligence, Covilha, Portugal, 5–8 December 2005*; IEEE: Covilha, Portugal, 2005; pp. 54–63. [CrossRef]
353. Kim, M.; Lee, H.S.; Park, J.W.; Jo, S.H.; Chung, M.J. Determining Color and Blinking to Support Facial Expression of a Robot for Conveying Emotional Intensity. In *RO-MAN 2008—The 17th IEEE International Symposium on Robot and Human Interactive Communication, Munich, Germany, 1–3 August 2008*; IEEE: Munich, Germany, 2008; pp. 219–224. [CrossRef]
354. Terada, K.; Yamauchi, A.; Ito, A. Artificial Emotion Expression for a Robot by Dynamic Color Change. In *2012 IEEE RO-MAN: The 21st IEEE International Symposium on Robot and Human Interactive Communication, Paris, France, 9–13 September 2012*; IEEE: Paris, France, 2012; pp. 314–321. [CrossRef]
355. Li, S.; Walters, G.; Packer, J.; Scott, N. Using Skin Conductance and Facial Electromyography to Measure Emotional Responses to Tourism Advertising. *Curr. Issues Tour.* **2018**, *21*, 1761–1783. [CrossRef]
356. Nakasone, A.; Prendinger, H.; Ishizuka, M. Emotion Recognition from Electromyography and Skin Conductance. In Proceedings of the 5th International Workshop on Biosignal Interpretation; 2005; pp. 219–222. Available online: https://citeseerx.ist.psu.edu/viewdoc/download?doi=10.1.1.64.7269&rep=rep1&type=pdf (accessed on 29 July 2022).
357. Val-Calvo, M.; Álvarez-Sánchez, J.R.; Ferrández-Vicente, J.M.; Díaz-Morcillo, A.; Fernández-Jover, E. Real-Time Multi-Modal Estimation of Dynamically Evoked Emotions Using EEG, Heart Rate and Galvanic Skin Response. *Int. J. Neur. Syst.* **2020**, *30*, 2050013. [CrossRef]
358. Minhad, K.N.; Ali, S.H.M.; Reaz, M.B.I. Happy-Anger Emotions Classifications from Electrocardiogram Signal for Automobile Driving Safety and Awareness. *J. Transp. Health* **2017**, *7*, 75–89. [CrossRef]
359. Orini, M.; Bailón, R.; Enk, R.; Koelsch, S.; Mainardi, L.; Laguna, P. A Method for Continuously Assessing the Autonomic Response to Music-Induced Emotions through HRV Analysis. *Med. Biol. Eng. Comput.* **2010**, *48*, 423–433. [CrossRef]
360. Hernando, A.; Lazaro, J.; Gil, E.; Arza, A.; Garzon, J.M.; Lopez-Anton, R.; de la Camara, C.; Laguna, P.; Aguilo, J.; Bailon, R. Inclusion of Respiratory Frequency Information in Heart Rate Variability Analysis for Stress Assessment. *IEEE J. Biomed. Health Inform.* **2016**, *20*, 1016–1025. [CrossRef]
361. Dasgupta, P.B. Detection and Analysis of Human Emotions through Voice and Speech Pattern Processing. *Int. J. Comput. Trends Technol.* **2017**, *52*, 1–3. [CrossRef]
362. Rüsch, N.; Corrigan, P.W.; Bohus, M.; Kühler, T.; Jacob, G.A.; Lieb, K. The Impact of Posttraumatic Stress Disorder on Dysfunctional Implicit and Explicit Emotions Among Women with Borderline Personality Disorder. *J. Nerv. Ment. Dis.* **2007**, *195*, 537–539. [CrossRef]
363. Yi, Q.; Xiong, S.; Wang, B.; Yi, S. Identification of Trusted Interactive Behavior Based on Mouse Behavior Considering Web User's Emotions. *Int. J. Ind. Ergon.* **2020**, *76*, 102903. [CrossRef]
364. Lozano-Goupil, J.; Bardy, B.G.; Marin, L. Toward an Emotional Individual Motor Signature. *Front. Psychol.* **2021**, *12*, 647704. [CrossRef]
365. Venture, G.; Kadone, H.; Zhang, T.; Grèzes, J.; Berthoz, A.; Hicheur, H. Recognizing Emotions Conveyed by Human Gait. *Int. J. Soc. Robot.* **2014**, *6*, 621–632. [CrossRef]
366. Bevacqua, E.; Mancini, M. Speaking with Emotions. In Proceedings of the AISB Symposium on Motion, Emotion and Cognition, Leeds, UK, 29 March–1 April 2004; pp. 58–65.

367. Maalej, A.; Kallel, I. Does Keystroke Dynamics Tell Us about Emotions? A Systematic Literature Review and Dataset Construction. In *2020 16th International Conference on Intelligent Environments (IE), Madrid, Spain, 20–23 July 2020*; IEEE: Madrid, Spain, 2020; pp. 60–67. [CrossRef]
368. Chanel, G.; Kierkels, J.J.M.; Soleymani, M.; Pun, T. Short-Term Emotion Assessment in a Recall Paradigm. *Int. J. Hum.-Comput. Stud.* **2009**, *67*, 607–627. [CrossRef]
369. Chanel, G.; Kronegg, J.; Grandjean, D.; Pun, T. Emotion Assessment: Arousal Evaluation Using EEG's and Peripheral Physiological Signals. In *Multimedia Content Representation, Classification and Security*; Gunsel, B., Jain, A.K., Tekalp, A.M., Sankur, B., Hutchison, D., Kanade, T., Eds.; Lecture Notes in Computer Science; Springer: Berlin/Heidelberg, Germany, 2006; Volume 4105, pp. 530–537. [CrossRef]
370. Peter, C.; Ebert, E.; Beikirch, H. A Wearable Multi-Sensor System for Mobile Acquisition of Emotion-Related Physiological Data. In *Affective Computing and Intelligent Interaction*; Tao, J., Tan, T., Picard, R.W., Hutchison, D., Kanade, T., Kittler, J., Kleinberg, J.M., Eds.; Lecture Notes in Computer Science; Springer: Berlin/Heidelberg, Germany, 2005; Volume 3784, pp. 691–698. [CrossRef]
371. Villon, O.; Lisetti, C. A User-Modeling Approach to Build User's Psycho-Physiological Maps of Emotions Using Bio-Sensors. In *ROMAN 2006—The 15th IEEE International Symposium on Robot and Human Interactive Communication, Hatfield, UK, 6–8 September 2006*; IEEE: Hatfield, UK, 2006; pp. 269–276. [CrossRef]
372. Lee, S.; Hong, C.-s.; Lee, Y.K.; Shin, H.-s. Experimental Emotion Recognition System and Services for Mobile Network Environments. In Proceedings of the 2010 IEEE Sensors; IEEE: Kona, HI, USA, 2010; pp. 136–140. Available online: https://ieeexplore.ieee.org/stamp/stamp.jsp?arnumber=5690570 (accessed on 29 July 2022). [CrossRef]
373. De Santos Sierra, A.; Ávila, C.S.; Casanova, J.G.; del Pozo, G.B. Real-Time Stress Detection by Means of Physiological Signals. In *Advanced Biometric Technologies*; IntechOpen: London, UK, 2011; pp. 23–44.
374. Hsieh, P.-Y.; Chin, C.-L. The Emotion Recognition System with Heart Rate Variability and Facial Image Features. In Proceedings of the 2011 IEEE International Conference on Fuzzy Systems (FUZZ-IEEE 2011); IEEE: Taipei, Taiwan, 2011; pp. 1933–1940. [CrossRef]
375. Zhang, J.; Chen, M.; Hu, S.; Cao, Y.; Kozma, R. PNN for EEG-Based Emotion Recognition. In Proceedings of the 2016 IEEE International Conference on Systems, Man, and Cybernetics (SMC); IEEE: Budapest, Hungary, 2016; pp. 2319–2323. [CrossRef]
376. Mehmood, R.; Lee, H. Towards Building a Computer Aided Education System for Special Students Using Wearable Sensor Technologies. *Sensors* **2017**, *17*, 317. [CrossRef]
377. Purnamasari, P.; Ratna, A.; Kusumoputro, B. Development of Filtered Bispectrum for EEG Signal Feature Extraction in Automatic Emotion Recognition Using Artificial Neural Networks. *Algorithms* **2017**, *10*, 63. [CrossRef]
378. Li, Y.; Huang, J.; Zhou, H.; Zhong, N. Human Emotion Recognition with Electroencephalographic Multidimensional Features by Hybrid Deep Neural Networks. *Appl. Sci.* **2017**, *7*, 1060. [CrossRef]
379. Hu, X.; Yu, J.; Song, M.; Yu, C.; Wang, F.; Sun, P.; Wang, D.; Zhang, D. EEG Correlates of Ten Positive Emotions. *Front. Hum. Neurosci.* **2017**, *11*, 26. [CrossRef]
380. Taneli, B.; Krahne, W. EEG Changes of Transcendental Meditation Practitioners. In *Advances in Biological Psychiatry*; Taneli, B., Perris, C., Kemali, D., Eds.; S. Karger AG: Basel, Switzerland, 1987; Volume 16, pp. 41–71. [CrossRef]
381. Si, Y.; Jiang, L.; Tao, Q.; Chen, C.; Li, F.; Jiang, Y.; Zhang, T.; Cao, X.; Wan, F.; Yao, D.; et al. Predicting Individual Decision-Making Responses Based on the Functional Connectivity of Resting-State EEG. *J. Neural Eng.* **2019**, *16*, 066025. [CrossRef] [PubMed]
382. Akash, K.; Hu, W.-L.; Jain, N.; Reid, T. A Classification Model for Sensing Human Trust in Machines Using EEG and GSR. *ACM Trans. Interact. Intell. Syst.* **2018**, *8*, 1–20. [CrossRef]
383. Tsao, Y.-C.; Huang, C.-M.; Miou, Y.-C. The Role of Opposing Emotions in Design Satisfaction and Perceived Innovation. *J. Sci. Des.* **2021**, *5*, 111–120.
384. Martin, O.; Kotsia, I.; Macq, B.; Pitas, I. The eNTERFACE' 05 Audio-Visual Emotion Database. In Proceedings of the 22nd International Conference on Data Engineering Workshops (ICDEW'06), Atlanta, GA, USA, 3–7 April 2006; IEEE: Atlanta, GA, USA, 2006; p. 8. [CrossRef]
385. McDermott, O.D.; Prigerson, H.G.; Reynolds, C.F.; Houck, P.R.; Dew, M.A.; Hall, M.; Mazumdar, S.; Buysse, D.J.; Hoch, C.C.; Kupfer, D.J. Sleep in the Wake of Complicated Grief Symptoms: An Exploratory Study. *Biol. Psychiatry* **1997**, *41*, 710–716. [CrossRef]
386. Rusalova, M.N.; Kostyunina, M.B.; Kulikov, M.A. Spatial Distribution of Coefficients of Asymmetry of Brain Bioelectrical Activity during the Experiencing of Negative Emotions. *Neurosci. Behav. Physiol.* **2003**, *33*, 703–706. [CrossRef] [PubMed]
387. Uyan, U. *EEG-Based Assessment of Cybersickness in a VR Environment and Adjusting Stereoscopic Parameters According to Level of Sickness to Present a Comfortable Vision*; Hacettepe University: Ankara, Turkey, 2020.
388. Yankovsky, A.E.; Veilleux, M.; Dubeau, F.; Andermann, F. Post-Ictal Rage and Aggression: A Video-EEG Study. *Epileptic Disord.* **2005**, *7*, 143–147.
389. Kim, S.-H.; Nguyen Thi, N.A. Feature Extraction of Emotional States for EEG-Based Rage Control. In Proceedings of the 2016 39th International Conference on Telecommunications and Signal Processing (TSP), Vienna, Austria, 27–29 June 2016; IEEE: Vienna, Austria, 2016; pp. 361–364. [CrossRef]
390. Cannon, P.A.; Drake, M.E. EEG and Brainstem Auditory Evoked Potentials in Brain-Injured Patients with Rage Attacks and Self-Injurious Behavior. *Clin. Electroencephalogr.* **1986**, *17*, 169–172.

391. Chen, X.; Lin, J.; Jin, H.; Huang, Y.; Liu, Z. The Psychoacoustics Annoyance Research Based on EEG Rhythms for Passengers in High-Speed Railway. *Appl. Acoust.* **2021**, *171*, 107575. [CrossRef]
392. Li, Z.-G.; Di, G.-Q.; Jia, L. Relationship between Electroencephalogram Variation and Subjective Annoyance under Noise Exposure. *Appl. Acoust.* **2014**, *75*, 37–42. [CrossRef]
393. Benlamine, M.S.; Chaouachi, M.; Frasson, C.; Dufresne, A. Physiology-Based Recognition of Facial Micro-Expressions Using EEG and Identification of the Relevant Sensors by Emotion. In Proceedings of the 3rd International Conference on Physiological Computing Systems; SCITEPRESS—Science and Technology Publications: Lisbon, Portugal, 2016; pp. 130–137. Available online: https://www.scitepress.org/Papers/2016/60027/60027.pdf (accessed on 9 July 2022). [CrossRef]
394. Aftanas, L.I.; Pavlov, S.V. Trait Anxiety Impact on Posterior Activation Asymmetries at Rest and during Evoked Negative Emotions: EEG Investigation. *Int. J. Psychophysiol.* **2005**, *55*, 85–94. [CrossRef]
395. Ragozinskaya, V.G. Features of Psychosomatic Patient's Aggressiveness. *Procedia-Soc. Behav. Sci.* **2013**, *86*, 232–235. [CrossRef]
396. Konareva, I.N. Correlation between Level of Aggressiveness of Personality and Characteristics of EEG Frequency Components. *Neurophysiology* **2006**, *38*, 380–388. [CrossRef]
397. Munian, L.; Wan Ahmad, W.K.; Xu, T.K.; Mustafa, W.A.; Rahim, M.A. An Aggressiveness Level Analysis Based On Buss Perry Questionnaire (BPQ) And Brain Signal (EEG). *J. Phys.: Conf. Ser.* **2021**, *2107*, 012045. [CrossRef]
398. Flores, A.; Münte, T.F.; Doñamayor, N. Event-Related EEG Responses to Anticipation and Delivery of Monetary and Social Reward. *Biol. Psychol.* **2015**, *109*, 10–19. [CrossRef]
399. Gorka, S.M.; Phan, K.L.; Shankman, S.A. Convergence of EEG and FMRI Measures of Reward Anticipation. *Biol. Psychol.* **2015**, *112*, 12–19. [CrossRef]
400. Alazrai, R.; Homoud, R.; Alwanni, H.; Daoud, M. EEG-Based Emotion Recognition Using Quadratic Time-Frequency Distribution. *Sensors* **2018**, *18*, 2739. [CrossRef]
401. Cai, J.; Chen, W.; Yin, Z. Multiple Transferable Recursive Feature Elimination Technique for Emotion Recognition Based on EEG Signals. *Symmetry* **2019**, *11*, 683. [CrossRef]
402. Chao, H.; Dong, L.; Liu, Y.; Lu, B. Emotion Recognition from Multiband EEG Signals Using CapsNet. *Sensors* **2019**, *19*, 2212. [CrossRef]
403. Gao, Z.; Cui, X.; Wan, W.; Gu, Z. Recognition of Emotional States Using Multiscale Information Analysis of High Frequency EEG Oscillations. *Entropy* **2019**, *21*, 609. [CrossRef]
404. Garg, D.; Verma, G.K. Emotion recognition in valence-arousal space from multi-channel EEG data and wavelet based deep learning framework. *Procedia Computer Science* **2020**, *171*, 857–867. [CrossRef]
405. Soleymani, M.; Lichtenauer, J.; Pun, T.; Pantic, M. A Multimodal Database for Affect Recognition and Implicit Tagging. *IEEE Trans. Affect. Comput.* **2012**, *3*, 42–55. [CrossRef]
406. Yogeeswaran, K.; Nash, K.; Jia, H.; Adelman, L.; Verkuyten, M. Intolerant of Being Tolerant? Examining the Impact of Intergroup Toleration on Relative Left Frontal Activity and Outgroup Attitudes. *Curr. Psychol.* **2021**, *41*, 7228–7239. [CrossRef]
407. Fan, C.; Peng, Y.; Peng, S.; Zhang, H.; Wu, Y.; Kwong, S. Detection of Train Driver Fatigue and Distraction Based on Forehead EEG: A Time-Series Ensemble Learning Method. *IEEE Trans. Intell. Transport. Syst.* **2021**, *23*, 13559–13569. [CrossRef]
408. Mück, M.; Ohmann, K.; Dummel, S.; Mattes, A.; Thesing, U.; Stahl, J. Face Perception and Narcissism: Variations of Event-Related Potential Components (P1 & N170) with Admiration and Rivalry. *Cogn. Affect. Behav. Neurosci.* **2020**, *20*, 1041–1055. [CrossRef]
409. Tolgay, B.; Dell'Orco, S.; Maldonato, M.N.; Vogel, C.; Trojano, L.; Esposito, A. EEGs as Potential Predictors of Virtual Agents' Acceptance. In *2019 10th IEEE International Conference on Cognitive Infocommunications (CogInfoCom), Naples, Italy, 23–25 October 2019*; IEEE: Naples, Italy, 2019; pp. 433–438. [CrossRef]
410. Tarai, S.; Mukherjee, R.; Qurratul, Q.A.; Singh, B.K.; Bit, A. Use of Prosocial Word Enhances the Processing of Language: Frequency Domain Analysis of Human EEG. *J. Psycholinguist Res.* **2019**, *48*, 145–161. [CrossRef]
411. Ho, V.A.; Nguyen, D.H.-C.; Nguyen, D.H.; Pham, L.T.-V.; Nguyen, D.-V.; Nguyen, K.V.; Nguyen, N.L.-T. Emotion Recognition for Vietnamese Social Media Text. In *Computational Linguistics*; Nguyen, L.-M., Phan, X.-H., Hasida, K., Eds.; Communications in Computer and Information Science; Springer: Singapore, 2020; Volume 1215, pp. 319–333. [CrossRef]
412. Hu, X.; Zhuang, C.; Wang, F.; Liu, Y.-J.; Im, C.-H.; Zhang, D. FNIRS Evidence for Recognizably Different Positive Emotions. *Front. Hum. Neurosci.* **2019**, *13*, 120. [CrossRef]
413. Khazankin, G.R.; Shmakov, I.S.; Malinin, A.N. Remote Facial Emotion Recognition System. In Proceedings of the 2019 International Multi-Conference on Engineering, Computer and Information Sciences (SIBIRCON), Novosibirsk, Russia, 21–27 October 2019; IEEE: Novosibirsk, Russia, 2019; pp. 0975–0979. [CrossRef]
414. Guo, J.; Lei, Z.; Wan, J.; Avots, E.; Hajarolasvadi, N.; Knyazev, B.; Kuharenko, A.; Jacques Junior, J.C.S.; Baro, X.; Demirel, H.; et al. Dominant and Complementary Emotion Recognition from Still Images of Faces. *IEEE Access* **2018**, *6*, 26391–26403. [CrossRef]
415. Mumenthaler, C.; Sander, D.; Manstead, A. Emotion Recognition in Simulated Social Interactions. *IEEE Trans. Affect. Comput.* **2018**, *11*, 308–312. [CrossRef]
416. Zheng, W.-L.; Lu, B.-L. A Multimodal Approach to Estimating Vigilance Using EEG and Forehead EOG. *J. Neural Eng.* **2017**, *14*, 026017. [CrossRef]
417. Tomar, D.; Agarwal, S. Multi-Label Classifier for Emotion Recognition from Music. In *Proceedings of 3rd International Conference on Advanced Computing, Networking and Informatics*; Nagar, A., Mohapatra, D.P., Chaki, N., Eds.; Smart Innovation, Systems and Technologies; Springer: New Delhi, India, 2016; Volume 43, pp. 111–123. [CrossRef]

418. Bhatti, A.M.; Majid, M.; Anwar, S.M.; Khan, B. Human Emotion Recognition and Analysis in Response to Audio Music Using Brain Signals. *Comput. Hum. Behav.* **2016**, *65*, 267–275. [CrossRef]
419. Shih, Y.-L.; Lin, C.-Y. The Relationship between Action Anticipation and Emotion Recognition in Athletes of Open Skill Sports. *Cogn. Process* **2016**, *17*, 259–268. [CrossRef]
420. Patwardhan, A.; Knapp, G. Aggressive Actions and Anger Detection from Multiple Modalities Using Kinect, 2016. *arXiv preprint* **2016**, arXiv:1607.01076.
421. Fernández-Alcántara, M.; Cruz-Quintana, F.; Pérez-Marfil, M.N.; Catena-Martínez, A.; Pérez-García, M.; Turnbull, O.H. Assessment of Emotional Experience and Emotional Recognition in Complicated Grief. *Front. Psychol.* **2016**, *7*, 126. [CrossRef]
422. Naji, M.; Firoozabadi, M.; Azadfallah, P. Classification of Music-Induced Emotions Based on Information Fusion of Forehead Biosignals and Electrocardiogram. *Cogn. Comput.* **2014**, *6*, 241–252. [CrossRef]
423. Wen, W.; Liu, G.; Cheng, N.; Wei, J.; Shangguan, P.; Huang, W. Emotion Recognition Based on Multi-Variant Correlation of Physiological Signals. *IEEE Trans. Affect. Comput.* **2014**, *5*, 126–140. [CrossRef]
424. Kamińska, D.; Pelikant, A. Recognition of Human Emotion from a Speech Signal Based on Plutchik's Model. *Int. J. Electron. Telecommun.* **2012**, *58*, 165–170. [CrossRef]
425. Furley, P.; Dicks, M.; Memmert, D. Nonverbal Behavior in Soccer: The Influence of Dominant and Submissive Body Language on the Impression Formation and Expectancy of Success of Soccer Players. *J. Sport Exerc. Psychol.* **2012**, *34*, 61–82. [CrossRef] [PubMed]
426. Wagner, J.; Kim, J.; Andre, E. From Physiological Signals to Emotions: Implementing and Comparing Selected Methods for Feature Extraction and Classification. In Proceedings of the 2005 IEEE International Conference on Multimedia and Expo, Amsterdam, The Netherlands, 6–8 July 2005; IEEE: Amsterdam, The Netherlands, 2005; pp. 940–943. [CrossRef]
427. Furman, J.M.; Wuyts, F.L. Vestibular Laboratory Testing. In *Aminoff's Electrodiagnosis in Clinical Neurology*; Elsevier: Philadelphia, PA, USA, 2012; pp. 699–723. [CrossRef]
428. Lord Mary, P.; Wright, W.D. The Investigation of Eye Movements. *Rep. Prog. Phys.* **1950**, *13*, 1–23. [CrossRef]
429. Landowska, A. Emotion Monitoring—Verification of Physiological Characteristics Measurement Procedures. *Metrol. Meas. Syst.* **2014**, *21*, 719–732. [CrossRef]
430. Skiendziel, T.; Rösch, A.G.; Schultheiss, O.C. Assessing the Convergent Validity between the Automated Emotion Recognition Software Noldus FaceReader 7 and Facial Action Coding System Scoring. *PLoS ONE* **2019**, *14*, e0223905. [CrossRef] [PubMed]
431. Frescura, A.; Lee, P.J. Emotions and Physiological Responses Elicited by Neighbours Sounds in Wooden Residential Buildings. *Build. Environ.* **2022**, *210*, 108729. [CrossRef]
432. Nikolova, D.; Petkova, P.; Manolova, A.; Georgieva, P. ECG-Based Emotion Recognition: Overview of Methods and Applications. In *ANNA '18; Advances in Neural Networks and Applications 2018*; VDE: Varna, Bulgaria, 2018; pp. 118–122.
433. Nakanishi, R.; Imai-Matsumura, K. Facial Skin Temperature Decreases in Infants with Joyful Expression. *Infant Behav. Dev.* **2008**, *31*, 137–144. [CrossRef]
434. Park, M.W.; Kim, C.J.; Hwang, M.; Lee, E.C. Individual Emotion Classification between Happiness and Sadness by Analyzing Photoplethysmography and Skin Temperature. In *2013 Fourth World Congress on Software Engineering*; IEEE: Hong Kong, China, 2013; pp. 190–194. [CrossRef]
435. Gouizi, K.; Bereksi Reguig, F.; Maaoui, C. Emotion Recognition from Physiological Signals. *J. Med. Eng. Technol.* **2011**, *35*, 300–307. [CrossRef]
436. Abadi, M.K.; Kia, S.M.; Subramanian, R.; Avesani, P.; Sebe, N. User-Centric Affective Video Tagging from MEG and Peripheral Physiological Responses. In Proceedings of the 2013 Humaine Association Conference on Affective Computing and Intelligent Interaction, Geneva, Switzerland, 2–5 September 2013; IEEE: Geneva, Switzerland, 2013; pp. 582–587. [CrossRef]
437. Aguiñaga, A.R.; Lopez Ramirez, M.; Alanis Garza, A.; Baltazar, R.; Zamudio, V.M. *Emotion Analysis through Physiological Measurements*; IOS Press: Amsterdam, The Netherlands, 2013; pp. 97–106.
438. Javaid, M.; Haleem, A.; Vaishya, R.; Bahl, S.; Suman, R.; Vaish, A. Industry 4.0 Technologies and Their Applications in Fighting COVID-19 Pandemic. *Diabetes Metab. Syndr. Clin. Res. Rev.* **2020**, *14*, 419–422. [CrossRef]
439. Kalhori, S.R.N.; Bahaadinbeigy, K.; Deldar, K.; Gholamzadeh, M.; Hajesmaeel-Gohari, S.; Ayyoubzadeh, S.M. Digital Health Solutions to Control the COVID-19 Pandemic in Countries with High Disease Prevalence: Literature Review. *J. Med. Internet Res.* **2021**, *23*, e19473. [CrossRef]
440. Rahman, M.S.; Peeri, N.C.; Shrestha, N.; Zaki, R.; Haque, U.; Hamid, S.H.A. Defending against the Novel Coronavirus (COVID-19) Outbreak: How Can the Internet of Things (IoT) Help to Save the World? *Health Policy Technol.* **2020**, *9*, 136–138. [CrossRef]
441. Rajeesh Kumar, N.V.; Arun, M.; Baraneetharan, E.; Stanly Jaya Prakash, J.; Kanchana, A.; Prabu, S. Detection and Monitoring of the Asymptotic COVID-19 Patients Using IoT Devices and Sensors. *Int. J. Pervasive Comput. Commun.* **2020**, *18*(4), 407–418. [CrossRef]
442. Stojanovic, R.; Skraba, A.; Lutovac, B. A Headset Like Wearable Device to Track COVID-19 Symptoms. In *2020 9th Mediterranean Conference on Embedded Computing (MECO), Budva, Montenegro, 8–11 June 2020*; IEEE: Budva, Montenegro, 2020; pp. 8–11. [CrossRef]
443. Xian, M.; Luo, H.; Xia, X.; Fares, C.; Carey, P.H.; Chiu, C.-W.; Ren, F.; Shan, S.-S.; Liao, Y.-T.; Hsu, S.-M.; et al. Fast SARS-CoV-2 Virus Detection Using Disposable Cartridge Strips and a Semiconductor-Based Biosensor Platform. *J. Vac. Sci. Technol. B* **2021**, *39*, 033202. [CrossRef]

444. Chamberlain, S.D.; Singh, I.; Ariza, C.; Daitch, A.; Philips, P.; Dalziel, B.D. Real-Time Detection of COVID-19 Epicenters within the United States Using a Network of Smart Thermometers. *Epidemiology* **2020**, 1–15. [CrossRef]
445. Cherry, K. The Role of Neurotransmitters. 2021. Available online: https://www.verywellmind.com/what-is-a-neurotransmitter-2795394 (accessed on 14 June 2022).
446. Ali Fahmi, P.N.; Kodirov, E.; Choi, D.-J.; Lee, G.-S.; Mohd Fikri Azli, A.; Sayeed, S. Implicit Authentication Based on Ear Shape Biometrics Using Smartphone Camera during a Call. In *2012 IEEE International Conference on Systems, Man, and Cybernetics (SMC)*, Seoul, Korea, 14–17 October 2012; IEEE: Seoul, Korea, 2012; pp. 2272–2276. [CrossRef]
447. Calvert, G. Everything You Need to Know about Implicit Reaction Time (IRTs). 2015. Available online: http://gemmacalvert.com/everything-you-need-to-know-about-implicit-reaction-time/ (accessed on 17 August 2022).
448. Harris, J.M.; Ciorciari, J.; Gountas, J. Consumer Neuroscience for Marketing Researchers. *J. Consum. Behav.* **2018**, *17*, 239–252. [CrossRef]
449. Fox, E. Perspectives from Affective Science on Understanding the Nature of Emotion. *Brain Neurosci. Adv.* **2018**, *2*, 239821281881262. [CrossRef]
450. Casado-Aranda, L.A.; Sanchez-Fernandez, J. Advances in neuroscience and marketing: Analyzing tool possibilities and research opportunities. *Span. J. Mark. – ESIC* **2022**, *26*, 3–22. [CrossRef]
451. Lantrip, C.; Gunning, F.M.; Flashman, L.; Roth, R.M.; Holtzheimer, P.E. Effects of Transcranial Magnetic Stimulation on the Cognitive Control of Emotion: Potential Antidepressant Mechanisms. *J. ECT* **2017**, *33*, 73–80. [CrossRef]
452. Catalino, M.P.; Yao, S.; Green, D.L.; Laws, E.R.; Golby, A.J.; Tie, Y. Mapping Cognitive and Emotional Networks in Neurosurgical Patients Using Resting-State Functional Magnetic Resonance Imaging. *Neurosurg. Focus* **2020**, *48*, E9. [CrossRef]
453. Grèzes, J.; Valabrègue, R.; Gholipour, B.; Chevallier, C. A Direct Amygdala-Motor Pathway for Emotional Displays to Influence Action: A Diffusion Tensor Imaging Study: A Direct Limbic Motor Anatomical Pathway. *Hum. Brain Mapp.* **2014**, *35*, 5974–5983. [CrossRef]
454. Alhargan, A.; Cooke, N.; Binjammaz, T. Affect Recognition in an Interactive Gaming Environment Using Eye Tracking. In *2017 Seventh International Conference on Affective Computing and Intelligent Interaction (ACII)*, San Antonio, TX, USA, 23–26 October 2017; IEEE: San Antonio, TX, USA, 2017; pp. 285–291. [CrossRef]
455. Szwoch, M.; Szwoch, W. Emotion Recognition for Affect Aware Video Games. In *Image Processing & Communications Challenges 6*; Choraś, R.S., Ed.; Advances in Intelligent Systems and Computing; Springer International Publishing: Cham, Switzerland, 2015; Volume 313, pp. 227–236. [CrossRef]
456. Krol, L.R.; Freytag, S.-C.; Zander, T.O. Meyendtris: A Hands-Free, Multimodal Tetris Clone Using Eye Tracking and Passive BCI for Intuitive Neuroadaptive Gaming. In Proceedings of the 19th ACM International Conference on Multimodal Interaction, Glasgow, UK, 13–17 November 2017; ACM: Glasgow, UK, 2017; pp. 433–437. [CrossRef]
457. Elor, A.; Powell, M.; Mahmoodi, E.; Teodorescu, M.; Kurniawan, S. Gaming Beyond the Novelty Effect of Immersive Virtual Reality for Physical Rehabilitation. *IEEE Trans. Games* **2022**, *14*, 107–115. [CrossRef]
458. Tiwari, S.; Agarwal, S. A Shrewd Artificial Neural Network-Based Hybrid Model for Pervasive Stress Detection of Students Using Galvanic Skin Response and Electrocardiogram Signals. *Big Data* **2021**, *9*, 427–442. [CrossRef]
459. Nakayama, N.; Arakawa, N.; Ejiri, H.; Matsuda, R.; Makino, T. Heart Rate Variability Can Clarify Students' Level of Stress during Nursing Simulation. *PLoS ONE* **2018**, *13*, e0195280. [CrossRef]
460. Tautchin, L.; Dussome, W. The Expanding Reach of Non-Traditional Marketing: A Discussion on the Application of Neuromarketing and Big Data Analytics in the Marketplace. Available online: https://lowelltautchin.ca/wp-content/uploads/2016/08/Neuromarketing-and-Big-Data-Analytics-Project.pdf (accessed on 14 June 2022).
461. Goyal, G.; Singh, J. Minimum Annotation Identification of Facial Affects for Video Advertisement. In Proceedings of the 2018 International Conference on Intelligent Circuits and Systems (ICICS), Phagwara, India, 20–21 April 2018; IEEE: Phagwara, India, 2018; pp. 300–305. [CrossRef]
462. Yadava, M.; Kumar, P.; Saini, R.; Roy, P.P.; Prosad Dogra, D. Analysis of EEG Signals and Its Application to Neuromarketing. *Multimed. Tools Appl.* **2017**, *76*, 19087–19111. [CrossRef]
463. Lakhan, P.; Banluesombatkul, N.; Changniam, V.; Dhithijaiyratn, R.; Leelaarporn, P.; Boonchieng, E.; Hompoonsup, S.; Wilaiprasitporn, T. Consumer Grade Brain Sensing for Emotion Recognition. *IEEE Sens. J.* **2019**, *19*, 9896–9907. [CrossRef]
464. Kong, W.; Wang, L.; Xu, S.; Babiloni, F.; Chen, H. EEG Fingerprints: Phase Synchronization of EEG Signals as Biomarker for Subject Identification. *IEEE Access* **2019**, *7*, 121165–121173. [CrossRef]
465. El-Amir, M.M.; Al-Atabany, W.; Eldosoky, M.A. Emotion Recognition via Detrended Fluctuation Analysis and Fractal Dimensions. In Proceedings of the 2019 36th National Radio Science Conference (NRSC), Port Said, Egypt, 16–18 April 2019; IEEE: Port Said, Egypt, 2019; pp. 200–208. [CrossRef]
466. Plassmann, H.; Kenning, P.; Deppe, M.; Kugel, H.; Schwindt, W. How Choice Ambiguity Modulates Activity in Brain Areas Representing Brand Preference: Evidence from Consumer Neuroscience. *J. Consum. Behav.* **2008**, *7*, 360–367. [CrossRef]
467. Perrachione, T.K.; Perrachione, J.R. Brains and Brands: Developing Mutually Informative Research in Neuroscience and Marketing. *J. Consum. Behav.* **2008**, *7*, 303–318. [CrossRef]
468. Gruter, D. Neuromarketing—New Science of Consumer Behavior. Available online: http://emarketingblog.nl/2014/12/neuromarketing-new-science-of-consumer-behavior/ (accessed on 14 June 2022).

469. Leon, E.; Clarke, G.; Callaghan, V.; Sepulveda, F. A User-Independent Real-Time Emotion Recognition System for Software Agents in Domestic Environments. *Eng. Appl. Artif. Intell.* **2007**, *20*, 337–345. [CrossRef]
470. Monajati, M.; Abbasi, S.H.; Shabaninia, F.; Shamekhi, S. Emotions States Recognition Based on Physiological Parameters by Employing of Fuzzy-Adaptive Resonance Theory. *Int. J. Intell. Sci.* **2012**, *02*, 166–175. [CrossRef]
471. Andrew, H.; Haines, H.; Seixas, S. Using Neuroscience to Understand the Impact of Premium Digital Out-of-Home Media. *Int. J. Mark. Res.* **2019**, *61*, 588–600. [CrossRef]
472. Kaklauskas, A.; Bucinskas, V.; Dzedzickis, A. Computer Implemented Neuromarketing Research Method. European Patent Application EP4016431, 7 February 2021.
473. Lajante, M.; Ladhari, R. The Promise and Perils of the Peripheral Psychophysiology of Emotion in Retailing and Consumer Services. *J. Retail. Consum. Serv.* **2019**, *50*, 305–313. [CrossRef]
474. Michael, I.; Ramsoy, T.; Stephens, M.; Kotsi, F. A Study of Unconscious Emotional and Cognitive Responses to Tourism Images Using a Neuroscience Method. *J. Islamic Mark.* **2019**, *10*, 543–564. [CrossRef]
475. Libert, A.; van Hulle, M.M. Predicting Premature Video Skipping and Viewer Interest from EEG Recordings. *Entropy* **2019**, *21*, 1014. [CrossRef]
476. Jiménez-Marín, G.; Bellido-Pérez, E.; López-Cortés, Á. Marketing Sensorial: El Concepto, Sus Técnicas y Su Aplicación En El Punto de Venta. *Vivat Acad.* **2019**, *148*, 121–147. [CrossRef]
477. Cherubino, P.; Martinez-Levy, A.C.; Caratù, M.; Cartocci, G.; Di Flumeri, G.; Modica, E.; Rossi, D.; Mancini, M.; Trettel, A. Consumer Behaviour through the Eyes of Neurophysiological Measures: State-of-the-Art and Future Trends. *Comput. Intell. Neurosci.* **2019**, *2019*, 1976847. [CrossRef]
478. Tichindelean, M.B.; Iuliana, C.; Tichindelean, M. Studying the User Experience in Online Banking Services: An Eye-Tracking Application. *Stud. Bus. Econ.* **2019**, *14*, 193–208. [CrossRef]
479. Soria Morillo, L.M.; Alvarez-Garcia, J.A.; Gonzalez-Abril, L.; Ortega Ramírez, J.A. Discrete Classification Technique Applied to TV Advertisements Liking Recognition System Based on Low-Cost EEG Headsets. *BioMed. Eng. OnLine* **2016**, *15*, 75. [CrossRef]
480. Pringle, H.; Field, P. Institute of Practitioners in Advertising. In *Brand Immortality: How Brands Can Live Long and Prosper*; Kogan Page: London, UK, 2008.
481. Takahashi, K. Remarks on Emotion Recognition from Bio-Potential Signals. In Proceedings of the 2nd International Conference on Autonomous Robots and Agents, Palmerston North, New Zealand, 13–15 December 2004.
482. Light, K.C.; Girdler, S.S.; Sherwood, A.; Bragdon, E.E.; Brownley, K.A.; West, S.G.; Hinderliter, A.L. High Stress Responsivity Predicts Later Blood Pressure Only in Combination with Positive Family History and High Life Stress. *Hypertension* **1999**, *33*, 1458–1464. [CrossRef]
483. Gray, M.A.; Taggart, P.; Sutton, P.M.; Groves, D.; Holdright, D.R.; Bradbury, D.; Brull, D.; Critchley, H.D. A Cortical Potential Reflecting Cardiac Function. *Proc. Natl. Acad. Sci. USA* **2007**, *104*, 6818–6823. [CrossRef]
484. Adrogué, H.J.; Madias, N.E. Sodium and Potassium in the Pathogenesis of Hypertension. *N. Engl. J. Med.* **2007**, *356*, 1966–1978. [CrossRef]
485. Blair, D.A.; Glover, W.E.; Greenfield, A.D.M.; Roddie, I.C. Excitation of Cholinergic Vasodilator Nerves to Human Skeletal Muscles during Emotional Stress. *J. Physiol.* **1959**, *148*, 633–647. [CrossRef]
486. Gasperin, D.; Netuveli, G.; Dias-da-Costa, J.S.; Pattussi, M.P. Effect of Psychological Stress on Blood Pressure Increase: A Meta-Analysis of Cohort Studies. *Cad. Saúde Pública* **2009**, *25*, 715–726. [CrossRef]
487. Sun, F.-T.; Kuo, C.; Cheng, H.-T.; Buthpitiya, S.; Collins, P.; Griss, M. Activity-Aware Mental Stress Detection Using Physiological Sensors. In *Mobile Computing, Applications, and Services*; Gris, M., Yang, G., Eds.; Lecture Notes of the Institute for Computer Sciences, Social Informatics and Telecommunications Engineering; Springer: Berlin/Heidelberg, Germany, 2012; Volume 76, pp. 211–230. [CrossRef]
488. Singh, R.R.; Conjeti, S.; Banerjee, R. A Comparative Evaluation of Neural Network Classifiers for Stress Level Analysis of Automotive Drivers Using Physiological Signals. *Biomed. Signal Processing Control* **2013**, *8*, 740–754. [CrossRef]
489. Palacios, D.; Rodellar, V.; Lázaro, C.; Gómez, A.; Gómez, P. An ICA-Based Method for Stress Classification from Voice Samples. *Neural Comput. Applic.* **2020**, *32*, 17887–17897. [CrossRef]
490. Oka, T.; Oka, K.; Hori, T. Mechanisms and Mediators of Psychological Stress-Induced Rise in Core Temperature. *Psychosom. Med.* **2001**, *63*, 476–486. [CrossRef] [PubMed]
491. Wu, C.-H.; Liang, W.-B. Emotion Recognition of Affective Speech Based on Multiple Classifiers Using Acoustic-Prosodic Information and Semantic Labels. *IEEE Trans. Affect. Comput.* **2011**, *2*, 10–21. [CrossRef]
492. Nilashi, M.; Mardani, A.; Liao, H.; Ahmadi, H.; Manaf, A.A.; Almukadi, W. A Hybrid Method with TOPSIS and Machine Learning Techniques for Sustainable Development of Green Hotels Considering Online Reviews. *Sustainability* **2019**, *11*, 6013. [CrossRef]
493. Kaklauskas, A.; Ubarte, I.; Kalibatas, D.; Lill, I.; Velykorusova, A.; Volginas, P.; Vinogradova, I.; Milevicius, V.; Vetloviene, I.; Grubliauskas, I.; et al. A Multisensory, Green, and Energy Efficient Housing Neuromarketing Method. *Energies* **2019**, *12*, 3836. [CrossRef]
494. Kaklauskas, A.; Dzitac, D.; Sliogeriene, J.; Lepkova, N.; Vetloviene, I. VINERS Method for the Multiple Criteria Analysis and Neuromarketing of Best Places to Live. *Int. J. Comput. Commun. Control* **2019**, *14*, 629–646. [CrossRef]
495. Etzold, V.; Braun, A.; Wanner, T. Eye Tracking as a Method of Neuromarketing for Attention Research—An Empirical Analysis Using the Online Appointment Booking Platform from Mercedes-Benz. In *Intelligent Decision Technologies 2019*; Czarnowski, I.,

Howlett, R.J., Jain, L.C., Eds.; Smart Innovation, Systems and Technologies; Springer: Singapore, 2019; Volume 143, pp. 167–182. [CrossRef]
496. Dedeoglu, B.B.; Bilgihan, A.; Ye, B.H.; Buonincontri, P.; Okumus, F. The Impact of Servicescape on Hedonic Value and Behavioral Intentions: The Importance of Previous Experience. *Int. J. Hosp. Manag.* **2018**, *72*, 10–20. [CrossRef]
497. Khan, S.N.; Mohsin, M. The Power of Emotional Value: Exploring the Effects of Values on Green Product Consumer Choice Behavior. *J. Clean. Prod.* **2017**, *150*, 65–74. [CrossRef]
498. Puustinen, P.; Maas, P.; Karjaluoto, H. Development and Validation of the Perceived Investment Value (PIV) Scale. *J. Econ. Psychol.* **2013**, *36*, 41–54. [CrossRef]
499. Shaw, C. What's Your Companies Emotion Score? *Introducing Net Emotional Value (Nev) and Its Relationship to NPS and CSAT*. 2012. Available online: https://beyondphilosophy.com/whats-your-companies-emotion-score-introducing-net-emotional-value-nev-and-its-relationship-to-nps-and-csat/ (accessed on 14 June 2022).
500. Shaw, C. New CX Measure to Compliment NPS: Net Emotional Value. 2016. Available online: https://customerthink.com/new-cx-measure-to-compliment-nps-net-emotional-value/ (accessed on 14 June 2022).
501. Shaw, C. How to Measure Customer Emotions. 2018. Available online: https://beyondphilosophy.com/measurecustomer-emotions/ (accessed on 14 June 2022).
502. Situmorang, S.H. Gen C and Gen Y: Experience, Net Emotional Value and Net Promoter Score. In Proceedings of the 1st International Conference on Social and Political Development (ICOSOP 2016), Medan, Indonesia, 21–22 November 2016; Atlantis Press: Medan, Indonesia, 2017; pp. 259–265. [CrossRef]
503. Williams, P.; Soutar, G.N. Value, Satisfaction and Behavioral Intentions in an Adventure Tourism Context. *Ann. Tour. Res.* **2009**, *36*, 413–438. [CrossRef]
504. Bailey, J.J.; Gremler, D.D.; McCollough, M.A. Service Encounter Emotional Value: The Dyadic Influence of Customer and Employee Emotions. *Serv. Mark. Q.* **2001**, *23*, 1–24. [CrossRef]
505. Zavadskas, E.K.; Bausys, R.; Kaklauskas, A.; Raslanas, S. Hedonic Shopping Rent Valuation by One-to-One Neuromarketing and Neutrosophic PROMETHEE Method. *Appl. Soft Comput.* **2019**, *85*, 105832. [CrossRef]
506. De Leersnyder, J.; Mesquita, B.; Boiger, M. What Has Culture Got to Do with Emotions?: (A Lot). In *Handbook of Advances in Culture and Psychology*; Oxford University Press: Oxford, UK, 2021; Volume 8, pp. 62–119. [CrossRef]
507. Frijda, N.H. *The Laws of Emotion*, 1st ed.; Psychology Press: New York, NY, USA, 2017. [CrossRef]
508. Levenson, R.W. Human Emotions: A Functional View. In *The Nature of Emotion: Fundamental Questions*; Oxford University Press: New York, NY, USA, 1994; pp. 123–126.
509. Nesse, R.M. Evolutionary Explanations of Emotions. *Hum. Nat.* **1990**, *1*, 261–289. [CrossRef]
510. Bonanno, G.A.; Colak, D.M.; Keltner, D.; Shiota, M.N.; Papa, A.; Noll, J.G.; Putnam, F.W.; Trickett, P.K. Context Matters: The Benefits and Costs of Expressing Positive Emotion among Survivors of Childhood Sexual Abuse. *Emotion* **2007**, *7*, 824–837. [CrossRef]
511. Coifman, K.G.; Bonanno, G.A. Emotion Context Sensitivity in Adaptation and Recovery. In *Emotion Regulation and Psychopathology: A Transdiagnostic Approach to Etiology and Treatment*; The Guilford Press: New York, NY, USA, 2010; pp. 157–173.
512. Pugh, Z.H.; Huang, J.; Leshin, J.; Lindquist, K.A.; Nam, C.S. Culture and Gender Modulate DlPFC Integration in the Emotional Brain: Evidence from Dynamic Causal Modeling. *Cogn. Neurodyn.* **2022**. Available online: https://link.springer.com/content/pdf/10.1007/s11571-022-09805-2.pdf (accessed on 14 June 2022). [CrossRef]
513. Tomasino, B.; Maggioni, E.; Bonivento, C.; Nobile, M.; D'Agostini, S.; Arrigoni, F.; Fabbro, F.; Brambilla, P. Effects of Age and Gender on Neural Correlates of Emotion Imagery. *Hum. Brain Mapp.* **2022**. [CrossRef]
514. Hampton, R.S.; Varnum, M.E.W. The Cultural Neuroscience of Emotion Regulation. *Cult. Brain* **2018**, *6*, 130–150. [CrossRef]
515. Rule, N.O.; Freeman, J.B.; Ambady, N. Culture in Social Neuroscience: A Review. *Soc. Neurosci.* **2013**, *8*, 3–10. [CrossRef]
516. Kraus, M.W.; Piff, P.K.; Keltner, D. Social Class, Sense of Control, and Social Explanation. *J. Personal. Soc. Psychol.* **2009**, *97*, 992–1004. [CrossRef]
517. Gallo, L.C.; Matthews, K.A. Understanding the Association between Socioeconomic Status and Physical Health: Do Negative Emotions Play a Role? *Psychol. Bull.* **2003**, *129*, 10–51. [CrossRef]
518. Choudhury, S.; Nagel, S.K.; Slaby, J. Critical Neuroscience: Linking Neuroscience and Society through Critical Practice. *BioSocieties* **2009**, *4*, 61–77. [CrossRef]
519. Goldfarb, M.G.; Brown, D.R. Diversifying Participation: The Rarity of Reporting Racial Demographics in Neuroimaging Research. *NeuroImage* **2022**, *254*, 119122. [CrossRef]
520. Lane, R.D. From Reconstruction to Construction: The Power of Corrective Emotional Experiences in Memory Reconsolidation and Enduring Change. *J. Am. Psychoanal. Assoc.* **2018**, *66*, 507–516. [CrossRef]
521. Nakamura, F. Creating or Performing Words? Observations on Contemporary Japanese Calligraphy. In *Creativity and Cultural Improvisation*; Routledge: Oxfordshire, UK, 2021; pp. 79–98.
522. Markus, H.R.; Kitayama, S. Cultural Variation in the Self-Concept. In *The Self: Interdisciplinary Approaches*; Strauss, J., Goethals, G.R., Eds.; Springer: New York, NY, USA, 1991; pp. 18–48. [CrossRef]
523. Mesquita, B.; Frijda, N.H. Cultural Variations in Emotions: A Review. *Psychol. Bull.* **1992**, *112*, 179–204. [CrossRef]
524. Mesquita, B.; Leu, J. The Cultural Psychology of Emotion. In *Handbook of Cultural Psychology*; The Guilford Press: New York, NY, USA, 2007; pp. 734–759.

525. Lim, N. Cultural Differences in Emotion: Differences in Emotional Arousal Level between the East and the West. *Integr. Med. Res.* **2016**, *5*, 105–109. [CrossRef]
526. Hareli, S.; Kafetsios, K.; Hess, U. A Cross-Cultural Study on Emotion Expression and the Learning of Social Norms. *Front. Psychol.* **2015**, *6*, 1501. [CrossRef]
527. Scollon, C.N.; Diener, E.; Oishi, S.; Biswas-Diener, R. Emotions Across Cultures and Methods. *J. Cross-Cult. Psychol.* **2004**, *35*, 304–326. [CrossRef]
528. Siddiqui, H.U.R.; Shahzad, H.F.; Saleem, A.A.; Khan Khakwani, A.B.; Rustam, F.; Lee, E.; Ashraf, I.; Dudley, S. Respiration Based Non-Invasive Approach for Emotion Recognition Using Impulse Radio Ultra Wide Band Radar and Machine Learning. *Sensors* **2021**, *21*, 8336. [CrossRef]
529. Houssein, E.H.; Hammad, A.; Ali, A.A. Human Emotion Recognition from EEG-Based Brain–Computer Interface Using Machine Learning: A Comprehensive Review. *Neural Comput Applic* **2022**, *34*, 12527–12557. [CrossRef]
530. Shi, Y.; Zheng, X.; Li, T. Unconscious Emotion Recognition Based on Multi-Scale Sample Entropy. In Proceedings of the 2018 IEEE International Conference on Bioinformatics and Biomedicine (BIBM), Madrid, Spain, 3–6 December 2018; IEEE: Madrid, Spain, 2018; pp. 1221–1226. [CrossRef]
531. Thomson, D.M.H.; Coates, T. Are Unconscious Emotions Important in Product Assessment? How Can We Access Them? *Food Qual. Prefer.* **2021**, *92*, 104123. [CrossRef]
532. Poria, S.; Cambria, E.; Bajpai, R.; Hussain, A. A Review of Affective Computing: From Unimodal Analysis to Multimodal Fusion. *Inf. Fusion* **2017**, *37*, 98–125. [CrossRef]
533. Caridakis, G.; Castellano, G.; Kessous, L.; Raouzaiou, A.; Malatesta, L.; Asteriadis, S.; Karpouzis, K. Multimodal Emotion Recognition from Expressive Faces, Body Gestures and Speech. In *Artificial Intelligence and Innovations 2007: From Theory to Applications*; Boukis, C., Pnevmatikakis, A., Polymenakos, L., Eds.; IFIP The International Federation for Information Processing; Springer: Boston, MA, USA, 2007; Volume 247, pp. 375–388. [CrossRef]
534. Cambria, E.; Das, D.; Bandyopadhyay, S.; Feraco, A. Affective Computing and Sentiment Analysis. In *A Practical Guide to Sentiment Analysis*; Cambria, E., Das, D., Bandyopadhyay, S., Feraco, A., Eds.; Socio-Affective Computing; Springer International Publishing: Cham, Switzerland, 2017; Volume 5, pp. 102–107. [CrossRef]
535. Dhanapal, R.; Bhanu, D. Electroencephalogram classification using various artificial neural networks. *J. Crit. Rev.* **2020**, *7*, 891–894. [CrossRef]
536. Gunawan, T.S.; Alghifari, M.F.; Morshidi, M.A.; Kartiwi, M. A Review on Emotion Recognition Algorithms Using Speech Analysis. *Indones. J. Electr. Eng. Inform.* **2018**, *6*, 12–20. [CrossRef]
537. Sánchez-Reolid, R.; García, A.; Vicente-Querol, M.; Fernández-Aguilar, L.; López, M.; González, A. Artificial Neural Networks to Assess Emotional States from Brain-Computer Interface. *Electronics* **2018**, *7*, 384. [CrossRef]
538. Nakisa, B.; Rastgoo, M.N.; Tjondronegoro, D.; Chandran, V. Evolutionary Computation Algorithms for Feature Selection of EEG-Based Emotion Recognition Using Mobile Sensors. *Expert Syst. Appl.* **2018**, *93*, 143–155. [CrossRef]
539. Saxena, A.; Khanna, A.; Gupta, D. Emotion Recognition and Detection Methods: A Comprehensive Survey. *J. Artif. Intell. Syst.* **2020**, *2*, 53–79. [CrossRef]
540. Ahmed, F.; Sieu, B.; Gavrilova, M.L. Score and Rank-Level Fusion for Emotion Recognition Using Genetic Algorithm. In Proceedings of the 2018 IEEE 17th International Conference on Cognitive Informatics & Cognitive Computing (ICCI*CC), IEEE, Berkeley, CA, USA, 7 October 2018; pp. 46–53. [CrossRef]
541. Slimani, K.; Kas, M.; El Merabet, Y.; Ruichek, Y.; Messoussi, R. Local Feature Extraction Based Facial Emotion Recognition: A Survey. *Int. J. Electr. Comput. Eng.* **2020**, *10*, 4080–4092. [CrossRef]
542. Maheshwari, D.; Ghosh, S.K.; Tripathy, R.K.; Sharma, M.; Acharya, U.R. Automated Accurate Emotion Recognition System Using Rhythm-Specific Deep Convolutional Neural Network Technique with Multi-Channel EEG Signals. *Comput. Biol. Med.* **2021**, *134*, 104428. [CrossRef]
543. Zatarain Cabada, R.; Rodriguez Rangel, H.; Barron Estrada, M.L.; Cardenas Lopez, H.M. Hyperparameter Optimization in CNN for Learning-Centered Emotion Recognition for Intelligent Tutoring Systems. *Soft Comput.* **2020**, *24*, 7593–7602. [CrossRef]
544. Poria, S.; Hazarika, D.; Majumder, N.; Naik, G.; Cambria, E.; Mihalcea, R. MELD: A Multimodal Multi-Party Dataset for Emotion Recognition in Conversations. In Proceedings of the 57th Annual Meeting of the Association for Computational Linguistics, Florence, Italy, 28 July–2 August 2019; Association for Computational Linguistics: Florence, Italy, 2019; pp. 527–536. [CrossRef]
545. Xu, Y.; Sun, Y.; Liu, X.; Zheng, Y. A Digital-Twin-Assisted Fault Diagnosis Using Deep Transfer Learning. *IEEE Access* **2019**, *7*, 19990–19999. [CrossRef]
546. Daneshfar, F.; Kabudian, S.J.; Neekabadi, A. Speech Emotion Recognition Using Hybrid Spectral-Prosodic Features of Speech Signal/Glottal Waveform, Metaheuristic-Based Dimensionality Reduction, and Gaussian Elliptical Basis Function Network Classifier. *Appl. Acoust.* **2020**, *166*, 107360. [CrossRef]
547. Shi, W.; Jiang, M. Fuzzy Wavelet Network with Feature Fusion and LM Algorithm for Facial Emotion Recognition. In Proceedings of the 2018 IEEE International Conference of Safety Produce Informatization (IICSPI), Chongqing, China, 10–12 December 2018; IEEE: Chongqing, China, 2018; pp. 582–586. [CrossRef]
548. Yildirim, S.; Kaya, Y.; Kılıç, F. A Modified Feature Selection Method Based on Metaheuristic Algorithms for Speech Emotion Recognition. *Appl. Acoust.* **2021**, *173*, 107721. [CrossRef]

549. Bellamkonda, S.S. Facial Emotion Recognition by Hyper-Parameter Tuning of Convolutional Neural Network Using Genetic Algorithm. 2021. Available online: http://urn.kb.se/resolve?urn=urn:nbn:se:bth-22308 (accessed on 14 June 2022).
550. Jalili, L.; Cervantes, J.; García-Lamont, F.; Trueba, A. Emotion Recognition from Facial Expressions Using a Genetic Algorithm to Feature Extraction. In *Intelligent Computing Theories and Application*; Huang, D.-S., Jo, K.-H., Li, J., Gribova, V., Bevilacqua, V., Eds.; Lecture Notes in Computer Science; Springer International Publishing: Cham, Switzerland, 2021; Volume 12836, pp. 59–71. [CrossRef]
551. Sun, L.; Li, Q.; Fu, S.; Li, P. Speech Emotion Recognition Based on Genetic Algorithm–Decision Tree Fusion of Deep and Acoustic Features. *ETRI J.* **2022**, *44*, 462–475. [CrossRef]
552. Madhoushi, Z.; Hamdan, A.R.; Zainudin, S. Sentiment Analysis Techniques in Recent Works. In Proceedings of the 2015 Science and Information Conference (SAI), London, UK, 28–30 August 2015; IEEE: London, UK, 2015; pp. 288–291. [CrossRef]
553. Li, G.; Zhou, X.; Cao, L. AI Meets Database: AI4DB and DB4AI. In Proceedings of the 2021 International Conference on Management of Data, Virtual Event, China; 2021; pp. 2859–2866. Available online: https://dbgroup.cs.tsinghua.edu.cn/ligl/papers/sigmod21-tutorial-paper.pdf (accessed on 14 June 2022). [CrossRef]
554. Arnau-Gonzalez, P.; Katsigiannis, S.; Arevalillo-Herraez, M.; Ramzan, N. BED: A New Data Set for EEG-Based Biometrics. *IEEE Internet Things J.* **2021**, *8*, 12219–12230. [CrossRef]
555. Stappen, L.; Schuller, B.; Lefter, I.; Cambria, E.; Kompatsiaris, I. Summary of MuSe 2020: Multimodal Sentiment Analysis, Emotion-Target Engagement and Trustworthiness Detection in Real-Life Media. In Proceedings of the 28th ACM International Conference on Multimedia, Seattle, WA, USA; ACM: Seattle, WA, USA, 2020; pp. 4769–4770. Available online: https://dl.acm.org/doi/pdf/10.1145/3394171.3421901 (accessed on 14 June 2022). [CrossRef]
556. Poria, S.; Majumder, N.; Mihalcea, R.; Hovy, E. Emotion Recognition in Conversation: Research Challenges, Datasets, and Recent Advances. *IEEE Access* **2019**, *7*, 100943–100953. [CrossRef]
557. Petta, P.; Pelachaud, C.; Cowie, R. (Eds.) *Emotion-Oriented Systems: The Humaine Handbook*; Cognitive Technologies; Springer: Berlin, Germany; London, UK, 2011.
558. Busso, C.; Bulut, M.; Lee, C.-C.; Kazemzadeh, A.; Mower, E.; Kim, S.; Chang, J.N.; Lee, S.; Narayanan, S.S. IEMOCAP: Interactive Emotional Dyadic Motion Capture Database. *Lang Resour. Eval.* **2008**, *42*, 335–359. [CrossRef]
559. Douglas-Cowie, E.; Campbell, N.; Cowie, R.; Roach, P. Emotional Speech: Towards a New Generation of Databases. *Speech Commun.* **2003**, *40*, 33–60. [CrossRef]
560. McKeown, G.; Valstar, M.; Cowie, R.; Pantic, M.; Schroder, M. The SEMAINE Database: Annotated Multimodal Records of Emotionally Colored Conversations between a Person and a Limited Agent. *IEEE Trans. Affect. Comput.* **2012**, *3*, 5–17. [CrossRef]
561. Koelstra, S.; Muhl, C.; Soleymani, M.; Lee, J.-S.; Yazdani, A.; Ebrahimi, T.; Pun, T.; Nijholt, A.; Patras, I. DEAP: A Database for Emotion Analysis; Using Physiological Signals. *IEEE Trans. Affect. Comput.* **2012**, *3*, 18–31. [CrossRef]
562. Katsigiannis, S.; Ramzan, N. DREAMER: A Database for Emotion Recognition Through EEG and ECG Signals from Wireless Low-Cost Off-the-Shelf Devices. *IEEE J. Biomed. Health Inform.* **2018**, *22*, 98–107. [CrossRef]
563. GitHub. EEG-Datasets. Available online: https://github.com/meagmohit/EEG-Datasets (accessed on 17 August 2022).
564. Olivas, E.S.; Guerrero, J.D.M.; Martinez-Sober, M.; Magdalena-Benedito, J.R.; Serrano, L. *Handbook Of Research On Machine Learning Applications and Trends: Algorithms, Methods and Techniques*; IGI global: Hershey, PA, USA, 2009.
565. Haerpfer, C.; Inglehart, R.; Moreno, A.; Welzel, C.; Kizilova, K.; Diez-Medrano, J.; Lagos, M.; Norris, P.; Ponarin, E.; Puranen, B. World Values Survey Wave 7 (2017–2022) Cross-National Data-Set. 2022. Available online: https://www.worldvaluessurvey.org/WVSDocumentationWV7.jsp (accessed on 14 June 2022). [CrossRef]
566. Sýkorová, K.; Flegr, J. Faster Life History Strategy Manifests Itself by Lower Age at Menarche, Higher Sexual Desire, and Earlier Reproduction in People with Worse Health. *Sci. Rep.* **2021**, *11*, 11254. [CrossRef]
567. Wlezien, C. Patterns of Representation: Dynamics of Public Preferences and Policy. *J. Politics* **2004**, *66*, 1–24. [CrossRef]
568. Kelley, K.; Preacher, K.J. On effect size. *Psychol. Methods* **2012**, *17*, 137–152. [CrossRef] [PubMed]
569. Wilkinson, L. Task Force on Statistical Inference, American Psychological Association, Science Directorate. Statistical methods in psychology journals: Guidelines and explanations. *Am. Psychol.* **1999**, *54*, 594–604. [CrossRef]
570. Durlak, J.A. How to select, calculate, and interpret effect sizes. *J. Pediatric Psychol.* **2009**, *34*, 917–928. [CrossRef] [PubMed]
571. Courville, T.; Thompson, B. Use of structure coefficients in published multiple regression articles: β is not enough. *Educ. Psychol. Meas.* **2001**, *61*, 229–248. [CrossRef]
572. Johnson, J.W. A heuristic method for estimating the relative weight of predictor variables in multiple regression. *Multivar. Behav. Res.* **2000**, *35*, 1–19. [CrossRef]
573. Depuydt, C.E.; Jonckheere, J.; Berth, M.; Salembier, G.M.; Vereecken, A.J.; Bogers, J.J. Serial type-specific human papillomavirus (HPV) load measurement allows differentiation between regressing cervical lesions and serial virion productive transient infections. *Cancer Med.* **2015**, *4*, 1294–1302. [CrossRef]
574. Funder, D.C.; Ozer, D.J. Evaluating effect size in psychological research: Sense and nonsense. *Adv. Methods Pract. Psychol. Sci.* **2019**, *2*, 156–168. [CrossRef]
575. Pogrow, S. How effect size (practical significance) misleads clinical practice: The case for switching to practical benefit to assess applied research findings. *Am. Stat.* **2019**, *73*, 223–234. [CrossRef]
576. Tabassi, E.; Wilson, C. A novel approach to fingerprint image quality. In *International Conference on Image Processing, ICIP'05*, Genoa, Italy, 11–14 September 2005; IEEE: Genova, Italy, 2005; pp. 37–40.

577. El-Abed, M.; Giot, R.; Charrier, C.; Rosenberger, C. Evaluation of biometric systems: An svm-based quality index. In Proceedings of the Third Norsk Information Security Conference, NISK; 2010; pp. 57–68. Available online: https://hal.archives-ouvertes.fr/hal-00995094/ (accessed on 14 June 2022).
578. iSO 13407:1999. Human Centred Design Process for Interactive Systems. Available online: https://www.iso.org/obp/ui/#iso:std:iso:13407:ed-1:v1:en (accessed on 10 May 2022).
579. Giot, R.; El-Abed, M.; Rosenberger, C. Fast computation of the performance evaluation of biometric systems: Application to multibiometrics. *Future Gener. Comput. Syst.* **2013**, *29*, 788–799. [CrossRef]
580. Mansfield, A. ISO/IEC 19795-1:2006; Information technology–biometric performance testing and reporting–part 1: Principles and framework. 2006. Available online: https://www.iso.org/standard/41447.html (accessed on 14 June 2022).
581. *iSO/IEC FCD 19792*; Information Technology—Security Techniques—Security Evaluation of Biometrics. Available online: https://webstore.iec.ch/preview/info_isoiec19792%7Bed1.0%7Den.pdf (accessed on 10 May 2022).
582. Rane, S. Standardization of biometric template protection. *IEEE MultiMedia* **2014**, *21*, 94–99. [CrossRef]
583. Dube, A.; Singh, D.; Asthana, R.K.; Walia, G.S. A Framework for Evaluation of Biometric Based Authentication System. In Proceedings of the 2020 3rd International Conference on Intelligent Sustainable Systems, ICISS, Thoothukudi, India, 3–5 December 2020; IEEE: Thoothukudi, India, 2020; pp. 925–932.
584. Mannepalli, K.; Sastry, P.N.; Suman, M. FDBN: Design and development of Fractional Deep Belief Networks for speaker emotion recognition. *Int. J. Speech Technol.* **2016**, *19*, 779–790. [CrossRef]
585. Al-Shayea, Q.; Al-Ani, M. Biometric face recognition based on enhanced histogram approach. *Int. J. Commun. Netw. Inf. Secur.* **2018**, *10*, 148–154. [CrossRef]
586. Valiyavalappil Haridas, A.; Marimuthu, R.; Sivakumar, V.G.; Chakraborty, B. Emotion recognition of speech signal using Taylor series and deep belief network based classification. *Evol. Intell.* **2020**, *15*, 1145–1158. [CrossRef]
587. Arora, M.; Kumar, M. AutoFER: PCA and PSO based automatic facial emotion recognition. *Multimed. Tools Appl.* **2021**, *80*, 3039–3049. [CrossRef]
588. Karmarkar, U.R.; Plassmann, H. Consumer neuroscience: Past, present, and future. *Organ. Res. Methods* **2019**, *22*, 174–195. [CrossRef]
589. Seitamaa-Hakkarainen, P.; Huotilainen, M.; Mäkelä, M.; Groth, C.; Hakkarainen, K. The Promise of Cognitive Neuroscience in Design Studies. Available online: https://dl.designresearchsociety.org/drs-conference-papers/drs2014/researchpapers/62 (accessed on 14 June 2022).
590. Su, F.; Xia, L.; Cai, A.; Ma, J. A dual-biometric-modality identification system based on fingerprint and EEG. In Proceedings of the IEEE 4th International Conference on Biometrics Theory, Applications and Systems, BTAS, Washington, DC, USA, 27–29 September 2010; IEEE: Washington, DC, USA, 2010; pp. 3–8.
591. Pal, S.; Mitra, M. Increasing the accuracy of ECG based biometric analysis by data modelling. *Measurement* **2012**, *45*, 1927–1932. [CrossRef]
592. Singh, Y.N.; Singh, S.K.; Gupta, P. Fusion of electrocardiogram with unobtrusive biometrics: An efficient individual authentication system. *Pattern Recognit. Lett.* **2012**, *33*, 1932–1941. [CrossRef]
593. Lourenço, A.; Silva, H.; Fred, A. Unveiling the biometric potential of finger-based ECG signals. *Comput. Intell. Neurosci.* **2011**, *2011*, 1–8. [CrossRef]
594. Wahabi, S.; Member, S.; Pouryayevali, S.; Member, S. On evaluating ECG biometric systems: Session-dependence and body posture. *IEEE Trans. Inf. Forensics Secur.* **2014**, *9*, 2002–2013. [CrossRef]
595. Havenetidis, K. Encryption and Biometrics: Context, methodologies and perspectives of biological data. *J. Appl. Math. Bioinform.* **2013**, *3*, 141.
596. Sanjeeva Reddy, M.; Narasimha, B.; Suresh, E.; Subba Rao, K. Analysis of EOG signals using wavelet transform for detecting eye blinks. In Proceedings of the 2010 International Conference on Wireless Communications & Signal Processing, WCSP 2010, Suzhou, China, 21–23 October 2010; pp. 1–3.
597. Punsawad, Y.; Wongsawat, Y.; Parnichkun, M. Hybrid EEG-EOG brain-computer interface system for practical machine control. In Proceedings of the 2010 Annual International Conference of the IEEE Engineering in Medicine Biology Society, EMBC 2010, Buenos Aires, Argentina, 31 August–4 September 2010; pp. 1360–1363.
598. Zapata, J.C.; Duque, C.M.; Rojas-Idarraga, Y.; Gonzalez, M.E.; Guzmán, J.A.; Botero, B. Data fusion applied to biometric identification–A review. In *Colombian Conference on Computing*; Springer: Cham, Switzerland, 2017; pp. 721–733.
599. Gutu, D. A Study of Facial Electromyography for Improving Image Quality Assessment. Ph.D. Thesis, University of Toyama, Toyama, Japan, 2015.
600. Jain, A.K.; Ross, A.; Prabhakar, S. An introduction to biometric recognition. *IEEE Transactions on circuits and systems for video technology* **2004**, *14*(1), 4–20. [CrossRef]
601. National Research Council. *Biometric Recognition: Challenges and Opportunities*; The National Academies Press: Washington, DC, USA, 2010; p. 182. [CrossRef]
602. Bhatia, R. Biometrics and face recognition techniques. *Int. J. Adv. Res. Comput. Sci. Softw. Eng.* **2013**, *3*, 93–99.
603. Sabhanayagam, T.; Venkatesan, V.P.; Senthamaraikannan, K. A comprehensive survey on various biometric systems. *Int. J. Appl. Eng. Res.* **2018**, *13*, 2276–2297.

604. Delac, K.; Grgic, M. A survey of biometric recognition methods. In *Proceedings Elmar-200, 46th International Symposium on Electronics in Marine, Zadar, Croatia, 16–18 June 2004*; IEEE: Zadar, Croatia, 2004; pp. 184–193.
605. Kataria, A.N.; Adhyaru, D.M.; Sharma, A.K.; Zaveri, T.H. A survey of automated biometric authentication techniques. In Proceedings of the 2013 Nirma University International Conference on Engineering (NUiCONE), Ahmedabad, India, 28–30 November 2013; IEEE: Ahmedabad, India, 2013; pp. 1–6.
606. Khairwa, A.; Abhishek, K.; Prakash, S.; Pratap, T. A comprehensive study of various biometric identification techniques. In Proceedings of the 2012 Third International Conference on Computing, Communication and Networking Technologies (ICCCNT'12), Coimbatore, India, 26–28 July 2012; IEEE: Coimbatore, India, 2012; pp. 1–6.
607. Ooms, K.; Dupont, L.; Lapon, L.; Popelka, S. Accuracy and precision of fixation locations recorded with the Low-cost Eye Tribe tracker in different experimental setups. *J. Eye Mov. Res.* **2015**, *8*, 1–24. [CrossRef]
608. Lopez-Basterretxea, A.; Mendez-Zorrilla, A.; Garcia-Zapirain, B. Eye/head tracking technology to improve HCI with iPad applications. *Sensors* **2015**, *15*, 2244–2264. [CrossRef]
609. Harinda, E.; Ntagwirumugara, E. Security & privacy implications in the placement of biometric-based ID card for Rwanda Universities. *J. Inf. Secur.* **2015**, *6*, 93. [CrossRef]
610. Ibrahim, D.R.; Tamimi, A.A.; Abdalla, A.M. Performance analysis of biometric recognition modalities. In Proceedings of the 2017 8th International Conference on Information Technology (ICIT); IEEE: Amman, Jordan, 2017; pp. 980–984. Available online: https://ieeexplore.ieee.org/stamp/stamp.jsp?tp=&arnumber=8079977 (accessed on 14 June 2022).
611. Vats, S.; Kaur, H. A Comparative Study of Different Biometric Features. *Int. J. Adv. Res. Comput. Sci.* **2016**, *7*, 30–35. [CrossRef]
612. Yu, F.X.; Suo, Y.N. Application of gesture recognition based on the somatosensory kinect sensor in human-computer interaction framework. *Rev. Fac. Ing.* **2017**, *32*, 580–585.
613. Meitram, R.; Choudhary, P. Palm vein recognition based on 2D Gabor filter and artificial neural network. *J. Adv. Inf. Technol.* **2018**, *9*, 68–72. [CrossRef]
614. Ahmed, A.A.E.; Traore, I. A new biometric technology based on mouse dynamics. *IEEE Trans. Dependable Secur. Comput.* **2007**, *4*, 165–179. [CrossRef]
615. Trewin, S.; Swart, C.; Koved, L.; Martino, J.; Singh, K.; Ben-David, S. Biometric authentication on a mobile device: A study of user effort, error and task disruption. In Proceedings of the 28th Annual Computer Security Applications Conference, ACSAC, New York, NY, USA, 3–7 December 2012; pp. 159–168. [CrossRef]
616. Haghighat, M.; Abdel-Mottaleb, M.; Alhalabi, W. Discriminant Correlation Analysis: Real-Time Feature Level Fusion for Multimodal Biometric Recognition. *IEEE Trans. Inf. Forensics Security. Wash. Bus. J.* **2016**, *11*, 1984–1996. [CrossRef]
617. Flook, B. This is the 'biometric war' Michael Saylor was talking about. *Wash. Bus. J.* **2013**, *9*, 91–98.
618. Islam, M. Feature and score fusion based multiple classifier selection for iris recognition. *Comput. Intell. Neurosci.* **2014**, *2014*, 380585. [CrossRef]
619. De Leersnyder, J.; Mesquita, B.; Kim, H.S. Where Do My Emotions Belong? A Study of Immigrants' Emotional Acculturation. *Pers. Soc. Psychol. Bull.* **2011**, *37*, 451–463. [CrossRef] [PubMed]
620. Vuong, Q.H.; Napier, N.K. Acculturation and Global Mindsponge: An Emerging Market Perspective. *Int. J. Intercult. Relat.* **2015**, *49*, 354–367. [CrossRef]
621. Vuong, Q.-H. Global Mindset as the Integration of Emerging Socio-Cultural Values through Mindsponge Processes: A Transition Economy Perspective. In *Global Mindsets: Exploration and Perspectives*; Routledge: London, UK, 2016; pp. 109–126.
622. Vuong, Q.-H.; Bui, Q.-K.; La, V.-P.; Vuong, T.-T.; Nguyen, V.-H.T.; Ho, M.-T.; Nguyen, H.-K.T.; Ho, M.-T. Cultural Additivity: Behavioural Insights from the Interaction of Confucianism, Buddhism and Taoism in Folktales. *Palgrave Commun.* **2018**, *4*, 143. [CrossRef]
623. Vuong, Q.-H.; Ho, M.-T.; Nguyen, H.-K.T.; Vuong, T.-T.; Tran, T.; Hoang, K.-L.; Vu, T.-H.; Hoang, P.-H.; Nguyen, M.-H.; Ho, M.-T.; et al. On How Religions Could Accidentally Incite Lies and Violence: Folktales as a Cultural Transmitter. *Palgrave Commun.* **2020**, *6*, 82. [CrossRef]
624. Ho, M.-T.; Mantello, P.; Nguyen, H.-K.T.; Vuong, Q.-H. Affective Computing Scholarship and the Rise of China: A View from 25 Years of Bibliometric Data. *Hum. Soc. Sci. Commun.* **2021**, *8*, 282. [CrossRef]
625. FaceReader. Reference Manual Version 7. Tool for Automatic Analysis of Facial Expressions. Available online: http://sslab.nwpu.edu.cn/uploads/1500604789-971697563f64.pdf (accessed on 2 March 2022).
626. Kaklauskas, A.; Abraham, A.; Milevicius, V. Diurnal Emotions, Valence and the Coronavirus Lockdown Analysis in Public Spaces. *Eng. Appl. Artif. Intell.* **2021**, *98*, 104122. [CrossRef]
627. Sun, Z.; Li, Q.; Liu, Y.; Zhu, Y. Opportunities and Challenges for Biometrics. *China's E-Sci. Blue Book* **2021**, 101–125.
628. Albuquerque, V.H.C.D.; Damaševičius, R.; Tavares, J.M.R.; Pinheiro, P.R. EEG-based biometrics: Challenges and applications. *Comput. Intell. Neurosci.* **2018**, *2018*, 5483921. [CrossRef]
629. Fierrez, J.; Morales, A.; Vera-Rodriguez, R.; Camacho, D. Multiple classifiers in biometrics. Part 2: Trends and challenges. *Inf. Fusion* **2018**, *44*, 103–112. [CrossRef]
630. Sivaraman, S. Top 10 Trending Biometric Technology for 2022. Available online: https://blog.mantratec.com/Top-10-trending-Biometric-technology-for-2022 (accessed on 2 March 2022).

 *sensors*

Article

Multispectral Facial Recognition in the Wild

Pedro Martins [1,2], José Silvestre Silva [1,3,4,*] and Alexandre Bernardino [2,5]

1. Military Electrical and Computer Engineering, Portuguese Military Academy, Rua Gomes Freire, 1169-203 Lisbon, Portugal; pedro.roque.martins@tecnico.ulisboa.pt
2. Instituto Superior Técnico, Universidade de Lisboa, 1049-001 Lisbon, Portugal; alex@isr.tecnico.ulisboa.pt
3. Military Academy Research Center (CINAMIL), Rua Gomes Freire, 1169-203 Lisbon, Portugal
4. Laboratory for Instrumentation, Biomedical Engineering and Radiation Physics (LIBPhys-UC), 3000-370 Coimbra, Portugal
5. Institute for Systems and Robotics (ISR), 1049-001 Lisbon, Portugal
* Correspondence: jose.silva@academiamilitar.pt

Abstract: This work proposes a multi-spectral face recognition system in an uncontrolled environment, aiming to identify or authenticate identities (people) through their facial images. Face recognition systems in uncontrolled environments have shown impressive performance improvements over recent decades. However, most are limited to the use of a single spectral band in the visible spectrum. The use of multi-spectral images makes it possible to collect information that is not obtainable in the visible spectrum when certain occlusions exist (e.g., fog or plastic materials) and in low- or no-light environments. The proposed work uses the scores obtained by face recognition systems in different spectral bands to make a joint final decision in identification. The evaluation of different methods for each of the components of a face recognition system allowed the most suitable ones for a multi-spectral face recognition system in an uncontrolled environment to be selected. The experimental results, expressed in Rank-1 scores, were 99.5% and 99.6% in the TUFTS multi-spectral database with pose variation and expression variation, respectively, and 100.0% in the CASIA NIR-VIS 2.0 database, indicating that the use of multi-spectral images in an uncontrolled environment is advantageous when compared with the use of single spectral band images.

Keywords: deep neural networks; multispectral imaging; face recognition; in the wild

1. Introduction

The sense of sight allows us to observe dangers, identify objects, and recognize people. This last task is fundamental for human beings as social beings. It enables us to differentiate the level of trust someone can give to a specific person, with this being at the base of the construction of communities. Such is the importance of this task that it has become one of the main topics of research with the emergence of machine learning, thus allowing machines to incorporate this biological capacity.

Multi-spectral images have several military applications, from detection of camouflaged people [1], classification of vegetation types in military regions [2], landmine detection [3] and face recognition [4]. The current face recognition systems operating in the visible (VIS) domain have reached a significant level of maturity. It is possible to observe their wide use nowadays, from security mechanisms to unlocking electronic devices such as smartphones and personal computers to population control systems [5].

However, most current face recognition systems [6] require the cooperation of the user to ensure that pictures are taken in favorable conditions (frontal postures, good illumination, no occlusion) and have trouble dealing with uncontrolled scenarios. Uncontrolled environment scenarios, such as riots and violent demonstrations, can often be used by criminals and terrorist cell members to move around and cause damage to Homeland Security, as this type of environment adds difficulty to their detection. The uncontrolled

Citation: Martins, P.; Silva, J.S.; Bernardino, A. Multispectral Facial Recognition in the Wild. *Sensors* **2022**, *22*, 4219. https://doi.org/10.3390/s22114219

Academic Editor: Mincheol Whang

Received: 5 April 2022
Accepted: 30 May 2022
Published: 1 June 2022

Publisher's Note: MDPI stays neutral with regard to jurisdictional claims in published maps and institutional affiliations.

Copyright: © 2022 by the authors. Licensee MDPI, Basel, Switzerland. This article is an open access article distributed under the terms and conditions of the Creative Commons Attribution (CC BY) license (https://creativecommons.org/licenses/by/4.0/).

environment is mainly characterized by a variety of lighting, pose, facial expressions and the existence of occlusions [5]. These features are challenges to face recognition systems due to the multiple intrapersonal variations they provide, making it difficult to correctly identify an individual's identity based on a collaborative image of the individual.

This work has as its main objective the development of a multi-spectral face recognition system in an uncontrolled environment. To achieve this goal, the solutions used by current recognition systems and the evaluation of the benefits of using multi-spectral images are explored. The developed face recognition system is evaluated in public multi-spectral image datasets with pose and expression variability.

This paper is organized into six sections. The Introduction section describes the motivation for the work, the objectives and the structure of the paper. The Background section explains important concepts, such as how a face recognition system works, what multi-spectral images are and what their advantages are. The Related Work section presents the study of the art of multispectral face recognition methods in an uncontrolled environment and of public multispectral databases. The Methodology section defines the proposed method in order to achieve the objectives. The Results and Discussion section describes the multispectral databases used, several experiments are also performed with the various modules proposed in the methodology, each experiment is accompanied by its respective analysis and discussion. The Conclusions section presents the conclusions of this work, thus consolidating the proposed objectives.

2. Background

2.1. Face Recognition

In general, a face recognition system is described in several phases. The first phase consists of acquiring the facial images and pre-processing them, such as locating the faces and cropping them. In a second phase, the features are extracted from the facial image, for instance, the position of facial landmarks, eye distance or even the face tones. Finally, these features are used in a classifier for identification or verification purposes.

Face recognition can be performed in a controlled or uncontrolled environment. The controlled environment, also known as consent recognition, is one in which the user cooperates in the recognition by facilitating it through correct and static posture in a place with good lighting. In the uncontrolled environment, recognition is dynamic, without the user cooperating in acquiring an image, making the face recognition process very difficult due to the diversity of the surrounding environment (e.g., low visibility), facial poses and expressions.

2.2. Multispectral Imaging in an Uncontrolled Environment

The databases of the VIS domain and the use of image synthesizers, which generate multiple poses and facial expressions from the obtained images, have allowed the difficulties associated with the variety of poses and facial expressions to be circumvented. However, two points have proved more difficult to overcome: the change of illumination and occlusions. This has led to the use of multiple spectral bands, with particular emphasis on the infrared (IR) spectral band, which can acquire images in environments with little or no brightness and overcome occlusions such as smoke and fog. In short, multispectral analysis allows a face recognition system to extract facial features that would be impossible to obtain with images from the VIS spectral band.

The IR bands can be categorized according to several spectral bands [7]. The active bands are the near-infrared (NIR) and short-wavelength infrared (SWIR). To acquire images in these bands, the object must receive illumination, even if scarce, because it is through reflection that the image is acquired. Such a fact means these images are commonly used in night vision devices. The NIR band allows the difficulties posed by the variation of illumination to be overcome, while the SWIR has the advantage of obtaining images through smoke and fog. The passive bands are the mid-wavelength infrared (MWIR) and long-wavelength infrared (LWIR). Unlike the active bands, the passive bands allow us to acquire images using only the thermal radiation emitted by a body, commonly known as thermal images.

The use of IR images for automatic face recognition is not without challenges, as these images are sensitive to the emotional, physical and health conditions of the individual, as well as the surroundings, and do not serve as an absolute alternative to the use of the VIS spectrum, but rather as a complement [8]. Another difficulty arises from the low number of public databases with images from both spectral ranges and in an uncontrolled environment [9], which limit the creation of rich classification models and the ability to characterize the performance of those systems in realistic conditions.

3. Related Work

Multi-spectral face recognition in an uncontrolled environment can be subdivided into two areas. The first is face recognition in an uncontrolled environment, which is already challenging. The second is multi-spectral face recognition, i.e., using different spectral bands in face recognition. This section briefly reviews the progress made in these two areas.

3.1. Face Recognition in an Uncontrolled Environment

The uncontrolled environment, strongly characterized by pose-light-expression factors, emerges as a problem for current recognition systems. A significant step was taken towards solving this type of problem by introducing very large databases to train Deep Convolutional Neural Networks (DCNN) in combination with the emergence of image synthesis methods [5]. The two main image synthesis methods are: (i) one-to-many augmentation, which consists of generating different poses of a face from a canonical face image; (ii) many-to-one normalization, which consists of normalizing any pose of the face to a canonical face pose [5]. The use of Generative Adversarial Networks (GAN), introduced by Goodfellow et al. [10], is characterized by the use of a generator and a discriminator (see Figure 1). The generator is responsible for producing samples given an input image so that the discriminator cannot discern which of the samples is real and which is false.

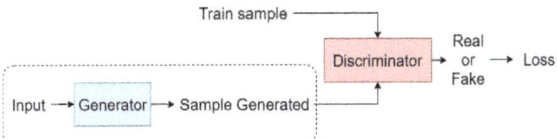

Figure 1. Schematic of the training of a GAN. The dashed line shows the process of sample generation.

Since their appearance in face normalization, with DR-GAN [11], GANs have taken the lead in solving the problem of pose and facial expression variation. As for one-to-many augmentation using GANs, as is the case with the DA-GAN network [12], their image production power also gives them an advantage compared to other algorithms.

Normalization of many-to-one images is an extreme image synthesis problem due to the pose differences of a face. Cao et al. [13] proposed HF-PIM, normalizing the face to a frontal pose through a texture fusion deformation procedure leveraging a dense matching field to interconnect the 2D and 3D surface spaces. Qian et al. [14] presented Face Normalization Module (FNM), which encodes images using a pre-trained network for feature extraction and generates realistic images.

One-to-many augmentation is another approach to achieve face recognition regardless of the pose. Tran et al. [15] synthesized different poses through 3D modeling and then trained a DCNN to perform face recognition with varied poses. The DA-GAN proposed by Zhao et al. [12] created 2D images through 3D modeling and then refined the obtained 2D images to be as realistic as possible, using a GAN to try to preserve the identity of the face. Thus, the DA-GAN network was also used to augment the training data.

3.2. Multispectral Face Recognition

The main multi-spectral face recognition methods can be characterized by three important features: Image Synthesis Methods, Fusion Methods and Loss Functions.

Fusion methods are subdivided into feature fusion and score fusion. In the first, a fusion of features from the different spectral bands of the facial image is performed, allowing the most relevant features to be extracted from the different bands and joining them in a vector. The second method combines the scores obtained from each classifier uni-band (e.g., a classifier operating only in the LWIR band and another operating only in the NIR band) [16].

The image synthesis methods allow an image of a spectral band to be transformed into another, helping to compare two images. The main advantage of image synthesis is that it enables an image to be passed from any spectral band to the VIS band, making it possible to use classifiers implemented to process images of the VIS spectrum [17]. One of the most recent works in this area synthesizes VIS images from NIR images using GANs [18].

Finally, all neural networks have cost functions for the training moment to update the network weights. However, certain cost functions have been proposed to proceed specifically to the classification of multi-spectral images. Examples of these cost functions are the Scatter Loss [19] and the Wasserstein Distance [20].

3.3. Gaps

Although several scientific works address multi-spectral face recognition, few of these demonstrate its power in an uncontrolled environment due to the limitations in current databases of multi-spectral face images. In existing datasets, the variations of conditions are not extreme, as they are usually semi-controlled environments and not *in the wild* (uncontrolled environment). For example, the most studied database in multi-spectral face recognition, CASIA NIR-VIS 2.0 [21], uses images in which the pose has few deviations from the frontal position, which does not reliably characterize the uncontrolled environment. Thus, the fact that these databases are incomplete (compared to those of the VIS band) is still a barrier to improving the capability of multi-spectral face recognition systems in an uncontrolled environment.

The present work proposes a system that integrates the capabilities of current face recognition systems in an uncontrolled environment in the VIS spectrum at the pose variation level and the capabilities of multi-spectral face recognition systems to surpass illumination variation.

4. Methodology

The proposed multi-spectral face recognition system consists of three tasks: Face Detection and Alignment, Image Synthesis and Face Recognition. In Figure 2, the general operation of the proposed face recognition system is shown, including the steps performed in each task.

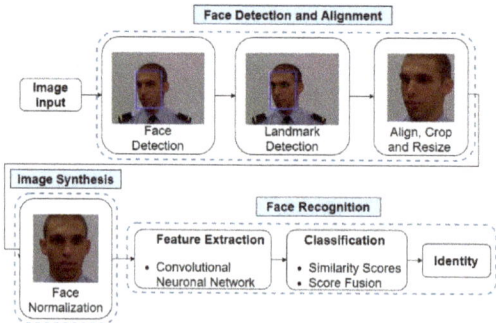

Figure 2. Schematic of the operation of the proposed face recognition system.

In the initial phase of the system, it is necessary to acquire multi-spectral images, which can be obtained through mono-spectral equipment (collects the image in only one spectral band) or multi-spectral (collects the image in different spectral bands). After image acquisition, the Face Detection and Alignment module aims to obtain an aligned and centered facial image with predefined dimensions. To achieve this goal, it is necessary to detect the presence of human faces and then perform a face marking, detecting essential landmarks of the face, such as eyes and nose, allowing a correct alignment of the face and clipping around it. The following task is Image Synthesis, which aims to obtain a frontal facial image. The next task is Face Recognition, where facial image features are extracted through a CNN and a one-shot learning methodology is followed for the classification task, obtaining similarity scores for each spectral band. These scores are combined using a score fusion method, and the predicted identity is the one with the highest combined score.

4.1. Face Detection and Alignment

Face detection, in conjunction with face alignment, aims to detect the faces presented in the input image and identify facial landmarks so that faces are centered, aligned and equally sized. Since face detection algorithms detect faces in rectangular areas without rotating the image, a face landmark detection algorithm is needed to apply a rotation so that the face is aligned on the horizontal plan, using the imaginary eye line. Thus, the procedure of face detection and alignment module (see Figure 3) does the following: is given an image, identifies the different faces present, extracts the facial landmarks and processes the image to produce facial images where the face is centered and aligned.

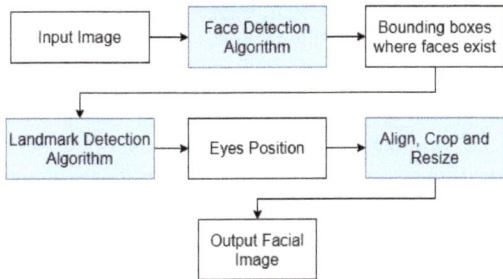

Figure 3. Flowchart of the steps of a facial detection and alignment module.

The face detection algorithms explored in this work are based on SSD (single-shot multibox detector), a deep learning architecture for object detection [22]. The basic idea of the SSD is to generate scores for the presence of each object category in each predefined box and produce adjustments to the box to match the shape of the object. In this work, three SSD based methods are tested: (i) the S3FD algorithm [23], (ii) the facial detection deep neural network of OpenCV [24] and (iii) the DSFD algorithm [25]. The S3FD has contributions to better cope with scaling variations with a single deep network. The DSFD uses a feature enhancement module to extend the single-shot detector to a dual-shot detector, obtaining more robust and discriminable features.

As for the facial landmark detection algorithms, the DLIB library's 68 landmark network, adapted from Khazemi and Sullivan [26], and Bulat's 2D-FAN [27], also with 68 landmarks, were tested. The latter one uses an Hour-Glass [28] based architecture to estimate the human pose. Both networks receive an image of a person and produce, as output, the position of the different facial landmarks around the face.

All the algorithms addressed in this subsection were trained in databases that only contain images in the spectral band of the VIS. To achieve data normalization, it is necessary to (i) rotate the image to align the eye line with the horizontal, (ii) crop the image to center the face image, and (iii) resize the image so that all output images have the specified dimensions.

4.2. Image Synthesis

To overcome the problems related with image acquisition in an uncontrolled environment, such as variation in lighting, occlusions and changes of poses, a face normalization module is used. This module aims to synthesize (create) an image of a face with frontal pose and neutral expression from a non-frontal face image.

To exemplify the expected behavior, Figure 4 shows an input face image in a non-frontal pose, with which the image synthesis module produces a frontal face image. Thus, it is intended that the image acquired helps obtain the identity features present in the facial image. The models FNM [14] and FFWM [29] are analyzed.

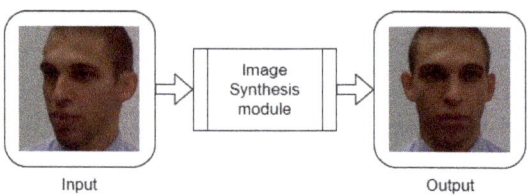

Figure 4. Input and output of the Image Synthesis module (intended function, not the result of a real experiment).

FNM is a GAN with two new features. First, it uses a network specialized in obtaining facial features to build the generator and provide the ability to preserve facial identity. Second, facial discriminators are used to refine local textures. Their authors claim that this model produces a face in the canonical pose without expression, which directly improves the performance of a face recognition system.

The normalization method of the FFWM model consists of using a deformation module, aiming to synthesize realistic frontal images with illumination preservation. For frontal image synthesis, it presents a module responsible for reducing pose discrepancy at the facial features level, thus preserving more details of profile images. The FFWM model uses pairs of face images for the training phase: one with a non-frontal pose and another with a frontal pose of the same person in the same conditions. Differently, the FNM model uses non-pair face images, where the images are not of the same person.

4.3. Face Recognition

This last module aims to identify the person present in an input face image, following the flowchart presented in Figure 5. For this purpose, it is necessary to perform two tasks: feature extraction and classification.

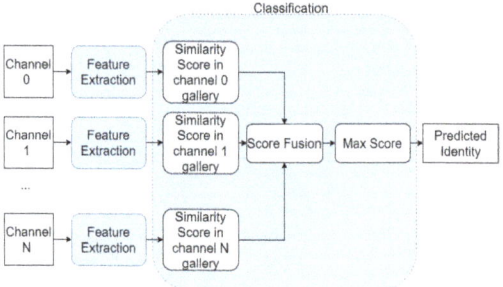

Figure 5. Schematic of the Face Recognition Module.

The extraction of representative features from a facial image is performed through a version of Light CNN [30] with 29 convolutional layers (Light CNN-29). To use this network for feature extraction in spectra other than VIS, transfer learning is used. According to [31], several models for biometric recognition are based on transfer learning when the databases are limited. Thus, one should use the Light CNN-29 model with the weights obtained by training on the VIS databases and fine tune with the facial image databases in spectra other than the VIS. At the end of the feature extraction phase, B vectors of 256 dimensions are generated, with B being the number of spectral bands in which the facial image was acquired.

The classification process applied by the one-shot learning technique determines the degree of similarity of the feature set extracted from the input image with the features sets extracted from the images of each class present in the support set, which is constituted by one example per class. The similarity functions to be used are the Euclidean distance and the cosine similarity. After obtaining the similarity values for each identity in the different spectral bands, a fusion of the obtained scores is performed, inspired by [27]:

$$S_{ic} = \sum_{b=1}^{B} S_{ib} W_b \tag{1}$$

where S_{ic} is the combined score for each identity i and S_{ib} is the score obtained for each band b for each identity i. W_b is the weight of each spectral band. The weights associated with each band are fixed, determined by the accuracy obtained when classifying with only that band. In this way, the band that usually obtains the most reliable similarity scores to classify will have a greater weight in the fusion of scores. The prediction is then made by choosing the identity i of the support set that has the highest combined similarity score:

$$prediction = max(S_{ic}) \; \forall i \in [1,\ldots,N] \tag{2}$$

5. Results and Discussion

5.1. Databases

We performed both qualitative and quantitative evaluations of the proposed methods. These images are in the VIS, NIR and LWIR bands. Two multi-spectral databases were used for quantitative evaluation: TUFTS [9] and CASIA NIR-VIS 2.0 [19]. The TUFTS database has facial images in the VIS, NIR and LWIR bands of 113 people with different poses and different illumination conditions. The TUFTS database has different subsets, divided into TUFTS-Pose (facial images with nine different poses per individual, in visible, NIR and LWIR) and TUFTS-Exp (four facial images with different expressions and one with sunglasses per individual, in visible and LWIR) to study pose variation and expression variation separately. CASIA NIR VIS 2.0 comprises 17,489 facial images of 715 people in VIS and NIR spectral bands under different light conditions.

5.2. Metrics

The metrics used are Rank-1, Rank-5 and TAR@FAR = 0.001. When using a generic expression Rank-n, given an image of a face as input, the classifier obtains the n most probable identities, one of which is the correct identity. TAR (true accept rate) is defined as the percentage of faces that, compared to the corresponding gallery identity, are identified as matches, while FAR (false accept rate) is the percentage of incorrect identities to which a face is matched.

5.3. Face Detection and Alignment

5.3.1. Face Detection

Regarding the qualitative results presented in Figure 6, all algorithms produced similar results in the VIS band. This was expected since they were all trained in databases of the spectral band of the VIS. In the LWIR spectral band, a failure of the OpenCV network was observed in the second facial pose, where it cannot detect any face. In addition, when OpenCV and S3FD detect the faces, there is a variation in the rectangle area compared to the VIS spectral band. The DSFD maintained the same results, which is a good indicator of its ability to extract characteristics even in the LWIR spectral band.

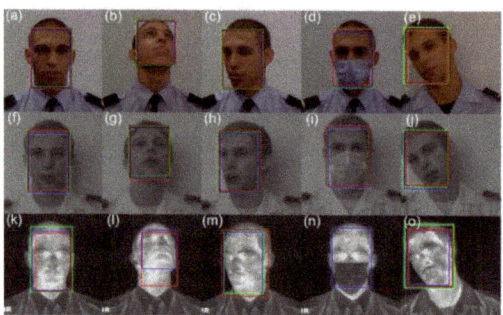

Figure 6. Results obtained by facial detection methods in the spectral bands of VIS (**a–e**), NIR (**f–j**) and LWIR (**k–o**). S3FD—red, DSFD—blue, OpenCV—green.

The quantitative results are presented in Table 1. It can be observed that the OpenCV network results are lower than the others, especially in infrared bands. Comparing results between the S3FD network and the DSFD, it is observed very similar results in the spectral band of the VIS and NIR. However, the results in LWIR are about 8 percentage points better. We observe that the DSFD maintains a very high accuracy for the different spectral bands, thus being the best network for face detection in a multispectral facial analysis system.

Table 1. Accuracy of the different face detection algorithms in the TUFTS database.

Method	Accuracy at Different Spectral Bands (%)		
	VIS	NIR	LWIR
OpenCV	99.2	90.4	77.7
S3FD	99.9	100.0	90.8
DSFD	99.9	100.0	98.8

5.3.2. Landmark Detection and Facial Alignment

The results for face landmark detection are shown in Figures 7 and 8. For the more challenging poses, we can see that the DLIB network fails, even in the VIS band (right eye, in Figure 7c), as it tends to maintain the shape of a near-frontal face. One possible cause of this behavior is that the face landmark detection model was trained in a dataset without significant variations at the pose level. The DLIB network reveals even more difficulties in the spectral band of LWIR.

2D-FAN reveals a good extraction of landmarks in any of the poses, including the LWIR band, where the results are somewhat like those obtained in the VIS band (Figure 8). In the case of Figure 8n, although it looks like there was a total failure, it is possible to observe that the eyes are correctly identified. 2D-FAN, unlike DLIB, was trained on a database with pronounced pose variations (including profile images), which is the justification for achieving better results.

Given the previous considerations, we decided to use the 2D-FAN over the DLIB's network due to two factors: (i) it shows better results with face pose variation, and (ii) it is the only one capable of producing positive results in the LWIR spectral band. After the face detection with DSFD and landmark face detection with 2D-FAN, the align, crop and resize phase took place, which aligned the imaginary eye line of all detected faces with the horizontal, centered the faces in the images, cropped them and resized to the same size, resulting in the results presented in Figure 9. The alignment effect is strongly noticeable on the rightmost facial image. This normalization of the facial images can help a multispectral face recognition system in an uncontrolled, where faces can be presented in several poses.

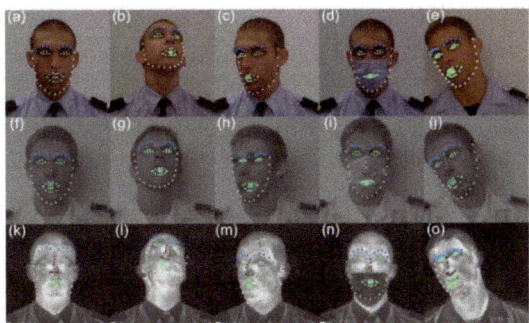

Figure 7. Results achieved by DLIB in the spectral bands of VIS (**a–e**), NIR (**f–j**) and LWIR (**k–o**). Yellow—jawline, green—eyes and mouth, purple—nose, blue—eyebrows.

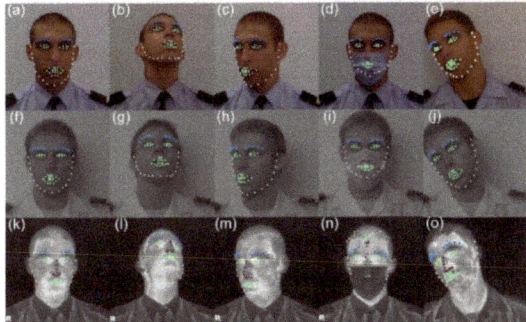

Figure 8. Results achieved by 2D-FAN in the spectral bands of VIS (**a–e**), NIR (**f–j**) and LWIR (**k–o**). Yellow—jawline, green—eyes and mouth, purple—nose, blue—eyebrows.

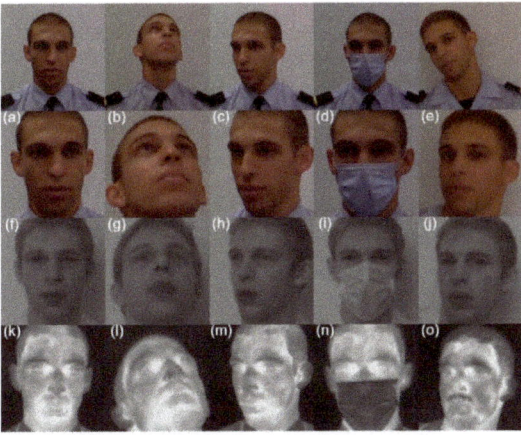

Figure 9. Results achieved by the proposed facial detection and alignment module in the different spectral bands. The images on the top are the originals in the VIS, before processing. Remain images correspond to facial alignment and detection in the spectral bands of VIS (**a–e**), NIR (**f–j**) and LWIR (**k–o**).

5.4. Image Synthesis

For all images used in the qualitative and quantitative evaluations, the images were previously processed to be properly centered, aligned and scaled. The FFWM model needs to receive the facial images with certain facial landmarks always in the same coordinates. Therefore, the face detection and alignment module provided by the authors of FFWM was used to obtain the results. The images used by the FNM model were processed by the face detection and alignment module developed by the authors of this work. The rightmost images used in the previous tasks were replaced by ones with a strong expression, to evaluate the capacity of the models to normalize expressions.

5.4.1. Selecting the Best Model

In Figure 10, the results obtained by the FFWM are shown. One of the images of the dataset could not be detected by the module provided by the authors of FFWM (see Figure 10n). It is possible to see that the performance of FFWM has a sharp drop as it moves away from the VIS band. Analyzing only the spectral band of the VIS and the images with pose variation (Figure 10b,c), a suitable normalization of the pose in Figure 10c is present.

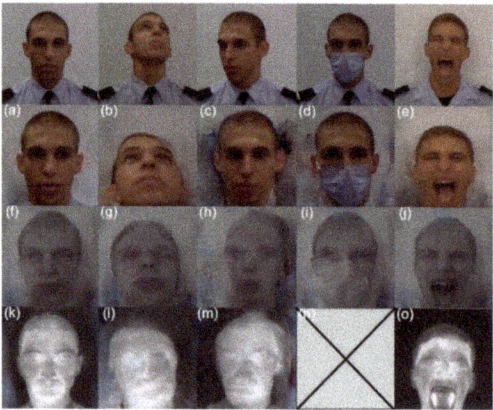

Figure 10. Results achieved by the FFWM in the different spectral bands. The images on the top are the originals in the VIS. The images (**a–e**), (**f–j**), and (**k–o**) were generated by the proposed methodology when it receives as input the images from the VIS, NIR and LWIR bands, respectively.

In Figure 10b, the FFWM produces a deformed face when the person looks upwards. The exclusive use of the Multi-PIE database [32] in training the FFWM means that it can only normalize the face where the pose varies along the horizontal plane.

The FNM presents more satisfactory results (see Figure 11) in the NIR spectral band, where the facial images are more realistic than those of the FFWM. It should be noted that with the FNM model, identities change, i.e., the person in the output face image appears to be different from the person in the input face image. However, the use of a face feature extractor by the FNM model allows the most relevant features in the output face image to be kept. It is also relevant to point out that the FNM normalizes pose and expression, eliminates face masks, as is the case of the surgical mask, and normalizes to the VIS spectral band. However, this normalization does not produce realistic results with the LWIR images due to the difference between the LWIR and VIS spectral bands.

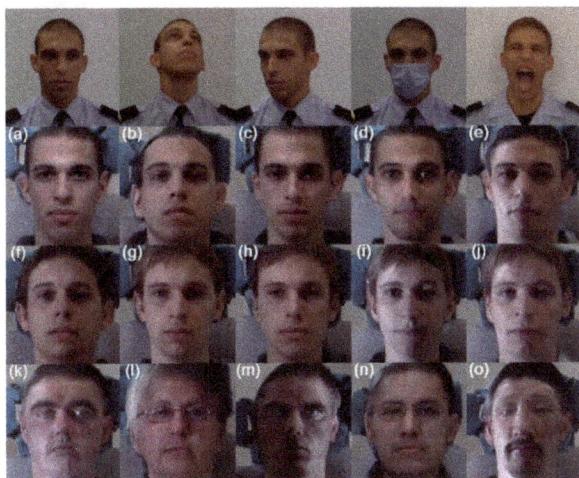

Figure 11. Results achieved by the FNM in the different spectral bands. The images on the top are the originals in the VIS. The images (**a–e**), (**f–j**), and (**k–o**) were generated by the proposed methodology when it receives as input the images from the VIS, NIR and LWIR bands, respectively.

Given the previous considerations, we decided to use the FNM instead of the FFWM due to two factors: (i) the FFWM requires a specific face detection and alignment module and that the face is perpendicular to the horizontal, while the FNM is more robust to pose variations in the input image; (ii) all images normalized by the FNM tend to maintain the face proportions, without deforming them, in the NIR and VIS spectral bands.

5.4.2. Evaluation of Selected Model

Identification with and without the use of FNM was performed to verify its advantage. For this purpose, the Light CNN-29 was used for feature extraction, and the identification was performed based on the score obtained by cosine similarity.

The results presented in Table 2 show that, without using the FNM, the use of the NIR spectral band produces better results than the VIS band in all metrics analyzed. A possible explanation is that the images obtained in the NIR band are not so affected by the illumination variation (due to pose variation), thus not causing as many occlusions as in the VIS band. The results improve with the use of the FNM in the VIS and NIR spectral bands, with increases in performance in Rank-1 of 15.9% and 0.7%, respectively. In the remaining metrics, it is also observed better values with the use of the normalization model. This shows that the apparent identity change in the qualitative tests (see Figure 11) does not have a negative impact. The results in the LWIR spectral band indicate that using the FNM does not improve the performance in any of the metrics.

Table 2. Results (in %) with and without FNM on the TUFTS-Pose database, using the Light CNN-29 and cosine similarity score.

	Rank-1		Rank-5		TAR @FAR = 0.001	
	w/o	w/	w/o	w/	w/o	w/
VIS	80.3	96.2	91.0	99.5	60.8	87.2
NIR	98.3	99.0	99.5	99.8	90.4	91.9
LWIR	41.8	34.9	58.2	57.8	28.7	14.0

Due to FNM's ability to normalize facial expression, tests were performed with TUFTS-Exp to verify whether normalization of expression allowed Light CNN-29 to extract more representative facial features. The results presented in Table 3 show that the sets of features extracted by Light CNN-29 without facial expression normalization are already representative enough, obtaining a Rank-1 of 99.6% in the VIS and 67.5% in the LWIR and a TAR@FAR = 0.001 of 99.4% in the VIS band and 57.0% in the LWIR band. The use of FNM impairs the feature extraction and consequently the results, especially in the LWIR spectral band, where FNM has more difficulties in generating realistic images. Analyzing the results obtained, the FNM model is used only to normalize facial images from the TUFTS-Pose database in the VIS and NIR spectral bands.

Table 3. Results (in %) with and without FNM on the TUFTS-Exp database, using the Light CNN-29 and cosine similarity score.

	Rank-1		Rank-5		TAR @FAR = 0.001	
	w/o	w/	w/o	w/	w/o	w/
VIS	99.6	93.3	100.0	98.5	99.4	82.9
LWIR	67.5	42.7	83.3	48.2	57.0	23.9

Table 4 presents the results obtained for Rank-1 with the variation of the quantized pose. The values achieved in the VIS band show a significant improvement in the Rank-1 metric with the use of the FNM, resulting in an increase from 77.5% to 97.7% with pose variations of 45° and from 43.3% to 87.4% with pose variations of 60. In the NIR, there is only an improvement when the pose variation is 60°, where the results go from 93.4% to 96.5%. The results obtained prove the ability of the FNM network regarding the pose normalization, where a higher pose variation results in a higher benefit of using it.

Table 4. Results (in %) of rank-1 with and without FNM on TUFTS-Pose database with quantification of pose variation, using the Light CNN-29 and cosine similarity score.

		Pose Variation			
		±60°	±45°	±30°	±15°
VIS	w/o	43.3	77.5	100.0	100.0
	w/	87.4	97.7	99.5	100.0
NIR	w/o	93.4	99.7	100.0	100.0
	w/	96.5	99.4	100.0	100.0

5.5. Face Recognition

5.5.1. Network Training

For the training phase, and considering the results presented above, it was decided to make only one fine adjustment to the LWIR band feature extraction network, because the results obtained in this band are considerably lower, due to the network having been trained in the visible. Thus, the fine-tuning aims for the network to learn to extract more representative features from facial images in the LWIR spectral band. In order to train the Light CNN-29 with identities (people) different from the test ones, a last connected layer was added for training purposes and the LWIR spectral band images from the IRIS database [33] were used. This last layer is used as the input of the softmax cost function and is simply set to the number of training set identities, as proposed by [30].

The optimization algorithms SGD and SGD with Nesterov were used, along with the Cross-Entropy loss function. Table 5 summarizes the parameters used during the training phase.

Table 5. Parameters used in the training procedure.

Parameter	Value
Batch Size	16
Learning Rate	10^{-4}
Momentum	0.9
Epoch Number	10

The objective of the training is that Light CNN-29 learns to extract representative features from facial images and not only to classify them. In this way, Light CNN-29 can be applied to other databases to extract features from facial images to be used as input for similarity functions. Thus, all the following processes make use of the 256-dimensional feature set obtained by Light CNN-29. Table 6 shows the results achieved by the original model and the models trained on the LWIR spectral band, using as similarity function the cosine similarity.

Table 6. Rank-1 results (in %) achieved by different models for extraction of LWIR band features.

	Original	SGD	SGD Nesterov
TUFTS-Pose	41.8	55.5	54.3
TUFTS-Exp	67.5	79.6	75.9

With the results achieved, it is seen that the fine-tuning allowed the network to learn to extract more representative features of facial images of the LWIR spectral band. It is also noticeable that the model that achieved the best results was the SGD without Nesterov, which was chosen for the remaining experiments.

5.5.2. Similarity Functions and Score-Level Fusion

At this stage, we have three Light CNN-29 models, each responsible for extracting features from a specific band. Only the Light CNN-29 responsible for the extraction of features from the LWIR spectral band underwent a fine-tuning. To proceed with classification, it was necessary to find the similarity function that best fits the face recognition task.

Table 7 present the results achieved with the similarity functions cosine similarity and Euclidean distance. The results show that the cosine similarity function is the one that obtains the best score, which is in agreement with [34,35].

Table 7. Rank-1 results (in %) achieved in the face recognition task with the cosine similarity (CSim) and Euclidean Distance (EDis).

	TUFTS-Pose		TUFTS-Exp		CASIA NIR-VIS 2.0	
	CSim	EDis	CSim	EDis	CSim	EDis
VIS	96.2	95.3	99.6	99.4	99.9	99.8
NIR	99.0	96.6	-	-	99.3	99.1
LWIR	55.5	42.0	79.6	69.6	-	-

It is now possible to use the scores obtained by each spectral band to proceed to the final classification. A fusion of the achieved scores was performed using (1). Two studies were conducted, with different weights of each band (W_b of Equation (1)) as shown in Tables 8 and 9.

Table 8. W_b values to be used for each spectral band in the different studies.

	Study 1	Study 2
VIS	1.0	1.0
NIR	1.0	1.0
LWIR	1.0	0.7

Table 9. Results (in %) obtained in the face recognition task, in the TUFTS-Pose database.

	Rank					TAR
	1	2	3	4	5	@FAR = 0.001
Study1	99.4	99.8	99.9	100.0	100.0	90.5
Study2	99.5	99.8	100.0	100.0	100.0	93.5
VIS	96.2	98.7	99.1	99.4	99.5	87.4
NIR	99.0	99.7	99.7	99.8	99.8	93.1
LWIR	55.6	62.2	66.7	69.9	72.6	30.5

In study 1, the previously obtained test results are not considered; thus, the same weight is used in all spectral bands. The final score is a simple arithmetic mean of the scores of the individual bands, which assumes that all spectral bands have the same classification capacity.

The W_b values in study 2 are derived from the mean of the Rank-1 average precision of each of the spectral bands in the tests performed on the TUFTS-Pose, TUFTS-Exp and CASIA NIR-VIS 2.0 databases (results obtained with the cosine similarity function in Table 7) rounded to tenths. Thus, in study 2, the final score was obtained as weighted arithmetic mean, where each band presents different weights reflecting its classification accuracy.

Tables 9–11 show our final face recognition results using both the individual bands and the combination of bands with the two different weight sets (Study 1 and Study 2).

Table 10. Results (in %) achieved in the face recognition task, using the TUFTS-Exp database.

	Rank					TAR
	1	2	3	4	5	@FAR = 0.001
Study1	99.6	100.0	100.0	100.0	100.0	98.7
Study2	99.6	100.0	100.0	100.0	100.0	99.3
VIS	99.6	99.6	99.8	100.0	100.0	99.4
LWIR	79.6	86.3	88.5	90.4	91.6	54.9

Table 11. Results (in %) achieved in the face recognition task, using the CASIA NIR-VIS 2.0 database.

	Rank					TAR
	1	2	3	4	5	@FAR = 0.001
Study1	100.0	100.0	100.0	100.0	100.0	100.0
VIS	99.9	100.0	100.0	100.0	100.0	100.0
NIR	99.6	99.7	99.9	99.9	99.9	99.1

Table 9 presents the results obtained with the TUFTS-Pose database. These results show that study 2 achieved better results than study 1, in the Rank-1 and Rank-3 metrics by 0.1 percentage points, and the TAR@FAR = 0.001 metric by 3 percentage points. The superiority of the results obtained by study 2 compared to study 1 shows that the weight assigned to the LWIR spectral band should be lower than the weight assigned to the others because the characteristics obtained in the LWIR spectral band are the least representative of the identity.

Analyzing the results of the different spectral bands separately, the NIR spectral band achieved the best results due to its robustness towards the variation of illumination present in the TUFTS-Pose database. Despite the promising results of the NIR band when used solo, study 2 obtained superior results in all metrics, with particular emphasis on Rank-1 (from 99.0% to 99.5%) and TAR@FAR = 0.001 (from 93.1% to 93.5%). It is relevant to point out that only the results obtained with score fusion reached the 100% accuracy rate in the assessed Ranks (Rank-4 for study 1 and Rank-3 for study 2).

Table 10 shows the results obtained with the TUFTS-Exp database. An analysis of the results allows us to see that the face recognition results obtained are better with score fusion, where both studies obtained the same result as the VIS spectral band in Rank-1 (99.6%) but managed to achieve a higher result in Rank-2 (100% against 99.6% of the VIS spectral band). However, the best result for TAR@FAR = 0.001 is obtained using only the VIS spectral band, with 99.4%, while the second-best result was obtained in study 2, with 99.3%.

The results achieved using the CASIA NIR-VIS 2.0 database (Table 11) show that study 1 reached a value of 100% in Rank-1. Using the VIS and NIR spectral bands separately, the results were 99.9% and 99.6%, respectively, using the same metric. It should be noted that study 2 was not performed for the CASIA NIR-VIS 2.0 database, as the difference between study 1 and study 2 is the weight assigned to the LWIR spectral band, which it does not have. In the TAR@FAR = 0.001 metric, study 1 matches the result for the VIS spectral band with 100%.

Performing a global analysis of all results, we can observe that the fusion of scores mainly favors cases where the results obtained by the different spectral bands separately were less satisfactory. Looking at the results obtained with the TUFTS-Exp and CASIA-NIR-VIS 2.0 databases (Tables 10 and 11), it is clear that the VIS spectral band already obtains very high values in all metrics. This fact makes the fusion of scores not so effective. However, despite a decrease of the TAR@FAR = 0.001 in Table 10, the results obtained by the fusion of scores, in general, were higher than those obtained by the spectral bands separately. The results obtained thus demonstrate the benefit of using multi-spectral images in a face recognition system.

6. Conclusions

In this paper, a multi-spectral face recognition system in an uncontrolled environment has been proposed, aiming to make a decision with the largest amount of data available, i.e., using the facial images obtained by the different spectral bands. The system is composed of three modules: (i) face detection and alignment, (ii) image synthesis and (iii) face recognition.

The state of the art regarding face recognition systems in an uncontrolled environment has led to the conclusion that image synthesis methods, mainly with GANs, have been used to combat intrapersonal variations, such as the difference in pose and facial expression. On the other hand, in the area of multispectral face recognition, with a plurality of solutions presented by the use of multispectral images, fusion methods are those that make the most use of images captured in different spectral bands in order to make a decision. The main problem encountered is the limited number of images (and people) in multispectral databases in an uncontrolled environment, which makes it challenging to train convolutional neural networks, which are the most used method for feature extraction.

Several techniques were implemented to validate them in different multi-spectral bands, since all of them were trained on visible databases, as well as to analyze the influence of facial image features (pose, illumination and expression). This analysis aimed to select the most appropriate technique for each module of the proposed face recognition system.

For the face detection task, three networks were evaluated qualitatively and quantitatively, which allowed us to conclude that the DSFD network was the most appropriate since it maintained a high accuracy in the different spectral bands. For the landmark detection task, three networks were evaluated qualitatively, and it was concluded that the 2D-FAN network was the best fit due to its ability to correctly identify facial landmarks in different spectral bands with a diversity of facial poses. Such evaluations allowed us to select the methods that are best suited for these tasks with multispectral images in an uncontrolled environment. Thus, this work presents an efficient face detection and face alignment module for a multispectral face recognition system in an uncontrolled environment.

The present work also performed evaluations of different face normalization methods, through image synthesis, to produce face images with a frontal pose. The FFWM and FNM models were analyzed, where the FNM model produced the most realistic facial images for the visible and NIR spectral bands, maintaining the proportions of the face and the most relevant facial features. Further analysis of the FNM model allowed us to conclude that: (i) the greater the pose variation, the greater the advantage in using the FNM model and (ii) the NIR images allow us to obtain a better identification/verification than the visible images because pose variation can entail variations in illumination, to which the NIR band is resistant.

The analysis of the performance of the different models allowed the selection of the most suitable one for a multispectral face recognition system in an uncontrolled environment, as well as the identification of the most advantageous situations for its use.

The extraction of the feature sets of the facial images from the different spectral bands is performed using Light CNN-29 [30], with a fine adjustment to the network weights for the LWIR spectral band since it was trained on the visible spectral band. For the classification phase, identification is performed in the different spectral bands, each producing different scores for each identity. These scores are computed by the similarity between the feature sets of each identity and the feature set of the input facial image. In this work, two different studies were performed for score fusion, which allowed us to conclude that: (i) simply using the different spectral bands to identify is advantageous (study 1) and (ii) a weighted average is beneficial when the different classifiers (of each spectral band) have different levels of reliability (study 2).

On the multi-spectral TUFTS database, with pose variation and expression variation, the results obtained in Rank-1 by the proposed system and with score fusion with a weighted average (study 2) were 99.5% and 99.6%, with the best results obtained using only one spectral band being 99.0% and 99.6%. On the TAR@FAR = 0.001 metric, the results obtained by weighted average are 93.5% and 99.3%, while with only one spectral band 93.1% and 99.4% were obtained. In the CASIA NIR-VIS 2.0 database, score fusion achieved the results of 100.0% in the Rank-1 and TAR@FAR = 0.001 metrics, where without score fusion, 99.9% and 100.0% in Rank-1 and TAR@FAR = 0.001, respectively, are obtained as the best result.

The original contributions of this work include the analysis of several techniques for different tasks, which allowed: (i) the presentation of an efficient face detection and alignment module to be used by any multi-spectral face analysis system, (ii) the identification of the situations in which the FNM model should be used to normalize facial images and (iii) the selection of a similarity function and the weights to be used in the fusion of scores to identify/verify identities. From the experimental results, it is also concluded that the proposed system allows us to obtain high results in multi-spectral face recognition in an uncontrolled environment, where the use of the scores obtained from different spectral bands allows us, in general, to achieve results that are superior to using only the scores obtained by one spectral band.

After performing the work described in this paper, the authors suggest as future work several relevant hypotheses. The first suggestion consists of the creation of a multispectral database to overcome the limitations in the public multispectral databases that currently exist. The second suggestion is to create a prototype and put it to work for access control in high security areas. The third suggestion for future work consists of the adaptation of the image input, to be able to process images obtained by drones with cameras in the spectrum of visible, NIR, SWIR and LWIR, having as an objective the processing of images in real time.

Author Contributions: J.S.S. and A.B. proposed the idea and concept; P.M. developed the software under the supervision of J.S.S. and A.B.; all authors revised and edited the manuscript. All authors have read and agreed to the published version of the manuscript.

Funding: This research was supported in part by the Military Academy Research Center (CINAMIL) under the project Multi-Spectral Facial Recognition, and by FCT with the projects UID/FIS/04559/2019, HAVATAR (PTDC/EEI-ROB/1155/2020) and LARSyS (UIDB/50009/2020).

Institutional Review Board Statement: Not applicable.

Informed Consent Statement: The Portuguese Military Academy (AM) database is a private database; the reproduction of the images present in this database without the explicit authorization of the authors is not allowed. During the image acquisition, a declaration of consent was presented to all participants for the use of the multispectral facial images in scientific works. As such, all persons present in the AM database gave informed consent for publication of identifying images in an online open-access publication. The image acquisition at the Military Academy was authorized by the Major-General (OF-7) Commander of the Military Academy according to the Information No. CINAMIL-2020-000224, Proc. 00.020.0181, of 28 February 2020. All methods were carried out following relevant guidelines and regulations and authorized by the Military Academy Commander.

Data Availability Statement: Not applicable.

Conflicts of Interest: The authors declare no conflict of interest.

References

1. Bento, N.A.; Silva, J.S.; Dias, J.B. Detection of Camouflaged People. *Int. J. Sens. Netw. Data Commun.* **2016**, *5*, 143–148.
2. Gonçalves, M.; Silva, J.S.; Bioucas-Dias, J. Classification of Vegetation Types in Military Region. In Proceedings of the SPIE Security and Defence 2015 Europe: Electro-Optical Remote Sensing, Photonic Technologies, and Applications, Toulouse, France, 21–22 September 2015; p. 9649.
3. Silva, J.S.; Guerra, I.F.L.; Bioucas-Dias, J.; Gasche, T. Landmine Detection Using Multispectral Images. *IEEE Sens. J.* **2019**, *19*, 9341–9351. [CrossRef]
4. Chambino, L.L.; Silva, J.S.; Bernardino, A. Multispectral Face Recognition Using Transfer Learning with Adaptation of Domain Specific Units. *Sensors* **2021**, *21*, 4520. [CrossRef] [PubMed]
5. Masi, I.; Wu, Y.; Hassner, T.; Natarajan, P. Deep face recognition: A survey. In Proceedings of the 31st SIBGRAPI Conference on Graphics, Patterns and Images, Foz do Iguaçu, Paraná, Brazil, 29 October–1 November 2018; pp. 471–478.
6. Chambino, L.L.; Silva, J.S.; Bernardino, A. Multispectral Facial Recognition: A Review. *IEEE Access* **2020**, *8*, 207871–207883. [CrossRef]
7. Munir, R.; Khan, R.A. An extensive review on spectral imaging in biometric systems: Challenges & advancements. *J. Vis. Commun. Image Represent.* **2019**, *65*, 102660. [CrossRef]
8. Zhang, W.; Zhao, X.; Morvan, J.; Chen, L. Improving Shadow Suppression for Illumination Robust Face Recognition. *IEEE Trans. Pattern Anal. Mach. Intell.* **2019**, *41*, 611–624. [CrossRef] [PubMed]
9. Panetta, K.; Wan, Q.; Agaian, S.; Rajeev, S.; Kamath, S.; Rajendran, R.; Rao, S.P.; Kaszowska, A.; Taylor, H.A.; Samani, A.; et al. A Comprehensive Database for Benchmarking Imaging Systems. *IEEE Trans. Pattern Anal. Mach. Intell.* **2020**, *42*, 509–520. [CrossRef] [PubMed]
10. Goodfellow, I.; Pouget-Abadie, J.; Mirza, M.; Xu, B.; Warde-Farley, D.; Ozair, S.; Courville, A.; Bengio, Y. Generative adversarial nets. In Proceedings of the Advances in Neural Information Processing Systems, Montreal, QC, Canada, 8–13 December 2014; Volume 27.
11. Tran, L.; Yin, X.; Liu, X. Disentangled representation learning gan for pose-invariant face recognition. In Proceedings of the IEEE Conference on Computer Vision and Pattern Recognition, San Juan, PR, USA, 21–26 July 2017; pp. 1415–1424.
12. Zhao, J.; Xiong, L.; Li, J.; Xing, J.; Yan, S.; Feng, J. 3d-aided dual-agent gans for unconstrained face recognition. *IEEE Trans. Pattern Anal. Mach. Intell.* **2018**, *41*, 2380–2394. [CrossRef] [PubMed]
13. Cao, J.; Hu, Y.; Zhang, H.; He, R.; Sun, Z. Towards high fidelity face frontalization in the wild. *Int. J. Comput. Vis.* **2019**, *128*, 1485–1504. [CrossRef]
14. Qian, Y.; Deng, W.; Hu, J. Unsupervised face normalization with extreme pose and expression in the wild. In Proceedings of the IEEE/CVF Conference on Computer Vision and Pattern Recognition, Long Beach, CA, USA, 15–20 June 2019; pp. 9851–9858.
15. Tuan Tran, A.; Hassner, T.; Masi, I.; Medioni, G. Regressing robust and discriminative 3D morphable models with a very deep neural network. In Proceedings of the IEEE Conference on Computer Vision and Pattern Recognition, San Juan, PR, USA, 21–26 July 2017; pp. 5163–5172.
16. Kanmani, M.; Narasimhan, V. Optimal fusion aided face recognition from visible and thermal face images. *Multimed. Tools Appl.* **2020**, *79*, 17859–17883. [CrossRef]
17. Peng, C.; Gao, X.; Wang, N.; Li, J. Graphical Representation for Heterogeneous Face Recognition. *IEEE Trans. Pattern Anal. Mach. Intell.* **2016**, *39*, 301–312. [CrossRef] [PubMed]
18. He, R.; Cao, J.; Song, L.; Sun, Z.; Tan, T. Adversarial cross-spectral face completion for NIR-VIS face recognition. *IEEE Trans. Pattern Anal. Mach. Intell.* **2019**, *42*, 1025–1037. [CrossRef] [PubMed]

19. Hu, W.P.; Hu, H.F.; Lu, X.L. Heterogeneous Face Recognition Based on Multiple Deep Networks with Scatter Loss and Diversity Combination. *IEEE Access* **2019**, *7*, 75305–75317. [CrossRef]
20. He, R.; Wu, X.; Sun, Z.N.; Tan, T.N. Wasserstein CNN: Learning Invariant Features for NIR-VIS Face Recognition. *IEEE Trans. Pattern Anal. Mach. Intell.* **2019**, *41*, 1761–1773. [CrossRef] [PubMed]
21. Li, S.; Yi, D.; Lei, Z.; Liao, S. The CASIA NIR-VIS 2.0 Face Database. In Proceedings of the IEEE Conference on Computer Vision and Pattern Recognition Workshops, Portland, OR, USA, 23–28 June 2013; pp. 348–353.
22. Liu, W.; Anguelov, D.; Erhan, D.; Szegedy, C.; Reed, S.; Fu, C.-Y.; Berg, A.C. SSD: Single shot multibox detector. *Lect. Notes Comput. Sci.* **2016**, *9905*, 21–37.
23. Zhang, S.; Zhu, X.; Lei, Z.; Shi, H.; Wang, X.; Li, S.Z. S3FD: Single shot scale-invariant face detector. In Proceedings of the IEEE International Conference on Computer Vision, Cambridge, MA, USA, 22–29 October 2017; pp. 192–201.
24. Bradski, G.; Kaehler, A. *Learning OpenCV: Computer Vision with the OpenCV Library*; O'Reilly Media, Inc.: Sevastopol, CA, USA, 2008.
25. Li, J.; Wang, Y.; Wang, C.; Tai, Y.; Qian, J.; Yang, J.; Wang, C.; Li, J.; Huang, F. DSFD: Dual shot face detector. In Proceedings of the Proceedings of the IEEE/CVF Conference on Computer Vision and Pattern Recognition, Long Beach, CA, USA, 27 October–2 November 2019; pp. 5060–5069.
26. Kazemi, V.; Sullivan, J. One Millisecond Face Alignment with an Ensemble of Regression Trees. In Proceedings of the Proceedings of the IEEE Conference on Computer Vision and Pattern Recognition, Columbus, OH, USA, 23–28 June 2014; pp. 1867–1874.
27. Bulat, A.; Tzimiropoulos, G. How Far Are We from Solving the 2d & 3d Face Alignment Problem? (And a Dataset of 230,000 3d Facial Landmarks). In Proceedings of the IEEE International Conference on Computer Vision, Cambridge, MA, USA, 22–29 October 2017; pp. 1021–1030.
28. Newell, A.; Yang, K.; Deng, J. Stacked hourglass networks for human pose estimation. In Proceedings of the European Conference on Computer Vision, Amsterdam, The Netherlands, 27–30 June 2016; pp. 483–499.
29. Wei, Y.; Liu, M.; Wang, H.; Zhu, R.; Hu, G.; Zuo, W. Learning flow-based feature warping for face frontalization with illumination inconsistent supervision. In Proceedings of the European Conference on Computer Vision, Glasgow, UK, 13–19 June 2020; pp. 558–574.
30. Wu, X.; He, R.; Sun, Z.; Tan, T. A Light CNN for Deep Face Representation with Noisy Labels. *IEEE Trans. Inf. Forensics Secur.* **2018**, *13*, 2884–2896. [CrossRef]
31. Minaee, S.; Abdolrashidi, A.; Su, H.; Bennamoun, M.; Zhang, D. Biometrics recognition using deep learning: A survey. *arXiv* **2019**, arXiv:1912.00271.
32. Gross, R.; Matthews, I.; Cohn, J.; Kanade, T.; Baker, S. Multi-pie. *Image Vis. Comput.* **2010**, *28*, 807–813. [CrossRef] [PubMed]
33. Dataset 02: IRIS Thermal/Visible Face Databases. 2005. Available online: http://vcipl-okstate.org/pbvs/bench/ (accessed on 27 March 2021).
34. Borade, S.N.; Deshmukh, R.R.; Shrishrimal, P. Effect of distance measures on the performance of face recognition using principal component analysis. In *Intelligent Systems Technologies and Applications*; Springer: Berlin/Heidelberg, Germany, 2016; pp. 569–577.
35. Liu, C. Discriminant analysis and similarity measure. *Pattern Recognit.* **2014**, *47*, 359–367. [CrossRef]

Article

A Robust Facial Expression Recognition Algorithm Based on Multi-Rate Feature Fusion Scheme

Seo-Jeon Park [1], Byung-Gyu Kim [1,*] and Naveen Chilamkurti [2]

1. Department of IT Engineering, Sookmyung Women's University, 100 Chungpa-ro 47 gil, Yongsna-gu, Seoul 04310, Korea; sj.park@ivpl.sookmyung.ac.kr
2. La Trobe Cybersecurity Research Hub, La Trobe University, Melbourne, VIC 3086, Australia; N.Chilamkurti@latrobe.edu.au
* Correspondence: bg.kim@sookmyung.ac.kr; Tel.: +82-2-2077-7293

Abstract: In recent years, the importance of catching humans' emotions grows larger as the artificial intelligence (AI) field is being developed. Facial expression recognition (FER) is a part of understanding the emotion of humans through facial expressions. We proposed a robust multi-depth network that can efficiently classify the facial expression through feeding various and reinforced features. We designed the inputs for the multi-depth network as minimum overlapped frames so as to provide more spatio-temporal information to the designed multi-depth network. To utilize a structure of a multi-depth network, a multirate-based 3D convolutional neural network (CNN) based on a multirate signal processing scheme was suggested. In addition, we made the input images to be normalized adaptively based on the intensity of the given image and reinforced the output features from all depth networks by the self-attention module. Then, we concatenated the reinforced features and classified the expression by a joint fusion classifier. Through the proposed algorithm, for the CK+ database, the result of the proposed scheme showed a comparable accuracy of 96.23%. For the MMI and the GEMEP-FERA databases, it outperformed other state-of-the-art models with accuracies of 96.69% and 99.79%. For the AFEW database, which is known as one in a very wild environment, the proposed algorithm achieved an accuracy of 31.02%.

Keywords: deep learning; facial expression recognition (FER); 3D convolutional neural network (3D CNN); multirate signal processing; minimum overlapped frame structure; self-attention; multi-depth network

1. Introduction

Communication skills have been developed based on the senses that play an important role in human interaction. There are five human senses: sight, sound, touch, taste, and smell. There is no doubt that sight is the most important one of the five senses for most people, since up to 80% of all senses are recognized through sight [1].

In recent years, the importance of human–computer interaction (HCI) grows larger as the artificial intelligence (AI) field develops. The basic goal of the HCI field is to improve the interaction between human and computer systems by making the computers more useful and accessible to humans. Additionally, the ultimate goal of the AI technology is to allow the machine to catch the user's intentions or emotions by itself, thereby reducing the burden of the user and making it more enjoyable. Therefore, understanding the feelings and the action of the human becomes important in various human-centric services. This technology based on the human face is called facial expression recognition (FER) technology.

FER technology has many applications in customer service [2], the automotive industry [3], entertainment [4], and home appliances [5]. There are good examples including: games with different modes based on classifications of the user's facial expression [6], identifying the driver's drowsiness and instructing an appropriate response [7,8], automatically collecting vast amounts of data necessary for the study of human emotional behavior patterns [9], detecting the emotional state of the patient and predicting the situation in need of help [10,11], and establishing an adaptive learning guidance strategy by grasping a student's psychological state using facial expressions and words that are used [12–14]. In recent years, interest and research on the development of intelligent home appliances and software that respond to the user's emotional state have been focused on.

One of the main technologies for emotion recognition is to recognize a user's emotional state from facial expression from an image sensor. Among various fields of biometrics, the face is a very important element that can be easily encountered in daily life. The emotional state that appears in the facial emotions is sufficient to be used as a human interface when sharing opinions with other people in the process of communicating with each other or conveying one's feelings. Reflecting this importance, many studies related to FER have been conducted. In the field of psychology, many studies on facial analysis and recognition have been done for many years.

According to a study by psychologist Ekman and Friesen, six emotions of a person, happiness, sadness, anger, surprise, disgust, and fear, have been classified as basic emotions that are perceived in common without being influenced by each culture [15,16]. Based on this, many studies have classified six emotions or seven emotions adding neutral expressions to identify emotional states. In recent years, research targets are expanding to expressions including not only depression, pain, and sleepiness but also expressions representing mental states such as agreement, concentration, interest, thinking, and confusion. In addition, research is also being conducted on the recognition of natural facial expressions and not only the research through an ideal database containing exaggerated expressions to the limited environment. However, despite these efforts, FER technology is still at a level that can be applied only in limited circumstances.

The FER system that recognizes facial expressions consists of three steps. The first step is to detect a person's face. This step is to detect a face area from an input image and to detect face elements such as the eyes, nose, or mouth. Representative algorithms include Adaboost [17], Haar-cascade [18,19], and the histogram of oriented gradients (HOG) [20]. Second, facial features are extracted from the recognized face using a geometric feature-based method or an appearance feature-based method. Finally, there is a classification step in which emotions are classified using the method based on the extracted features.

Facial expression recognition is a field with high dependency on datasets. There are two types of factors that influence facial expression recognition. The first type of external factors is uniqueness of each person such as gender, race, and age. The second type of external factors is the environment such as lighting, poses, resolution, and noise. However, many facial-expression datasets were created in controlled environments, so the second type of external factors were affected less than the first type. To overcome this problem, the dataset must be rich enough to accommodate these factors. Therefore, we used data augmentation to supply various information. Another method is to create a cross dataset that uses multiple datasets. This is to learn and test by combining different datasets under the same conditions. Through this, there is an advantage that facial expressions in a more diverse environment can be generalized.

Datasets used for FER are largely divided into two types according to the type of dataset. A static dataset consists of static images, and a dynamic dataset consists of dynamic images, which are called videos. In order to apply the FER in practice, we need to use a dynamic dataset, which is found in real life. In general, the accuracy of a dynamic dataset is lower than a static dataset, because dynamic images have different features such as facial movements over time. Therefore, temporal dynamics must be considered. Through the 2D convolutional neural network (2D CNN) [21], only spatial features can be identified within an image. Therefore, to classify the facial expressions in dynamic images through this 2D CNN, there is a limitation in processing temporal motion.

To solve this problem, a network dealing with the time axis is needed. Recurrent neural networks (RNN) are a type of artificial neural network that forms a circular structure in which hidden nodes are connected by directional edges. Data appearing sequentially can be usefully processed through RNN [22]. However, if the distance between the relevant information and the point where the information is used is long, the gradient gradually decreases during back-propagation, leading to a problem that the learning ability is greatly degraded. This is called the vanishing gradient problem. The long short-term memory (LSTM) [23] was devised to overcome this problem. LSTM is a structure in which the cell state is added to the hidden state of the RNN. Another method is the use of 3-dimensional convolution neural networks (3D CNN) [24]. Unlike conventional 2D CNN, 3D CNN uses a 3D convolution kernel to extract features not only for the space domain but also for the time domain.

In [25,26], they used geometric features such as landmarks, and the reference of a facial expression such as neutral expression was required while extracting features. However, in the case of real-life FER, no reference is given, and it cannot be guaranteed that the face of the neutral expression will be given. Therefore, a model that can recognize facial expressions without a reference is needed to use FER in practice.

We proposed a new facial expression recognition model to solve these problems. First, a 3D CNN structure that can simultaneously extract spatial and temporal features was used to obtain more accurate facial expression recognition results. Second, we used multi-networks with different frame rates to extract various features. The frames used for inputs entering each network should not overlap as much as possible, so we can utilize more spatio-temporal information. Third, we applied self-attention to the features that were extracted from each network, to make more reinforced features. The facial expressions were classified by combining these features through a joint fusion classifier.

In order to make a facial expression recognition model, the most relevant contributions are as follows:

- We defines a multi-depth network based on multi-frame rate input.
- A structure that minimizes the overlapping between input frames to each model was designed.
- The proposed scheme reinforced the features that are the result of the networks by self-attention, and it showed a better result than each network's result.
- We verified the robustness of the multi-depth network on the variation of dataset and different facial expression acquisition conditions.

The rest of the article is organized as follows. The related works for facial expression recognition are introduced in Section 2. Section 3 introduces a detailed description of the facial expression recognition algorithm composed of five main steps. Section 4 provides several experimental results and the performance comparison results with the latest models. Finally, the concluding remarks of this article are given in Section 5.

2. Related Works

2.1. The Facial Expression Recognition Methods

2.1.1. Classical Feature-Based Approaches

Features representing facial expressions are divided into the permanent facial features (PFF), which expresses permanent facial features such as the eyes and nose, and the transient facial features (TFF), which expresses wrinkles or protrusions that occur temporarily as facial muscles move [27]. In face recognition, the proportion of the PFF is large, but in the field of facial expression recognition, the TFF also plays an important role as well as the PFF. Representative methods of expressing these facial features in an image include a geometric feature-based method and an appearance feature-based method. Analyzing the existing studies in terms of expressing facial features is as follows:

Geometric Feature-Based Method

Systems based on geometric features express changes in the shape and expression of a face by using the positions of various facial elements and the relationships between them. Since the positions and movements of the mechanical features of the face are changed according to the difference between the shape of the face and the facial expression, an intuitive expression recognition method can be used by using dynamic information obtained by tracking these features from a video image. The difficult point of the geometric feature-based method is that because each person has a different face shape, the location of the feature cannot be used as it is. To solve this problem, the facial parts are modeled with the active appearance model (AAM) [28] or action unit (AU) [29] according to facial expressions, and based on the information extracted from the image, they are tracked to obtain the relative distance between the parts.

The geometric feature has the advantage of being able to implement a system that requires less memory and can easily adapt to changes in the position, size, and orientation of the face because the motion of the feature can be simply expressed with a few factors. On the other hand, since it is difficult to express the TFF that appears temporarily while the expression occurs, the geometrical features are similar, but there is a limit to distinguishing expressions with different facial textures such as wrinkles.

Appearance Feature-Based Method

The facial expression recognition method based on appearance features can accommodate both permanent features such as the eyes and mouth according to facial expressions and temporary features such as wrinkles for the entire image or the regional image. The appearance feature-based method is divided into a holistic image-based approach and a local image-based approach according to the size of the image used for feature extraction.

Holistic Feature-Based Method The holistic feature-based method considers each pixel constituting a face image as one feature element and expresses the entire image as one feature vector. Therefore, when the number of pixels constituting the face image is large, the size of the feature vector becomes excessively large, and the amount of calculation increases accordingly. As a solution to this problem, the linear subspace method (LSM) was proposed. LSM [30] improved the overall processing speed and accuracy by expressing the feature vector composed of the pixels of the face image as a low-dimensional spatial vector through linear transformation. Representative LSMs include principal component analysis (PCA) [31], linear discriminant analysis (LDA) [32], and independent component analysis (ICA) [33].

This holistic feature-based method is simple because it targets the entire image without going through a separate feature extraction process, but it has a disadvantage in that its performance is poor in a dynamic environment in which the face pose, lighting, and facial expressions move.

Local Feature-Based Method The regional feature-based method constructs a feature vector representing the overall face shape by setting a regional window in a region where changes can occur due to facial expressions in a face image and extracting features based on the brightness distribution within the window. In general, since the lighting of an image or changes in facial expressions appear in a part of the facial image, the regional feature-based method sets a local window only in the area where changes in the face can occur. Therefore, it has the advantage of being relatively less sensitive to these changes compared to the global feature-based method. Representative methods based on regional features include the Gabor filter [34], the Haar-like feature [18], and the local binary pattern (LBP) [35].

2.1.2. Deep-Learning-Based Approaches

Most facial expression recognition algorithms used in recent studies use deep learning-based methods. When AlexNet showed a performance improvement in the ImageNet challenge [36], many researchers began to apply the 2D CNN structure to various fields, and it was also applied to the FER [37,38]. There have been many attempts to apply 2D

CNN to the video frames. However, 2D CNN has structural limitations because they cannot provide temporal information to the neural network.

Many studies use two architectures to overcome this problem. First, 3D CNN was designed by transforming the structure of 2D CNN [24]. 3D CNN uses a 3D convolution operation, which has three-dimensional convolution filters. Therefore, the feature map generated by one filter is also three-dimensional, and 3D CNN can learn temporal learning of successive frames from the convolution filter. This structure enabled spatio-temporal-feature learning for short-term input frames. Second, a hybrid method that combines multiple networks was also used. A CNN-RNN or CNN-LSTM [39,40] structure is one of the examples. It learns spatial features with CNN and then learns temporal features by RNN or LSTM.

A hybrid method is also used for improving accuracy as well as solving 2D CNN problems. In [26,41], they used two networks to extract temporal appearance and geometric features from image sequences and facial landmark points. They combined these two networks with a new integration method to make the two models cooperate with each other and improve the performance. Based on these methods, a hybrid method that combines multiple-depth networks based on 3D CNN is suggested.

2.2. Multirate Filter Bank

In [42], multi-rate filter banks produced multiple output signals by filtering and sub-sampling a single input signal or, conversely, generating a single output by up-sampling and interpolating multiple inputs. An analysis filter bank divides the signal into M-filtered and sub-sampled versions. A synthesis filter bank generates a single signal from M-up-sampled and interpolated signals. The proposed algorithm looks like a sub-band coder, which was combined by an analysis filter bank and a synthesis filter bank.

We divided the input video (dynamic image) into multiple outputs, which have different frame rates, and put them into networks, which have different network-depth models. By using this structure, we could construct various spatio-temporal features. These features were combined into one feature, and we classified it by a joint fusion classifier.

2.3. Self-Attention

Attention is a methodology that started from the perspective of "let the model learn even the parts that need to be learned intensively for better performance." It makes network-to-weight features and uses the weighted features to help achieve the task. It is widely used in natural language processing (NLP), multivariate time series, and machine translation.

The attention mechanism was first devised for sequence learning [43]. It figures out which output sequence of the encoder is most associated with the particular output sequence of the decoder.

The attention itself is almost similar to the transformer [44]. The transformer can be divided into self-supervision and self-attention. By self-supervision, it is possible to train a model with an unlabeled dataset and learn generalizable representations. Self-attention calculates the attention by itself, and it assumes a minimum inductive bias unlike models such as CNN and RNN.

The self-attention method has been applied in computer vision tasks such as [45–49]. In [45], they inserted an attention block between convolutional layers to improve image feature creation performance. In [46], the attention was performed per channel through a dot product on the channel characteristic vector, and the authors used a channel and spatial attention block in [47]. Figure 1 shows some examples of the visual attention.

(a) The butterfly is sitting on the flower.

(b) The postal delivery officer delivers mail by motorcycle.

(c) A baby is sleeping in the crib.

(d) A rainbow hangs in the sky.

Figure 1. The examples of attending to the correct object (*white* indicates the attended regions, and *underlining* indicates the corresponding objects).

3. Proposed Scheme

This section introduces the proposed method in detail. Section 3.1 introduces the method of how we pre-processed the input images before feeding them into the networks. Additionally, we describe the data augmentation process in Section 3.2. Section 3.3 elaborates the network that was used to extract the feature maps. Section 3.4 goes into detail about how to reinforce the features and the joint fusion classifier, which classifies the facial expressions with the reinforced features.

Figure 2 shows the overall structure of the proposed algorithm based on multirate inputs and multi-depth networks to make a robust scheme.

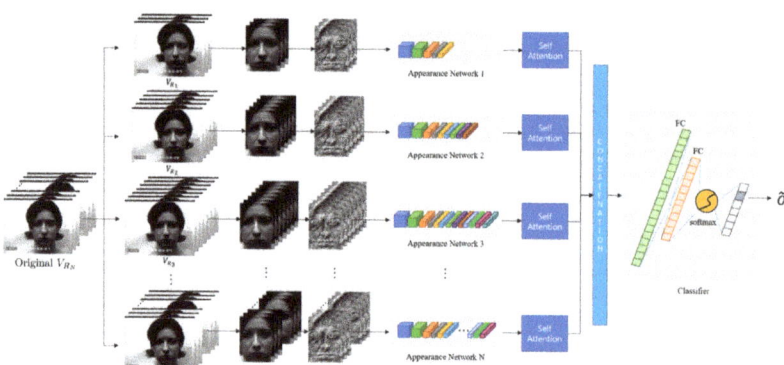

Figure 2. The process of the proposed facial expression recognition scheme.

3.1. Data Pre-Processing

The environments of each database such as resolution, brightness, and pose are changeable. In order to have a general environment, a data pre-processing step is required, and Figure 3 shows the entire process of input with one sequence.

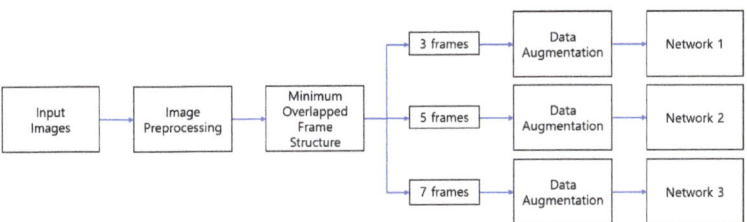

Figure 3. The entire process of making input dataset with a sequence.

We augmented the pre-processed dataset to avoid the overfitting problem. Then, each network received those dataset as input since CNN requires the fixed size of the input. Through this process, unnecessary sequence parts were removed, and important features were highlighted, so that the network can extract informative features efficiently.

3.1.1. Image Pre-Processing

In order to have a general condition of the input, we went through four steps. The flowchart of the image pre-processing algorithm is shown in Figure 4.

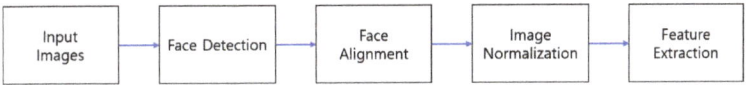

Figure 4. Architecture of data pre-processing algorithm.

3.1.1.1. Face Detection

For FER, we needed to detect the face area first. Then, we cropped the detected face area not to be affected by unnecessary parts such as hair or accessories.

We used the FaceBoxes module [50] to detect the face region. It consists of the rapidly digested convolution layers (RDCL), the multiple scale convolution layers (MSCL), and the divide and conquer Head (DCH).

3.1.1.2. Face Alignment

Through facial landmarks, we checked whether the face is frontal or not and aligned the askew frontal face in order to fix the posture. We used the style-aggregated network (SAN) module [51] to extract the landmark of the face. We tilted the face by aligning the x axis of the tip of the nose and the x axis of the center of the eyes vertically. The tip of the nose was the 34-th landmark, and the center of the eyes was the average of the 37-th to 46-th landmark—refer to Figure 5a.

After alignment, the face was judged to be front if the 34-th landmark, which is the tip of the nose, was between the 40-th landmark and the 43-th landmark, which are the nearest points from the nose of the left and right eye. The example of this part is shown in Figure 5b. After the face alignment process, we cropped the minimized face area without empty data again. Then, we resized the image into 128 × 128 in order to make the same resolution. This alignment process can be considered as a kind of affine transformation based on two points. This had two constraints as: (1) the images of the two lines were also parallel, and (2) translations are isometries.

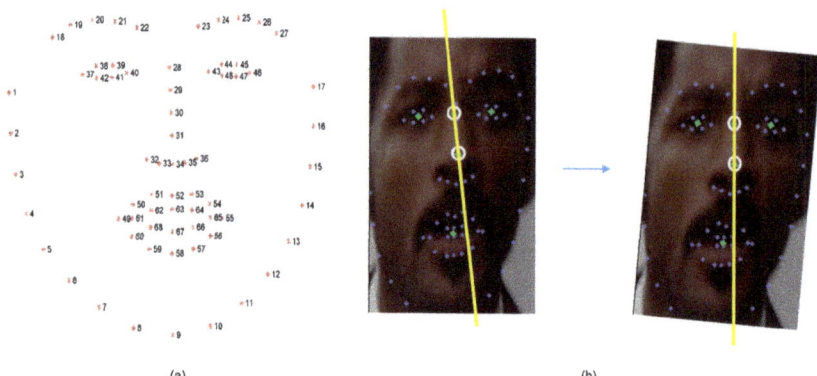

Figure 5. Aligning the face: (**a**) 68 facial landmarks; (**b**) face alignment with the landmarks.

3.1.1.3. Image Normalization

There are two ways to normalize an image. The first is to normalize the size of the image. In general, when using CNN, the dimension of an input image or feature needs to be fixed. Therefore, we resized all the input images into the same size 128 × 128. This accelerates the convergence of the network. The second is to normalize the image numerically. It means we normalized the pixel distribution of the original image. Through Equation (1), which has been reported in [35], the values followed the standard normal distribution standardized by the Z-score. The standard conversion formula for this is as follows:

$$x' = \frac{x - \mu}{\sigma}. \tag{1}$$

Here, x is the pixel value of the original image, and x' is the new value of the converted image. In addition, μ is the average pixel value of the image through calculation, and σ is the pixel standard deviation value of the image through calculation. The data subjected to Z-score standardization showed a normal distribution with an average of 0 and a deviation of 1 approximately. This intensity normalization can give better features than using one by 255.

In most of the deep learning approaches, an input image is given into the designed deep neural network after normalizing it by 255, to make robustness in illumination change. However, it always gives an intensity range as (0, 1.0). That is, this normalization by 255 compresses into too small an intensity range. However, the suggested Z-score maintains a larger range as (−1.0, +1.0) by the standard deviation of illumination in the given image. Through experiment, we verified the suggested normalization to be more effective to make features in convolution neural networks.

3.1.1.4. Feature Extraction Using LBP

We extracted features from the resized image to reduce the computational complexity and to emphasize facial characteristics. In this study, facial features were extracted through an LBP. The LBP classifies the texture of the image and is widely used in fields such as facial recognition and gender, race, and age classification [52,53]. Additionally, the LBP function was used to eliminate the effect of lighting.

In [54], Timo et al. proposed a method of applying LBP to facial recognition problems for the first time. This showed a better result than many of the existing approaches.

In order to have the LBP feature value for one pixel, a 3 × 3 size block was used, and it is shown in Figure 6. Each pixel value except the center was compared with the pixel value located in the center, and if it was brighter than the center, it was encoded as 1; if it was darker than the center, it was encoded as 0. The formula is as follows:

$$LBP(x,y) = \sum_{n=0}^{N-1} s(p_o - p_c) \times 2^n, \qquad (2)$$

where

$$s(x) = \begin{cases} 1, & \text{if } x > 0, \\ 0, & \text{otherwise.} \end{cases} \qquad (3)$$

The value of the center point $s(x)$ was converted to a binary number 0 and 1 through Equation (3) where x refers to the difference between the center pixel p_c and the other pixel p_o. As value of the center pixel p_c, a different 8-bit binary string was generated if N is 8. Then, the binary code was converted to decimal $LBP(x,y)$ by Equation (2). The LBP's capabilities help reduce computational complexity compared to the original image. It also emphasizes the main texture of the face in the image.

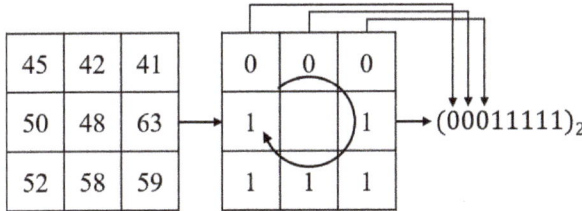

Figure 6. The LBP feature extraction.

3.1.2. Minimum Overlapped Frame Structure

The proposed model extracts features from multiple networks, whose inputs are various using different input frame rates, and classifies facial expression by combining extracted features. Therefore, we thought that it would be more efficient to learn if various information is given.

In the conventional structure, frames are extracted with regular intervals. This assumes that the expression of the sequence goes from the neutral to the peak. When the number of the sequence is n, then the structure of the number of N input frames $S(N)$ is made from the X sequence as follows:

$$S(N) = \{X[1], X[2], ..., X[N]\}, \qquad (4)$$

where

$$X[i] = X[round(\frac{n-1}{N-1} \times (i-1))]. \qquad (5)$$

However, in this case, the first $X[1]$ and the last $X[N]$ images are always given as an input into each network. Additionally, middle part of the input can be overlapped. Then, the same information is overlapped into each network. As a result, the same spatial features are extracted. This is not good situation to learn the given input sequences. The example of the original structure of picking 3, 5, and 7 input frames is in Figure 7.

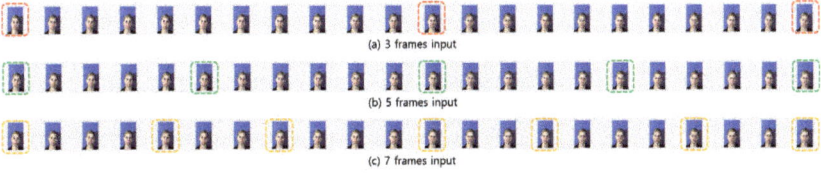

Figure 7. The example of the original structure of selecting input frames.

As in Figure 7, when n = 22, which means the sequence has 22 image frames, 3 frames of input are selected as S(3) = {X[0], X[11], X[21]}. In the case of 5 frames of input, S(5) is chosen as {X[0], X[5], X[11], X[16], X[21]}, and 7 frames of input sequence are constructed as S(7) = {X[0], X[4], X[7], X[11], X[14], X[18], X[21]}. All of them have the same images of X[0], X[11], and X[21] when constructing input sequences. In terms of information, the overlapped portion is not desirable to make reliable learning.

In order to solve this problem, we designed an input frame structure that can make a minimized overlapped between the generated input sequences. We extracted frames with regular intervals the same as the existing structure, but it made a different condition by making the start and end points different. The equation for the structure of the number of 3, 5, and 7 input frames $S(N)$ from the original X sequence where the number of the sequence is n as follows:

$$S(N) = X[1], X[2], ..., X[N],\qquad(6)$$

where

$$X[i] = \begin{cases} X[0 + round(\frac{(n-1)-2}{3-1} \times (i-1)]), & N = 3 \text{ input frames.} \\ X[2 + round(\frac{(n-1)-2}{5-1} \times (i-1)]), & N = 5 \text{ input frames.} \\ X[1 + round(\frac{(n-1)-2}{7-1} \times (i-1)]), & N = 7 \text{ input frames.} \end{cases} \qquad(7)$$

The start and end point of the seven input frames were set between the start and end points of the three and five input frames. In our example, seven frames was the largest number of the selected frames in a sequence with the number of n. If the start and the end point of seven input frames is shifted by one order from other input frames, the probability of overlap may be decreased. The example of the designed minimum overlapped frame structure, which selects three, five, and seven of input frames, is shown in Figure 8.

Figure 8. The example of selecting input frames through the minimum overlapped frame structure.

When n = 22, three frames of input have S(3) = {X[0], X[10], X[19]}. Five frames of the input sequence can be selected as S(5) = {X[2], X[7], X[12], X[16], X[21]}, and seven frames of input are constructed by S(7) = {X[1], X[4], X[8], X[11], X[14], X[17], X[20]}. None of the input images overlap as shown in Figure 8. The proposed structure can give more spatio-temporal information to extract features in the neural network. With the suggested three multi-depth network, full frames cannot be utilized. However, if we add one or two more different depth networks, then we can utilize almost-full frames with a larger frame rate for our FER task.

3.2. Data Preparation

3.2.1. Data Augmentation

For FER, we needed enough datasets of human faces. However, most of the FER databases have been labeled with a well-controlled environment, and it needs a high-cost task. Therefore, there are not enough datasets for the experiment in most cases. When training through deep learning with insufficient datasets, the network can be easily overfitted. Therefore, most researchers use data augmentation to solve this overfitting problem.

Data augmentation is largely divided into two types. The first method is to utilize some deep learning technologies such as autoencoder (AE) [22,55] or generative adversarial networks (GAN) [56]. Usually, autoencoder (AE) [22,55] with generative adversarial networks (GAN) [56] together could be used for input data augmentation. The second method is augmentation through image pre-processing like rotation, skewing, and scaling. Flipping horizontally is also effective in increasing the dataset. This is effective to increase the number of data while maintaining the geometric relationship between the eyes of the face image and important parts of the face such as the nose and mouth. Another method is to add noise to the image. This method includes salt and pepper noise, speckle noise, and Gaussian noise. In [57], the amount of the dataset was increased by 14 times through horizontal flipping and rotation. Figure 9 shows data augmentation using image pre-processing.

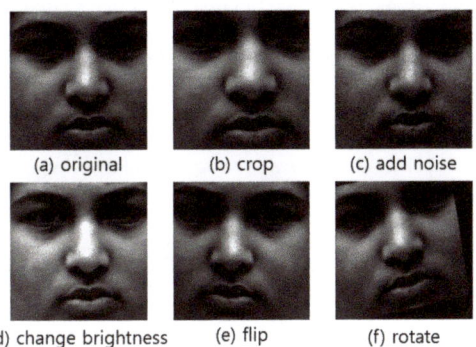

Figure 9. The example of data augmentation.

In this experiment, the second data augmentation method was used to increase the amount of the dataset. Table 1 shows the number of the original input dataset in each database. The CK+ database contains images labeled with "contempt," but other databases do not have this label [58,59]. Therefore, we excluded sequences labeled with "contempt" to establish the same experimental conditions. For the MMI dataset [60], we separated frames for each emotion before making inputs.

In the case of the GEMEP-FERA database [61], the total number of the emotion class was 5. However, one of them was not "neutral" but had a label of "relief." We changed the label of "relief" into "neutral." In Table 1, the first row is an abbreviation for the emotion classes such as neutral, anger, disgust, fear, happiness, sadness, and surprise in that order. The AFEW database [62] has already been divided into training, validation, and test datasets. However, the test dataset had no annotation about the expression. Therefore, we used the provided train dataset for the train, and the validation dataset was used for the test stage.

Table 1. The number of the original input datasets.

	Neu.	Ang.	Dis.	Fea.	Hap.	Sad.	Sur.	Total
CK+	0	45	56	24	69	28	78	300
MMI	0	33	32	28	42	32	41	208
FERA	31	32	—	30	31	31	—	155
AFEW(Train+Val.)	143	146	77	80	156	119	74	795
AFEW(Test)	63	61	39	41	60	54	43	361

The expression input data set was constructed in the following way. First, several frames were extracted from the input sequence through the minimum overlapped fame structure. If there was a separate sequence with the neutral label in the database, the neutral label dataset was also configured in the same way. However, in the case of a database where the neutral label was not specified, the neutral dataset was created through the first three frames of each sequence.

In the case of CK+ databases, there were no neutral labeled sequences. Therefore, a labeled emotion dataset was created through the minimum overlapped frame structure method, and an neutral dataset was created through the first three frames of each sequence. Because of this, datasets labeled with neutral existed in all sequences. Since each emotion-labeled dataset can only be created in a specific labeled sequence, the difference between the amount of neutral datasets and the other emotion datasets became large. In order to avoid the overfitting problem that can be caused by insufficient and biased distribution of the datasets, it was necessary to increase the emotion-labeled dataset.

Data augmentation was mainly performed to increase the amount of the emotion-labeled dataset so that the dataset was evenly distributed. For the created neutral expression dataset, each image was flipped horizontally to increase two times. For the other dataset, each image was flipped horizontally and rotated by {−7.5°, −5°, −2.5°, 2.5°, 5°, and 7.5°}. Through this process, the emotion-labeled dataset increased 14 times. Table 2 shows the specific values of the increased datasets for the CK+, MMI, and GEMEP-FERA datasets. In particular, we augmented two times for the neutral dataset, which was created from all emotion-labeled sequences due to no neutral emotion in the MMI dataset. The '−' symbol in Table 2 means that the class does not exist in the GEMEP-FERA dataset.

Table 2. The number of the augmented input datasets.

	Neu.	Ang.	Dis.	Fea.	Hap.	Sad.	Sur.	Total
CK+	600	630	784	336	966	392	1092	4800
MMI	416	462	448	392	588	448	574	3328
FERA	434	448	–	420	434	434	–	2170
AFEW(Train+Val.)	572	584	308	320	624	476	296	3180
AFEW(Test)	126	122	78	82	120	108	86	722

The provided AFEW train dataset, which was used as a train and validation dataset in our experiment, was augmented four times. We flipped horizontally and rotated by {−2.5°, 2.5°}. The provided AFEW validation dataset, which was used as a test dataset in our experiment, was augmented two times by flipping horizontally. The result of the augmented AFEW dataset is in the fifth and sixth rows of Table 2.

3.2.2. Making Neutral Label of Dataset

The CK+ database is composed of images to go from neutral to the peak of expression. Thus, the neutral sequence in the CK+ database is at the beginning of the video. To make three consecutive frames as inputs, the first three frames were assigned to the frames labeled as neutral. Input consisted by five frames was made by using the first frame once, the second frame twice, and the third frame twice among the first three consecutive frames. Input consisted by seven frames was created by using the first frame twice, the second frame twice, and the third frame three times among the first three consecutive frames. Figure 10 shows an example of a "neutral" label frame extracted from a sequence.

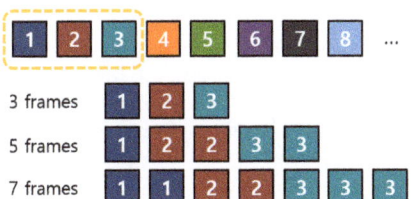

Figure 10. The method of creating a neutral dataset where the neutral label is not specified.

Unlike the CK+ database, the MMI database had an emotion flow, which was "neutral" to one of the peaks of expression and then to "neutral." We judged that the peak of the emotion was in the middle of the video. Therefore, the dataset was created using only the half that was the first to middle sequence out of the total sequence. Then, it had the same emotion flow like in the CK+ database as the "neutral" emotion to the peak of one expression. The dataset for the neutral expression was made through the same method, which created the neutral dataset from the CK+ database.

On the other hand, the GEMEP-FERA database did not have a label for "neutral" but a label for "relief." In order to match the conditions with other databases, we defined the "relief" as the "neutral" label.

3.3. 3D Convolutional Neural Network (3D CNN)

Spatial and temporal information was simultaneously captured using a 3D convolution and a 3D input dataset. Unlike the kernel used in 2D CNN, 3D CNN has a 3D cube-shaped convolution kernel, which has one more depth in the time axis. This preserves the time information of the input sequence and creates an output that forms the volume. Therefore, motion information can be obtained by connecting the feature map of the convolutional layer from multi-frames as input. Additionally, it considers adjacent pixels within the frame like the operation of 2D convolution at the same time. Therefore, spatial and temporal information can be simultaneously extracted through 3D convolution.

Shuiwang et al. [24] have explained 3D CNN mathematically. The value at position (x, y, z) on the j-th feature map in the i-th layer is given by:

$$v_{ij}^{xyz} = tanh\left(b_{ij} + \sum_{m} \sum_{p=0}^{P_i-1} \sum_{q=0}^{Q_i-1} \sum_{r=0}^{R_i-1} w_{ijm}^{pqr} v_{(i-1)m}^{(x+p)(y+q)(z+r)}\right), \quad (8)$$

where (p, q) is the spatial dimension index, r is the temporal dimension index of the kernel, w_{ijm}^{pqr} is the (p, q, r)-th value of the kernel connected to the m-th feature map in the previous layer, and R_i is the size of the 3D kernel. $tanh()$ assumed that activation function is the hyperbolic tangent, so other activation function can also be used.

In this study, we used a 3D CNN from study [26], which is called an "Appearance Network," as a basic model to capture spatio-temporal information. Figure 11 shows the detailed configuration of the network.

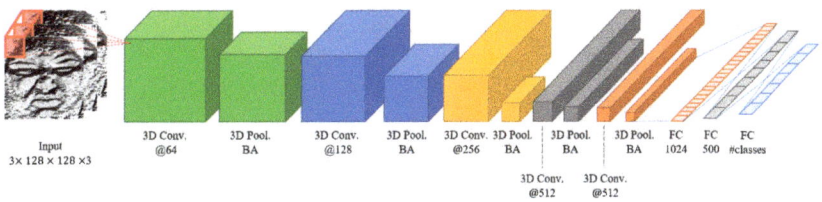

Figure 11. The architecture of 3D CNN from study [26], which has five 3D convolution layers.

First, the 3D convolutional layer extracts spatial and temporal features. All convolutional layers use a 5 × 5 × 3 kernel and a restricted linear unit (ReLU) activation function. In addition, 3D pooling is applied to reduce the number of parameters and cope with changes in the position of image elements. In this case, the pooling layer is max pooling that transfers only the maximum value of the volume area. After the maxpooling operation, the size of the feature map is reduced. Due to 3D pooling, dimension reduction on the time axis also occurs. The maximum value of 2 × 2 × 2 blocks is mapped to a single pixel of the output 3D feature map.

After max pooling layers, a batch normalization layer follows. Batch normalization is one of the ideas for preventing the disappearance or explosion of the gradient [63]. During deep learning training, if the hierarchy is deep and the number of epochs increases, the slope may explode or disappear. This problem arises because the scale of the parameters is different. This means the distribution of input to each layer or activation function of the network would be better to be controlled in the signal scale. To solve this problem, the input distribution needs to be normalized. However, this method is very complicated because the covariance matrix and the inverse matrix must be calculated. Instead, through batch normalization, the mean and standard deviation are obtained from each feature rather than the entire dataset, and they are normalized for each feature.

At the end of the network, emotions are classified as consecutive values through the softmax function. However, this classification module is not used in this study because we designed different joint fusion classifier based on the self-attention.

3.4. Joint Fusion Classifier Using Self-Attention

In this section, a joint fusion classifier is designed for a combination of multiple networks. This classifier serves to classify facial expressions based on various pieces of information by combining features extracted from each different input frame. In other words, it is possible to obtain more accurate results by supplementing the results of each network. Here, feature vector 1, ..., N were extracted to make the final 3D features from each depth network in Figure 11. In this study, there were three features since we employed three depth networks. When we extended this up to the N depths network, we could obtain N number of features before the classification module.

Additionally, we employed a squeeze-and-excitation network (SENet) for self-attention [46]. For any given transformation $F_{tr} : X \to U, X \in R^{H' \times W' \times C'}, U \in R^{H \times W \times C}$ (e.g., a convolution or a set of convolutions), we employed the squeeze-and-excitation (SE) block [46] to perform feature re-calibration as follows. In this structure, the features U are first passed through a squeeze operation, which aggregates the feature maps across spatial dimensions $H \times W$ to produce a channel descriptor. This descriptor embeds the global distribution of channel-wise feature responses, enabling information from the global receptive field of the network to be leveraged by its lower layers. This is followed by an excitation operation, in which sample-specific activation, learned for each channel by a self-gating mechanism based on channel dependence, govern the excitation of each channel. The feature maps U are then re-weighted to generate the output of the SE block, which can then be fed directly into subsequent layers.

Therefore, we could obtain emphasized and reinforced features through self-attention. Those features were concatenated in one-dimension and fed into the joint fusion classifier, which is depicted in Figure 12.

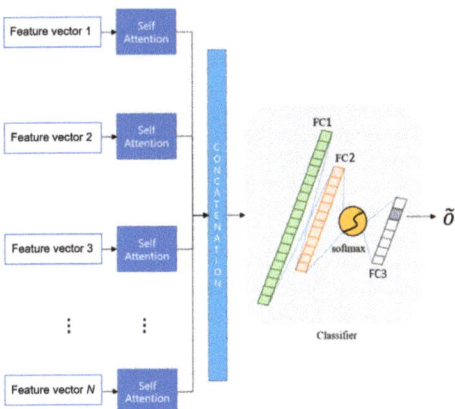

Figure 12. The architecture of the joint fusion classifier using self-attention.

In Figure 12, a joint fusion classifier was composed as follows: fully connected (FC) layer one and fully connected (FC) layer two of each network use ReLU. Fully connected (FC) layer three uses the softmax as an activation function. Additionally, cross entropy was used as the loss function, and loss was reduced by using the Adam optimizer. This determined the final emotion and used the same training dataset for each network to use it.

As mentioned in the above, we designed a multi-depth network based on multi-rate feature fusion for efficient facial expression recognition. Additionally, we developed a new image normalization and different depth networks as frame rates to give more robustness for various datasets. We verified the robustness and effectiveness of the proposed algorithm through experiments.

4. Experimental Results and Discussion

This section introduces the experiment and its environment in detail. We present and analyze the performance through several experimental results. Additionally, we compare the proposed FER algorithm with other latest algorithms. To train this network, the Adam optimizer was used with the default parameter setting [64]. We implemented all methods on a GPU server with Intel i-7 CPU and GTX 1080 Ti 11G memory.

4.1. Ablation Study

4.1.1. Performance of Image Normalization

This experiment confirmed the better performance when the image was normalized as described in Section 3.1.1.3. In the AFEW dataset, most of the image sequences were not taken from the controlled environment but were the same as in real-life conditions. Therefore, the brightness of the images varied, even being too dark or too bright. By using image normalization, we could overcome such problems, and the result of using image normalization is shown in Figure 13.

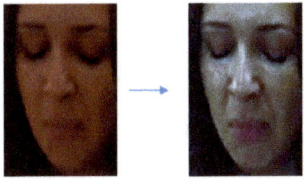

Figure 13. Examples of applying image normalization in AFEW dataset.

The results of image normalization in CK+, MMI, GEMEP-FERA, and AFEW datasets is in Table 3. In MMI and GEMEP-FERA datasets, this method mostly showed a better result than not using image normalization. The bold letter in the Table 3 means a better or same accuracy than not using image normalization. In the CK+ database, the employed image normalization was 0.14% better on average. However, in MMI, GEMEP-FERA, and AFEW datasets, most of the results using image normalization showed better performances of 0.7%, 0.61%, and 0.23% on average.

Table 3. The results of image normalization.

Datasets	Input Frames	Depth of Network	Image Normalization	
			Not Used	Used
CK+	3	5	98.65	98.02
		10	98.02	**98.33**
		15	97.92	97.81
MMI	3	5	95.58	**96.19**
		10	94.97	**97.1**
		15	94.40	93.75
FERA	3	5	98.85	**100**
		10	99.77	**99.77**
		15	99.31	**100**
AFEW	3	5	28.32	**28.95**
		10	26.59	**27.49**
		15	23.41	**23.89**

4.1.2. Correlation between Depth of the Network and Frame Rate of Input

This experiment was to find out the correlation between the depth of the network model and the number of the input frame. We took the experiment with applying and transforming the depth of the base model (5 layers) based on the 3D appearance network [26]. We gave three, five, and seven frames input into the 3D CNN with 5 (base model), 10, 15, 20, and 25 layers. As mentioned, we gave the depths of the models as 5 layers, 10 layers, 15 layers, 20 layers, and 25 layers to check on the relationship.

We used CK+, MMI, and GEMEP-FERA datasets to deduce the relationship. The results of the experiment by combining each depth of the network and input frame rate are shown in Table 4. The bold face denotes the maximum accuracy for each network depth according to input frame rate. In Figure 14, it was converted into a graph to visually show the results of Table 4. The dotted lines indicate the trend line. The result shows that if the depth of the model and the frame rate of the input are proportional, then the accuracy is inclined to increase. This means the accuracy is higher as the depth of the model is large and the number of frames of the input increases. Additionally, as the depth of the model is shallow and the number of frames of the input is smaller, the accuracy tends to be high. We utilize this observation to design our multirate-based network model.

Table 4. The results of correlation between the depth of network and frame rate of the input.

Datasets	Depth of Network	Input Frames	Image Normalization
CK+	5	3	98.45
		5	98.34
		7	98.23
	10	3	98.89
		5	97.90
		7	98.31
	15	3	97.90
		5	97.23
		7	98.45
MMI	10	3	96.65
		5	95.88
		7	93.75
	15	3	89.84
		5	92.99
		7	94.67
	20	3	90.85
		5	89.94
		7	92.84
FERA	5	3	100
		5	99.54
		7	99.54
	15	3	100
		5	100
		7	99.77
	25	3	98.85
		5	99.77
		7	99.77

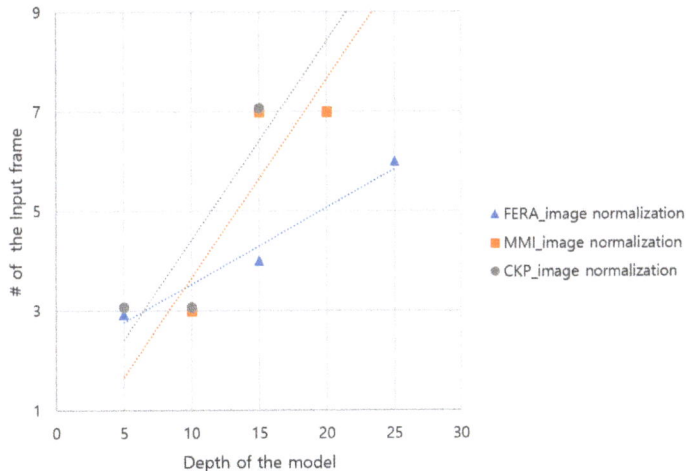

Figure 14. The graph of correlation between the depth of network and frame rate of the input.

4.1.3. Performance of the Minimum Overlapped Frame Structure

In this experiment, when creating the input dataset structure that is used in multiple networks, we verified that more various temporal information is helpful for learning. The minimum number of frames in the dataset was set to nine frames. Previously, the input dataset entering each network was determined as follows. If there is an arbitrary sequence of images, the total number of images is divided by equal intervals to obtain the required number of input frames. In this case, the beginning and end of three frames of input, five-frames of input, and seven frames of input always contained the same image. It means that the probability of overlapping the intermediate image was also high. In order to compensate for this problem, the method described in Section 3.1.2 was designed to create an input frame that does not overlap as much as possible. Because of the minimum overlapped frame structure, it was possible to give more various information when the network was learning.

Based on the correlation between the depth of the network and the frame rate of the input in Section 4.1.2, we fed 3 frames of input into the 3D CNN with 5 layers, 5 frames input into the 3D CNN with 10 layers, and 7 frames input into the 3D CNN with 15 layers. We obtained the feature from the networks without using image normalization and LBP feature extraction. Through Table 5, it can be seen that providing a variety of information to the network improves the performance in all of the databases. In the CK+, MMI, and GEMEP-FERA datasets, better performances of about 1.97%, 1.53%, and 0.46%, respectively, were shown. Moreover, the network using a minimum overlapped frame structure showed an improvement of 0.97% in the AFEW database.

Table 5. The results of using minimum overlapped frame structure.

Datasets	Minimum Overlapped Frame Structure	
	Not Used	Used
CK+	96.88	98.85
MMI	89.48	91.01
FERA	99.31	99.77
AFEW	27.70	28.67

4.1.4. Performance of Self-Attention Module

We also fed 3 frames into the 3D CNN with 5 layers, 5 frames into the 3D CNN with 10 layers, and 7 frames into the 3D CNN with 15 layers using a minimum overlapped frame structure, and we did not use the image normalization and LBP feature extraction. When the features came out from each network, we reinforced the feature using the self attention. Then, we concatenated the reinforced features into one-dimension and fed them into the joint fusion classifier.

We checked whether the self-attention module reinforced the features or not by comparing between the concatenated feature without the self-attention module and the concatenated feature with the self-attention module. The result is shown in Table 6. We can see that the self-attention module reinforced the feature and improved the FER performance in most of the databases. In the CK+, MMI, and GEMEP-FERA databases, the proposed scheme showed about 0.21%, 0.91%, and 0.23% better performances, respectively. Additionally, in the AFEW database, it showed a 0.42% better performance with the self-attention module.

Table 6. The results of the performance using self-attention.

Datasets	Self-Attention	
	Not Used	Used
CK+	98.85	99.06
MMI	91.01	91.92
FERA	99.77	100.00
AFEW	28.67	29.09

4.1.5. Effectiveness of Multi-Depth Network Structure

To show the effectiveness of the proposed multi-depth network structure, we tested a single layer network, which was from [26], as shown in Figure 11. We set three frames as the input sequence. For obtaining the results, we used a 10-fold validation approach.

Table 7 summarizes the number of trainable parameters of the proposed three-depths network model. It was assumed that three frames were given as input. It is also showed only the layers with trainable parameters in the entire network. As shown in the table, the number of layers in the individual network increased to a multiple of fie according to the number of frames given as input, and the number of parameters increased significantly accordingly. The outputs of each network were finally combined into the last three FC-layers, with the total number of parameters including them reaching about 237 million. If we extend the proposed network with more depths, then the complexity will be further increased.

Table 7. Summary of trainable parameters of the proposed multi-depth network (*input shape = (3, 128, 128, 1)*).

Structure	Depth Network 1		Depth Network 2		Depth Network 3	
	Layers	Params	Layers	Params	Layers	Params
Conv3d +MaxPool +BatchNorm	5	14,720 615,040 2,458,880 9,832,960 19,663,360	10	14,720 307,520 615,040 1,229,440 2,458,880 4,916,480 9,832,960 19,663,360 19,663,360 19,663,360	15	14,720 307,520 307,520 615,040 1,229,440 1,229,440 2,458,880 4,916,480 4,916,480 9,832,960 19,663,360 19,663,360 19,663,360 19,663,360 19,663,360
Self-attention	1	32,065	1	32,065	1	32,065
Fully connected				1,537,024 512,500 3,507		
Total params				237,244,586		

For the CK+ dataset, the proposed multi-depth network gave slightly better accuracy than the single network in Table 8. Additionally, we observed up to 7% of the accuracy in the MMI and GEMEP-FERA datasets. From these results, we can conclude that the proposed multi-depth network was effective for the facial expression recognition task.

Table 8. The performance results of the proposed network (multi-depth network) and single network (%).

Datasets	CK+	MMI	GEMEP-FERA
Single Network	96.11	92.52	93.38
Proposed (Multi-depth network)	96.23	96.69	99.79

4.2. Overall Accuracy Performance of the Proposed Scheme

In this section, we demonstrate that the proposed scheme shows competitive performance compared with the recent existing methods. Among various techniques for facial expression recognition, we compared with spatio-temporal network approaches or hybrid network approaches. Table 9 shows input construction and model setting of the recent existing methods, which were compared with the proposed method.

Table 9. Analysis of the recent existing methods for performance comparison.

Method	Datasets	Input Construction	Models
3DIR [65]	CK+, MMI, GEMEP-FERA	Multiple frames, facial landmarks	3D CNN, LSTM, Inception-ResNet
STCNN-CRF [66]	CK+, MMI, GEMEP-FERA	Multiple frames	2D CNN, CRF, Inception-ResNet
CNN-CTSLSTM [67]	CK+, MMI, AFEW	Multiple frames, facial landmarks	VGG-CTSLSTM, LEMHI-VGG
DDL [68]	CK+, MMI	Multiple frames	DFEM, DDM
DJSTN [26]	CK+, MMI, GEMEP-FERA	Multiple frames, facial landmarks	Hybrid network (App., Geo.)
FDRL [69]	CK+, MMI	Multiple frames	ResNet-18, FDN, FRN
MC-DCN [70]	CK+, MMI, GEMEP-FERA	Multiple frames	Hybrid network (C3Ds)

For experiments, we used three datasets: the CK+, MMI, and GEMEP-FERA datasets. The number of image sequences in each dataset was listed in Table 2. We used 3 frames, 5 frames, and 7 frames as input, and the multi-depth network was composed of 5 layers, 10 layers, and 15 layers. We used self-attention to reinforce the features, which came from each network and fed into the joint fusion classifier.

The results from the 10-times trial on the CK+ dataset are in Table 10. "Without Pre-processing" means that we did not use the image normalization, the LBP feature extraction, the minimum overlapped frame structure, and the self-attention module. In contrast, "With Pre-processing" means that we used all proposed image pre-processing methods, including the minimum overlapped frame structure and the self-attention module. The average accuracy was shown as the bold face in each processing. For the CK+ dataset, the network performance of "Without Pre-processing" showed better results—about 1.11% on average. This CK+ dataset is a very static one. However, the proposed scheme was based on several video frames to extract more temporal information. This means that the proposed algorithm works well for more dynamic video sequences.

The accuracy comparisons of each method using the CK+ database is shown in Table 11. For the CK+ database, the proposed scheme which was denoted as the bold face, did not get the best result compared with some existing methods [26,68–70].

Table 10. Overall accuracy and improvement on the CK+ dataset (%).

	Without Preprocessing	With Preprocessing
1	96.88	95.31
2	97.08	95.52
3	97.60	95.94
4	97.71	96.15
5	97.60	96.25
6	97.81	96.77
7	97.71	97.40
8	97.08	95.63
9	96.88	96.67
10	97.08	96.67
Average	**97.33**	**96.23**

Table 11. Comparison results of accuracy in the CK+ dataset (%).

Methods	Accuracy
3DIR [65]	93.21
STCNN-CRF [66]	93.04
CNN-CTSLSTM [67]	93.90
DDL [68]	99.16
DJSTN [26]	99.21
FDRL [69]	99.54
MC-DCN [70]	95.50
Proposed Scheme	**96.23**

The results from the 10-times trial on the MMI dataset are in Table 12. The proposed scheme showed a better result by about 4.79% on average (as the bold face) than "Without Preprocessing." The comparison of experimental results showed the outperformed results for the MMI dataset in Table 13. Here, the bold face denotes the performance of the proposed scheme.

Table 12. Overall accuracy and improvement on the MMI dataset (%).

	without Preprocessing	with Preprocessing
1	89.48	97.87
2	92.07	98.02
3	92.23	95.27
4	92.53	96.95
5	93.29	97.26
6	91.31	95.12
7	92.53	95.43
8	89.18	95.88
9	92.99	97.25
10	93.45	94.35
Average	**91.91**	**96.69**

Table 13. Comparison results of accuracy in the MMI dataset (%).

Methods	Accuracy
3DIR [65]	77.50
STCNN-CRF [66]	68.51
CNN-CTSLSTM [67]	78.4
DDL [68]	83.67
DJSTN [26]	87.88
FDRL [69]	85.23
MC-DCN [70]	78.6
Proposed Scheme	**96.69**

Additionally, the proposed method outperformed in the GEMEP-FERA dataset. The result from the 10-times trial on the GEMEP-FERA dataset is displayed in Table 14. The network performance of "With Pre-processing" showed better results of about 0.64% in average (as the bold face) than "Without Pre-processing." Table 15 shows the comparison of experimental results in the GEMEP-FERA dataset. The proposed scheme (the bold face in average) achieved an improvement of 8%, at least compared to the recent methods.

Table 14. Overall accuracy and improvement on the GEMEP-FERA dataset (%).

	Without Preprocessing	With Preprocessing
1	99.31	100.00
2	99.77	100.00
3	98.62	99.54
4	99.54	99.77
5	99.54	100.00
6	98.16	99.77
7	97.24	99.54
8	99.54	100.00
9	99.77	99.54
10	100.00	99.77
Average	**99.15**	**99.79**

Table 15. Comparison results of accuracy in the GEMEP-FERA dataset (%).

Methods	Accuracy
3DIR [65]	77.42
STCNN-CRF [66]	66.66
DJSTN [26]	91.83
MC-DCN [70]	78.3
Proposed Scheme	**99.79**

The proposed method showed a little weak performance on the CK+ dataset. However, in the MMI and GEMEP-FERA datasets, it showed the highest performance. According to the results of the CK+, MMI, and GEMEP-FERA datasets, the proposed model showed better performance in the more complex dataset.

In the AFEW dataset, the result is shown in Table 16. The AFEW dataset is well known as data capture in a very wild environment. The network performance of "With Pre-processing" showed a result that was about 3.32% better than "Without Pre-processing" by using only video data. From this result, we can expect that the proposed scheme can improve the recognition accuracy of the facial expression in real environments.

Table 16. Overall accuracy and improvement on the AFEW dataset (%).

	Without Preprocessing	With Preprocessing
Accuracy	27.70	31.02

For the processing time of the proposed scheme, the inference time was measured. This inference time contained the consumed time of the data pre-processing, the construction of frame structures, and the prediction for the final decision. When testing the proposed multi-depth network (three layers and three, five, and seven frames of input), the inference time was measured by about 102.0 ms on our GPU server with an Intel i7 CPU and GTX 1080 Ti 11G memory. In terms of the frame processing rate, a value of 9.8 frames per second (FPS) was obtained. When we used a single0layer network with an input of three frames, as shown in Figure 11, 49.3 ms was measured due to a very small network structure.

5. Conclusions

We proposed a robust facial expression recognition algorithm on the variation of datasets and different facial expression acquisition conditions. The proposed scheme extracted various features by combining several networks based on external features and classified them by putting them in a joint fusion classifier. This network simultaneously extracted spatial and temporal features using 3D CNN to overcome the problem of the existing 2D CNN model trained only with spatial features. In addition, in order to obtain a various features, we designed a multi-depth network structure by multiple input frames which were the least overlapped and composed of LBP features. The features extracted from each network were reinforced through the self-attention module. Then, these were combined and fed into the joint fusion network to newly learn and classify the emotions.

Through experiments, we found the correlation between the number of input frames and the depth of the network. When the number of frames increases, the network depth increases. When the number of frames decreases, the shallower the network depth, which showed the better performance. Through comparative analysis, we proved that the proposed multirate feature fusion scheme could achieve more accurate results than the state-of-the-art methods. The performance of the proposed model enhanced by 96.23%, 96.69%, and 99.79% the average accuracy of the CK +, MMI, and GEMEP-FERA datasets, respectively. Additionally, a 31.02% accuracy was achieved in the AFEW dataset through the features enhanced by the self-attention module and the proposed multi-depth network structure.

Author Contributions: Conceptualization, B.-G.K.; methodology, B.-G.K. and S.-J.P.; formal analysis, B.-G.K. and S.-J.P.; investigation, B.-G.K. and S.-J.P.; writing—original draft preparation, S.-J.P.; writing—review and editing, B.-G.K. and N.C.; supervision, B.-G.K. and N.C. All authors have read and agreed to the published version of the manuscript.

Funding: This research received no external funding.

Institutional Review Board Statement: Not applicable.

Informed Consent Statement: Not applicable.

Data Availability Statement: All sources and data can be found at https://github.com/smu-ivpl/MultiRateFeatureFusion_FER (accessed on 18 October 2021).

Acknowledgments: Authors thank the reviewers for valuable suggestion and comments for improving this paper.

Conflicts of Interest: The authors declare no conflict of interest.

References

1. Rosenblum, L.D. *See What I'm Saying: The Extraordinary Powers of Our Five Senses*, 1st ed.; W. W. Norton Company: New York, NY, USA, 2010.
2. Dang, L.T.; Cooper, E.W.; Kamei, K. Development of facial expression recognition for training video customer service representatives. In Proceedings of the 2014 IEEE International Conference on Fuzzy Systems (FUZZ-IEEE), Beijing, China, 6–11 July 2014; pp. 1297–1303.
3. Tischler, M.A.; Peter, C.; Wimmer, M.; Voskamp, J. Application of emotion recognition methods in automotive research. In Proceedings of the 2nd Workshop on Emotion and Computing—Current Research and Future Impact, Osnabrück, Germany, 9–13 October 2007; Volume 1, pp. 55–60.
4. Takahashi, K.; Mitsukura, Y. An entertainment robot based on head pose estimation and facial expression recognition. In Proceedings of the SICE Annual Conference (SICE), Akita, Japan, 20–23 August 2012; pp. 2057–2061.
5. Khowaja, S.A.; Dahri, K.; Kumbhar, M.A.; Soomro, A.M. Facial expression recognition using two-tier classification and its application to smart home automation system. In Proceedings of the 2015 International Conference on Emerging Technologies (ICET), Peshawar, Pakistan, 19–20 December 2015; pp. 1–6.
6. Blom, P.M.; Bakkes, S.; Tan, C.; Whiteson, S.; Roijers, D.; Valenti, R.; Gevers, T. Towards personalised gaming via facial expression recognition. In Proceedings of the AAAI Conference on Artificial Intelligence and Interactive Digital Entertainment, Raleigh, NC, USA, 3–7 October 2014; Volume 10.
7. Assari, M.A.; Rahmati, M. Driver drowsiness detection using face expression recognition. In Proceedings of the 2011 IEEE International Conference on Signal and Image Processing Applications (ICSIPA), Kuala Lumpur, Malaysia, 16–18 November 2011; pp. 337–341.
8. Kavitha, K.R.; Lakshmi, S.V.; Reddy, P.B.K.; Reddy, N.S.; Chandrasekhar, P.; Sisindri, Y. Driver Drowsiness Detection Using Face Expression Recognition. *Ann. Rom. Soc. Cell Biol.* **2021**, *25*, 2785–2789.
9. Sajjad, M.; Zahir, S.; Ullah, A.; Akhtar, Z.; Muhammad, K. Human behavior understanding in big multimedia data using CNN based facial expression recognition. *Mob. Netw. Appl.* **2020**, *25*, 1611–1621. [CrossRef]
10. Bediou, B.; Krolak-Salmon, P.; Saoud, M.; Henaff, M.-A.; Burt, M.; Dalery, J.; D'Amato, T. Facial expression and sex recognition in schizophrenia and depression. *Can. J. Psychiatry* **2005**, *50*, 525–533. [CrossRef] [PubMed]

11. Csukly, G.; Czobor, P.; Szily, E.; Takács, B.; Simon, L. Facial expression recognition in depressed subjects: The impact of intensity level and arousal dimension. *J. Nerv. Mental Dis.* **2009**, *197*, 98–103. [CrossRef] [PubMed]
12. Whitehill, J.; Bartlett, M.; Movellan, J. Automatic facial expression recognition for intelligent tutoring systems. In Proceedings of the 2008 IEEE Computer Society Conference on Computer Vision and Pattern Recognition Workshops, Anchorage, AK, USA, 23–28 June 2008; pp. 1–6.
13. Yang, M.T.; Cheng, Y.J.; Shih, Y.C. Facial expression recognition for learning status analysis. In *Human-Computer Interaction. Users and Applications, Proceedings of the 14th International Conference on Human-Computer Interaction, Orlando, FL, USA, 9–14 July 2011*; Jacko, J.A., Ed.; Springer: Berlin/Heidelberg, Germany, 2011; pp. 131–138.
14. Khalfallah, J.; Slama, J.B.H. Facial expression recognition for intelligent tutoring systems in remote laboratories platform. *Procedia Comput. Sci.* **2015**, *73*, 274–281. [CrossRef]
15. Ekman, P.; Friesen, W.V. Constants across cultures in the face and emotion. *J. Personal. Soc. Psychol.* **1971**, *17*, 124–129. [CrossRef] [PubMed]
16. Ekman, P.; Friesen, W.V. A new pan-cultural facial expression of emotion. *Motiv. Emot.* **1986**, *10*, 159–168. [CrossRef]
17. Freund, Y.; Schapire, R.E. A decision-theoretic generalization of on-line learning and an application to boosting. *J. Comput. Syst. Sci.* **1997**, *55*, 119–139. [CrossRef]
18. Viola, P.; Jones, M. Rapid object detection using a boosted cascade of simple features. In Proceedings of the 2001 IEEE Computer Society Conference on Computer Vision and Pattern Recognition (CVPR 2001), Kauai, HI, USA, 8–14 December 2001; Volume 1.
19. Viola, P.; Jones, M. Robust real-time object detection. *Int. J. Comput. Vision* **2001**, *4*, 4.
20. Dalal, N.; Triggs, B. Histograms of oriented gradients for human detection. In Proceedings of the 2005 IEEE Computer Society Conference on Computer Vision and Pattern Recognition (CVPR'05), San Diego, CA, USA, 20–25 June 2005; Volume 1, pp. 886–893.
21. LeCun, Y.; Haffner, P.; Bottou, L.; Bengio, Y. Object recognition with gradient-based learning. In *Shape, Contour and Grouping in Computer Vision*; Springer: Berlin/Heidelberg, Germany, 1999; pp. 319–345.
22. Rumelhart, D.E.; Hinton, G.E.; Williams, R.J. *Learning Internal Representations by Error Propagation*; Technical Report; California Univ San Diego La Jolla Inst for Cognitive Science: San Diego, CA, USA, 1985.
23. Hochreiter, S.; Schmidhuber, J. Long short-term memory. *Neural Comput.* **1997**, *9*, 1735–1780. [CrossRef]
24. Ji, S.; Xu, W.; Yang, M.; Yu, K. 3D convolutional neural networks for human action recognition. *IEEE Trans. Pattern Anal. Mach. Intell.* **2012**, *35*, 221–231. [CrossRef] [PubMed]
25. Kim, J.H.; Kim, B.G.; Roy, P.P.; Jeong, D.M. Efficient facial expression recognition algorithm based on hierarchical deep neural network structure. *IEEE Access* **2019**, *7*, 41273–41285. [CrossRef]
26. Jeong, D.; Kim, B.G.; Dong, S.Y. Deep Joint Spatiotemporal Network (DJSTN) for Efficient Facial Expression Recognition. *Sensors* **2020**, *20*, 1936. [CrossRef] [PubMed]
27. Tian, Y.I.; Kanade, T.; Cohn, J.F. Recognizing action units for facial expression analysis. *IEEE Trans. Pattern Anal. Mach. Intell.* **2001**, *23*, 97–115. [CrossRef] [PubMed]
28. Cootes, T.F.; Edwards, G.J.; Taylor, C.J. Active appearance models. *IEEE Trans. Pattern Anal. Mach. Intell.* **2001**, *23*, 681–685. [CrossRef]
29. Ekman, P.; Rosenberg, E.L. *What the Face Reveals: Basic and Applied Studies of Spontaneous Expression Using the Facial Action Coding System (FACS)*; Oxford University Press: New York, NY, USA, 1997.
30. Jepson, A.; Heeger, D. Linear subspace methods for recovering translational direction. In *Spatial Vision in Humans and Robots*; Cambridge University Press: Cambridge, UK, 1994.
31. Pearson, K. LIII. On lines and planes of closest fit to systems of points in space. *Lond. Edinb. Dublin Philos. Mag. J. Sci.* **1901**, *2*, 559–572. [CrossRef]
32. Mika, S.; Ratsch, G.; Weston, J.; Scholkopf, B.; Mullers, K.R. Fisher discriminant analysis with kernels. In Proceedings of the Neural Networks for Signal Processing IX: Proceedings of the 1999 IEEE Signal Processing Society Workshop (Cat. No. 98th8468), Madison, WI, USA, 25 August 1999; pp. 41–48.
33. Comon, P. Independent component analysis, a new concept? *Signal Process.* **1994**, *36*, 287–314. [CrossRef]
34. Gabor, D. Theory of communication. Part 1: The analysis of information. *J. Inst. Electr. Eng. Part III Radio Commun. Eng.* **1946**, *93*, 429–441. [CrossRef]
35. Ojala, T.; Pietikainen, M.; Maenpaa, T. Multiresolution gray-scale and rotation invariant texture classification with local binary patterns. *IEEE Trans. Pattern Anal. Mach. Intell.* **2002**, *24*, 971–987. [CrossRef]
36. Krizhevsky, A.; Sutskever, I.; Hinton, G.E. Imagenet classification with deep convolutional neural networks. *Adv. Neural Inf. Process. Syst.* **2012**, *25*, 1097–1105. [CrossRef]
37. Ouellet, S. Real-time emotion recognition for gaming using deep convolutional network features. *arXiv* **2014**, arXiv:1408.3750.
38. Mollahosseini, A.; Chan, D.; Mahoor, M.H. Going deeper in facial expression recognition using deep neural networks. In Proceedings of the 2016 IEEE Winter Conference on Applications of Computer Vision (WACV), Lake Placid, NY, USA, 7–10 March 2016; pp. 1–10.
39. Kim, D.H.; Baddar, W.J.; Jang, J.; Ro, Y.M. Multi-objective based spatio-temporal feature representation learning robust to expression intensity variations for facial expression recognition. *IEEE Trans. Affect. Comput.* **2017**, *10*, 223–236. [CrossRef]
40. Baddar, W.J.; Lee, S.; Ro, Y.M. On-the-Fly Facial Expression Prediction using LSTM Encoded Appearance-Suppressed Dynamics. *IEEE Trans. Affect. Comput.* **2019**. [CrossRef]

41. Jung, H.; Lee, S.; Yim, J.; Park, S.; Kim, J. Joint fine-tuning in deep neural networks for facial expression recognition. In Proceedings of the IEEE International Conference on Computer Vision, Santiago, Chile, 7–13 December 2015; pp. 2983–2991.
42. Vetterli, M. A theory of multirate filter banks. *IEEE Trans. Acoust. Speech Signal Process.* **1987**, *35*, 356–372. [CrossRef]
43. Sutskever, I.; Vinyals, O.; Le, Q.V. Sequence to sequence learning with neural networks. In Proceedings of the Advances in Neural Information Processing Systems, Montreal, QC, Canada, 8–13 December 2014; pp. 3104–3112.
44. Vaswani, A.; Shazeer, N.; Parmar, N.; Uszkoreit, J.; Jones, L.; Gomez, A.N.; Kaiser, Ł.; Polosukhin, I. Attention is all you need. In Proceedings of the Advances in Neural Information Processing Systems, Long Beach, CA, USA, 4–9 December 2017; pp. 5998–6008.
45. Xu, K.; Ba, J.; Kiros, R.; Cho, K.; Courville, A.; Salakhudinov, R.; Zemel, R.; Bengio, Y. Show, attend and tell: Neural image caption generation with visual attention. In Proceedings of the International Conference on Machine Learning, PMLR, Lille, France, 6–11 July 2015; pp. 2048–2057.
46. Hu, J.; Shen, L.; Sun, G. Squeeze-and-excitation networks. In Proceedings of the IEEE Conference on Computer Vision and Pattern Recognition, Salt Lake City, UT, USA, 18–22 June 2018; pp. 7132–7141.
47. Woo, S.; Park, J.; Lee, J.Y.; Kweon, I.S. Cbam: Convolutional block attention module. In Proceedings of the European Conference on Computer Vision (ECCV), Munich, Germany, 8–14 September 2018; pp. 3–19.
48. Wang, X.; Girshick, R.; Gupta, A.; He, K. Non-local neural networks. In Proceedings of the IEEE Conference on Computer Vision and Pattern Recognition, Salt Lake City, UT, USA, 18–22 June 2018; pp. 7794–7803.
49. Ramachandran, P.; Parmar, N.; Vaswani, A.; Bello, I.; Levskaya, A.; Shlens, J. Stand-alone self-attention in vision models. *arXiv* **2019**, arXiv:1906.05909.
50. Zhang, S.; Zhu, X.; Lei, Z.; Shi, H.; Wang, X.; Li, S.Z. Faceboxes: A CPU real-time face detector with high accuracy. In Proceedings of the 2017 IEEE International Joint Conference on Biometrics (IJCB), Denver, CO, USA, 1–4 October 2017; pp. 1–9.
51. Dong, X.; Yan, Y.; Ouyang, W.; Yang, Y. Style aggregated network for facial landmark detection. In Proceedings of the IEEE Conference on Computer Vision and Pattern Recognition, Salt Lake City, UT, USA, 18–22 June 2018; pp. 379–388.
52. Canedo, D.; Neves, A.J. Facial expression recognition using computer vision: A systematic review. *Appl. Sci.* **2019**, *9*, 4678. [CrossRef]
53. Zhong, L.; Liu, Q.; Yang, P.; Liu, B.; Huang, J.; Metaxas, D.N. Learning active facial patches for expression analysis. In Proceedings of the 2012 IEEE Conference on Computer Vision and Pattern Recognition, Providence, RI, USA, 16–21 June 2012; pp. 2562–2569.
54. Ahonen, T.; Hadid, A.; Pietikäinen, M. Face recognition with local binary patterns. In Proceedings of the European Conference on Computer Vision, Prague, Czech Republic, 11–14 May 2004; pp. 469–481.
55. Ballard, D.H. Modular learning in neural networks. *AAAI* **1987**, *647*, 279–284.
56. Goodfellow, I.J.; Pouget-Abadie, J.; Mirza, M.; Xu, B.; Warde-Farley, D.; Ozair, S.; Courville, A.; Bengio, Y. Generative adversarial networks. *arXiv* **2014**, arXiv:1406.2661.
57. Yu, Z.; Liu, G.; Liu, Q.; Deng, J. Spatio-temporal convolutional features with nested LSTM for facial expression recognition. *Neurocomputing* **2018**, *317*, 50–57. [CrossRef]
58. Lucey, P.; Cohn, J.F.; Kanade, T.; Saragih, J.; Ambadar, Z.; Matthews, I. The extended cohn-kanade dataset (ck+): A complete dataset for action unit and emotion-specified expression. In Proceedings of the 2010 IEEE Computer Society Conference on Computer Vision and Pattern Recognition-Workshops, San Francisco, CA, USA, 13–18 June 2010; pp. 94–101.
59. Kanade, T.; Cohn, J.F.; Tian, Y. Comprehensive database for facial expression analysis. In Proceedings of the 2000 IEEE International Conference on Automatic Face and Gesture Recognition, Grenoble, France, 28–30 March 2000; pp. 46–53.
60. Pantic, M.; Valstar, M.; Rademaker, R.; Maat, L. Web-based database for facial expression analysis. In Proceedings of the 2005 IEEE International Conference on Multimedia and Expo, Amsterdam, The Netherlands, 6–8 July 2005; p. 5.
61. Bänziger, T.; Scherer, K.R. Introducing the geneva multimodal emotion portrayal (gemep) corpus. *Bluepr. Affect. Comput. Sourceb.* **2010**, *2010*, 271–294.
62. Dhall, A.; Goecke, R.; Lucey, S.; Gedeon, T. Collecting large, richly annotated facial-expression databases from movies. *IEEE Ann. Hist. Comput.* **2012**, *19*, 34–41. [CrossRef]
63. Ioffe, S.; Szegedy, C. Batch normalization: Accelerating deep network training by reducing internal covariate shift. In Proceedings of the International Conference on Machine Learning, PMLR, Lille, France, 7–9 July 2015; pp. 448–456.
64. Diederik, P.; Kingma, J.B. Adam: A Method for Stochastic Optimization. In Proceedings of the 3rd International Conference for Learning Representations, San Diego, CA, USA, 7–9 May 2015; pp. 1–15.
65. Hasani, B.; Mahoor, M.H. Facial expression recognition using enhanced deep 3D convolutional neural networks. In Proceedings of the IEEE Conference on Computer Vision and Pattern Recognition Workshops, Honolulu, HI, USA, 21–26 July 2017; pp. 30–40.
66. Hasani, B.; Mahoor, M.H. Spatio-temporal facial expression recognition using convolutional neural networks and conditional random fields. In Proceedings of the 2017 12th IEEE International Conference on Automatic Face & Gesture Recognition (FG 2017), Washington, DC, USA, 30 May–3 June 2017; pp. 790–795.
67. Hu, M.; Wang, H.; Wang, X.; Yang, J.; Wang, R. Video facial emotion recognition based on local enhanced motion history image and CNN-CTSLSTM networks. *J. Vis. Commun. Image Represent.* **2019**, *59*, 176–185. [CrossRef]
68. Ruan, D.; Yan, Y.; Chen, S.; Xue, J.H.; Wang, H. Deep disturbance-disentangled learning for facial expression recognition. In Proceedings of the 28th ACM International Conference on Multimedia, Seattle, WA, USA, 12–16 October 2020; pp. 2833–2841.

69. Ruan, D.; Yan, Y.; Lai, S.; Chai, Z.; Shen, C.; Wang, H. Feature Decomposition and Reconstruction Learning for Effective Facial Expression Recognition. In Proceedings of the IEEE/CVF Conference on Computer Vision and Pattern Recognition, Nashville, TN, USA, 19–25 June 2021; pp. 7660–7669.
70. Wu, H.; Lu, Z.; Zhang, J.; Li, X.; Zhao, M.; Ding, X. Facial Expression Recognition Based on Multi-Features Cooperative Deep Convolutional Network. *Appl. Sci.* **2021**, *11*, 1428. [CrossRef]

Article

Changes in Computer-Analyzed Facial Expressions with Age

Hyunwoong Ko [1,2,3,†], Kisun Kim [2,†], Minju Bae [1,2], Myo-Geong Seo [2], Gieun Nam [2], Seho Park [4], Soowon Park [5], Jungjoon Ihm [1,3] and Jun-Young Lee [1,2,*]

1. Interdisciplinary Program in Cognitive Science, Seoul National University, Seoul 08826, Korea; powerzines@snu.ac.kr (H.K.); minju1222@snu.ac.kr (M.B.); ijj127@snu.ac.kr (J.I.)
2. Department of Psychiatry, SMG-SNU Boramae Medical Center, Seoul National University College of Medicine, Seoul 03080, Korea; imkisun@snu.ac.kr (K.K.); coieyu102@cau.ac.kr (M.-G.S.); genam1006@cau.ac.kr (G.N.)
3. Dental Research Institute, School of Dentistry, Seoul National University, Seoul 08826, Korea
4. Behavioral Neuroscience Program, School of Medicine, Boston University, Boston, MA 02101, USA; sehopark@bu.edu
5. Division of Teacher Education, College of Liberal Arts and Interdisciplinary Studies, Kyonggi University, Suwon 16200, Korea; swpark1@kyonggi.ac.kr
* Correspondence: benji@snu.ac.kr
† These authors contributed equally to this work.

Abstract: Facial expressions are well known to change with age, but the quantitative properties of facial aging remain unclear. In the present study, we investigated the differences in the intensity of facial expressions between older (n = 56) and younger adults (n = 113). In laboratory experiments, the posed facial expressions of the participants were obtained based on six basic emotions and neutral facial expression stimuli, and the intensities of their faces were analyzed using a computer vision tool, OpenFace software. Our results showed that the older adults expressed strong expressions for some negative emotions and neutral faces. Furthermore, when making facial expressions, older adults used more face muscles than younger adults across the emotions. These results may help to understand the characteristics of facial expressions in aging and can provide empirical evidence for other fields regarding facial recognition.

Keywords: facial action unit; facial aging; facial expression; posed emotion

Citation: Ko, H.; Kim, K.; Bae, M.; Seo, M.-G.; Nam, G.; Park, S.; Park, S.; Ihm, J.; Lee, J.-Y. Changes in Computer-Analyzed Facial Expressions with Age. *Sensors* **2021**, *21*, 4858. https://doi.org/10.3390/s21144858

Academic Editor: Wataru Sato

Received: 17 April 2021
Accepted: 15 July 2021
Published: 16 July 2021

Publisher's Note: MDPI stays neutral with regard to jurisdictional claims in published maps and institutional affiliations.

Copyright: © 2021 by the authors. Licensee MDPI, Basel, Switzerland. This article is an open access article distributed under the terms and conditions of the Creative Commons Attribution (CC BY) license (https://creativecommons.org/licenses/by/4.0/).

1. Introduction

Expression and recognition of emotions through facial expressions are fundamental functions of basic communication. Facial expressions are critical for communicating with one's surroundings in terms of their role to convey the primary meaning of social information [1,2]. People can communicate and convey their emotions in diverse manners; however, facial expressions can be used in the most flexible way [3]. Investigating how facial movements are controlled and how people recognize others' facial expressions, therefore, is an essential way to understand the nature of human beings as social beings and can also facilitate emotional functioning.

It has been well established that emotional expression and recognition skills through facial expressions change with age [4,5]. A previous study showed older and young people a variety of facial expressions and confirmed how they recognized them [6]. Young and old people were both aware of expressions of positive emotion, while older people were less aware of negative facial expressions. In addition, the performance of the older group declined in sadness facial expression recognition but improvement in disgust facial expression recognition [7–9]. The older people were also more inclined to think that they felt happy when they were shown smiles [10]. A recent meta-analysis demonstrated that older adults showed lower performance on emotional face identification than a younger group of adults [11].

Owing to physical aging, sarcopenia, such as atrophy of facial skeleton, malposition of fatty muscles, and loss of soft tissue happen most commonly in the areas of the maxilla, mandible, and anterior nasal spine [12]. A previous study showed that human facial aging demonstrated a common pattern of morphological, chronological, and dermatological changes in various biomedical studies [13]. In an aspect of neuromuscular mechanism, voluntary facial expressions (i.e., posed facial expressions) using the lower part of the face are prominently controlled by the left hemisphere and vice versa [14–16]. Specifically, aging of the orofacial motor cortex, which involves involuntary facial expressions, can cause a decline in cognitive control for the lower part of the face [17,18]. While facial aging is natural and inevitable for most people, multiple studies have suggested there are several markers of facial expression and recognition in neuropathological changes including epilepsy [19], Parkinson's disease [20], Alzheimer's disease [21], and other neurocognitive disorders [22]. Despite this, identifying the quantitative characteristics of facial aging is still limited.

The posed facial expression, which is commonly exhibited on portrayal of other's facial expression, has distinct characteristics compared to spontaneous facial expression in aspects of neuromotor system and display rules. Whereas posed facial expression is generated cognitively within the pyramidal system, spontaneous facial expression exhibits independent motor control and is driven by extrapyramidal system [15,23]. The movements inherent to posed facial expression tend to display intended emotions in the context of social interactions (i.e., display rules), while spontaneous facial expression correspond to a primary emotional system [15,24]. Although, several studies have pointed out the limitations of the characteristic of the posed facial expression for its artificiality by actor's and variability by experimental conditions [25–27], research leveraging posed facial expression has clear advantages. For interpretability, posed facial expression is less ambiguous than spontaneous facial expression [28] and is also universal across the basic emotion [29]. Such universality has also been identified in recent study for East Asian population [27]. Since cumulative literatures have studied the pose facial expression [30], posed facial expression is may expected to be a valid indicator for investigating aging.

Quantitative measurements of facial expressions and their analyses has been an active research topic in behavioral science. Among several studies, a facial action coding system (FACS) [31,32] is the most widely used in this area. A series of facial muscle movements that represent facial expressions, termed as action units (AUs), can help a facial recognition-based analysis to be more standardized [33]. Since AUs were originally developed from basic emotion theory and manually rated by highly trained coders, the FACS-based AUs have had limited accessibility for standardization. Recently, automated computer vision and multidiscipline study for facial expression analysis have emerged [34]. These studies enable scaling facial expressions more feasible; facial aging study remains in three-dimensional (3D) morphometric [13,35] or electromyography (EMG) studies [36,37]. In that regard, little is known about quantitative facial aging.

Given that facial expressions are crucial indicators of human health status [38,39], applying machine learning algorithm techniques to facial expressions, such as computer-aided diagnosis (CAD) in the biomedical signal [40], and the medical imaging field [41], can contribute to digital health. This technique is often used in facial paralysis [42,43], face transplant [44], pain detection through facial expression [45], and neurologic studies such as those involving autism [46], Turner syndrome [47], and Parkinson's disease [48]. Since language production and discourse decrease with aging [49], identifying the characteristics of facial expressions in the older adults is a promising and challenging research area in gerontology, which can diagnose disease regardless of patient communication skills. Moreover, the uniqueness of facial expressions has led to consistent studies in the area of personal identification for health records [50], to improve performances on CAD and identification using facial expressions, to develop the algorithm, and to provide interpretable results for facial expressions with aging. Although there has been much work on automatic facial expression recognition in computer vision research, the algorithms have been experimentally validated primarily on younger faces. For facial expressions to be better used as digital markers related to aging, finding quantitative differences in facial changes with aging should be studied.

The aim of this study was to identify the characteristics of facial expressions based on the basic emotion theory and to compare the differences in facial expressions between younger and older adults for each basic emotion and AU, respectively. Additionally, a feature-selection approach was used to identify multivariate patterns of the changes in facial expressions related to aging. Finally, the predictive ability for selected AUs was evaluated.

2. Materials and Methods

2.1. Ethics Statement

This study was approved by the Institutional Review Board of the SMG-SNU Boramae Medical Center (IRB No. 30-2017-63), and all participants submitted written consent for participating in the study.

2.2. Participants

A total of 61 older adults and 115 younger adults were recruited for this study. The older adults were between 62 and 84 years old and recruited from the Alzheimer's disease research center of the SMG-SNU Boramae hospital. Healthy young participants were recruited from the university student participant pool and aged between 18 to 39. None of them had a history of psychiatric disorder. Major medical diseases, severe head injury, and visual impairment were excluded in all groups. Especially, all the older adults were free from the diagnosis criteria of Alzheimer's disease and depressive spectrum disorder with DSM-IV [51]. All medical judgements were determined by a board-certified psychiatrist (J.-Y.L.).

To screen the potential emotion related problems such as depression, anxiety, and alexithymia, participants were asked to answer self-reported measures: Beck Depression Inventory (BDI), Beck Anxiety Inventory (BAI), and Toronto Alexithymia Scale (TAS). The Korean version of BDI involves 21 questions to evaluate the severity of depression, with scores ranging from 0 to 63 [52,53]. A higher score indicates severe depressive symptoms, and the cutoff score is 18 in the Korean version [54]. The Korean version of BAI utilizes 21 questions to measure the severity of anxiety, with scores ranging from 0 to 63 [55]. A higher BAI score indicates severe anxiety symptoms with a cutoff score of 19 [56]. A twenty-item TAS was developed and validated to measure the severity of alexithymia. A score ranging from 20 to 100 [57,58], with a cutoff score at 61 was used for the Korean version [59]. The TAS is made up of three subscales: Difficulty identifying feeling, difficulty describing feeling, and externally oriented thinking. Neither group had an abnormal level of emotional problems (Table 1).

Table 1. Demographic characteristics across the groups.

		Younger Adults (n = 113)	Older Adults (n = 56)
		Mean ± SD	Mean ± SD
Age		21.9± 2.91	72.2± 4.72
Education		14.5± 1.10	9.8± 4.47
Sex, n (%)			
	Male	57 (50.4)	27 (48.2)
	Female	56 (49.6)	29 (51.8)
Usage of botox, n (%)		2 (1.8)	1 (1.7)
Left-handed, n (%)		8 (7.1)	1 (1.7)

Table 1. Cont.

	Younger Adults (n = 113)	Older Adults (n = 56)
	Mean ± SD	Mean ± SD
BDI	10.7±6.88	14.4± 11.04
BAI	25.1 4.28	25.3 ± 6.22
TAS	45.3± 10.52	50.3± 8.95

Note. Botox, botulinum toxin; BDI, Beck Depression Inventory; BAI, Beck Anxiety Inventory; TAS, Toronto Alexithymia Scale; SD, standard deviation; BOLD indicates statistically significant differences.

Since data for five older adults and two younger adults failed to pass the quality check, 169 of 176 participants were included in the analysis. Table 1 summarizes the demographic and clinical characteristics of the participants. Significant differences were found in age, education, left-handed, BDI score, and TAS score. Except for age, these variables were adjusted in further analyses.

2.3. Procedures

A series of photos containing six basic emotions and a neutral facial expression were presented to participants, which consisted of seven stimuli and had been selected by researchers from a photography dataset used in a previous study [50]. Instructions were given in both verbal and visual form, and the participants were asked to answer verbally for stimuli. Then, participants performed posed facial expressions for the given list of six basic emotions and the neutral emotion. For example, for happy facial expression, a photograph of a person with a happy face was presented; participants were asked to identify the emotion conveyed; and "make a happy face for 15 s towards the camera" to be video recorded. The facial stimuli were given once participants were fully aware of the instruction of the study. Examples of stimuli are shown in Figure 1. Each facial stimulus was presented for a maximum of 7 s; the researcher moved on to the next stimulus when the participant made a verbal response. Facial expressions were acquired for a total of 105 s for each emotion.

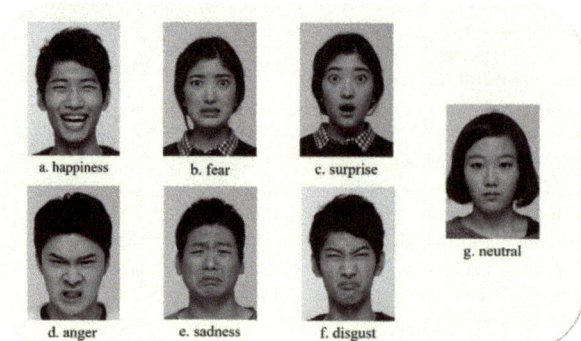

Figure 1. The facial stimuli representing the six basic emotions and the neutral emotion, adapted from [60].

2.4. Data Acquisition

The participants' video recordings of posed facial expressions were administered with a Canon EOS 70D DSLR Camera with a 50 mm prime lens, 720 p resolution, and 60 fps frame rate. The camera was positioned on a fixed stand approximately 120–140 cm above the floor to correctly capture the entire face of the participants. The posed facial expressions were recorded for 15 s after a clear instruction to imitate a previously recognized emotional face.

For each frame of the recorded videos, the presence and intensity were estimated using OpenFace 2.0, an open-source toolkit for facial behavior analysis, which consists of four pipelines: (1) facial landmark detection and tracking, (2) head pose estimation, (3) eye gaze estimation, and (4) facial expression recognition [34]. For analyzing facial expressions, OpenFace 2.0 recognizes facial expressions by detecting AU intensity and presence according to FACS [31]. Without using all the AUs listed in FACS, OpenFace 2.0 offers a subset of 18 AUs by cross-dataset learning, specifically, 01, 02, 04, 05, 06, 07, 09, 10, 12, 14, 15, 17, 20, 23, 25, 26, 28, and 45. The occurrences and intensities in AUs are estimated by using machine learning algorithms. The methods for AU estimation and analysis are described in more detail elsewhere [61]. In the present study, AU intensities were used to derive measures of individual emotional facial expression and six basic emotions were created according to emotional FACS (EMFACS) [62]. The EMFACS were based on the FACS that have been proven to have significant reliability for the assessment of human facial movements [63,64]. The highest intensity for each AU was calculated as the maximum score across all the video frames, which is validated in prior work [65]. Examples of each AU and emotion are shown in Table 2.

Table 2. Action unit descriptions and combination of each emotion.

No.	FACS Name	Facial Muscle (Location)
1	Inner brow raiser	Frontalis, pars medialis (U)
2	Outer brow raiser	Frontalis, pars lateralis (U)
4	Brow lowering	Depressor glabellae, depressor supercilli, currugator (U)
5	Upper lid raiser	Levator palpebrae superioris (U)
6	Cheek raiser	Orbicularis oculi, pars orbitalis (U)
7	Lid tightener	Orbicularis oculi, pars palpebralis (U)
9	Nose wrinkle	Levator labii superioris alaquae nasi (L)
10	Upper lip raiser	Levator labii superioris, caput infraorbitalis (L)
11	Nasolabial deepener	Zygomatic minor (L)
12	Lip corner puller	Zygomatic major (L)
14	Dimpler	Buccinator (L)
15	Lip corner depressor	Depressor anguli oris (triangularis) (L)
17	Chin raiser	Mentalis (L)
20	Lip stretcher	Risorius (L)
23	Lip tightener	Orbicularis oris (L)
25	Lip parting	Depressor labii, relaxation of mentalis, orbicularis oris (L)
26	Jaw drop	Masseter, temporal and internal pterygoid relaxed (L)
45	Blink	Levator palpebrae superioris, orbicularis oculi (U)
Emotion	AU combination	
Angry	04 + 05 + 07 + 23	
Disgust	09 + 15	
Fear	01 + 02 + 04 + 05 + 20 + 26	
Happy	06 + 12	
Sad	01 + 04 + 15	
Surprise	01 + 02 + 05 + 26	

Note. AU, action unit; FACS, facial action coding system; L, lower face; U, upper face.

2.5. Statistical Analysis

Descriptive statistics for demographic variables were calculated as mean scores and standard deviations. The difference in AU was compared, applying for multiple comparisons (followed by Bonferroni correction). Chi-squared tests were used to compare categorical outcomes such as sex and usage of botulinum toxin (botox). The correlation between age and the AU intensity was investigated. To explain multivariate profiles with respect to input features that were accurately distinguished from the older group, the adaptive least absolute shrinkage and selection operator (LASSO) ML algorithm were applied to the dataset [66]. The adaptive LASSO, which is a regularized regression method with L1-norm penalty [67] is a popular technique for simultaneous estimation and consistent variable selection [66]. It is a powerful model that performs regularization and feature selection, and it can provide model interpretability by excluding irrelevant features that are not related to the class from the model. L1 regularization, which penalizes elements of redundant complexity, focuses on the most significant features, and thus prevents overfitting of the data and is supported by well-grounded theoretical analysis [68]. The regression coefficients of unimportant variables shrank to 0 upon implementing the adaptive LASSO. In that regard, the adaptive LASSO algorithm provided interpretable results related to the older adults. Due to its high accessibility and low computational complexity as compared with other feature selection models, recently, this approach has been highly recommended in behavioral science [69].

In order to avoid the overfitting issue and to evaluate the generalizability of the results from the ML algorithms, 10-fold cross-validation was applied during the variable selection process [70]. First, the data were randomly split into a training set (66.7% of the data) and a test set (33.4% of the data). All the ML models were fitted using the training set, and classifications were separately made on the test and training datasets. The optimal parameter, lambda, was determined across 1000 iterations of 10-fold cross-validation to minimize the deviance of the model. Then, predictions were made on the test set based on the ML models trained in the training set. All reported p values have been adjusted for multiple comparison analyses.

3. Results

3.1. The Differences in Facial Expression between the Older Adults and Younger Adtuls

Figures 2 and 3 demonstrate the AU values of the older and younger adults for the neutral and emotional face. The results applied for multiple comparisons are presented in Table 3. In AU 06, 07, 12, and 14, older adults showed higher intensity compared to younger adults. For AU 45, older adults showed lower intensity than younger adults.

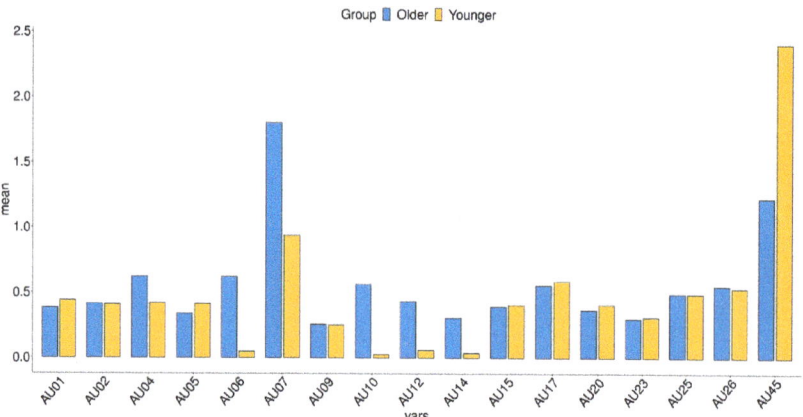

Figure 2. Prevalence of AU values by groups for neutral face. AU, action unit.

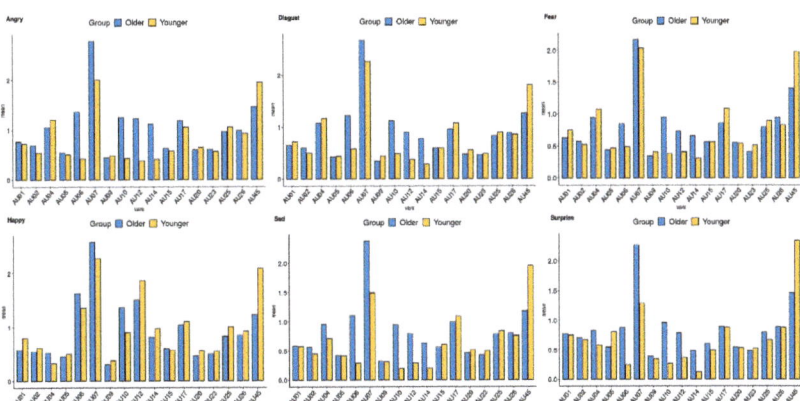

Figure 3. Prevalence of emotional AU values by groups for emotional face. AU, action unit.

Table 3. AU comparisons by groups for six basic emotions.

Variables	Younger Adults	Older Adults	Direction	Location	p-Value
	Mean ± SD	Mean ± SD			
AU06 (ang)	0.42 ± 0.58	1.36 ± 0.85	Y < O	U	<0.001
AU06 (dis)	0.58 ± 0.54	1.22 ± 0.73	Y < O	U	0.0276
AU06 (neu)	0.05 ± 0.14	0.62 ± 0.45	Y < O	U	<0.001
AU06 (sad)	0.29 ± 0.45	1.11 ± 0.59	Y < O	U	<0.001
AU06 (sur)	0.25 ± 0.50	0.88 ± 0.59	Y < O	U	<0.001
AU07 (neu)	0.94 ± 0.70	1.80 ± 0.83	Y < O	U	<0.001
AU07 (sad)	1.50 ± 0.94	2.37 ± 1.03	Y < O	U	<0.001
AU07 (sur)	1.29 ± 0.94	2.27 ± 0.97	Y < O	U	0.0105
AU10 (ang)	0.43 ± 0.58	1.25 ± 0.61	Y < O	L	<0.001
AU10 (dis)	0.49 ± 0.50	1.11 ± 0.62	Y < O	L	<0.001
AU10 (fea)	0.38 ± 0.49	0.95 ± 0.58	Y < O	L	<0.001
AU10 (neu)	0.03 ± 0.13	0.57 ± 0.47	Y < O	L	<0.001
AU10 (sad)	0.20 ± 0.34	0.95 ± 0.55	Y < O	L	<0.001
AU10 (sur)	0.26 ± 0.46	0.96 ± 0.61	Y < O	L	<0.001
AU12 (ang)	0.38 ± 0.56	1.23 ± 0.83	Y < O	L	<0.001
AU12 (neu)	0.06 ± 0.15	0.43 ± 0.40	Y < O	L	<0.001
AU12 (sad)	0.29 ± 0.43	0.79 ± 0.65	Y < O	L	<0.001
AU14 (ang)	0.41 ± 0.63	1.12 ± 0.81	Y < O	L	0.0255
AU14 (neu)	0.04 ± 0.15	0.31 ± 0.38	Y < O	L	<0.001
AU14 (sad)	0.20 ± 0.41	0.64 ± 0.60	Y < O	L	0.0036
AU45 (hap)	2.09 ± 0.70	1.23 ± 0.63	Y > O	U	0.0029
AU45 (neu)	2.41 ± 0.69	1.22 ± 0.55	Y > O	U	<0.001
AU45 (sad)	1.95 ± 0.75	1.18 ± 0.61	Y > O	U	0.0495
AU45 (sur)	2.34 ± 0.77	1.46 ± 0.73	Y > O	U	0.0022

Note: AU, action unit; BOLD, indicates significant p-values; ang, angry; dis, disgust; fea, fear; hap, happy; neu, neutral; sur, surprise; L, lower face; U, upper face. Comparisons were adjusted for covariates. p-values were adjusted for multiple comparisons.

To explore the relationship between age and each AU, a correlation analysis was conducted. The patterns of the results were similar to differences in group comparisons (Figure 4). For AU 06, 07, 12, 10, and 14, positive correlations between AU and age were found, while negative correlation were found in AU 45 across the emotions.

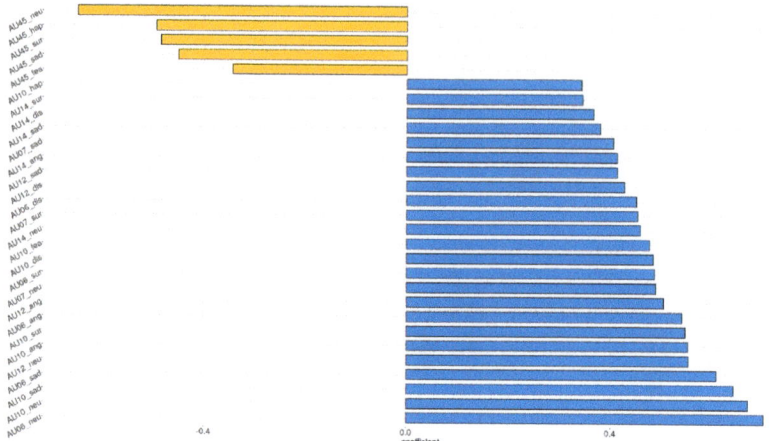

Figure 4. Correlation plot for age and AUs. AU, action unit; ang, angry; dis, disgust; fea, fear; hap, happy; neu, neutral; sur, surprise. *p*-values were adjusted for multiple comparisons.

3.2. Feature Selection for Predicting Age

The adaptive LASSO model was implemented to identify significant features for distinguishing the older group among the input variables. Demographics (education, sex, left-handed, and botox), self-reported measure (TAS and BDI), and all AUs were assessed for their ability to classify the older adults. Figure 5 shows the multivariate profiles for distinguishing the older adults from the participants in the current study. Demographics and self-reported measure were not significant in the adaptive LASSO result. Among the total 119 AUs, only 11 AUs remain significant: AU 10 in angry; AU 02, 10, 14, and 45 in sad; AU 05 and 14 in surprise; AU 06, 10, 20, and 45 in neutral, respectively. The receiver operating characteristic (ROC) demonstrated an AUC of 0.924 for the adaptive LASSO model.

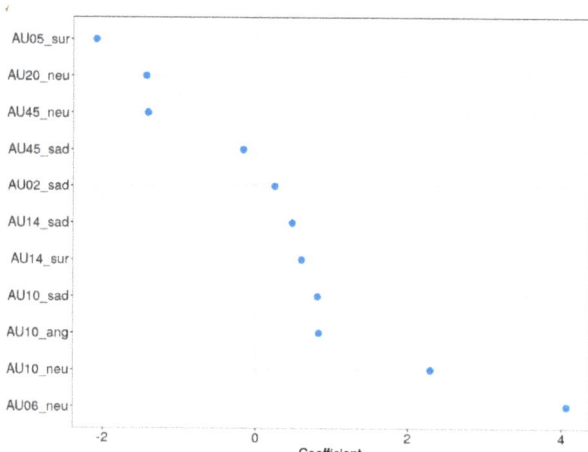

Figure 5. The adaptive LASSO results. AU, action unit; ang, angry; neu, neutral; sur, surprise.

4. Discussion and Conclusions

The purpose of the present study was to investigate the differences in facial expressions of older and younger adults and to examine how facial muscles contributed to aging through AUs for six basic emotion and neutral facial expression. Throughout the emotions and AUs, the older adults appeared to exhibit greater intensity in facial expression than the younger adults. In some area, the older adults showed lower facial intensity than the younger adults.

4.1. Degenerative Changes in Facial Expression Differences with Age

The main findings show that the older adults have higher AU values than young people for neutral and negative emotion (i.e., angry and sad). An increasing amount of the literatures has demonstrated that aging is associated with dramatic reductions in muscle strength (i.e., dynapenia) and motor control [71–73]. With advancing age, decreased neuromuscular changes may result in deficits in voluntary activation for facial activities [73,74]. In that regard, the facial expressions of older adults can naturally differ from those of younger adults [75].

Given that the cortex, spinal cord, and neuromuscular junction are functionally correlated, and they influence voluntary activation of muscle fibers [76], voluntary facial expressions can be addressed by neurological evidence [77]. For older adults to make facial expressions as intended, therefore, it is necessary to utilize their brain in the top-down processing format to ensure that the commands from the brain are correctly delivered to the facial muscles. In addition to facial aging due to sarcopenia, this suggests that changes in the motor cortex with aging can cause changes in facial expressions in the older adults [78,79].

Regarding the expression of strong negative emotions in the older adults representing our results, age differences are reported between the older and the young adults when they discriminate negative emotion. A previous study demonstrated that older adults had more difficulty distinguishing low intensity negative emotions [80]. They may tend to make facial expressions excessively because the older adults themselves may not be able to identify low intensity negative emotions.

Previous studies well support the differences in AUs intensity between the two groups. On upper facial expression, namely AU 06 and 07, the older adults can show greater intensity than the younger adults. Increased activity in orbicularis oculi muscle [81], deeply set of eye [82], and changes in eyelid due to poor visual acuity [83] may have affected the changes in upper facial expression. For lower facial expression, AU 10, 12, 14, the strength of the face may have been further tapped due to the highlighted facial contour caused by loss of subcutaneous fill around the nose and mouth in the older adults [84]. In AU 45, the older adults rather showed reduced AUs than the younger people. Elevated duration of eye blink may explain this reason. Duration of the eye-blinking decreases with aging, apparently reflecting decreased intensities in AU 45 [85], since the deterioration of the orbicularis oculi muscle can affect the complete eye closure rate [86].

As for the adaptive LASSO, the result was shown to be similar to the comparisons between two groups, expect for the AU 02, 05, and 20. The increase in AU 02 in sad condition, as previously mentioned, may have resulted in increased activity in the eyebrow and strong representation of negative representations [80,81]. For the AU 05 in surprise condition, the reduction of muscles may also involve in eye activity have affected the weaker construction of surprise facial expressions [85,86]. For the AU 20, aging may lead to the relaxation of the lip stretcher owing to decreased muscle around the mouth [17,87].

4.2. Limitations and Future Direction

There are several limitations in the current study. First, we employed only posed emotions. Given that the mechanisms of the posed emotions and the spontaneous facial expressions differ [88], further studies are needed to compare the difference between two distinct facial expressions. Secondly, we did not employ physiological assessment. The OpenFace software, unlike EMG, could not measure sensitive intensities in facial muscles at a physiological level. However, since the OpenFace library is based on FACS and provides reliable results along with recent technological advances, measurement errors are not likely to be a problem. In addition, recent study on the difference between computer vision and EMG has demonstrated only a few differences among the two techniques with respect to accessing overt facial expressions, and that computer vision showed better performance as compared with human [89]. Thirdly, age group is less continuous. Thus, future studies should be designed for providing normative data for facial aging with respect to demographics, such as age and sex. Lastly, the presence of the imbalanced class between the younger group and older group can be a potential limitation of the current study. This issue may not be critical, if the ratios between two classes are not too different. An experimental study showed that low class imbalance ratios do not cause significant performance loss [90], where the class ratio of 40:60, which is similar to our study (Table 2), seemed to converge to nearly zero with respect to performance loss. Another study used metabolomics data and showed that a false positive ratio even decreases as the class-imbalanced ratio rises, due to the prevention of over selection in identifying biomarker features with the LASSO algorithm [91]. Despite these studies, our findings should be interpreted with caution.

With the above limitations, our study has the following strengths. Our findings regarding posed emotions, which require conscious effort of facial muscles, can be used as an evidence to censor individuals who deliberately deceive others, especially for lie detection [92]. In situations where biophysiological assessment is limited, computer vision-based face recognition tools would be beneficial. In a clinical setting, our findings can be used for detecting frailty and other senile changes in muscle. For computer vision-based facial recognition, our findings may also provide researchers with empirical evidence for the characteristics of a human aging face, which would help develop the service and/or product for recognizing the faces of older adults. Notably, so far, there has been little attempt for facial expression recognizing study that compares the characteristics between the younger and the older. Our findings can provide interpretable evidence and explainable features for aging faces. This could provide an important basis for CAD studies for older people in the future.

4.3. Conclusions

Taken together, the present study is the first to investigate the differences in posed facial expressions between older adults and younger adults using a computer analysis method. Our findings provide evidence for implications in facial expression intensity based on FACS-AU-derived emotional faces. The older adults expressed more intense expressions in neutral and negative emotions than younger adults and tended to use more muscles when they were making facial expressions. In some part of the facial expression, the older adults showed weaker intensity than the younger adults. Our findings may suggest that changes in the muscles around the eyes and mouth due to aging can be indicators of the characteristics for identifying the aging face. The results of this study were obtained quantitatively from a normal population, which has several strengths as compared with previous studies of facial expression based on EMG, 3D morphometry, or subjective rating. They can be used as a basic methodology for analyzing and for identification of the characteristics of facial aging. We hope that the various features of the posed emotions of the older adults in this study can be a significant contribution to other scientific fields with respect to facial expressions, such as criminological research using lie detection, behavioral medicine, and computer vision research based on facial recognition. Future studies are needed for investigating other attributes in facial expressions regarding dynamic emotions, natural environments, and diverse groups.

Author Contributions: J.-Y.L. and S.P. (Soowon Park) designed the study; S.P. (Soowon Park) and J.-Y.L. recruited participants and collected facial and clinical data; M.B., M.-G.S., G.N. and J.I. wrote the protocol and performed interpretation of data; H.K. and S.P. (Seho Park) contributed to facial behavioral data analyses and wrote the methodology; K.K. and H.K. undertook statistical data analyses; K.K. and H.K. wrote the manuscript. All authors have read and agreed to the published version of the manuscript.

Funding: This research was funded by the Ministry of Education through the National Research Foundation of Korea (NRF), grant number (NRF-2017R1D1A1A02018479).

Institutional Review Board Statement: This study was conducted in accordance with the Declaration of Helsinki and the protocol was approved by the Institutional Review Board of SMG-SNU Boramae Medical Center (IRB No. 30-2017-63).

Informed Consent Statement: Informed consent was obtained from all subjects involved in the study.

Data Availability Statement: The data presented in this study are available on request from the corresponding author.

Acknowledgments: We would like to thank the anonymous reviewers for their time and constructive comments.

Conflicts of Interest: The authors declare no conflict of interest.

References

1. Buck, R. Nonverbal communication of affect in children. *J. Personal. Soc. Psychol.* **1975**, *31*, 644–653. [CrossRef]
2. Buck, R.W.; Savin, V.J.; Miller, R.E.; Caul, W.F. Communication of affect through facial expressions in humans. *J. Personal. Soc. Psychol.* **1972**, *23*, 362–371. [CrossRef] [PubMed]
3. Crivelli, C.; Fridlund, A.J. Facial displays are tools for social influence. *Trends Cogn. Sci.* **2018**, *22*, 388–399. [CrossRef]
4. Malatesta, C.Z.; Izard, C.E.; Culver, C.; Nicolich, M. Emotion communication skills in young, middle-aged, and older women. *Psychol. Aging* **1987**, *2*, 193–203. [CrossRef]
5. Sullivan, S.; Ruffman, T. Emotion recognition deficits in the elderly. *Int. J. Neurosci.* **2004**, *114*, 403–432. [CrossRef] [PubMed]
6. Ebner, N.C.; Johnson, M.K. Young and older emotional faces: Are there age group differences in expression identification and memory? *Emotion* **2009**, *9*, 329–339. [CrossRef]
7. Calder, A.J.; Keane, J.; Manly, T.; Sprengelmeyer, R.; Scott, S.; Nimmo-Smith, I.; Young, A.W. Facial expression recognition across the adult life span. *Neuropsychologia* **2003**, *41*, 195–202. [CrossRef]
8. MacPherson, S.E.; Phillips, L.H.; Della Sala, S. Age, executive function and social decision making: A dorsolateral prefrontal theory of cognitive aging. *Psychol. Aging* **2002**, *17*, 598–609. [CrossRef]
9. Suzuki, A.; Hoshino, T.; Shigemasu, K.; Kawamura, M. Decline or improvement? Age-related differences in facial expression recognition. *Biol. Psychol.* **2007**, *74*, 75–84. [CrossRef]
10. Slessor, G.; Miles, L.K.; Bull, R.; Phillips, L.H. Age-related changes in detecting happiness: Discriminating between enjoyment and nonenjoyment smiles. *Psychol. Aging* **2010**, *25*, 246–250. [CrossRef] [PubMed]
11. Gonçalves, A.R.; Fernandes, C.; Pasion, R.; Ferreira-Santos, F.; Barbosa, F.; Marques-Teixeira, J. Effects of age on the identification of emotions in facial expressions: A meta-analysis. *PeerJ* **2018**, *6*, e5278. [CrossRef]
12. Fedok, F.G. The aging face. *Facial Plast. Surg.* **1996**, *12*, 107–115. [CrossRef]
13. Windhager, S.; Mitteroecker, P.; Rupić, I.; Lauc, T.; Polašek, O.; Schaefer, K. Facial aging trajectories: A common shape pattern in male and female faces is disrupted after menopause. *Am. J. Phys. Anthropol.* **2019**, *169*, 678–688. [CrossRef]
14. Müri, R.M. Cortical control of facial expression. *J. Comp. Neurol.* **2016**, *524*, 1578–1585. [CrossRef]
15. Ross, E.D.; Prodan, C.I.; Monnot, M. Human facial expressions are organized functionally across the upper-lower facial axis. *Neuroscience* **2007**, *13*, 433–446. [CrossRef]
16. Ross, E.D.; Pulusu, V.K. Posed versus spontaneous facial expressions are modulated by opposite cerebral hemispheres. *Cortex* **2013**, *49*, 1280–1291. [CrossRef] [PubMed]
17. Bilodeau-Mercure, M.; Kirouac, V.; Langlois, N.; Ouellet, C.; Gasse, I.; Tremblay, P. Movement sequencing in normal aging: Speech, oro-facial, and finger movements. *Age* **2015**, *37*, 1–13. [CrossRef] [PubMed]
18. Avivi-Arber, L.; Sessle, B.J. Jaw sensorimotor control in healthy adults and effects of ageing. *J. Oral Rehabil.* **2018**, *45*, 50–80. [CrossRef] [PubMed]
19. Balestrini, S.; Lopez, S.M.; Chinthapalli, K.; Sargsyan, N.; Demurtas, R.; Vos, S.; Altmann, A.; Suttie, M.; Hammond, P.; Sisodiya, S.M. Increased facial asymmetry in focal epilepsies associated with unilateral lesions. *Brain Commun.* **2021**, *3*, fcab068. [CrossRef]
20. Sonawane, B.; Sharma, P. Review of automated emotion-based quantification of facial expression in Parkinson's patients. *Vis. Comput.* **2021**, *37*, 1151–1167. [CrossRef]
21. Burton, K.W.; Kaszniak, A.W. Emotional experience and facial expression in Alzheimer's disease. *Aging Neuropsychol. Cogn.* **2006**, *13*, 636–651. [CrossRef] [PubMed]
22. Zeghari, R.; König, A.; Guerchouche, R.; Sharma, G.; Joshi, J.; Fabre, R.; Robert, P.; Manera, V. Correlations between facial expressivity and apathy in elderly people with neurocognitive disorders: Exploratory study. *JMIR Form. Res.* **2021**, *5*, e24727. [CrossRef]
23. Borod, J.C.; Haywood, C.S.; Koff, E. Neuropsychological aspects of facial asymmetry during emotional expression: A review of the normal adult literature. *Neuropsychol. Rev.* **1997**, *7*, 41–60. [CrossRef] [PubMed]

24. Namba, S.; Makihara, S.; Kabir, R.S.; Miyatani, M.; Nakao, T. Spontaneous facial expressions are different from posed facial expressions: Morphological properties and dynamic sequences. *Curr. Psychol.* **2017**, *36*, 593–605. [CrossRef]
25. Galati, D.; Scherer, K.R.; Ricci-Bitti, P.E. Voluntary facial expression of emotion: Comparing congenitally blind with normally sighted encoders. *J. Personal. Soc. Psychol.* **1997**, *73*, 1363. [CrossRef]
26. Gosselin, P.; Kirouac, G.; Doré, F.Y. Components and recognition of facial expression in the communication of emotion by actors. *J. Personal. Soc. Psychol.* **1995**, *68*, 83. [CrossRef]
27. Sato, W.; Hyniewska, S.; Minemoto, K.; Yoshikawa, S. Facial expressions of basic emotions in Japanese laypeople. *Front. Psychol.* **2019**, *10*, 259. [CrossRef] [PubMed]
28. Van Der Zant, T.; Nelson, N. Motion increases recognition of naturalistic postures but not facial expressions. *J. Nonverbal Behav.* **2021**, 1–14.
29. Elfenbein, H.A.; Ambady, N. On the universality and cultural specificity of emotion recognition: A meta-analysis. *Psychol. Bull.* **2002**, *128*, 203. [CrossRef]
30. Aviezer, H.; Ensenberg, N.; Hassin, R.R. The inherently contextualized nature of facial emotion perception. *Curr. Opin. Psychol.* **2017**, *17*, 47–54. [CrossRef]
31. Ekman, P.; Friesen, W. *Facial Action Coding System (FACS): Manual*; Consulting Psychologists Press: Palo Alto, CA, USA, 1978.
32. Hamm, J.; Kohler, C.G.; Gur, R.C.; Verma, R. Automated facial action coding system for dynamic analysis of facial expressions in neuropsychiatric disorders. *J. Neurosci. Methods* **2011**, *200*, 237–256. [CrossRef]
33. Kar, N.B.; Babu, K.S.; Sangaiah, A.K.; Bakshi, S. Face expression recognition system based on ripplet transform type II and least square SVM. *Multimed. Tools Appl.* **2019**, *78*, 4789–4812. [CrossRef]
34. Baltrusaitis, T.; Zadeh, A.; Lim, Y.C.; Morency, L.P. Openface 2.0: Facial behavior analysis toolkit. In Proceedings of the 2018 13th IEEE International Conference on Automatic Face & Gesture Recognition (FG 2018), Xi'an, China, 15–19 May 2018; pp. 59–66.
35. Cotofana, S.; Assemi-Kabir, S.; Mardini, S.; Giunta, R.E.; Gotkin, R.H.; Moellhoff, N.; Avelar, L.E.T.; Mercado-Perez, A.; Lorenc, P.Z.; Frank, K. Understanding facial muscle aging: A surface electromyography study. *Aesthetic Surg. J.* **2021**, sjab202. [CrossRef]
36. Bailey, P.E.; Henry, J.D. Subconscious facial expression mimicry is preserved in older adulthood. *Psychol. Aging* **2009**, *24*, 995–1000. [CrossRef]
37. Labuschagne, I.; Pedder, D.J.; Henry, J.D.; Terrett, G.; Rendell, P.G. Age differences in emotion regulation and facial muscle reactivity to emotional films. *Gerontology* **2020**, *66*, 74–84. [CrossRef]
38. Wang, F.; Chen, H.; Kong, L.; Sheng, W. Real-time facial expression recognition on robot for healthcare. In Proceedings of the 2018 IEEE International Conference on Intelligence and Safety for Robotics (ISR), Shenyang, China, 24–27 August 2018; pp. 402–406.
39. Stephen, I.D.; Hiew, V.; Coetzee, V.; Tiddeman, B.P.; Perrett, D.I. Facial shape analysis identifies valid cues to aspects of physiological health in Caucasian, Asian, and African populations. *Front. Psychol.* **2017**, *8*, 1883. [CrossRef] [PubMed]
40. Khan, M.A.; Kim, Y. Cardiac arrhythmia disease classification using LSTM deep learning approach. *CMC Comput. Mater. Contin.* **2021**, *67*, 427–443.
41. Giger, M.L.; Suzuki, K. Computer-aided diagnosis. In *Biomedical Information Technology*; Academic Press: Cambridge, MA, USA, 2008; pp. 359–374.
42. Parra-Dominguez, G.S.; Sanchez-Yanez, R.E.; Garcia-Capulin, C.H. Facial paralysis detection on images using key point analysis. *Appl. Sci.* **2021**, *11*, 2435. [CrossRef]
43. Guarin, D.L.; Yunusova, Y.; Taati, B.; Dusseldorp, J.R.; Mohan, S.; Tavares, J.; Jowett, N. Toward an automatic system for computer-aided assessment in facial palsy. *Facial Plast. Surg. Aesthetic Med.* **2020**, *22*, 42–49. [CrossRef] [PubMed]
44. Dorante, M.I.; Kollar, B.; Obed, D.; Haug, V.; Fischer, S.; Pomahac, B. Recognizing emotional expression as an outcome measure after face transplant. *JAMA Netw. Open* **2020**, *3*, e1919247. [CrossRef]
45. Roy, S.D.; Bhowmik, M.K.; Saha, P.; Ghosh, A.K. An approach for automatic pain detection through facial expression. *Procedia Comput. Sci.* **2016**, *84*, 99–106. [CrossRef]
46. De Belen, R.A.J.; Bednarz, T.; Sowmya, A.; Del Favero, D. Computer vision in autism spectrum disorder research: A systematic review of published studies from 2009 to 2019. *Transl. Psychiatry* **2020**, *10*, 1–20. [CrossRef]
47. Chen, S.; Pan, Z.X.; Zhu, H.J.; Wang, Q.; Yang, J.J.; Lei, Y.; Li, J.Q.; Pan, H. Development of a computer-aided tool for the pattern recognition of facial features in diagnosing Turner syndrome: Comparison of diagnostic accuracy with clinical workers. *Sci. Rep.* **2018**, *8*, 9317. [CrossRef]
48. Jin, B.; Qu, Y.; Zhang, L.; Gao, Z. Diagnosing Parkinson disease through facial expression recognition: Video analysis. *J. Med. Internet Res.* **2020**, *22*, e18697. [CrossRef]
49. Ardila, A.; Rosselli, M. Spontaneous language production and aging: Sex and educational effects. *Int. J. Neurosci.* **1996**, *87*, 71–78. [CrossRef]
50. Jayanthy, S.; Anishkka, J.B.; Deepthi, A.; Janani, E. Facial Recognition and Verification System for Accessing Patient Health Records. In Proceedings of the 2019 International Conference on Intelligent Computing and Control Systems (ICCS), Madurai, India, 15–17 May 2019; pp. 1266–1271.
51. Association, A.P. *Diagnostic and Statistical Manual of Mental Disorder: DSM-IV-TR*; American Psychiatric Association: Washington, DC, USA, 2000.
52. Beck, A.T.; Ward, C.H.; Mendelson, M.; Mock, J.; Erbaugh, J. An inventory for measuring depression. *Arch. Gen. Psychiatry* **1961**, *4*, 561–571. [CrossRef] [PubMed]

53. Sung, H.; Kim, J.; Park, Y.; Bai, D.; Lee, S.; Ahn, H. A study on the reliability and the validity of Korean version of the Beck Depression Inventory (BDI). *J. Korean Soc. Biol. Ther. Psychiatry* **2008**, *14*, 201–212.
54. Lim, S.Y.; Lee, E.J.; Jeong, S.W.; Kim, H.C. The validation study of Beck Depression Scale 2 in Korean version. *Anxiety Mood* **2011**, *7*, 48–53.
55. Beck, A.T.; Epstein, N.; Brown, G.; Steer, R.A. An inventory for measuring clinical anxiety: Psychometric properties. *J. Couns. Clin. Psychol.* **1988**, *56*, 893–897. [CrossRef]
56. Julian, L.J. Measures of anxiety: State-Trait Anxiety Inventory (STAI), Beck Anxiety Inventory (BAI), and Hospital Anxiety and Depression Scale-Anxiety (HADS-A). *Arthritis Care Res.* **2011**, *63*, S467–S472. [CrossRef]
57. Bagby, R.M.; Parker, J.D.A.; Taylor, G.J. The twenty-item Toronto Alexithymia Scale-I. Item selection and cross-validation of the factor structure. *J. Psychosom. Res.* **1994**, *38*, 23–32. [CrossRef]
58. Lee, J.Y.; Rim, Y.H.; Lee, H.D. Development and validation of a Korean version of the 20-item Toronto Alexithymia Scale (TAS-20K). *J. Korean Neuropsychiatr. Assoc.* **1996**, *35*, 888–899.
59. Seo, S.S.; Chung, U.S.; Rim, H.D.; Jeong, S.H. Reliability and validity of the 20-item Toronto Alexithymia Scale in Korean adolescents. *Psychiatry Investig.* **2009**, *6*, 173. [CrossRef] [PubMed]
60. Park, S.; Kim, T.; Shin, S.A.; Kim, Y.K.; Sohn, B.K.; Park, H.J.; Youn, J.H.; Lee, J.Y. Behavioral and neuroimaging evidence for facial emotion recognition in elderly korean adults with mild cognitive impairment, Alzheimer's disease, and frontotemporal dementia. *Front. Aging Neurosci.* **2017**, *9*, 389. [CrossRef] [PubMed]
61. Baltrušaitis, T.; Mahmoud, M.; Robinson, P. Cross-dataset learning and person-specific normalisation for automatic action unit detection. In Proceedings of the 2015 11th IEEE International Conference and Workshops on Automatic Face and Gesture Recognition (FG), Ljubljana, Slovenia, 4–8 May 2015; pp. 1–6.
62. Friesen, W.; Ekman, P. *EMFACS-7: Emotional Facial Action Coding System*; University of California at San Francisco: San Francisco, CA, USA, 1983; Unpublished manuscript.
63. Sayette, M.A.; Cohn, J.F.; Wertz, J.M.; Perrott, M.A.; Parrott, D.J. A psychometric evaluation of the facial action coding system for assessing spontaneous expression. *J. Nonverbal Behav.* **2001**, *25*, 167–185. [CrossRef]
64. Scherer, K.R. *Handbook of Methods in Nonverbal Behavior Research*; Cambridge University Press: Cambridge, UK, 1985.
65. Olderbak, S.; Hildebrandt, A.; Pinkpank, T.; Sommer, W.; Wilhelm, O. Psychometric challenges and proposed solutions when scoring facial emotion expression codes. *Behav. Res. Methods* **2014**, *46*, 992–1006. [CrossRef]
66. Zou, H. The adaptive lasso and its oracle properties. *J. Am. Stat. Assoc.* **2006**, *101*, 1418–1429. [CrossRef]
67. Tikhonov, A.N. On the stability of inverse problems. *Dokl. Akad. Nauk SSSR* **1943**, *39*, 195–198.
68. Vidaurre, D.; Bielza, C.; Larranaga, P. A survey of L1 regression. *Int. Stat. Rev.* **2013**, *81*, 361–387. [CrossRef]
69. McNeish, D.M. Using lasso for predictor selection and to assuage overfitting: A method long overlooked in behavioral sciences. *Multivar. Behav. Res.* **2015**, *50*, 471–484. [CrossRef]
70. Lever, J.; Krzywinski, M.; Altman, N. Points of significance: Model selection and overfitting. *Nat. Methods* **2016**, *13*, 703–704. [CrossRef]
71. Clark, B.C.; Manini, T.M. Sarcopenia ≠ dynapenia. *J. Gerontol. Ser. A* **2008**, *63*, 829–834. [CrossRef] [PubMed]
72. Enoka, R.M.; Christou, E.A.; Hunter, S.K.; Kornatz, K.W.; Semmler, J.G.; Taylor, A.M.; Tracy, B.L. Mechanisms that contribute to differences in motor performance between young and old adults. *J. Electromyogr. Kinesiol.* **2003**, *13*, 1–12. [CrossRef]
73. Clark, B.C. Neuromuscular changes with aging and sarcopenia. *J. Frailty Aging* **2019**, *8*, 7–9. [PubMed]
74. Klass, M.; Baudry, S.; Duchateau, J. Voluntary activation during maximal contraction with advancing age: A brief review. *Eur. J. Appl. Physiol.* **2007**, *100*, 543–551. [CrossRef]
75. Oliviero, A.; Profice, P.; Tonali, P.A.; Pilato, F.; Saturno, E.; Dileone, M.; Ranieri, F.; Di Lazzaro, V. Effects of aging on motor cortex excitability. *Neurosci. Res.* **2006**, *55*, 74–77. [CrossRef]
76. Gandevia, S.C. Spinal and supraspinal factors in human muscle fatigue. *Physiol. Rev.* **2001**, *81*, 1725–1789. [CrossRef]
77. Manini, T.M.; Clark, B.C. Dynapenia and aging: An update. *J. Gerontol. Ser. A* **2012**, *67*, 28–40. [CrossRef]
78. Morecraft, R.J.; Stilwell-Morecraft, K.S.; Rossing, W.R. The motor cortex and facial expression: New insights from neuroscience. *Neurol.* **2004**, *10*, 235–249. [CrossRef]
79. Salat, D.H.; Buckner, R.L.; Snyder, A.Z.; Greve, D.N.; Desikan, R.S.; Busa, E.; Morris, J.C.; Dale, A.M.; Fischl, B. Thinning of the cerebral cortex in aging. *Cereb. Cortex* **2004**, *14*, 721–730. [CrossRef]
80. Mienaltowski, A.; Johnson, E.R.; Wittman, R.; Wilson, A.T.; Sturycz, C.; Norman, J.F. The visual discrimination of negative facial expressions by younger and older adults. *Vis. Res.* **2013**, *81*, 12–17. [CrossRef]
81. Yun, S.; Son, D.; Yeo, H.; Kim, S.; Kim, J.; Han, K.; Lee, S.; Lee, J. Changes of eyebrow muscle activity with aging: Functional analysis revealed by electromyography. *Plast. Reconstr. Surg.* **2014**, *133*, 455e–463e. [CrossRef]
82. Hennekam, R.C. The external phenotype of aging. *Eur. J. Med. Genet.* **2020**, *63*, 103995. [CrossRef]
83. Moon, J.H.; Oh, Y.H.; Kong, M.H.; Kim, H.J. Relationship between visual acuity and muscle mass in the Korean older population: A cross-sectional study using Korean National Health and Nutrition Examination Survey. *BMJ Open* **2019**, *9*, e033846. [CrossRef]
84. Coleman, S.R.; Grover, R. The anatomy of the aging face: Volume loss and changes in 3-dimensional topography. *Aesthetic Surg. J.* **2006**, *26*, S4–S9. [CrossRef]
85. Sun, W.S.; Baker, R.S.; Chuke, J.C.; Rouholiman, B.R.; Hasan, S.A.; Gaza, W.; Stava, M.W.; Porter, J.D. Age-related changes in human blinks. Passive and active changes in eyelid kinematics. *Investig. Ophthalmol. Vis. Sci.* **1997**, *38*, 92–99.

86. Sforza, C.; Rango, M.; Galante, D.; Bresolin, N.; Ferrario, V.F. Spontaneous blinking in healthy persons: An optoelectronic study of eyelid motion. *Ophthalmic Physiol. Opt.* **2008**, *28*, 345–353. [CrossRef] [PubMed]
87. Cecílio, F.; Regalo, S.; Palinkas, M.; Issa, J.; Siéssere, S.; Hallak, J.; Machado-de-Sousa, J.; Semprini, M. Ageing and surface EMG activity patterns of masticatory muscles. *J. Oral Rehabil.* **2010**, *37*, 248–255. [CrossRef] [PubMed]
88. Motley, M.T.; Camden, C.T. Facial expression of emotion: A comparison of posed expressions versus spontaneous expressions in an interpersonal communication setting. *West. J. Commun. (Incl. Commun. Rep.)* **1988**, *52*, 1–22. [CrossRef]
89. Perusquia-Hernández, M.; Ayabe-Kanamura, S.; Suzuki, K.; Kumano, S. The invisible potential of facial electromyography: A comparison of EMG and computer vision when distinguishing posed from spontaneous smiles. In Proceedings of the 2019 CHI Conference on Human Factors in Computing Systems, New York, NY, USA, 4–9 May 2019.
90. Prati, R.C.; Batista, G.E.; Silva, D.F. Class imbalance revisited: A new experimental setup to assess the performance of treatment methods. *Knowl. Inf. Syst.* **2015**, *45*, 247–270. [CrossRef]
91. Fu, G.H.; Yi, L.Z.; Pan, J. LASSO-based false-positive selection for class-imbalanced data in metabolomics. *J. Chemom.* **2019**, *33*, e3177. [CrossRef]
92. Avola, D.; Cinque, L.; Foresti, G.L.; Pannone, D. Automatic deception detection in RGB videos using Facial Action Units. In Proceedings of the 13th International Conference on Distributed Smart Cameras, New York, NY, USA, 9–11 September 2019; pp. 1–6.

Article

The Analysis of Emotion Authenticity Based on Facial Micromovements

Sung Park [1,*], Seong Won Lee [2] and Mincheol Whang [2]

1 School of Design, Savannah College of Art and Design, Savannah, GA 31401, USA
2 Department of Human Centered Artificial Intelligence, Sangmyung University, Jongno-gu, Seoul 03016, Korea; wonii0916@gmail.com (S.W.L.); whang@smu.ac.kr (M.W.)
* Correspondence: spica7601@gmail.com

Abstract: People tend to display fake expressions to conceal their true feelings. False expressions are observable by facial micromovements that occur for less than a second. Systems designed to recognize facial expressions (e.g., social robots, recognition systems for the blind, monitoring systems for drivers) may better understand the user's intent by identifying the authenticity of the expression. The present study investigated the characteristics of real and fake facial expressions of representative emotions (happiness, contentment, anger, and sadness) in a two-dimensional emotion model. Participants viewed a series of visual stimuli designed to induce real or fake emotions and were signaled to produce a facial expression at a set time. From the participant's expression data, feature variables (i.e., the degree and variance of movement, and vibration level) involving the facial micromovements at the onset of the expression were analyzed. The results indicated significant differences in the feature variables between the real and fake expression conditions. The differences varied according to facial regions as a function of emotions. This study provides appraisal criteria for identifying the authenticity of facial expressions that are applicable to future research and the design of emotion recognition systems.

Keywords: facial micromovement; emotion recognition; emotion authenticity

1. Introduction

Humans utilize both verbal and nonverbal communication channels. The latter category includes facial expressions, gestures, posture, gait, gaze, distance, and tone and manner of voice [1]. Facial expressions, which account for up to 30% of nonverbal expressions, are the most rapidly processed type of expression by visual recognition [2]. Facial expressions project the communicator's intentions and emotions [3]. However, people may conceal their true feelings and produce fake expressions [4]. Such false expressions are exhibited for a very short time with only subtle changes [5], and it is extremely difficult to detect their authenticity with eyesight [6]. Identifying fake expressions is paramount to counter deception and recognize users' true intent in advanced intelligent systems (e.g., social robots and assistive systems).

Early research involving facial expressions focused on establishing a quantitative classification framework to recognize emotions. Ekman built a facial action coding system (FACS), a computation system that encodes facial features' movements to taxonomize emotions from facial expressions. Analysis of facial expressions also spurred interest in the authenticity of expressions.

Researchers have found asymmetric intensity in facial expressions. Dopson revealed that the intensity of expressions in the left face was stronger than that in the right face in the case of voluntary expressions [7]. Conversely, the intensity was weaker than that in the right face in the case of involuntary expressions. These results suggest that the comparison of both sides may identify the authenticity of expressions. The sensitivity of left-face expressions is because facial movements are connected to the right hemisphere of the brain. Patients with

right-brain injuries are reported to experience significant degradation in recognizing emotions from facial expressions compared with patients with left-brain injuries [8].

Studies have also found differential activation of facial muscles between real and fake expressions. Duchenne experimented on facial muscular contractions with electrical probes to understand how the human face produces expressions [9]. He observed that participants produced a genuine smile with a unique contraction of the Orbicularis oculi muscle [10]. This "smiling with the eyes" is called the Duchenne smile, in his honor.

Ekman analyzed human false expressions and identified minute vibrations or spontaneous changes in the facial muscles responsible for emotional expression [11]. Such micromovements are observed in false (e.g., deception) or pretended (e.g., to be polite) expressions [5]. Facial micromovement is also called microexpression. Micromovement occurs with less than a second of movement and with vibration lasting between 0.04 and 0.5 s [12–14]. Simultaneously, in a typical interaction, an emotional expression begins and ends with a macroexpression that occurs in less than 4 s [15]. The degree of movement or the vibration of the facial muscles between real and fake expressions can be significantly different [11].

Recent advances in AI technology have led to research on identifying the authenticity of facial expressions using repetitive training with paired data of facial expressions and visual content (an image and a video clip) [16,17]. Microexpression recognition (MER) researchers have put massive effort into open innovation (e.g., facial microexpressions grand challenge [18,19]) to improve the state-of-the-art algorithm. Academic challenges include all aspects of MER sequences such as data collection, preprocessing (face detection and landmark detection), feature extraction, microexpression recognition, and emotion classification within the computer vision domain (for a comprehensive review, see [20] and [21]). Similar to other AI domains, convolutional neural networks (CNNs) have been used the most for MER [22]. A generative adversarial network (GAN), with a generator and an adversarial discriminator model, has been used for feature extraction [23] and facial image synthesis [24]. Most recently, extended local binary patterns on three orthogonal plans (ELBPTOP) were introduced to counter information loss and computational burden of the previous dominant descriptors, LBPTOP [25].

While researchers continue to pursue better algorithms to improve MER accuracy and reliability, in the most recent survey of facial microexpression analysis [20], Xie observed that MER literature on facial asymmetrical phenomena is scarce and limited. While researchers have found an asymmetric intensity in facial expressions, less is known regarding where in the facial region such microexpressions are the most salient and how they interact with different emotions. Specifically, feature variables (i.e., the degree and variance of movement, and vibration level) of emotions that are primarily expressed with the relaxation of facial muscles (e.g., contentment, sadness) may have weaker intensity in the real condition. Systematic research identifying reliable indicators of authenticity per facial region as a function of emotion is imperative.

Physiological data, including electrocardiogram (ECG), are powerful signals for emotion identification [26]. ECG correlates with the contraction of the heart muscles and varies as a function of emotion [27]. In order to achieve a deeper understanding of MER, facial vision data should be fused with cardio signals [20]. To the best of our knowledge, no research has combined the two.

In summary, the study hypothesized that (1) there is a significant difference in the micromovements at the onset of expression between real and fake conditions, and (2) such differences vary by representative emotions (happiness, sadness, contentment, anger). The findings were cross-validated with neurological measurements (ECG).

2. Methods

2.1. Experiment Design

The present study used a 2 × 4 within-subject design. The authenticity factor had two levels (real and fake), and the emotion factor had four levels (happiness, sadness, anger, and contentment). The visual stimulus consisted of a still photo and a video clip. The still photo depicted a facial expression of the target emotion. The video clip, which was shown after the still photo, was a recording that was designed to induce either the target emotion shown in the still photo or a neutral emotion.

The participants were then asked to produce a facial expression that the participant felt while watching the still photo. The real condition was manipulated by showing the two materials, the still photo and the video clip, congruently. The false condition was manipulated by having the video clip induce a neutral emotion. In this case, participants were forced to produce a facial expression based on the photo that they viewed earlier. If a different emotion was induced other than neutral, the participant's emotion may have been compounded, which made the measurements difficult to explain. After every 30 s during the video, participants were signaled with a visual cue to produce a facial expression.

The dependent measurements involved micromovements in the face. That is, the average movement, standard deviation, and variance of the facial muscle movements were measured. Facial vibration was analyzed with the dominant frequency elicited by the fast Fourier transform (FFT).

2.2. Participants

Fifty university students were recruited as participants. The participants' average age was 22.5 years (SD = 2.13) with an even ratio in gender. We selected participants with corrective vision of 0.7 or above to ensure the participants' reliable recognition of visual stimuli. The participants were not allowed to wear glasses. All participants were briefed on the purpose and procedure of the experiment and signed a consent form. Participants were compensated with participation fees.

2.3. Procedure and Materials

Figure 1 illustrates the experimental procedure. Each participant's neutral facial expression was captured for 210 s before the main task. This was considered as the individual's reference expression. Participants were then exposed to eight combinations of visual stimuli—four sets (happiness, sadness, anger, and contentment) to elicit real emotions and four sets to elicit fake emotions. The order was randomized to counter order and learning effects. A set of stimuli consisted of a still photo and a video clip. A set used to induce real emotion had congruent emotions between the two materials. Conversely, a set to induce false emotions had inconsistent emotions between the two materials. In this case, the video clip induced a neutral emotion.

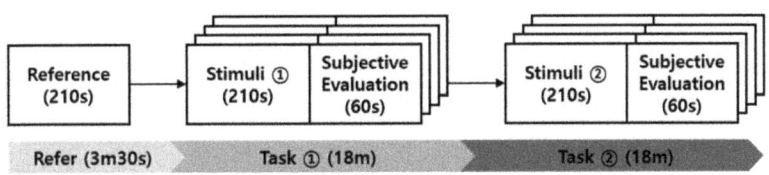

Figure 1. Experimental procedure.

After viewing the visual stimuli, the participants were given a resting period of 60 s. During this period, participants reported their current emotional state with a subjective evaluation. Participants reported their (1) emotional state (happiness, sadness, anger, disgust, fear, surprise, and contentment), (2) degree of arousal, and (3) degree of pleasantness. The latter two were rated on a five-point Likert scale. (1) We provided a comprehensive set of seven emotions to select from to exclude any data from participants who felt nothing or had a different emotion from the target emotion. The exclusion was determined for each condition, even for neutral video clips, to eliminate any compounding factors from the data.

The participant's facial data were acquired using a webcam. A Logitech c920 webcam (Logitech, Lausanne, Switzerland) was used to obtain image data with a resolution of 1280 × 980 at 30 frames per second. To analyze the activation level of the autonomic nervous system (ANS) when participants were exposed to visual stimuli, participants' heart rate variability (HRV) and electrocardiogram (ECG) data were acquired. The latter was obtained through a Biopac (Biopac, Goleta, CA, USA) system with a frequency of 500 Hz.

Figure 2 shows the experimental setup. Participants were asked to sit and view the experiment monitor at a distance of 60 cm. A webcam, which acquired facial data from the participant, was placed on top atop the monitor.

Figure 2. Experimental setting.

2.4. Statistical Analysis

The present study compared the differences in the micromovement of facial expressions between real and fake emotions. From the participant's expression data, feature variables (i.e., the degree and variance of movement, as well as vibration level) obtained at 4 s (macromovement), 1 s, and 0.5 s (micromovements) after the onset (t) of facial expression were analyzed. For each representative emotion (happiness, contentment, anger, and sadness), a t-test was used to compare the differences between the feature variables in the two conditions (real and fake) for all 11 AUs responsible for emotional expression. The following section explains how the feature variables were extracted and how the ECG data were obtained.

3. Analysis

Figure 3 outlines the analysis process. To analyze the data, we established an operational definition of facial expression muscles and extracted facial movement data for such muscles. In total, 40 datasets were analyzed; participants who had excessive facial movements or participants who did not display emotion were excluded. That is, the experimenter screened each recorded video clip and excluded participants who had turned their faces, clearly looking at an object outside of the screen, or when the system had failed to track their faces. To minimize the exclusion, we had instructed the participants to reduce the facial movement and look straight ahead.

The expression onset segment was defined (② in Figure 3) based on the threshold of facial movement. Feature variables were then extracted by comparing the rate of change in action units (AUs) between data frames. The effective feature variables were selected by comparing the feature variables of real and fake expressions for each emotion.

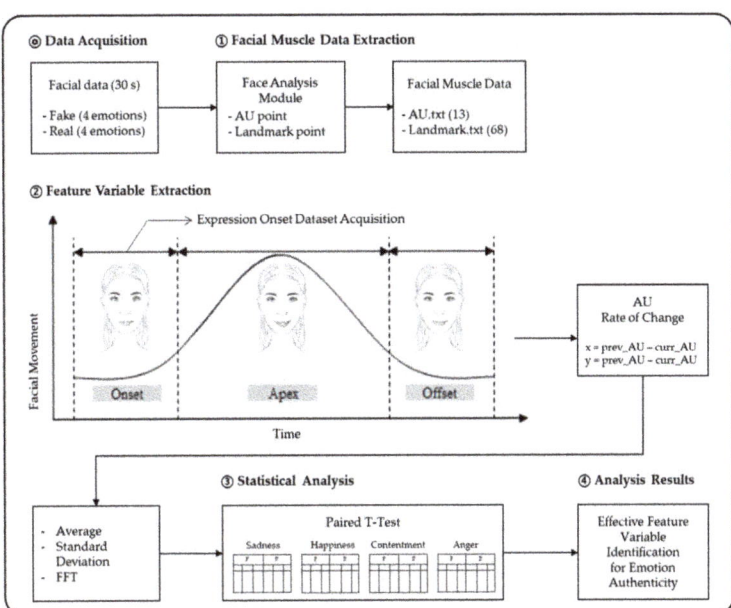

Figure 3. Data analysis process.

3.1. Operational Definition of Facial Muscles

The present study recognized the participants' emotions by identifying the activation of anatomical regions that represent a particular emotion. The AUs were extracted using facial landmarks through a Python program. Figure 4 depicts the extraction process.

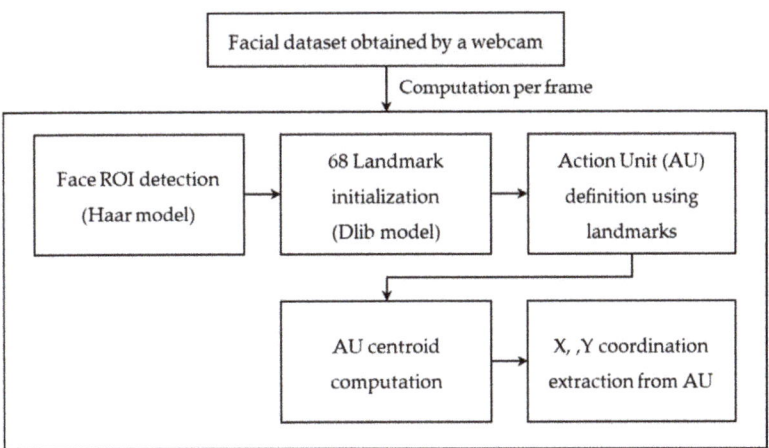

Figure 4. Facial muscle extraction process.

Each frame obtained from the webcam was analyzed. First, the location of the face in the image was identified using a face detection model, the Haar cascade classifier [28]. Face detection models extracted the target object's features from the dataset and compared the features from the pretrained data to identify the object. Specifically, the present system used the Haar-like feature to detect the region of a face (region of interest (ROI)) by identifying the location of the nose and eyes. The system then identified 68 facial landmarks by tracking the eye, eyebrows, nose, lips, and chin line using the Dlib library [29], which was trained with a massive quantity of data. The differential facial muscles per facial expression were predefined and utilized to extract 11 muscle areas (i.e., coordination). Eleven facial muscle units (AUs) involving the brow, eyes, cheeks, chin, and lips responsible for facial expressions were predefined and extracted from the participant's dataset (see Table 1). Figure 5 visualizes the relative locations of action units.

Table 1. Action Unit Definitions.

Action Units	Description	Muscular Basis	Landmark
AU4_M	Brow depressor	Depressor glabella Depressor supercilii Corrugator supercilii	21, 22, 27
AU5_L AU5_R	Upper Lip raiser	Lavator palpebrae superioris	23, 25, 44 18, 20, 37
AU6_L AU6_R	Cheek raiser Lip tightener	Orbicularis oculi	15, 26, 45 1, 17, 36
AU12_L AU12_R	Lip corner puller	Zygomaticus major	14, 35, 54 2, 31, 48
AU15_L AU15_R	Lip corner depressor	Depressor anguli oris	10, 11, 54 5, 6, 48
AU23_L AU23_R	Lip tightener	Orbicularis oris	52, 53, 63 49, 50, 61

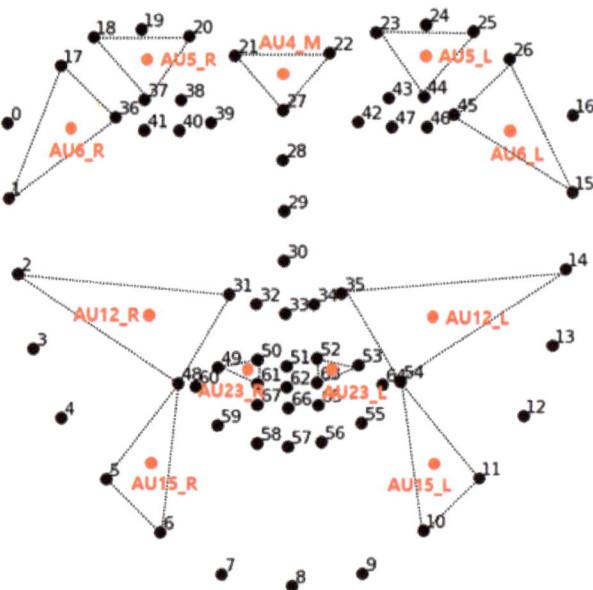

Figure 5. The relative locations of action units.

These 11 AUs are the centroid values of the three corresponding facial landmarks, computed as follows:

$$A(x_1, y_1), B(x_2, y_2), C(x_3, y_3)$$
$$P(\tfrac{x_1+x_2+x_3}{3}, \tfrac{y_1+y_2+y_3}{3})$$

For further analysis, facial data from the last 30 s were extracted and analyzed. That is, we defined the first 180 s as time for the visual content to sufficiently "sink in" for the participants.

3.2. Feature Variable Extraction

To extract feature variables involving facial micromovement, we developed a micromovement extraction program built by LabVIEW 2016 for massive data processing. From the last 30 s of the participant's dataset, 11 AUs (Table 1) were calculated. A threshold was used, the average movement of an AU, using the following min-max algorithm to determine the onset of facial expressions. The micromovement section before the onset was extracted.

$$Threshold = \frac{(Max + Min)}{2}$$

The expression section after the onset consisted of one macromovement section (4 s) and two micromovement sections (1 s, 0.5 s). These three sections may overlap. The movement data from the three sections were extracted. That is, the degree of change (delta) in the coordination of an AU between the current and previous frames was computed as follows, which was performed to analyze the degree of facial vibration.

$$x_n = prevAU[n] \cdot x - currAU[n] \cdot x$$
$$y_n = prevAU[n] \cdot y - currAU[n] \cdot y$$

Finally, we extracted feature values by analyzing the delta value. That is, the average and standard deviation of the delta and FFT values were extracted. The former two were used to analyze the degree and variance of the change. The latter was used to analyze the degree of facial vibration through the dominant frequency obtained by the FFT.

3.3. Heart Rate Variability Analysis

In addition to the facial data, ECG data were measured while the visual stimuli were shown for 210 s. The participants' time-series data were transformed into a frequency band using FFT. This enabled measurement of the ANS responses of participants exposed to emotion-inducing stimuli [30,31]. Table 2 outlines the HRV variables used in this study. To measure the change in the serial heart rate data, a 180-s sliding window was used.

Table 2. Heart Rate Variability Variables.

Variable	Unit	Definition	Frequency Range
VLF	ms^2	The power value in the VLF frequency	0.003~0.04 Hz
LF	ms^2	The power value in the LF frequency	0.04~0.15 Hz
HF	ms^2	The power value in the HF frequency	0.15~0.4 Hz
VLF (%)	%	VLF divided by the overall power value	
LF (%)	%	LF divided by the overall power value	
HF (%)	%	HLF divided by the overall power value	

4. Results

The current study analyzed changes in facial micromovements between real and fake expressions of representative emotions. A *t*-test was used to compare the differences between the participants' facial expressions in the two conditions. The feature variables obtained at 4 s (macromovement), 1 s, and 0.5 s (micromovements) after the onset (*t*) of facial expression were analyzed.

Figure 6 shows the template used to visualize the results. The blank squares on the right indicate the 11 AUs (Table 1) representing the facial muscles responsible for facial expressions. The statistical difference between the real and fake conditions are color coded in Figures 7, 9, 11 and 13 in three levels: $p < 0.001$: *** ■; $p < 0.01$: ** ■; $p < 0.05$: * ■.

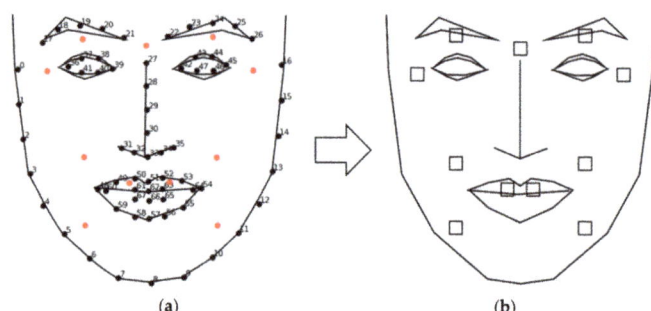

Figure 6. (a) Eleven AU regions (red dots) for feature variable extraction; (b) visualization framework for reporting the results.

In the HRV analysis, we compared the difference in ANS activation between the real and fake conditions.

4.1. Authenticity of Happiness

The results of the analysis of micromovement involving expressions of happiness are as follows. Figure 7 depicts the differential movement of the facial regions between the two conditions through the visualization of a face. All 11 AUs had at least one significant difference in the dependent variables (dominant peak frequency, average, and standard deviation of movement).

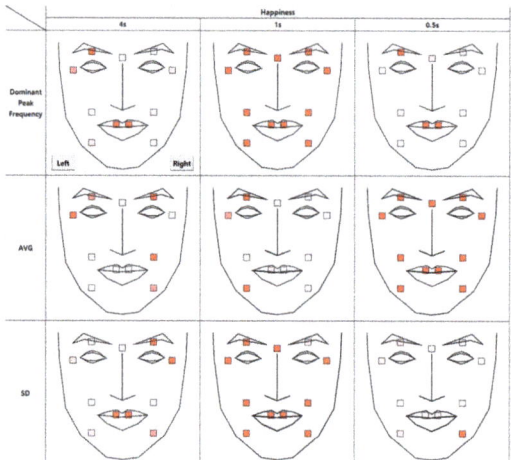

Figure 7. Statistical differences between real and fake happiness expressions (AVG = Average, SD = Standard Deviation).

The average at t + 0.5 (0.5 s after the onset) showed a significant difference in all AUs, whereas only partially significant differences appeared at t + 1, mostly in the left face.

This implies that expressions of happiness may be most prominent in the early stage (0.5 s) of a microexpression but persist until t + 1 in the left face. Further regression analysis on average movement showed that the time segment factor enters the regression equation ($R^2 = 0.97$), $p < 0.001$, along with the authenticity factor, $p < 0.05$.

However, for the standard deviation, the values at t + 1 significantly differed in all 11 AUs. The domain peak frequency also showed a significant difference at t + 1 in all AUs. The domain peak frequency at t + 0.5 showed a significant difference in the lips, left eyebrows, and brow.

Figure 8 presents a statistical comparison between dependent variables for each AU, collapsing data from the three sections (t + 0.5, t + 1, and t + 4). The measured values were higher in real expressions in almost all regions.

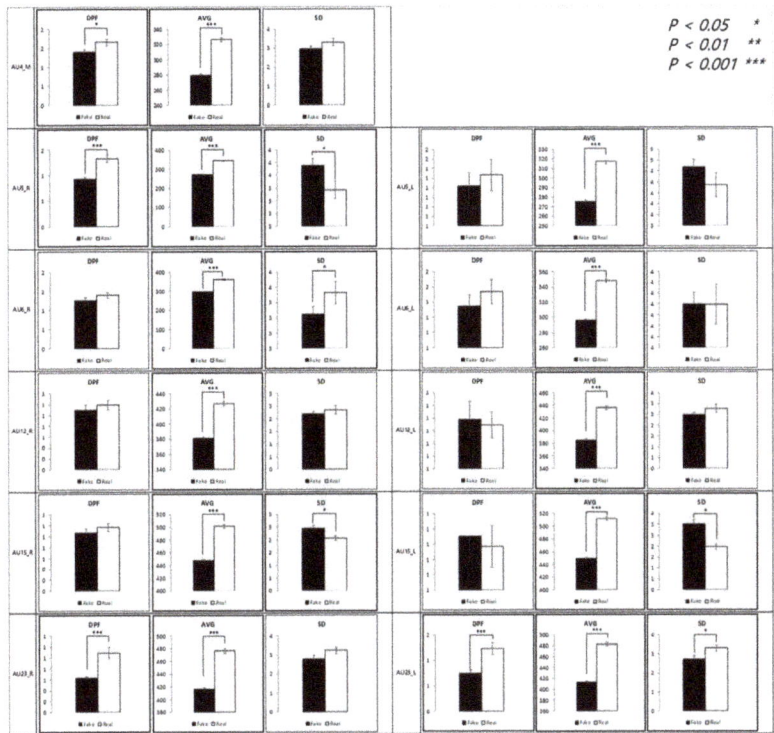

Figure 8. Comparison between feature variables of happiness expressions.

4.2. Authenticity of Contentment

The results of the analysis of micromovements involving expressions of contentment are as follows. Figure 9 depicts the differences in the movement of facial regions between the two conditions.

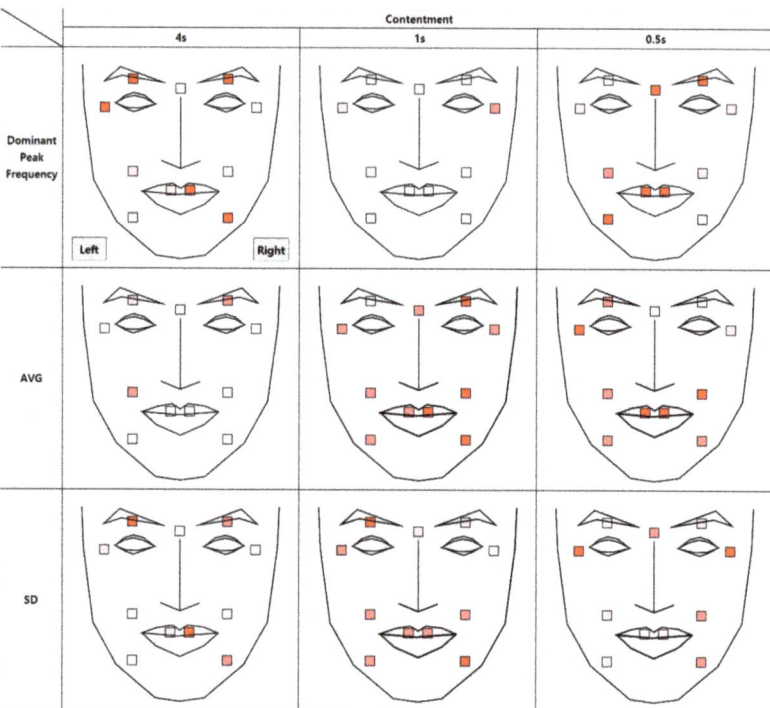

Figure 9. Statistical differences between real and fake contentment expressions (AVG = Average, SD = Standard Deviation).

At t + 1, except for the left eyelid, all 10 AUs were found to have a significant average difference. Similar results were observed for the standard deviation. At t + 0.5, nine AUs were reported to have a significant average difference. This indicates that the microexpression of contentment, compared to happiness, may persist longer. Further regression analysis on average movement showed that the time segment factor enters the regression equation ($R^2 = 0.97$), $p < 0.001$, along with the authenticity factor, $p < 0.001$ and the face side factor, $p < 0.001$.

The vibration of the macromovement (dominant peak frequency at t + 4) was significantly different in many facial regions, including the mouth tail and eyelid of the right side and the eye tail, eyelid, and mouth tail of the left side. Similar results were observed for the standard deviation in the same regions.

As shown in Figure 10, similar to the happiness condition, the average was significantly higher in the real condition, but the dominant peak frequency was significantly higher in the fake condition. That is, there was more facial movement in the real condition but more facial vibration in the fake condition.

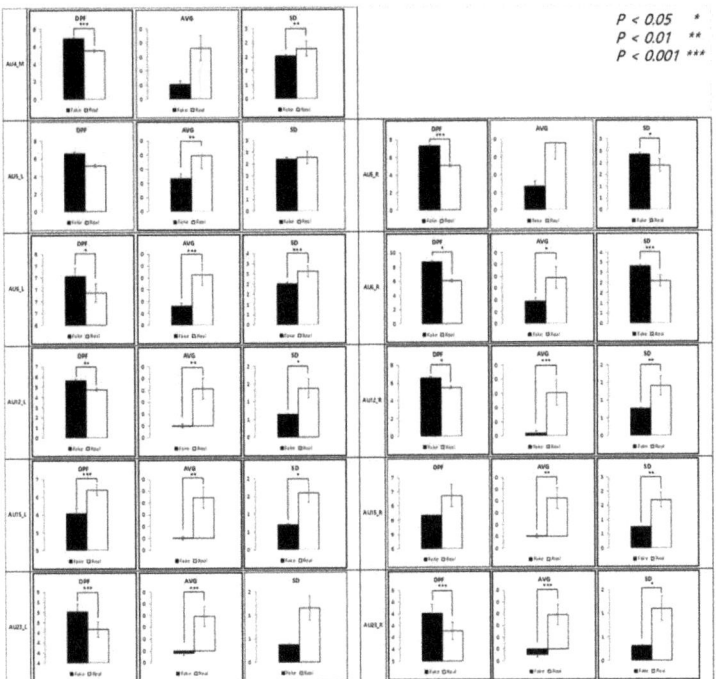

Figure 10. Comparison between feature variables of contentment expressions.

4.3. Authenticity of Anger

The results of the analysis of micromovement involving expressions of anger are as follows. Figure 11 depicts the differences in the movement of facial regions between the two conditions.

Similar to the results in the happiness condition, micromovements at t + 0.5 had a statistical difference in all regions, 11 of them at $p < 0.001$. Unlike with happiness, however, the differential micromovements of anger persisted through t + 1, except for in two of the facial regions. Further regression analysis on average movement showed that the time segment factor entered the regression equation ($R^2 = 0.96$), $p < 0.001$, along with the authenticity factor, $p < 0.001$ and the face side factor, $p < 0.05$.

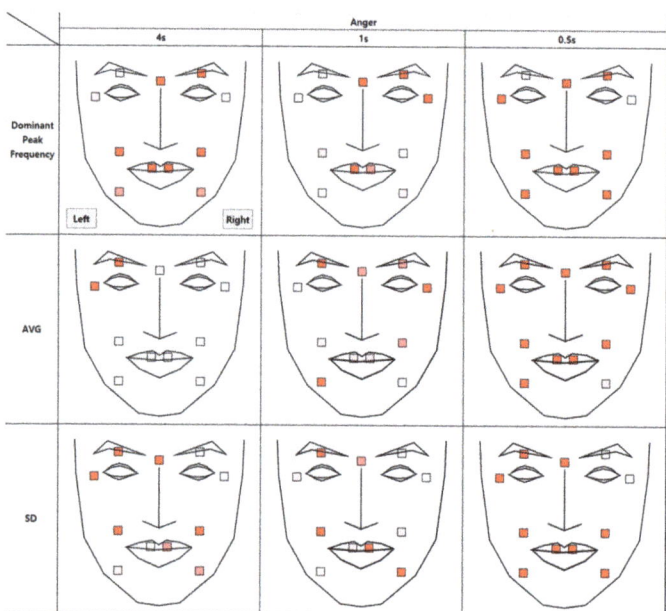

Figure 11. Statistical differences between real and fake anger expressions (AVG = Average, SD = Standard Deviation).

A significant difference in dominant peak frequency was found in all regions except for the right eye tail and the left eyelid in all time segments.

As shown in Figure 12, similar to the happiness condition but unlike the contentment condition, all three measurements were higher in the real condition than in the fake condition.

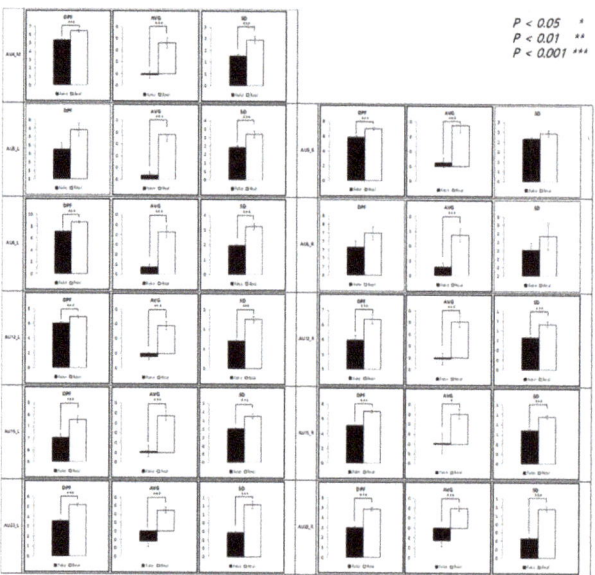

Figure 12. Comparison between feature variables of anger expression.

4.4. Authenticity of Sadness

The results of the analysis of micromovement involving the expression of sadness are as follows. Figure 13 depicts the differential movement of the facial regions between the two conditions.

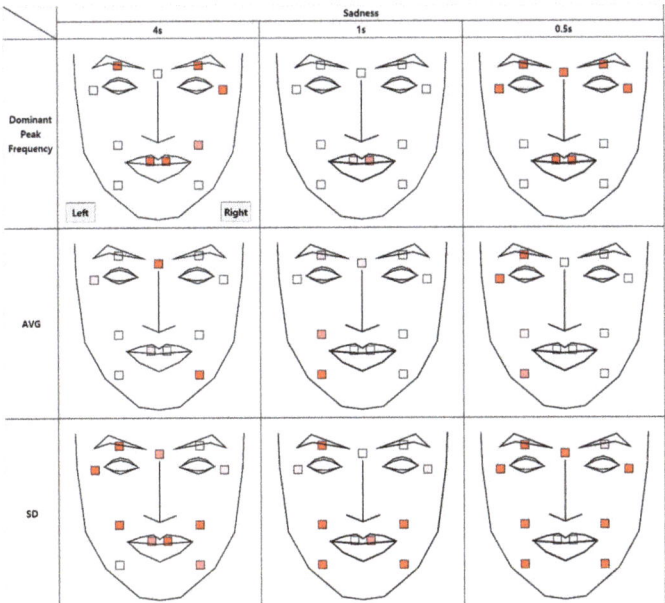

Figure 13. Statistical differences between real and fake sadness expression (AVG = Average, SD = Standard Deviation).

The significant differences were not dominant in all facial regions compared to other emotion conditions, but instead concentrated on the left side of the face. Specifically, similar results were found in the micromovements (t + 1 and t + 0.5) in the left eyelid and mouth tail. Further regression analysis on average movement showed that the face side (left or right) factor enters the regression equation ($R^2 = 0.96$), $p < 0.001$, along with the authenticity factor, $p < 0.001$ and the time segment factor, $p < 0.001$.

A significant difference was found in the mouth region in all segments with respect to the dominant peak frequency. However, the difference in vibration was prominent and salient at t + 4 and t + 0.5.

As shown in Figure 14, when the data are collapsed, similar to the contentment condition, the standard deviation and the dominant peak frequency were higher in the fake condition than in the real condition.

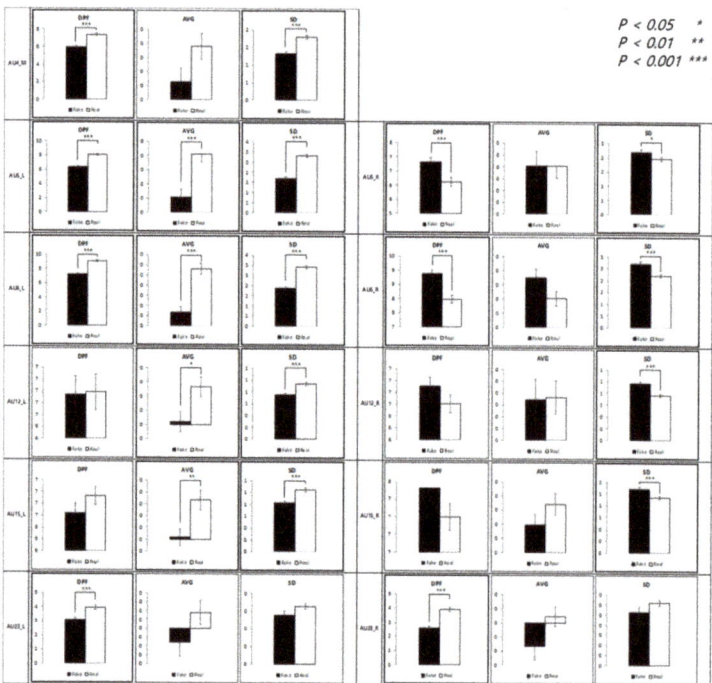

Figure 14. Comparison between feature variables of sadness expressions.

4.5. Analysis of Heart Rate Variability

The HRV data of the fake condition were compared to those of the real condition of the three frequency bands (very low, low, and high) (see Figure 15). This was performed to compare the ANS response, independent of emotions. Except for the LF (%) variable, a significant difference was found in all variables ($p < 0.001$). Specifically, VLF and VLF (%) were higher in the real condition than in the fake condition. Conversely, HF and HF (%) were higher in the fake condition than in the real condition. LF was significantly higher in the fake condition.

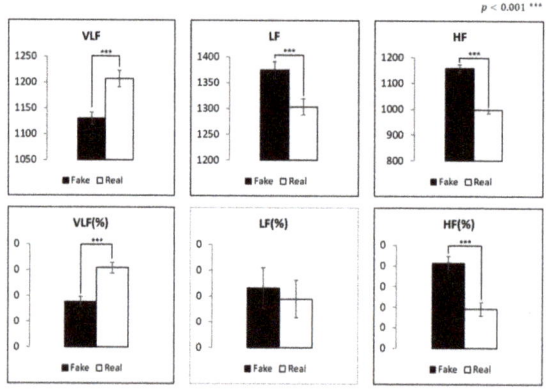

Figure 15. Comparison of frequency domain.

5. Conclusions and Discussion

The present study compared the differences in the micromovement of facial expressions between real and fake emotions. The study utilized 11 AUs based on anatomical muscle location responsible for emotional expression. That is, we identified the difference in the feature variables (average and standard deviation of movement, as well as dominant peak frequency) between the real and fake conditions by facial regions for each representative emotion (happiness, contentment, anger, and sadness). In conclusion, the study showed that the degree of activation is higher if the expression is authentic, implying more micromovement.

The study analyzed the feature variables in three time segments (0.5, 1, and 4 s) after the onset (t) of facial expression for each representative emotion. Results indicated that micromovements are more informative at an early stage (less than a second) of expression. In the case of $t + 1$ and $t + 0.5$, a significant difference between the real and fake conditions was observed in the left face than the right in the happiness condition. The asymmetric difference in the activation of the face can be explained by activation of the right brain region [32]. Campbell found that the left face expresses more than the right in voluntary expressions. Conversely, the left face expresses less than the right in involuntary expressions [33]. In the anger condition, compared with other emotions, the brow had the highest number of feature variables that were significantly different between the real and fake conditions. This was a result of muscle movement from the participant's frowning.

At $t + 4$, compared to the time segments in which less than a second had elapsed, less statistical differences were observed between the two conditions for all four emotions. This confirms that measurements at $t + 4$ cannot reliably capture the differential micromovements between real and fake expressions. The data at $t + 4$ also include the macromovements of facial muscles and hence may not be sensitive enough to identify abrupt changes in facial movements (i.e., micromovements).

Collapsing the data across time segments, all three feature variables (average, standard deviation, and dominant peak frequency) of the real condition were significantly higher than those of the fake condition in the happiness and anger conditions. Conversely, in the contentment and sadness conditions, the standard deviation and dominant peak frequency of the fake condition were significantly higher. That is, emotions that are primarily expressed with the relaxation of facial muscles, such as contentment and sadness, were observed with weaker intensity in the real condition. The results support the hypothesis that the degree of expression differs between the real and fake conditions as a function of emotions.

Our findings were cross-validated with neurological measurements involving the PSNS and ANS. In the HRV analysis, both HF and HF (%) indicators for the parasympathetic nervous system (PSNS) were higher in the fake condition than in the real condition. Conversely, both VLF and VLF (%) indicators for the ANS were higher in the real condition than in the fake condition. LF (%), an indicator that involves both the PSNS and ANS, did not show a significant difference. In conclusion, the stimuli in the real condition led to the activation of the ANS, which implies an increase in the participant's arousal. In addition, the stimuli in the fake condition led to the activation of the PSNS, which implies the participant's relaxation.

The study acknowledges the individual variance in participants' emotions when they were exposed to visual stimuli. To minimize this difference, a target facial expression was provided. In the fake condition, to ensure that other emotions did not interfere, visual content inducing a neutral emotion was used. That is, participants had to pretend an expression while the stimuli conveyed neutrality. We acknowledge the limitations of this experimental design, which may lower the ecological validity. However, future studies may investigate when a real emotion is replaced by another emotion and study the change in microexpressions.

Follow-up studies may introduce experimental treatments that are congruent with real-world settings. Specifically, micromovements of expressions in complex emotions merit further analysis. In addition, the study was limited to four representative emotions. Although not related to emotion authenticity, Adegun and Vadapalli analyzed microexpressions to recognize seven universal emotions with machine learning [34].

Another limitation of the study involves facial landmark detection. Proper landmark detection is necessary to secure recognition accuracy [20]. We have identified 68 facial landmarks by tracking the eye, eyebrows, nose, lips, and chin line using the Dlib library [29]. However, recent state-of-the-art methods, including tweaked convolutional neural networks (TCNN), may improve the robustness of facial landmark detection [35].

The breakdown of feature variables may be used as an appraisal criterion to authenticate facial data with emotional expressions. This study identified that data at less than one second is critical for analysis of the authenticity of an expression, which may not be reportable by the participants.

Systems capable of recognizing human emotions (e.g., social robots, recognition systems for the blind, monitoring systems for drivers) may use the authenticity of the user's facial expression to provide a useful and practical response. Recognizing fake expressions is imperative in security interfaces and systems that counter crime. For a social robot to provide effective services, identifying the user's intent is paramount. A recent human-robot interaction study applied deep neural networks to recognize a user's facial expressions in real time [36]. Further recognition of the user's false (e.g., deception) or pretended (e.g., to be polite) expressions might introduce more social, rich, and effective interactions.

Supplementary Materials: The following are available online at https://www.mdpi.com/article/10.3390/s21134616/s1, Python Code_AU Extraction.

Author Contributions: S.P.: methodology, validation, formal analysis, investigation, writing, review, editing, S.W.L.: conceptualization, methodology, software, validation, formal analysis, investigation, resources, data curation, writing, visualization, project administration, M.W.: conceptualization, methodology, writing, review, supervision, funding acquisition. All authors have read and agreed to the published version of the manuscript.

Funding: This work was supported by the National Research Foundation of Korea (NRF) grant funded by the Korean government (MSIT) (NRF-2020R1A2B5B02002770).

Institutional Review Board Statement: The study was conducted according to the guidelines of the Declaration of Helsinki, and approved by the Institutional Review Board of Sangmyung University (protocol code BE2018-31, approved at 10 August 2018).

Informed Consent Statement: Informed consent was obtained from all subjects involved in the study. Written informed consent has been obtained from the subjects to publish this paper.

Conflicts of Interest: The authors declare no conflict of interest.

References

1. Patterson, M.L. Nonverbal communication. *Corsini Encycl. Psychol.* **2010**, *30*, 1–2.
2. Mehrabian, A.; Williams, M. Nonverbal concomitants of perceived and intended persuasiveness. *J. Pers. Soc. Psychol.* **1969**, *13*, 37–58. [CrossRef] [PubMed]
3. Patterson, M.L. Invited article: A parallel process model of nonverbal communication. *J. Nonverbal Behav.* **1995**, *19*, 3–29. [CrossRef]
4. Ekman, P.; Friesen, W.V. Detecting deception from the body or face. *J. Pers. Soc. Psychol.* **1974**, *29*, 288–298. [CrossRef]
5. Frank, M.G.; Ekman, P. The ability to detect deceit generalizes across different types of high-stake lies. *J. Pers. Soc. Psychol.* **1997**, *72*, 1429–1439. [CrossRef] [PubMed]
6. Ekman, P. Darwin, deception, and facial expression. *Ann. N. Y. Acad. Sci.* **2003**, *1000*, 205–221. [CrossRef]
7. Dopson, W.G.; Beckwith, B.E.; Tucker, D.M.; Bullard-Bates, P.C. Asymmetry of Facial Expression in Spontaneous Emotion. *Cortex* **1984**, *20*, 243–251. [CrossRef]
8. Cicone, M.; Wapner, W.; Gardner, H. Sensitivity to Emotional Expressions and Situations in Organic Patients. *Cortex* **1980**, *16*, 145–158. [CrossRef]
9. Duchenne, G.B.; de Boulogne, G.B. *The Mechanism of Human Facial Expression*; Cambridge University Press: Cambridge, UK, 1990.
10. Reincke, H.; Nelson, K.R. Duchenne de boulogne: Electrodiagnosis of poliomyelitis. *Muscle Nerve* **1990**, *13*, 56–62. [CrossRef]
11. Ekman, P.; Friesen, W.V.; O'Sullivan, M.; Rosenberg, E.L. Smiles When Lying. In *What the Face Reveals: Basic and Applied Studies of Spontaneous Expression Using the Facial Action Coding System (FACS)*; Oxford University Press: Oxford, UK, 2005; pp. 201–215. [CrossRef]

12. Porter, S.; Ten Brinke, L. Reading between the lies: Identifying concealed and falsified emotions in universal facial expressions. *Psychol. Sci.* **2008**, *19*, 508–514. [CrossRef]
13. Endres, J.; Laidlaw, A. Micro-expression recognition training in medical students: A pilot study. *BMC Med. Educ.* **2009**, *9*, 47. [CrossRef]
14. Matsumoto, D.; Hwang, H.S. Evidence for training the ability to read microexpressions of emotion. *Motiv. Emot.* **2011**, *35*, 181–191. [CrossRef]
15. Ramachandran, V.S. *The Tell-Tale Brain: A Neuroscientist's Quest for What Makes Us Human*; WW Norton & Company: New York, NY, USA, 2012.
16. Sebe, N.; Cohen, I.; Gevers, T.; Huang, T.S. Emotion Recognition Based on Joint Visual and Audio Cues. In Proceedings of the 18th International Conference on Pattern Recognition (ICPR'06), Hong Kong, China, 20–24 August 2006; Volume 1, pp. 1136–1139.
17. Tarnowski, P.; Kołodziej, M.; Majkowski, A.; Rak, R.J. Emotion recognition using facial expressions. *Procedia Comput. Sci.* **2017**, *108*, 1175–1184. [CrossRef]
18. See, J.; Yap, M.H.; Li, J.; Hong, X.; Wang, S.-J. MEGC 2019—The Second Facial Micro-Expressions Grand Challenge. In Proceedings of the 2019 14th IEEE International Conference on Automatic Face & Gesture Recognition (FG 2019), Lille, France, 14–18 May 2019; pp. 1–5.
19. Liu, Y.; Du, H.; Zheng, L.; Gedeon, T. A Neural Micro-Expression Recognizer. In Proceedings of the 2019 14th IEEE International Conference on Automatic Face & Gesture Recognition (FG 2019), Lille, France, 14–18 May 2019; pp. 1–4.
20. Xie, H.X.; Lo, L.; Shuai, H.H.; Cheng, W.H. An Overview of Facial Micro-Expression Analysis: Data, Methodology and Challenge. *arXiv* **2020**, arXiv:2012.11307.
21. Pan, H.; Xie, L.; Wang, Z.; Liu, B.; Yang, M.; Tao, J. Review of micro-expression spotting and recognition in video sequences. *Virtual Real. Intell. Hardw.* **2021**, *3*, 1–17. [CrossRef]
22. Choi, D.Y.; Song, B.C. Facial Micro-Expression Recognition Using Two-Dimensional Landmark Feature Maps. *IEEE Access* **2020**, *8*, 121549–121563. [CrossRef]
23. Liong, S.-T.; Gan, Y.S.; Zheng, D.; Li, S.-M.; Xu, H.-X.; Zhang, H.-Z.; Lyu, R.-K.; Liu, K.-H. Evaluation of the Spatio-Temporal Features and GAN for Micro-Expression Recognition System. *J. Signal Process. Syst.* **2020**, *92*, 705–725. [CrossRef]
24. Zhang, F.; Zhang, T.; Mao, Q.; Xu, C. Joint Pose and Expression Modeling for Facial Expression Recognition. In Proceedings of the 2018 IEEE/CVF Conference on Computer Vision and Pattern Recognition, Salt Lake City, UT, USA, 18–23 June 2018; pp. 3359–3368.
25. Guo, C.; Liang, J.; Zhan, G.; Liu, Z.; Pietikainen, M.; Liu, L. Extended Local Binary Patterns for Efficient and Robust Spontaneous Facial Micro-Expression Recognition. *IEEE Access* **2019**, *7*, 174517–174530. [CrossRef]
26. Nikolova, D.; Petkova, P.; Manolova, A.; Georgieva, P. ECG-based Emotion Recognition: Overview of Methods and Applications. In *ANNA'18; Advances in Neural Networks and Applications*; VDE: Berlin, Germany, 2018; pp. 1–5.
27. Brás, S.; Ferreira, J.H.T.; Soares, S.C.; Pinho, A.J. Biometric and Emotion Identification: An ECG Compression Based Method. *Front. Psychol.* **2018**, *9*, 467. [CrossRef]
28. Wilson, P.I.; Fernandez, J. Facial feature detection using Haar classifiers. *J. Comput. Sci. Coll.* **2006**, *21*, 127–133.
29. King, D.E. Dlib-ml: A machine learning toolkit. *J. Mach. Learn. Res.* **2009**, *10*, 1755–1758.
30. Park, K.-J.; Jeong, H. Assessing Methods of Heart Rate Variability. *Korean J. Clin. Neurophysiol.* **2014**, *16*, 49–54. [CrossRef]
31. Kleiger, R.E.; Stein, P.K.; Bigger, J.T., Jr. Heart rate variability: Measurement and clinical utility. *Ann. Noninvasive Electrocardiol.* **2005**, *10*, 88–101. [CrossRef]
32. Fusar-Poli, P.; Placentino, A.; Carletti, F.; Landi, P.; Allen, P.; Surguladze, S.; Benedetti, F.; Abbamonte, M.; Gasparotti, R.; Barale, F.; et al. Functional atlas of emotional faces processing: A voxel-based meta-analysis of 105 functional magnetic resonance imaging studies. *J. Psychiatry Neurosci.* **2009**, *34*, 418–432.
33. Campbell, R. Asymmetries in Interpreting and Expressing a Posed Facial Expression. *Cortex* **1978**, *14*, 327–342. [CrossRef]
34. Adegun, I.P.; Vadapalli, H. Facial micro-expression recognition: A machine learning approach. *Sci. Afr.* **2020**, *8*, e00465. [CrossRef]
35. Wu, Y.; Hassner, T.; Kim, K.; Medioni, G.; Natarajan, P. Facial Landmark Detection with Tweaked Convolutional Neural Networks. *IEEE Trans. Pattern Anal. Mach. Intell.* **2017**, *40*, 3067–3074. [CrossRef]
36. Melinte, D.O.; Vladareanu, L. Facial Expressions Recognition for Human–Robot Interaction Using Deep Convolutional Neural Networks with Rectified Adam Optimizer. *Sensors* **2020**, *20*, 2393. [CrossRef]

Article

Multi-Path and Group-Loss-Based Network for Speech Emotion Recognition in Multi-Domain Datasets

Kyoung Ju Noh *, Chi Yoon Jeong, Jiyoun Lim, Seungeun Chung, Gague Kim, Jeong Mook Lim and Hyuntae Jeong

Artificial Intelligence Research Laboratory, Electronics and Telecommunications Research Institute, Daejeon 34129, Korea; iamready@etri.re.kr (C.Y.J.); kusses@etri.re.kr (J.L.); schung@etri.re.kr (S.C.); ggkim@etri.re.kr (G.K.); jmlim21@etri.re.kr (J.M.L.); htjeong@etri.re.kr (H.J.)
* Correspondence: kjnoh@etri.re.kr; Tel.: +82-42-860-1764

Abstract: Speech emotion recognition (SER) is a natural method of recognizing individual emotions in everyday life. To distribute SER models to real-world applications, some key challenges must be overcome, such as the lack of datasets tagged with emotion labels and the weak generalization of the SER model for an unseen target domain. This study proposes a multi-path and group-loss-based network (MPGLN) for SER to support multi-domain adaptation. The proposed model includes a bidirectional long short-term memory-based temporal feature generator and a transferred feature extractor from the pre-trained VGG-like audio classification model (VGGish), and it learns simultaneously based on multiple losses according to the association of emotion labels in the discrete and dimensional models. For the evaluation of the MPGLN SER as applied to multi-cultural domain datasets, the Korean Emotional Speech Database (KESD), including KESDy18 and KESDy19, is constructed, and the English-speaking Interactive Emotional Dyadic Motion Capture database (IEMOCAP) is used. The evaluation of multi-domain adaptation and domain generalization showed 3.7% and 3.5% improvements, respectively, of the F1 score when comparing the performance of MPGLN SER with a baseline SER model that uses a temporal feature generator. We show that the MPGLN SER efficiently supports multi-domain adaptation and reinforces model generalization.

Keywords: speech emotion recognition; domain adaptation; SER generalization; Korean Emotional Speech Database; ensemble model; multi-path; group-loss; BLSTM network

1. Introduction

Human speech is a natural communication method in human–computer interaction (HCI) and human–robot interaction (HRI). Speech emotion recognition (SER), which is based on natural human language, is a key method used to recognize individual emotions in everyday speech. SER uses the acoustic features of a speech segment, not the lexical features having the semantic information of the segment [1]. Hence, it recognizes subjects' emotions from "how" they speak rather than the content of their words. The predicted emotional context of a target speaker can then be used as an important factor for decision making in intelligent HCI and HRI services [2,3].

Prior to deploying SER models in real applications, the lack of SER databases tagged with emotion labels must be addressed, because they are not sufficient for training deep-SER models. Another challenge is the limited generality of the SER model, owing to the high variability of the acoustic signals of the emotional speech samples.

Emotions have characteristics of high subjectivity and diversity, depending on the individual or culture. Therefore, it is time-consuming and expensive to build a large-scale emotional database annotated with reliable gold-standard emotion labels via human observation. Most SER datasets having gold-standard labels contain thousands of speech samples collected from a limited number of speakers in a specific environment [4–7]. Therefore, the performance of an SER model trained on single-domain samples is inherently degraded when applied to unseen domain samples that reflect different languages, cultures, speakers, genders, microphone types, positions, and signal-to-noise ratios [8–10]. This study defines a single SER domain dataset collected using one collection procedure at one place using the same collection device.

Many studies have effectively utilized limited emotion databases to improve the SER performance. In addition to the typical augmentation methods of speech samples [11,12], there exists a domain adaptation method that utilizes speech datasets already established in the unknown target domain [8–10,13–16]. In comparison with the results of data augmentation in a single domain, it is difficult to guarantee good performance because of the high variability of the acoustic features of the emotional speech samples in the domain [8–10,13,14]. However, domain adaptation based on multi-domain datasets can be used to construct better SER models to support such generalities without overfitting.

We propose a multi-path and group-loss-based network (MPGLN) for SER, which supports supervised domain adaptation in multi-domain datasets acquired from multiple environments. The proposed MPGLN for SER (MPGLN SER) is based on an ensemble learning structure for multi-level embedding vector learning for speech segments. It includes a temporal embedding feature generator, transferred feature extractor, and prediction function network that classifies the emotion labels based on the generated and extracted feature vectors. The bidirectional long short-term memory (BLSTM)-based temporal feature generator network learns an embedding vector as a 74-D input of handcrafted low-level descriptions (LLD) of a speech segment. The transferred feature extractor creates feature vectors from the pre-trained VGG-like audio classification model (VGGish) [17], and the proposed MPGLN SER is trained based on multiple losses by the association between the discrete and continuous dimensional emotion labels [1] of the multi-domain samples.

The proposed MPGLN SER is evaluated over five multi-domain SER datasets: the benchmark English Interactive Emotional Dyadic Motion Capture (IEMOCAP) dataset [7], which was widely used in previous studies for SER model evaluation, and the four Korean Emotional Speech Database (KESD) datasets that are built for this study.

In our evaluation, we use an SER model comprising a BLSTM-based temporal feature generator and the MPGLN predicting network, excluding transferred features, as our baseline model. We then verify the reliability of the baseline SER model using the IEMOCAP dataset. Comparing it with the performance of the baseline SER model, it is confirmed that the proposed MPGLN SER is effective in supporting supervised multi-domain adaptations and reinforcing generalizations [18] of the SER model in multi-domain datasets.

This paper is organized as follows. In Section 2, we present a brief overview of related SER and domain adaptation works. Section 3 describes the proposed MPGLN, which supports multi-domain adaption of SER in multi-domain datasets. Section 4 details the evaluation results of the MPGLN SER, and Section 5 concludes this study and suggests future works.

2. Related Works

Recent SER models based on deep-learning architectures [19–30] have demonstrated state-of-the-art performance with an attention mechanism [19,20,22,23,25,26]. The deep-learning architectures adopted in previous studies included recurrent neural networks (RNN) [19], convolutional neural networks (CNN) [24], and convolutional RNNs (CRNN) [20,26]. Liu et al. [21] presented an SER model of a decision tree for an extreme learning machine having a single hidden-layer feed-forward neural network, using a mixture of deep learning and typical classification techniques.

The input features for deep-learning-based SER models are generally extracted from the time or spectrum axis in units of speech segments or frames. There are various LLDs and high-level statistical functions of the LLD single features [19,20,31–33]. The spectrum LLD features of speech signals include logMel filter-banks and mel-frequency cepstral coefficients (MFCC). Zero-crossing rates and signal energies are representative time-domain features [27–30], whereas spectral roll-off and spectral centroid are classified as spectral parameters [33]. A set of multiple single features for acoustic signal processing, such as the extended Geneva Minimalistic Acoustic Parameter Set [34] and the INTERSPEECH 2010 Paralinguistic Challenge (IS10) dataset [35], is now accessible from open-source frameworks, such as OpenSmile [36]. Some studies have investigated the mechanism of modeling and integrating of temporal acoustic features to improve the performance of speech emotion recognition or audio classification [31,32]. Jing et al. [37] presented an evaluation of multiple acoustic feature sets that combined features generated from the pre-trained acoustic model [15,17,38,39].

A typical deep-learning model requires large-scale samples for training. Unfortunately, SER datasets annotated with emotion labels are scarce. Furthermore, collecting SER speech samples and tagging them with emotion labels is time-consuming and expensive. Thus, to overcome the limitations of volume and diversity of labeled speech samples for deep-learning SER models, studies have been performed using data augmentation [11,12,40–42], active learning [12,43] based on collected datasets, and domain adaptation [8–10,13–16] to adapt the existing SER datasets to the target domains.

Park et al. [11] presented a data augmentation experiment for speech samples using warping and masking in a frequency channel with a time step. Chatziagapi et al. [40] proposed a method that used generative adversarial networks [44] to extract artificial spectrograms of augmented data to balance each emotion class.

Active-learning methods have been used to present greedy selection methods of speech samples to construct an initial SER model suitable for a target speaker based on limited samples [12,43]. Abdelwahab et al. [43] proposed the active learning of greedy sampling to select the most informative samples to improve the performance of DNN-based SER models. In a study by Bang et al. [12], samples that were close to the target speaker's samples in the embedding space were selected; the synthetic minority oversampling technique was applied to increase the number of samples of the minority class.

Domain adaptation techniques are actively being studied in the field of visual classification [18,45]. Metric-based learning is a representative method of learning distances containing the features of inter-domain and -class samples to minimize domain mismatches between the source and target domains. Gao et al. [46] proposed an acoustic model based on ResNet [47] for acoustic scene classification; its learning process is such that it is difficult to distinguish the domain to which a sample belongs.

The domain adaptation for SER models based on multi-domain datasets has the purpose of building an SER model that is not overfitted to a specific dataset and is generalized for unknown target-domain speech data. However, the SER model based on multi-domain datasets has a different applicability from the case that applies data augmentation by oversampling a single domain dataset. It does not guarantee the SER performance improvement, even if several multi-domain speech samples are used to train the SER model, because there is high domain discrepancy in the speech signal, which depends on the collection environments [8–10,13,14].

Liang et al. [9] proposed a structure that learned emotion-salient features based on audio and video data through an adversarial learning framework, generating embedding features for the purpose of reducing domain discrepancies. Huang et al. [13] presented a network model that aligned the distribution shift in the intermediate feature space between the source and target domains. Neumann et al. [14] introduced an adaptive technique to fine-tune the weights of SER neural networks trained in the source domain using a small number of samples from the target. By using the transferred features from the pre-trained model, Li et al. [15] demonstrated improvements in the SER performance using additional embedding vectors extracted from the pretrained VGGish in AudioSet [48]. Lee et al. [16] presented the generalization effect of emotion recognition by applying dropout and normalization methods in multilingual heterogeneous datasets.

3. Ensemble Learning Model for SER in Multi-Domain Datasets

We propose an ensemble learning model to improve the performance of SER generalization in multi-domain datasets. The operational flow of the supervised multi-domain adaptation of the proposed MPGLN SER is shown in Figure 1. We denote speech-input samples and class-label spaces as X and Y, respectively, and the domain datasets are $D = \{D_1, D_2, \ldots, D_k\}$. This study assumes a supervised learning environment wherein each domain sample has common emotion labels. In this study, each domain dataset consists of pairs $D_k = \left\{ \left(X_i^k, (y_{i\_d}^k, y_{i\_v}^k) \right) \right\}_{i=1}^{N_k}$, where N_k is the number of speech samples of the k-th domain dataset, and datasets in each speech sample have multiple Y labels. The discrete emotion label is $y_{i\_d}^k$ (e.g., "happy" and "sad"), and that of the valence-level is $y_{i\_v}^k$ in the continuous dimensional emotion model.

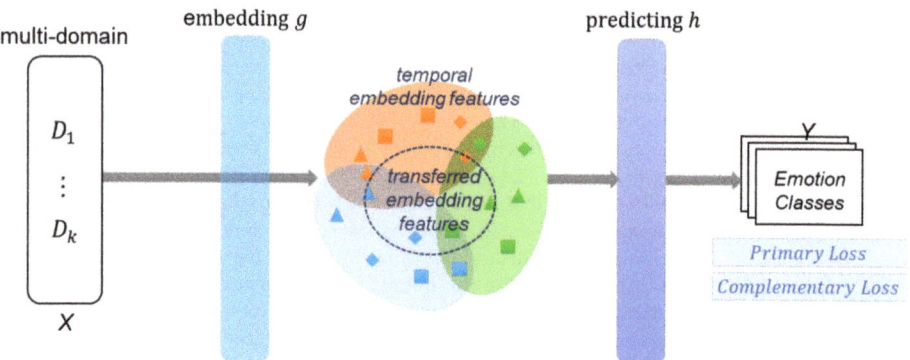

Figure 1. Supervised multi-domain adaptation of the multi-path and group-loss-based network (MPGLN) speech emotion recognition (SER). The model generates the temporal embedding feature and the transferred embedding feature for the speech segment and learns based on multiple losses.

The source-domain dataset used for model training is domain D_s, and the domain to which test samples to be predicted belong is the target domain, D_t. There are variant shifts and domain discrepancies of the feature distribution, $d(X^S)$ and $d(X^T)$, of data samples of different domain datasets, D_s and D_t, respectively [45].

The goal of the SER model is to learn the classifier function, $f : X \to Y$, in the target domain. Function f consists of the composition of two functions, $f = h \circ g$, where g is an embedding feature generator from the input data space, X, to an embedding feature space, and h is the function used to predict the embedding feature to label-space Y.

Figure 2 shows the architecture of the proposed MPGLN SER, which generates the multi-level embedding vectors from the multi-path generators. The BLSTM-based feature generator, g_{BLSTM}, generates a temporal embedding vector, and the transferred feature extractor, g_{vgg}, extracts a transferred embedding vector from the pre-trained VGGish model [17].

In the prediction function, h, of the proposed ensemble structure, discrete emotional labels are classified based on the fusion of multi-path embedding vectors from g_{BLSTM} and g_{vgg}. It also includes a dimensional valence-level classification function based on the temporal embedding feature generated by g_{BLSTM}.

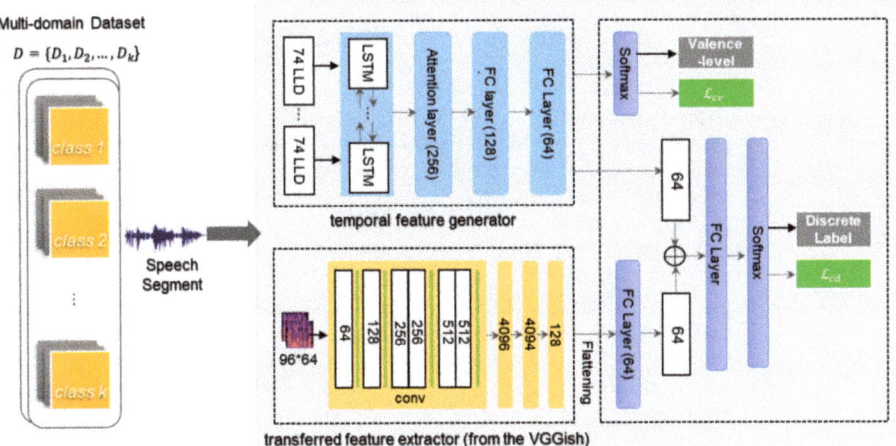

Figure 2. Architecture of the multi-path and group-loss-based network for SER. The MPGLN SER model comprises a bidirectional long short-term memory (BLSTM)-based temporal embedding generator and a transferred feature extractor from the VGG-like audio classification model (VGGish) and its prediction function.

3.1. Multi-Path Embedding Features

In this study, the speech segments of an utterance unit are embedded in the feature space through g_{BLSTM}, a temporal feature generator of the ensemble structure, and g_{vgg}, a transmitted feature extractor. In Figure 2, the temporal feature generator, g_{BLSTM}, of the BLSTM architecture reflects a characteristic of the temporal relevance of before-and-after speech features. The 74-D LLD-per-frame speech segment comprises a 13-D MFCC and 40-D Mel-spectrogram, along with 21-D time- and frequency-domain LLDs such as zero-crossing rate, energy, spectral centroid, and spectral roll-off. The 74-D LLD are extracted by the frame that applies sliding windows of 200 ms with a 50% shift in the speech segment. Each speech segment is padded with a zero value to have a fixed number of 100 frames, and the sequence of 100 × 74 per segment is input to g_{BLSTM}. The padded input sequence is fed into the g_{BLSTM}, comprising 128 cells in each direction, and g_{BLSTM} produces a 256-D feature vector.

The feature generator, g_{BLSTM}, adopts an attention mechanism and focuses on those more discriminative parts of the BLSTM output sequence before activation of the final emotion classification. The attention mechanism for SER assumes that there are certain words and salient parts that express emotions well in the speech segment. Using the attention method, it gives more weight to relevant speech frames of an utterance-level segment for emotion recognition.

The attention layer focuses on relevant parts of the output sequence of the BLSTM by giving different weight scores and generates the high-level features (hf). It computes weight α_t using the softmax function via the attention layer (see Equation (1)), where the BLSTM output vector is $h_t = [\overrightarrow{h_t}, \overleftarrow{h_t}]$ at time t. It produces the high-level feature, hf, which is the weighted sum, h_t, obtained by multiplying the weights, α_t (see Equation (2)). The generated hf is transited again to an embedding feature vector of $\mathbb{R}^{64}$ through the two fully-connected (FC) layers in the MPGLN.

$$\alpha_t = \frac{\exp(W \cdot h_t)}{\sum_{t=1}^{T} \exp(W \cdot h_t)} \quad (1)$$

$$hf = \sum_{t=1}^{T} \alpha_t \cdot h_t, \quad (2)$$

The temporal feature generator, $g_{BLSTM} : X \rightarrow \mathbb{R}^{64}$, generates a 64-D embedding vector from the input of the 74-D LLD in units of speech-segment frames. The feature generator, g_{BLSTM}, in the MPGLN SER can operate as an SER model alone by combining the prediction function, $h_d^{baseline} : \mathbb{R}^{64} \rightarrow Y(y_{i\_d}^k)$, without using the transferred features from the VGGish. This study uses the BLSTM-based SER model as a baseline for the evaluation of the MPGLN SER.

The transferred feature extractor, $g_{vgg} : X \rightarrow \mathbb{R}^{VGGish}$, extracts the transferred feature vector of data-sample X using the VGGish model. The input speech segment is divided into non-overlapping 960 ms time-unit frames, and 64 mel-spaced spectrogram features that apply a 25 ms window every 10 ms in each frame are extracted using the VGGish model [17]. Using the transferred feature extractor, g_{vgg}, it generates a 128-D embedding feature vector from the VGGish model for the speech segment by inputting a frame-by-frame spectrogram in units of 96 × 64. The extracted 128-D embedding vector passes through the fattening and FC layers and is transited to a 64-D embedding vector.

3.2. Group Loss

Equation (3) shows how classifier f is trained on the classification loss, $\mathcal{L}_c(f)$, of the emotion labels Y of the speech samples X, where ℓ is an appropriate loss function similar to cross-entropy for multi-class classification [45,49].

$$\mathcal{L}_c(f) = \ell(f(X), Y) \quad (3)$$

The proposed MPGLN SER is trained to simultaneously minimize multiple losses, which are induced by the association of multi-dimensional emotion labels. The discrete emotion labels are intuitive for expressing the emotion, but it has difficulty in expressing complex emotions. The dimensional emotion labels are capable of normalized expressions of complex emotions. However, doing so, it is difficult to intuitively distinguish emotions at similar positions (e.g., "fear" and "anger") in the arousal-valence axis [1]. This study derives an association between discrete and dimensional valence-level labels based on real SER domain datasets and applies a method of simultaneously learning the loss for each emotion-label classification in the MPGLN model.

As shown in Figure 2, the MPGLN SER learns simultaneously based on the two losses: $\mathcal{L}_{cv}$ for the valence-level label using the $\mathbb{R}^{64}$ feature vector generated from g_{BLSTM} and $\mathcal{L}_{cd}$ for predicting the discrete emotion label.

The primary loss, $\mathcal{L}_{cd}$, is used for the predicting function, $f_d = h_d \circ (g_{BLSTM} \oplus g_{VGGish})$, where $h_d : \left(\mathbb{R}^{64} \oplus \mathbb{R}^{VGGish}\right) \rightarrow Y(y_{i\_d}^k)$ predicts the discrete emotion label of $y_{i\_d}^k$ via the combination of two embedding vectors. The complementary loss, $\mathcal{L}_{cv}$, is that of the predicting function, $f_v = h_v \circ g_{BLSTM}$, which classifies the valence-level labels, where $h_V : \mathbb{R}^{64} \rightarrow Y(y_{i\_v}^k)$. Equation (4) shows that the proposed MPGLN SER is trained to minimize group loss $\mathcal{L}_g$ about the prediction functions, f_d and f_v:

$$\mathcal{L}_g = \text{Group}(\mathcal{L}_{cd}(f_d), \mathcal{L}_{cv}(f_v)). \quad (4)$$

4. Evaluation

4.1. Datasets

We evaluated the proposed model using five multi-domain datasets contained in three real SER databases. For the evaluation of the MPGLN SER based on multi-cultural datasets, two KESD databases (i.e., KESDy18 and KESDy19) constructed for this study, and the IEMOCAP are used. KESDy18 and KESDy19 comprise two domain datasets based on heterogeneous microphone devices.

In the IEMOCAP dataset, data were collected from the scenarios for inducing the five target emotions ("happy", "sad", "neutral", "angry", and "frustration"), and annotators selected one of the six basic emotions ("angry", "sad", "happy", "disgust", "fear", and "surprise") [50] along with "frustration", "excited", and "neutral" as the discrete emotion labels. Numerous data were annotated with the emotion categories such as "fear" and "disgust", which do not belong to the target emotions in IEMOCAP [7]. Even in the KESD database, considering the subjectivity and diversity of human emotion perception, the categorical emotion label was tagged as one of the six basic emotion labels along with "neutral".

The KESDy18 comprises speech samples in which 30 voice actors uttered 20 sentences while expressing the four given emotions of "angry", "happy", "neutral", and "sad". The six external taggers evaluated the speech segments while listening to the recorded utterances as shown in Figure 3a. The annotators tagged one of the seven categorical emotion labels comprising the six basic emotions [50] in addition to "neutral", whose tagged labels are more diverse than the classification of the actor's expressed emotion. They tagged labels of arousal and valence-level on a five-point scale for each segment. The final categorical emotion label was determined by majority vote. The label of arousal and the valence-level were determined from the average value of the levels tagged by the evaluators. KESDy18 simultaneously collected speech data from two heterogeneous microphones (i.e., a cell-phone's built-in microphone (PM) and an external microphone (EM) connected to a computer). According to the type of microphone devices, KESDy18 comprised the KESDy18_PM dataset plus the KESDy18_EM dataset.

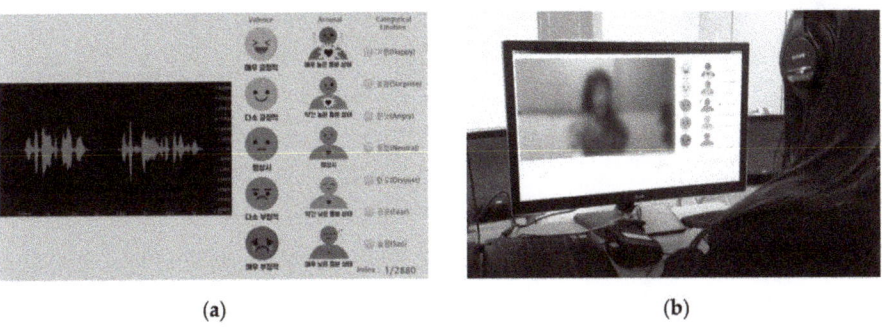

Figure 3. External annotator tags the emotion labels for speech segments using the tagging application while watching the recorded video and listening to the Emotional Speech Database (KESD) speech segments: (**a**) evaluating emotional labels of KESDy18 via the tagging application; (**b**) evaluation of the KESDy19 speech segments looking at the recorded video.

The KESDy19 includes the speech samples of 40 voice-actors who speak Korean as their native language using collection scenarios similar to those of the IEMOCAP. KESDy19 consists of 20 sessions collected from speech and electrocardiogram signals produced during the dyadic acting of two voice actors, the process of acting was recorded. Each session consists of 10 plays having lengths of 4–10 min. Six plays were based on scenarios written to induce specific emotions, and the other four were improvised during the dyadic interactions. Each speech segment per speaker was tagged using one of seven categorical emotion labels, and the average value of the five-point scale of arousal and valence-level was annotated by 10 external taggers using the same tagging application as shown in Figure 3b. KESDy19 comprises a KESDy19_EM dataset that used an external microphone and a KESDy19_PM dataset that simulated the KESDy19_EM dataset via a cell-phone's microphone.

The IEMOCAP is a widely used SER performance evaluation model organized into five sessions of multi-modal audio, visual, and textual data taken from interactive dyadic interactions performed by 10 voice actors. In each session, two voice actors emotionally performed improvisations or scripted scenarios. The speech segments of their utterance-levels were tailored to discrete emotion labels of "happy," "sad," "neutral," "angry," "surprise," "frustration," "excited," "disgust," or "fear" based on the majority opinions of three external human annotators. The IEMOCAP data were also tagged with labels of arousal and valence based on a five-point dimensional emotion scale [39,51]. The IEMOCAP database provides the re-rounded average score of the evaluations of arousal and valence-levels according to the five-point scale based on evaluations by six external evaluators. Many prior studies evaluated SER performance using the IEMOCAP database to classify the four emotion categories of "happy," "sad," "neutral," and "angry."

Figure 4 shows the distribution of four discrete emotion and arousal/valence-level labels on the five-point scales of IEMOCAP, KESDy18, and KESDy19. As shown in Figure 4a–c, the speech samples of the "happy" class are distributed at the highest valence level, and the "neutral" samples are in the middle. The speech data labeled with "sad" and "angry" classes show a distribution of low-level valences across all three SER databases. The association between discrete emotion labels and those of arousal-level shows more irregularities in Figure 4d–f. The speech samples tagged with the "sad" class are distributed in the overall arousal-level, and the samples of the IEMOCAP with the "happy" label are distributed in the overall level of arousal, unlike the other two KESD.

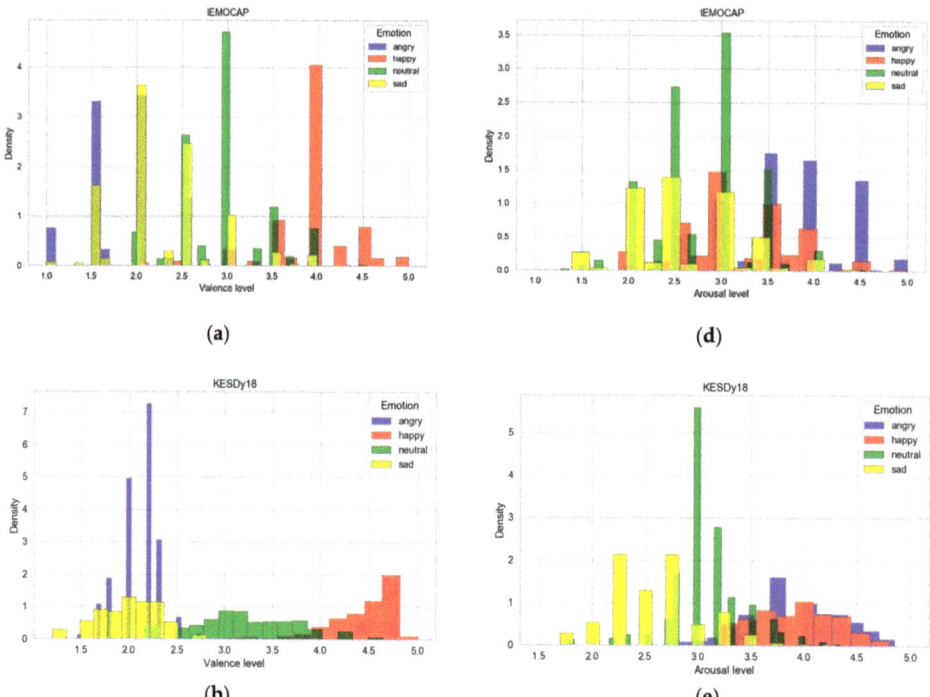

Figure 4. Cont.

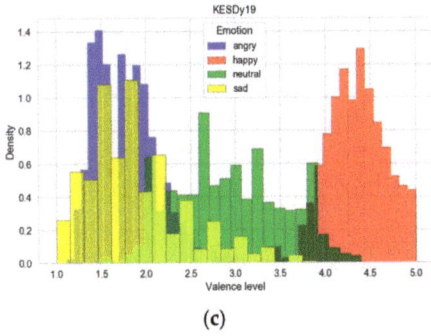
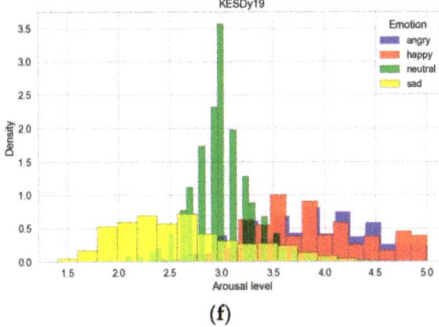

Figure 4. Distribution between discrete and dimensional emotion labels of the five-point scale: (**a**) distribution of discrete and valence-level labels of Interactive Emotional Dyadic Motion Capture database (IEMOCAP); (**b**) distribution of discrete and valence-level of KESDy18; (**c**) distribution of discrete and valence-level of KESDy19; (**d**) distribution of discrete and arousal-level of IEMOCAP; (**e**) distribution of discrete and arousal-level of KESDy18; and (**f**) distribution of discrete and arousal-level of KESDy19.

In Figure 4, the speech samples corresponding to the discrete emotion classes constitute roughly three distribution groups across the label of valence-level. The three distribution groups are "happy," "neutral," and "sad" or "angry."

In this study, we mapped the valence-level labels of the five-point scale to a three-point scale using the induced association between discrete and dimensional emotion labels, as shown in Table 1 and Figure 4. Each valence-level (i.e., 1, 2, and 3) of the three-point scale represents "negative", "neutral", and "positive" emotional states, respectively. For the conversion to the valence-level of the three-point scale, this study assigned sample labels of valences less than 2.5 to the first valence-level, samples of 4.0 or higher to the third, and the others to the second, respectively. Table 1a shows the mean and standard variation of arousal and valence-levels on a five-point scale for each discrete emotion category. Table 1b shows the confidences of association [52] of the speech samples of four discrete emotion classes included in the valence levels of the three-point scale. The confidence $Conf.(C_i \rightarrow V_j) = \frac{N_{C_i \cup V_j}}{N_{C_i}}$, where C_i is the discrete emotion label, $1 \leq i \leq 4$, and V_j denotes the valence-level, $1 \leq j \leq 3$.

Table 1. Association properties of discrete emotion labels and valence-levels in multi-domain SER datasets: (**a**) Mean and standard variation of arousal and valence levels on a five-point scale for each discrete emotion category; (**b**) Confidence of discrete emotion labels and valence-level of three-point scale.

Index	Association Property		IEMOCAP	KESDy18	KESDy19
(a)	Valence Mean ± variation	angry	1.89 ± 0.52	2.11 ± 0.21	1.78 ± 0.37
		happy	3.94 ± 0.47	4.42 ± 0.34	4.33 ± 0.36
		neutral	2.95 ± 0.49	3.23 ± 0.53	2.94 ± 0.60
		sad	2.24 ± 0.57	2.00 ± 0.33	1.89 ± 0.52
	Arousal Mean ± variation	angry	3.69 ± 0.66	3.93 ± 0.46	3.81 ± 0.58
		happy	3.16 ± 0.61	3.92 ± 0.36	3.90 ± 0.53
		neutral	2.79 ± 0.53	3.08 ± 0.38	2.99 ± 0.33
		sad	2.61 ± 0.61	2.60 ± 0.44	2.63 ± 0.64
(b)	Confidence	$Conf.(\{C_i = angry\} \rightarrow \{V_1\})$	0.8	0.95	0.95
		$Conf.(\{C_i = sad\} \rightarrow \{V_1\})$	0.58	0.9	0.86
		$Conf.(\{C_i = neutral\} \rightarrow \{V_2\})$	0.85	0.83	0.71
		$Conf.(\{C_i = happy\} \rightarrow \{V_3\})$	0.77	0.93	0.86

Table 2 shows properties of the five domain datasets of three SER databases used for the evaluation, where we used speech segments having lengths of 2 s or longer as one of four categories of emotion labels, "angry", "happy", "neutral", and "sad."

Table 2. Properties of multi-domain SER datasets.

Property	IEMOCAP	KESDy18	KESDy19 [2]
Language	English	Korean	Korean
Speakers	10 (5 male, 5 female)	30 (15 male, 15 female)	40 (20 male, 20 female)
Utterance type	Acted (Scripted/Improvised)	Acted (Scripted)	Acted (Scripted/Improvised)
Datasets (Mic.)	IEMOCAP (2 Mic. of the same type)	KESDy18_PM (Galaxy S6), KESDy18_EM [1] (Shure S35)	KESDy19_PM (Galaxy S8), KESDy19_EM (AKG C414)
angry	947	431	1628
happy	507	157	1121
neutral	1320	1193	2859
sad	966	467	694
Total	3740	2248	6302

[1] KESDy18_EM is available online at https://nanum.etri.re.kr/share/kjnoh/SER-DB-ETRIv18?lang=eng (accessed on 7 January 2021).
[2] The collecting process of the KESDy19 was approved by the Institutional Review Board of Korea National Institute for Bioethics Policy (approval number P01-201907-22-010 and 22 July 2019).

4.2. Evaluation of the BLSTM-Based Baseline SER

As shown in Table 2, the five domain SER datasets used for evaluation were unbalanced in the number of samples of the discrete emotion classes. We did not apply oversampling, data augmentation [11], or weighted loss methods [46] to minority classes for objective verification of the proposed MPGLN SER.

Speech samples of each class in the multi-domain datasets were trained in the SER model by the units of the speech segment, which consisted of the voiced part of the vocal-cord vibrations and unvoiced parts such as a silence section between voiced parts [53]. This study did not remove the unvoiced region from any speech segment. However, it framed the entire voiced and unvoiced parts of the segment as input to the model.

We present four performance metrics in consideration of the sample imbalance of each emotion class: weighted accuracy (WA), unweighted accuracy (UA), precision (PR), and F1 score. WA is the overall accuracy, calculated as the ratio of the total number of test data and the number of samples accurately predicted by the actual label. UA is calculated as the average of the recall values of four classes and is an important performance indicator in the evaluation of the SER model based on imbalanced datasets [19,20,26].

This study applied z-normalization [1] of the means and standard deviations of each dataset to reduce the fluctuations of the speaker and speech signals. We evaluated the speaker-independent leave-p-subjects-out (LpSO) validation technique, where p is the number of subjects to leave out when training the model. For training, we used separated samples belonging to speakers accounting for 80% of the total number in each dataset; samples of the remaining 20% were evaluated as test data.

For the evaluation of IEMOCAP, we used a leave-two-subjects-out evaluation that applied speech data from two speakers participating in one session as the test data, which was the leave-one-session-out (LOSO) validation. KESDy18 was evaluated as a leave-six-subjects-out sample from the set of 30 speakers. The evaluation of KESDy19 was conducted as a leave-eight-subjects-out sample for four sessions of the 20 sessions played in pairs by 40 speakers. The training and test data separated for speaker-independent evaluation in each dataset were equally applied to the evaluation of a single domain, multi-domain, or domain generalization, as shown in Tables 3 and 4 and Tables 6–8.

Table 3. Performance of the baseline BLSTM-based SER model according to the input low-level descriptions (LLD) feature set in SER datasets.

Model	Dataset	Input LLDs	WA	UA	PR	F1
Our baseline (SPSL: single-path-single-loss)	IEMOCAP	MFCC	0.616	0.588	0.576	0.559
		Mel-spec	0.534	0.525	0.504	0.491
		MFCC + Mel-spec	0.608	0.58	0.574	0.562
		MFCC + Mel-spec + TimeSpectral	0.611	0.59	0.58	**0.575**
	KESDy18_EM	MFCC	0.742	0.712	0.715	0.71
		Mel-spec	0.62	0.57	0.553	0.556
		MFCC + Mel-spec	0.762	0.736	0.719	0.724
		MFCC + Mel-spec + TimeSpectral	0.774	0.738	0.737	**0.734**
	KESDy19_EM	MFCC	0.613	0.563	0.581	0.567
		Mel-spec	0.56	0.483	0.518	0.491
		MFCC + Mel-spec	0.617	0.562	0.579	0.568
		MFCC + Mel-spec + TimeSpectral	0.643	0.595	0.608	**0.599**

Table 4. Performance of the baseline BLSTM-based SER model.

Model	Dataset	WA	UA	PR	F1
Our baseline (SPSL)	IEMOCAP	0.611	0.59	0.58	0.575
	KESDy18_PM	0.776	0.739	0.739	0.736
	KESDy18_EM	0.774	0.738	0.737	0.734
	KESDy19_PM	0.624	0.574	0.589	0.58
	KESDy19_EM	0.643	0.595	0.608	0.599

In the evaluation of this study, a model based on the temporal embedding features and the learning loss, $\mathcal{L}_{cd}$, without the transferred embedding feature was assumed to be the baseline SER model. It can be seen that this baseline operated using a single-path-single-loss (SPSL) scheme. In the evaluation, the proposed MPGLN and the baseline SPSL SER model were trained with a batch size of 200 samples at 25 epochs using an Adam optimizer and a drop rate of 0.6 to the last two FC layers. The learning rate of the optimizer was 1.10^{-3}. The model was evaluated over 10 iterations of training and testing, and the final value of each performance metric was calculated as the average value.

The baseline SPSL SER model uses the 74-D LLD integration per-frame of speech segment, which comprises 13-D MFCC and 40-D Mel-spectrogram (Mel-spec), along with 21-D time- and spectral-domain (TimeSpectral) LLDs such as zero-crossing rate, energy, spectral centroid, and spectral roll-off. We evaluated the performance of each combination of LLDs with our baseline SER model based on multiple SER datasets. Table 3 summarizes the performance evaluation according to the input feature set of the LLDs used in this study, as shown in the evaluation results based on the IEMOCAP, KESDy18_EM, and KESDy19_EM datasets. It can be observed that MFCC is the dominant feature of SER from the results in Table 3. The SER performance improved from 1.6% to 3.2% based on the F1 score in comparison with the single input of MFCC when using the input combination of MFCC and Mel-spectrogram, along with TimeSpectral LLDs.

Table 4 shows the results of the speaker-independent evaluation of the BLSTM baseline SPSL when classifying the four discrete emotion labels in each of the five domain datasets. The evaluation based on KESDy19 showed similar performance results as IEMOCAP. In the evaluation of KESDy18, it showed higher performance results than the other two databases.

A previous study by Zheng et al. [54] demonstrated the performance of 40% WA of the CNN-based SER model for the five emotion classes based on IEMOCAP. For a fair comparison of the SER performance, this study performed a comparison with the previous RNN-based SER models that presented the UA performance of the four emotion classes based on IEMOCAP, which was the test environment in many previous SER studies.

In Table 5, we compare the performance results of previous RNN-based SER models and the SPSL baseline model in the LOSO evaluation to classify the four emotion labels based on the IEMOCAP dataset. These studies present a UA metric of the average recall for each emotion class, considering the imbalance of the number of samples. As shown in Table 5, our baseline BLSTM SER model achieved a competitive performance of UA 59% in the LOSO validation based on IEMOCAP.

Table 5. Performance results reported in previous recurrent neural networks (RNN)-based studies of SER model and our baseline model based on IEMOCAP.

Researches	Features	Network	UA	Emotions
Mirsamadi [19]	32 LLD	RNN	0.585	4
Chen [1] [20]	logMel	CRNN	0.647 ± 0.054	4
Mu [26]	Spectrogram	CRNN	0.564	4
Our baseline (SPSL)	74 LLD	RNN	0.59 ± 0.08	4

[1] This study used only the improvisation data of female speakers as test data.

4.3. Evaluation of Multi-Domain Adaptation

As shown in Tables 6–8, evaluations were performed using a single-domain evaluation, a multi-domain adaptation, and a multi-domain generalization according to the source and target domains participating in training and evaluation. The division of training and testing data separated for speaker-independent evaluation in each dataset used the same configurations as those used in Tables 3–8. In Tables 6–8, the highest F1 scores are highlighted.

Table 6. Evaluation results in a single domain dataset. Single-path-single-loss (SPSL) is the baseline SER model that learns by the temporal embedding features and the loss $\mathcal{L}_{cd}$; Multi-path-single-loss (MPSL) is that model learns using the multi-path embedding vectors and loss $\mathcal{L}_{cd}$ without the loss $\mathcal{L}_{cv}$; MPGL is the model that learns based on multi-path embedding vectors and the group loss $\mathcal{L}_g$.

Index	Domain	Model	WA	UA	PR	F1
(a)	IEMOCAP	SPSL	0.611	0.59	0.58	0.575
		MPSL	0.611	0.606	0.576	0.583
		MPGL	0.619	0.607	0.582	**0.588**
(b)	KESDy18_PM	SPSL	0.776	0.739	0.739	0.736
		MPSL	0.781	0.753	0.747	0.746
		MPGL	0.814	0.778	0.771	**0.773**
(c)	KESDy18_EM	SPSL	0.774	0.738	0.737	0.734
		MPSL	0.788	0.756	0.732	0.741
		MPGL	0.797	0.768	0.761	**0.762**
(d)	KESDy19_PM	SPSL	0.624	0.574	0.589	0.58
		MPSL	0.625	0.581	0.594	0.586
		MPGL	0.637	0.586	0.607	**0.594**
(e)	KESDy19_EM	SPSL	0.643	0.595	0.608	**0.599**
		MPSL	0.629	0.581	0.591	0.584
		MPGL	0.642	0.592	0.608	0.598

Table 7. Evaluation results of multi-domain adaptation.

Index	Multi-Domain	Model	WA	UA	PR	F1
(a)	KESDy18_PM, KESDy18_EM	SPSL	0.774	0.749	0.722	0.731
		MPSL	0.799	0.764	0.753	0.756
		MPGL	0.806	0.773	0.766	**0.768**
(b)	KESDy19_PM, KESDy19_EM	SPSL	0.618	0.581	0.584	0.581
		MPSL	0.626	0.58	0.589	0.584
		MPGL	0.631	0.585	0.595	**0.589**
(c)	KESDy18_PM, KESDy18_EM, KESDy19_PM, KESDy19_EM	SPSL	0.653	0.628	0.63	0.628
		MPSL	0.664	0.639	0.642	**0.639**
		MPGL	0.663	0.63	0.639	0.634
(d)	KESDy18_PM, KESDy18_EM, IEMOCAP	SPSL	0.683	0.649	0.63	0.637
		MPSL	0.706	0.675	0.654	0.66
		MPGL	0.713	0.677	0.656	**0.664**
(e)	KESDy19_PM, KESDy19_EM, IEMOCAP	SPSL	0.599	0.577	0.575	0.573
		MPSL	0.602	0.583	0.576	0.578
		MPGL	0.616	0.587	0.59	**0.588**

Table 8. Evaluation results of multi-domain generalization.

Index	Source Domain	Target Domain	Model	WA	UA	PR	F1
(a)	KESDy18_PM, KESDy18_EM, KESDy19_EM	KESDy19_PM	SPSL	0.594	0.532	0.563	0.539
			MPSL	0.592	0.53	0.559	0.536
			MPGL	0.606	0.543	0.573	**0.551**
(b)	KESDy18_EM, IEMOCAP	KESDy18_PM	SPSL	0.682	0.69	0.652	0.658
			MPSL	0.688	0.704	0.643	0.658
			MPGL	0.718	0.74	0.677	**0.693**
(c)	KESDy19_EM, IEMOCAP	KESDy19_PM	SPSL	0.572	0.55	0.538	0.538
			MPSL	0.577	0.552	0.545	0.542
			MPGL	0.596	0.555	0.561	**0.554**

Table 6 shows the evaluation results when classifying four discrete emotion classes based on each of the five domain datasets. The evaluation was conducted in three experimental environments according to the type of SER model: The baseline SPSL model learns from the temporal embedding features and the single-loss $\mathcal{L}_{cd}$. Multi-path-single-loss (MPSL) uses multi-path embedding vectors and is trained only on $\mathcal{L}_{cd}$ without the complementary loss, $\mathcal{L}_{cv}$, for valence-level classification. Multi-path-group-loss (MPGL) learns from multi-path embedding vectors and the group loss, $\mathcal{L}_g$, consisting of $\mathcal{L}_{cd}$ and $\mathcal{L}_{cv}$.

When compared with the harmonic-mean F1 score based on the KESDy18_PM dataset shown in Table 6b, the performance of the SER of the MPSL using a single-loss $\mathcal{L}_{cd}$ showed an improvement of 1% over that of the baseline SPSL. The SER MPGL model trained on the loss group, $\mathcal{L}_g$, showed an F1 improvement of up to 3.7% over the SPSL's F1.

Table 7 shows the results of multi-domain adaptation evaluation when the SER model was trained with samples aggregated from multiple-domain SER datasets collected from various environments. The separated test samples for about 20% of the speakers were evaluated for speaker-independent evaluation. As shown in Table 7a, regarding KESDy18, which consisted of two datasets collected simultaneously via heterogeneous devices, the proposed SER model trained on the group-loss $\mathcal{L}_g$ of MPGL achieved an F1 improvement of up to 3.7% over the baseline SPSL.

Table 8 presents the evaluation results of the proposed MPGLN SER for supporting multi-domain generalization. In the evaluation of Table 8a, the SER model was trained with the aggregated samples of KESDy18_PM, KESDy18_EM, and KESDy19_EM datasets and was evaluated against the separated test samples of the KESDy19_PM domain, which was not used for training but was collected from the same language culture. The evaluation results of Table 8a shows that the F1 score of the MPGL model improved by 1.2% compared with the baseline SPSL. In the evaluation of Table 8b, when the SER model was trained on KESDy18_EM and IMEOCAP datasets, which were from different language cultures, the model was evaluated using the Korean KESDy18_PM domain dataset. The proposed MPGLN SER showed an F1-score improvement of about 3.5% over the baseline model.

Figure 5 shows the changes in losses from Table 8b, including the loss, $\mathcal{L}_{cd}$, of the baseline SPSL model and losses $\mathcal{L}_{cd}$ and $\mathcal{L}_{cv}$ of the MPGL SER model. These losses were measured every 25 epochs during training using aggregated KESDy18_EM and IMEOCAP samples. The loss, $\mathcal{L}_{cd}$, of the MPGL model, which learned two losses simultaneously, trained faster than did the $\mathcal{L}_{cd}$ of the baseline SER model. This shows that the other complementary loss, $\mathcal{L}_{cv}$, of the proposed MPGLN, used to predict the valence-level label, decreased similarly to the loss, $\mathcal{L}_{cd}$, of the baseline SPSL.

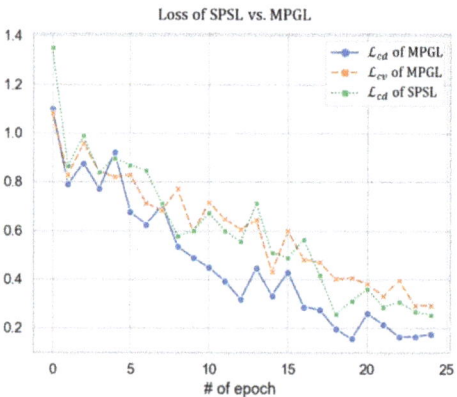

Figure 5. Change in losses of the baseline SER and the proposed MPGLN SER in Table 8b. The loss, $\mathcal{L}_{cd}$, of the baseline SPSL model and losses $\mathcal{L}_{cd}$ and $\mathcal{L}_{cv}$ of the SER model of MPGL.

Figure 6 shows the distribution of the 64-D embedding vectors of the test data reduced to a 2-D embedding space via t-stochastic neighbor embedding (t-SEN). The 64-D embedding vectors were generated in the FC layer just prior to the MPSL and MPGL softmax activations of the evaluation in Table 8b.

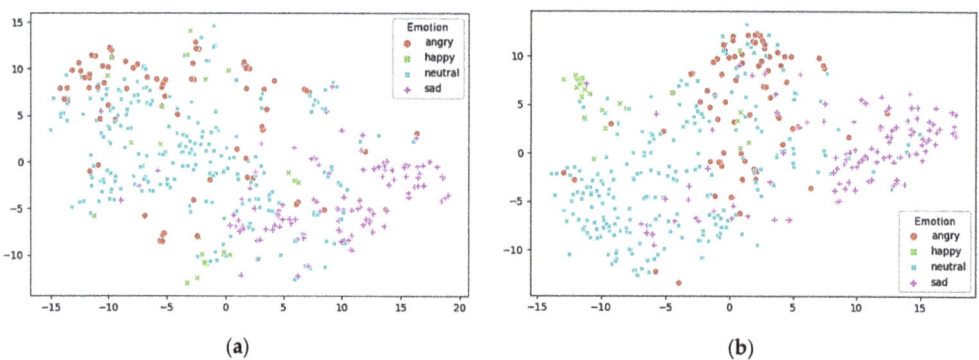

Figure 6. Distribution of reduced embedding vectors (the 64-D embedding vectors of the test data in the last fully-connected (FC) layer in the ensemble network) that are reduced to 2-D via t-stochastic neighbor embedding (t-SEN) dimension reduction: (a) embedding space for MPSL in Table 8b; (b) embedding space for MPGL in Table 8b.

Figure 6a shows the distribution of the embedding feature vector in the MPSL trained by the loss, $\mathcal{L}_{cd}$, only without the complementary loss, $\mathcal{L}_{cv}$. Figure 6b displays the distribution of the MPGL model based on the loss group, $\mathcal{L}_g$, of the two losses: $\mathcal{L}_{cd}$ and $\mathcal{L}_{cv}$. Figure 6b shows the MPGLN SER model that learns from multi-path embedding vectors and the loss group, $\mathcal{L}_g$, where the samples belonging to the "happy" class were more closely grouped, and the samples of the "angry" and "sad" classes are located closer together compared with the MPSL distribution shown in Figure 6a.

5. Conclusions

We determined that it is essential to improve the generalization of the SER model for deployment to real applications. This paper proposed the MPGLN for SER in support of supervised multi-domain adaptation and generalization based on multi-domain datasets. The proposed MPGLN SER includes a temporal feature generator for the BLSTM network using the input of handcrafted LLD features of a speech sample. Additionally, we leveraged the transferred feature extractor from the pre-trained VGGish model for the MPGLN. The proposed MPGLN SER learned simultaneous multiple losses induced by associations between discrete emotion and dimension labels.

The proposed MPGLN SER was evaluated using five real SER datasets of various speaker domains, language cultures, collecting devices, and procedural environments. This included KESDy18 and KESDy19 databases. KESDy18 comprised speech samples delivered by voice actors who uttered Korean short sentences by expressing specific discrete emotions. The KESDy18 database consisted of KESDy18_PM and KESDy18_EM datasets from heterogeneous devices and environments with different device locations. The KESDy19 database comprised KESDy19_EM and KESDy19_PM, which contained the collected speech sample voices acted using a similar procedure as that of the IEMOCAP and that of the simulated dataset based on the cell-phone's built-in microphone, respectively.

This study assumed that the SER model was trained only with the BLSTM-based temporal embedding feature generator included with MPGLN without transferred feature as the baseline SER model. We verified the performance reliability of the baseline SER model using the IEMOCAP. The BLSTM-baseline SER model showed competitive UA results of 59% when classifying the four categorical emotion labels. The multi-domain adaptation and domain generalization evaluation of the proposed MPGLN SER was performed using the English-speaking IEMOCAP and the Korean KESDy18 and KESDy19 datasets by comparing the performances of the baseline model according to various evaluation environments.

The proposed MPGLN SER model trained on multiple losses showed an F1 performance improvement of up to 3.7% over the baseline model when classifying four emotion labels in a single domain dataset. The performance evaluation of the MPGLN SER for supervised multi-domain adaptation, which trained and tested on the SER model using the aggregated speech samples of the multi-domain datasets, also showed an improvement of up to 3.7% over the baseline F1 score. From the evaluation of the multi-domain generalization of the proposed MPGLN SER, the F1 score enjoyed an improvement of 3.5% over the baseline SER when using samples from other language cultures not used for training. From these results, we found that our MPGLN SER, which supports supervised multi-domain adaptations, is also effective in reinforcing the generalization of the SER model based on multi-domain datasets.

For future works, we plan to derive the differences in acoustic features of emotional expressions based on multi-cultural SER datasets and study the learning method for the deep-learning-based SER model considering the domain discrepancy. Furthermore, we will continue enhancing our model's generalizability through evaluations of speech data in the wild by deploying the proposed MPGLN SER to real applications.

Author Contributions: Conceptualization, K.J.N., C.Y.J., J.L., S.C., G.K., J.M.L., and H.J.; methodology, K.J.N. and C.Y.J.; software, K.J.N.; validation, K.J.N.; formal analysis, K.J.N.; investigation, K.J.N., C.Y.J., J.L., S.C., G.K., J.M.L., and H.J.; resources, H.J.; data curation, K.J.N.; writing—original draft preparation, K.J.N.; writing—review and editing, K.J.N., C.Y.J., and S.C.; visualization, K.J.N.; supervision, H.J.; project administration, H.J.; funding acquisition, H.J. All authors have read and agreed to the published version of the manuscript.

Funding: This work was supported by Electronics and Telecommunications Research Institute (ETRI) grant funded by the Korean government. (21ZS1100, Core Technology Research for Self-Improving Integrated Artificial Intelligence System).

Institutional Review Board Statement: The collecting process of the KESDy19 database was conducted according to the guidelines of the Declaration of Helsinki, and was approved by the Institutional Review Board of Korea national Institute for Bioethics Policy (approval number P01-201907-22-010 and 22 July 2019). The study did not require additional ethical approval.

Informed Consent Statement: Not applicable.

Data Availability Statement: Statistical results are contained within the article. The KESDy18_EM dataset collected in this study is available online at https://nanum.etri.re.kr/share/kjnoh/SER-DB-ETRIv18?lang=eng (accessed on 7 January 2021).

Conflicts of Interest: The authors declare no conflict of interest.

References

1. Akçay, M.B.; Oğuz, K. Speech Emotion Recognition: Emotional Models, Databases, Features, Preprocessing Methods, Supporting Modalities, and Classifiers. *Speech Commun.* **2020**, *116*, 56–76. [CrossRef]
2. Hazer-Rau, D.; Meudt, S.; Daucher, A.; Spohrs, J.; Hoffmann, H.; Schwenker, F.; Traue, H.C. The UulmMAC Database—A Multimodal Affective Corpus for Affective Computing in Human-Computer Interaction. *Sensors* **2020**, *20*, 2308. [CrossRef]
3. Marín-Morales, J.; Llinares, C.; Guixeres, J.; Alcañiz, M. Emotion Recognition in Immersive Virtual Reality: From Statistics to Affective Computing. *Sensors* **2020**, *20*, 5163. [CrossRef]
4. Haq, S.; Jackson, P.J.; Edge, J. Speaker-Dependent Audio-Visual Emotion Recognition. In Proceedings of the International Conference on Auditory-Visual Speech Processing (AVSP), Norwich, UK, 10–13 September 2009; pp. 53–58.
5. Vryzas, N.; Kotsakis, R.; Liatsou, A.; Dimoulas, C.A.; Kalliris, G. Speech Emotion Recognition for Performance Interaction. *J. Audio Eng. Soc.* **2018**, *66*, 457–467. [CrossRef]
6. Livingstone, S.R.; Russo, F.A. The Ryerson Audio-Visual Database of Emotional Speech and Song (RAVDESS): A Dynamic, Multimodal Set of Facial and Vocal Expressions in North American English. *PLoS ONE* **2018**, *13*, e0196391. [CrossRef] [PubMed]
7. Busso, C.; Bulut, M.; Lee, C.; Kazemzadeh, A.; Mower, E.; Kim, S.; Chang, J.N.; Lee, S.; Narayanan, S.S. IEMOCAP: Interactive Emotional Dyadic Motion Capture Database. *Lang. Resour. Eval.* **2008**, *42*, 335–359. [CrossRef]
8. Abdelwahab, M.; Busso, C. Supervised Domain Adaptation for Emotion Recognition from Speech. In Proceedings of the IEEE International Conference on Acoustics, Speech and Signal Processing (ICASSP), Brisbane, Australia, 19–24 April 2015; pp. 5058–5062.
9. Liang, J.; Chen, S.; Zhao, J.; Jin, Q.; Liu, H.; Lu, L. Cross-Culture Multimodal Emotion Recognition with Adversarial Learning. In Proceedings of the ICASSP 2019 IEEE International Conference on Acoustics, Speech and Signal Processing (ICASSP), Brighton, UK, 12–17 May 2019; pp. 4000–4004.
10. Schuller, B.; Vlasenko, B.; Eyben, F.; Wöllmer, M.; Stuhlsatz, A.; Wendemuth, A.; Rigoll, G. Cross-Corpus Acoustic Emotion Recognition: Variances and Strategies. *IEEE Trans. Affect. Comput.* **2010**, *1*, 119–131. [CrossRef]
11. Park, D.S.; Chan, W.; Zhang, Y.; Chiu, C.-C.; Zoph, B.; Cubuk, E.D.; Le, Q.V. Specaugment: A Simple Data Augmentation Method for Automatic Speech Recognition. In Proceedings of the INTERSPEECH, Graz, Austria, 15–19 September 2019. [CrossRef]
12. Bang, J.; Hur, T.; Kim, D.; Lee, J.; Han, Y.; Banos, O.; Kim, J.-I.; Lee, S. Adaptive Data Boosting Technique for Robust Personalized Speech Emotion in Emotionally-Imbalanced Small-Sample Environments. *Sensors* **2018**, *18*, 3744. [CrossRef] [PubMed]
13. Huang, Z.; Xue, W.; Mao, Q.; Zhan, Y. Unsupervised Domain Adaptation for Speech Emotion Recognition Using PCANet. *Multimed. Tools Appl.* **2017**, *76*, 6785–6799. [CrossRef]
14. Neumann, M. Cross-Lingual and Multilingual Speech Emotion Recognition on English and French. In Proceedings of the IEEE International Conference on Acoustics, Speech and Signal Processing (ICASSP), Calgary, AB, Canada, 15–20 April 2018; pp. 5769–5773.
15. Li, Y.; Yang, T.; Yang, L.; Xia, X.; Jiang, D.; Sahli, H. A Multimodal Framework for State of Mind Assessment with Sentiment Pre-Classification. In Proceedings of the 9th International on Audio/Visual Emotion Challenge and Workshop, Nice, France, 21 October 2019; The Association for Computing Machinery: New York, NY, USA, 2019; pp. 13–18.
16. Lee, S. The Generalization Effect for Multilingual Speech Emotion Recognition across Heterogeneous Languages. In Proceedings of the ICASSP 2019 IEEE International Conference on Acoustics, Speech and Signal Processing (ICASSP), Brighton, UK, 12–17 May 2019; pp. 5881–5885.

17. Hershey, S.; Chaudhuri, S.; Ellis, D.P.; Gemmeke, J.F.; Jansen, A.; Moore, R.C.; Plakal, M.; Platt, D.; Saurous, R.A.; Seybold, B. CNN Architectures for Large-Scale Audio Classification. In Proceedings of the IEEE International Conference on Acoustics, Speech and Signal Processing (ICASSP), New Orleans, LA, USA, 5–9 March 2017; pp. 131–135.
18. Motiian, S.; Piccirilli, M.; Adjeroh, D.A.; Doretto, G. Unified Deep Supervised Domain Adaptation and Generalization. In Proceedings of the IEEE International Conference on Computer Vision, Venice, Italy, 22–29 October 2017; pp. 5715–5725.
19. Mirsamadi, S.; Barsoum, E.; Zhang, C. Automatic Speech Emotion Recognition Using Recurrent Neural Networks with Local Attention. In Proceedings of the IEEE International Conference on Acoustics, Speech and Signal Processing (ICASSP), New Orleans, LA, USA, 5–9 March 2017; pp. 2227–2231.
20. Chen, M.; He, X.; Yang, J.; Zhang, H. 3-D Convolutional Recurrent Neural Networks with Attention Model for Speech Emotion Recognition. *IEEE Signal Process. Lett.* **2018**, *25*, 1440–1444. [CrossRef]
21. Liu, Z.-T.; Wu, M.; Cao, W.-H.; Mao, J.-W.; Xu, J.-P.; Tan, G.-Z. Speech Emotion Recognition Based on Feature Selection and Extreme Learning Machine Decision Tree. *Neurocomputing* **2018**, *273*, 271–280. [CrossRef]
22. Huang, C.-W.; Narayanan, S.S. Attention Assisted Discovery of Sub-Utterance Structure in Speech Emotion Recognition. In Proceedings of the INTERSPEECH, San Francisco, CA, USA, 8–12 September 2016; pp. 1387–1391.
23. Chorowski, J.K.; Bahdanau, D.; Serdyuk, D.; Cho, K.; Bengio, Y. Attention-Based Models for Speech Recognition. *Adv. Neural Inf. Process. Syst.* **2015**, *28*, 577–585.
24. Anvarjon, T.; Kwon, S. Deep-Net: A Lightweight CNN-Based Speech Emotion Recognition System Using Deep Frequency Features. *Sensors* **2020**, *20*, 5212. [CrossRef] [PubMed]
25. Yeh, S.-L.; Lin, Y.-S.; Lee, C.-C. An Interaction-Aware Attention Network for Speech Emotion Recognition in Spoken Dialogs. In Proceedings of the ICASSP 2019 IEEE International Conference on Acoustics, Speech and Signal Processing (ICASSP), Brighton, UK, 12–17 May 2019; pp. 6685–6689.
26. Mu, Y.; Gómez, L.A.H.; Montes, A.C.; Martínez, C.A.; Wang, X.; Gao, H. Speech Emotion Recognition Using Convolutional-Recurrent Neural Networks with Attention Model. In Proceedings of the International Conference on Computer Engineering, Information Science and Internet Technology (CII), Sanya, China, 11–12 November 2017; pp. 341–350.
27. Yao, Z.; Wang, Z.; Liu, W.; Liu, Y.; Pan, J. Speech Emotion Recognition Using Fusion of Three Multi-Task Learning-Based Classifiers: HSF-DNN, MS-CNN and LLD-RNN. *Speech Commun.* **2020**, *120*, 11–19. [CrossRef]
28. Jin, Q.; Li, C.; Chen, S.; Wu, H. Speech Emotion Recognition with Acoustic and Lexical Features. In Proceedings of the IEEE International Conference on Acoustics, Speech and Signal Processing (ICASSP), Brisbane, Australia, 19–24 April 2015; pp. 4749–4753.
29. Glodek, M.; Tschechne, S.; Layher, G.; Schels, M.; Brosch, T.; Scherer, S.; Kächele, M.; Schmidt, M.; Neumann, H.; Palm, G. Multiple Classifier Systems for the Classification of Audio-Visual Emotional States. In Proceedings of the International Conference on Affective Computing and Intelligent Interaction, Memphis, TN, USA, 9–12 October 2011; Springer: Berlin/Heidelberg, Germany, 2011; pp. 359–368.
30. Hong, I.S.; Ko, Y.J.; Shin, H.S.; Kim, Y.J. Emotion Recognition from Korean Language Using MFCC HMM and Speech Speed. In Proceedings of the 12th International Conference on Multimedia Information Technology and Applications (MITA2016), Luang Prabang, Laos, 4–6 July 2016; pp. 12–15.
31. Ntalampiras, S.; Fakotakis, N. Modeling the Temporal Evolution of Acoustic Parameters for Speech Emotion Recognition. *IEEE Trans. Affect. Comput.* **2011**, *3*, 116–125. [CrossRef]
32. Vrysis, L.; Tsipas, N.; Thoidis, I.; Dimoulas, C. 1d/2d Deep CNNs vs. Temporal Feature Integration for General Audio Classification. *J. Audio Eng. Soc.* **2020**, *68*, 66–77. [CrossRef]
33. Sandhya, P.; Spoorthy, V.; Koolagudi, S.G.; Sobhana, N.V. Spectral Features for Emotional Speaker Recognition. In Proceedings of the Third International Conference on Advances in Electronics, Computers and Communications (ICAECC), Bengaluru, India, 11–12 December 2020; pp. 1–6.
34. Eyben, F.; Scherer, K.R.; Schuller, B.W.; Sundberg, J.; André, E.; Busso, C.; Devillers, L.Y.; Epps, J.; Laukka, P.; Narayanan, S.S. The Geneva Minimalistic Acoustic Parameter Set (GeMAPS) for Voice Research and Affective Computing. *IEEE Trans. Affect. Comput.* **2015**, *7*, 190–202. [CrossRef]
35. Schuller, B.; Steidl, S.; Batliner, A.; Burkhardt, F.; Devillers, L.; Müller, C.; Narayanan, S.S. The INTERSPEECH 2010 Paralinguistic Challenge. In Proceedings of the Eleventh Annual Conference of the International Speech Communication Association, Makuhari, Japan, 26–30 September 2010.
36. Eyben, F.; Wullmer, M.; Schuller, B.O. OpenSMILE - The Munich Versatile and Fast Open-Source Audio Feature Extractor. In Proceedings of the ACM International Conference on Multimedia (MM), Firenze, Italy, 25–29 October 2010; pp. 1459–1462.
37. Jing, S.; Mao, X.; Chen, L. Prominence Features: Effective Emotional Features for Speech Emotion Recognition. *Digit. Signal Process.* **2018**, *72*, 216–231. [CrossRef]
38. Sahoo, S.; Kumar, P.; Raman, B.; Roy, P.P. A Segment Level Approach to Speech Emotion Recognition Using Transfer Learning. In Proceedings of the Asian Conference on Pattern Recognition, Auckland, New Zealand, 26–29 November 2019; Springer: Berlin/Heidelberg, Germany, 2019; pp. 435–448.
39. Jiang, W.; Wang, Z.; Jin, J.S.; Han, X.; Li, C. Speech Emotion Recognition with Heterogeneous Feature Unification of Deep Neural Network. *Sensors* **2019**, *19*, 2730. [CrossRef]

40. Chatziagapi, A.; Paraskevopoulos, G.; Sgouropoulos, D.; Pantazopoulos, G.; Nikandrou, M.; Giannakopoulos, T.; Katsamanis, A.; Potamianos, A.; Narayanan, S. Data Augmentation Using GANs for Speech Emotion Recognition. In Proceedings of the INTERSPEECH, Graz, Austria, 15–19 September 2019; pp. 171–175.
41. Salamon, J.; Bello, J.P. Deep Convolutional Neural Networks and Data Augmentation for Environmental Sound Classification. *IEEE Signal Process. Lett.* **2017**, *24*, 279–283. [CrossRef]
42. Vryzas, N.; Vrysis, L.; Matsiola, M.; Kotsakis, R.; Dimoulas, C.; Kalliris, G. Continuous Speech Emotion Recognition with Convolutional Neural Networks. *J. Audio Eng. Soc.* **2020**, *68*, 14–24. [CrossRef]
43. Abdelwahab, M.; Busso, C. Active Learning for Speech Emotion Recognition Using Deep Neural Network. In Proceedings of the 8th International Conference on Affective Computing and Intelligent Interaction (ACII), Cambridge, UK, 3–6 September 2019; pp. 1–7.
44. Goodfellow, I.; Pouget-Abadie, J.; Mirza, M.; Xu, B.; Warde-Farley, D.; Ozair, S.; Courville, A.; Bengio, Y. Generative Adversarial Nets. In Proceedings of the Advances in Neural Information Processing Systems, Montreal, QC, Canada, 8–12 December 2014; pp. 2672–2680.
45. Kang, G.; Jiang, L.; Yang, Y.; Hauptmann, A.G. Contrastive Adaptation Network for Unsupervised Domain Adaptation. In Proceedings of the IEEE Conference on Computer Vision and Pattern Recognition, Long Beach, CA, USA, 16–20 June 2019; pp. 4893–4902.
46. Gao, W.; McDonnell, M.; UniSA, S. *Acoustic Scene Classification Using Deep Residual Networks with Focal Loss and Mild Domain Adaptation*; Technical Report; Detection and Classification of Acoustic Scenes and Event: Mawson, Australia, 2020.
47. He, K.; Zhang, X.; Ren, S.; Sun, J. Deep Residual Learning for Image Recognition. In Proceedings of the IEEE Conference on Computer Vision and Pattern Recognition, Las Vegas, NV, USA, 26 June–1 July 2016; pp. 770–778.
48. Gemmeke, J.F.; Ellis, D.P.; Freedman, D.; Jansen, A.; Lawrence, W.; Moore, R.C.; Plakal, M.; Ritter, M. Audio Set: An Ontology and Human-Labeled Dataset for Audio Events. In Proceedings of the IEEE International Conference on Acoustics, Speech and Signal Processing (ICASSP), New Orleans, LA, USA, 5–9 March 2017; pp. 776–780.
49. Dou, Q.; Coelho de Castro, D.; Kamnitsas, K.; Glocker, B. Domain Generalization via Model-Agnostic Learning of Semantic Features. *Adv. Neural Inf. Process. Syst.* **2019**, *32*, 6450–6461.
50. Ekman, P.; Friesen, W.V.; Ellsworth, P. *Emotion in the Human Face: Guidelines for Research and an Integration of Findings*; Elsevier: Amsterdam, The Netherlands, 2013; Volume 11.
51. Povolny, F.; Matejka, P.; Hradis, M.; Popková, A.; Otrusina, L.; Smrz, P.; Wood, I.; Robin, C.; Lamel, L. Multimodal Emotion Recognition for AVEC 2016 Challenge. In Proceedings of the 6th International Workshop on Audio/Visual Emotion Challenge, Amsterdam, The Netherlands, 15–19 October 2016; pp. 75–82.
52. Verykios, V.S.; Elmagarmid, A.K.; Bertino, E.; Saygin, Y.; Dasseni, E. Association Rule Hiding. *IEEE Trans. Knowl. Data Eng.* **2004**, *16*, 434–447. [CrossRef]
53. Kumar, S. Real-Time Implementation and Performance Evaluation of Speech Classifiers in Speech Analysis-Synthesis. *ETRI J.* **2020**, *43*, 82–94. [CrossRef]
54. Zheng, W.Q.; Yu, J.S.; Zou, Y.X. An Experimental Study of Speech Emotion Recognition Based on Deep Convolutional Neural Networks. In Proceedings of the 2015 International Conference on Affective Computing and Intelligent Interaction (ACII), Xi'an, China, 21–24 September 2015; pp. 827–831.

Article

Recognition of Emotion by Brain Connectivity and Eye Movement

Jing Zhang [1,†], Sung Park [1,†], Ayoung Cho [1] and Mincheol Whang [2,*]

1. Department of Emotion Engineering, Sangmyung University, Seoul 03016, Korea
2. Department of Human-Centered Artificial Intelligence, Sangmyung University, Seoul 03016, Korea
* Correspondence: whang@smu.ac.kr; Tel.: +82-2-2287-5293
† These authors contributed equally to this work.

Abstract: Simultaneous activation of brain regions (i.e., brain connection features) is an essential mechanism of brain activity in emotion recognition of visual content. The occipital cortex of the brain is involved in visual processing, but the frontal lobe processes cranial nerve signals to control higher emotions. However, recognition of emotion in visual content merits the analysis of eye movement features, because the pupils, iris, and other eye structures are connected to the nerves of the brain. We hypothesized that when viewing video content, the activation features of brain connections are significantly related to eye movement characteristics. We investigated the relationship between brain connectivity (strength and directionality) and eye movement features (left and right pupils, saccades, and fixations) when 47 participants viewed an emotion-eliciting video on a two-dimensional emotion model (valence and arousal). We found that the connectivity eigenvalues of the long-distance prefrontal lobe, temporal lobe, parietal lobe, and center are related to cognitive activity involving high valance. In addition, saccade movement was correlated with long-distance occipital-frontal connectivity. Finally, short-distance connectivity results showed emotional fluctuations caused by unconscious stimulation.

Keywords: emotion recognition; attention; eye movement; brain connectivity

1. Introduction

Studies have shown that different brain regions participate in various perceptual and cognitive processes. For example, the frontal lobe is related to thinking and consciousness, whereas the temporal lobe is associated with processing complex stimulus information, such as faces, scenes, smells, and sounds. The parietal lobe integrates a variety of sensory inputs and the operational control of objects, while the occipital lobe is related to vision [1].

The brain is an extensive network of neurons. Brain connectivity refers to the synchronous activity of neurons in different regions and may provide useful information on neural activity [2]. Mauss and Robinson [3] suggested that emotion processing occurs in distributed circuits, rather than in specific isolated brain regions. Analysis of the simultaneous activation of brain regions is a robust pattern-based analysis method for emotional recognition [4]. Researchers have developed methods to capture asymmetric brain activity patterns that are important for emotion recognition [5].

Users search massive amounts of information until they find something useful [6]. However, although the information is presented visually, users do not recognize it, because of a lack of attention. The cortical area known as the frontal eye field (FEF) plays a vital role in the control of visual attention and eye movements [7].

Eye tracking is the process of measuring eye movements. Eye tracking signals imply the user's subconscious behaviors and provide essential clues to the context of the subject's current activity [8], which allow us to determine what elicits users' attention.

The brain activity is significantly related to eye movement features involving pupil, saccade, and fixation. Our pupils change their size accordingly [9] when one is stimulated from resting to emotional states. The saccade is a decision made every time we move our eyes [10,11]. Decisions are influenced by one's expectations, goals, personalities, memories, and intentions [12].

A gaze is a potent social cue. For example, mutual gaze often implies threat or evasion, signaling submission or avoidance [13–16]. Eye gaze processing is one of the bases for social interactions, because the neural substrate for gaze processing is an essential step in developing neuroscience for social cognition [17,18].

By analyzing eye movement data, such as gaze position and gaze time, researchers can obtain explanations for multiple cognitive operations involving multiple behaviors [19]. For example, language researchers can use eye-tracking to analyze how people read and understand spoken language. Consumer researchers can study how shoppers make purchases. Researchers can gain a better cognitive understanding by integrating eye tracking with neuroimaging technologies (e.g., fMRI and EEG) [20].

Table 1 compares the few studies on eye movement features and EEG signals with an interest in producing a robust emotion-recognition model [21]. Wu et al. [22] integrated functional features from EEG and eye movements with deep canonical correlation analysis (DCCA). Their classification achieved 95.08% ± 6.42% accuracy on SEED public emotion EEG datasets [23]. Zheng et al. [24] used a multimodal depth neural network to incorporate eye movement and EEG signals to improve recognition performance. The results demonstrated that modality fusion with deep neural networks significantly enhances the performance compared with a single modality. Soleymani [25] learned that the decision-level fusion strategy is more adaptive than feature-level fusion when incorporating EEG signals and eye movement data. They also found that user-independent emotion recognition can perform better than individual self-reports for arousal assessment. While studies focused on improving recognition accuracy, currently, there is a lack of understanding of the relationship between brainwave connectivity and eye movement features (fixation, saccade, and left and right pupils). Specifically, we do not know how the functional relationship varies according to visual content's emotional characteristics (valence, arousal).

Table 1. Comparison of previous and proposed methods.

Methods	Strengths	Weaknesses
Deep canonical correlation analysis (DCCA) of integrated functional features [22]	Applied machine learning and incorporated and analyzed brain connectivity and eye movement data.	The statistical significance of brain connectivity and eye movement feature variables was not analyzed.
Designed a six-electrode placement to collect EEG and combined them with eye movements to integrate internal cognitive states and external behaviors [24].	Demonstrated the effect of modality fusion with a multimodal deep neural network. The mean accuracy was 85.11% for four emotions (happy, sad, fear, and neutral).	The study did not analyze the functional relationship between brainwave connectivity and eye movements.
User-independent emotion recognition method to identify affective tags for videos using gaze distance, pupillary response, and EEG [25].	Investigated pupil diameter, gaze distance, eye blinking, and EEG and applied modality fusion strategy at both feature and decision levels.	The experimental session limited the number of videos shown to participants. The study did not investigate brainwave connectivity.
Recognition of emotion by brain connectivity and eye movement (proposed method).	Explored the characteristics of brainwave connectivity and eye movement eigenvalues and the relationship between the two in a two-dimensional emotional model.	Did not apply machine learning to formulate a model. The analysis was based on one stimulus for each of the four quadrants in the two-dimensional model.

In this study, our research question involves the functional characteristics of brainwave connectivity and eye movement eigenvalues in valence-arousal emotions in a two-dimensional emotional model. We hypothesized that when viewing video content, the activation features of brain connections are significantly related to eye movement characteristics. We divided and analyzed brainwave connectivity into three groups: (1) long-distance occipital-frontal connectivity, (2) long-distance prefrontal and temporal, parietal, and central connectivity, and (3) short-distance connectivity, including frontal-temporal, frontal-central, temporal-parietal, and parietal-central connectivity. We applied k-means clustering to distinguish emotional feature responses, and eye movement eigenvalues were further differentiated. We then analyzed the relationship between eye movements and brain wave connectivity, depicting the differential characteristics of a two-dimensional emotional model.

2. Materials and Methods

We adopted Russell's two-dimensional model [26], where emotional states can be defined at any valence or arousal level. We invited participants to view emotion-eliciting videos with varying valences (i.e., from unpleasant to pleasant) and arousal levels (i.e., from relaxed to aroused). To understand brain connectivity and causality of brain regions according to different emotions, we used supervised learning to classify emotional and non-emotional states, and extract eye movement feature values associated with such different emotional states to analyze the relationship between brain activity and eye movement.

2.1. Stimuli Selection

We edited 6-min video clips (e.g., dramas or films) to elicit emotions from the participants. The content used to induce emotional conditions (valence and arousal) was collected in a two-dimensional model. To ensure that the emotional videos were effective, we conducted a stimulus selection experiment prior to the main experiment. We selected 20 edited dramas or movies containing emotions; five video clips were used for each quadrant in the two-dimensional model. Thirty participants viewed the emotional videos and responded to a subjective questionnaire. They received USD 20 for their participation in the study. Among the five video clips, the most representative video for each of the four quadrants in the two-dimensional model was selected (see Figure 1). Four stimuli were selected for the main experiment.

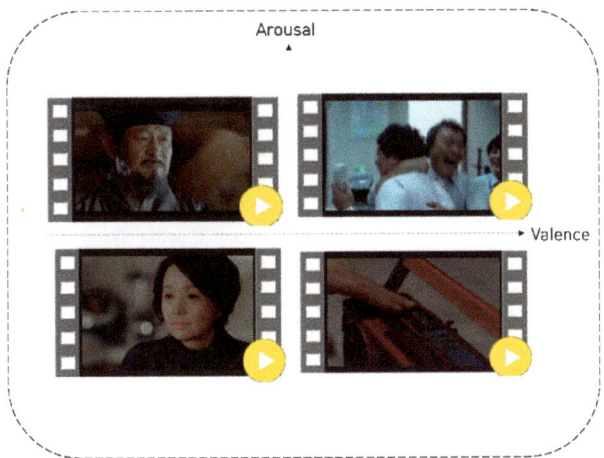

Figure 1. Video stimulus for each quadrant on a two-dimensional model.

2.2. Experiment Design

The main experiment had a factorial design of two (valence: pleasant and unpleasant) × two (arousal: aroused and relaxed) independent variables. The dependent variables included participants' brainwaves, eye movements (fixation, saccade, and left and right pupils), and subjective responses to a questionnaire.

2.3. Participants

We conducted an a priori power analysis using the program G*Power with the power set at 0.8 and α = 0.05, d = 0.6 (independent *t*-test), two-tailed. These results suggest that an N value of approximately 46 is required to achieve appropriate statistical power. Therefore, 47 university students were recruited for the study. Participants' ages ranged from 20 to 30 years (mean = 28, STD = 2.9), with 20 (44%) men and 27 (56%) women. We selected participants with a corrective vision ≥ 0.8, without any vision deficiency, to ensure reliable recognition of visual stimuli. We recommended that the participants sleep sufficiently and refrain from smoking and consuming alcohol and caffeine the day before the experiment. As the experiment required valid recognition of the participant's facial expression, we limited the use of glasses and cosmetic makeup. All participants were briefed on the purpose and procedure of the experiment, and signed consent was obtained from them. They were then compensated for their participation by payment of a fee.

2.4. Experimental Protocol

Figure 2 outlines the experimental process and the environment used in this study. The participants were asked to sit 1 m away from a 27-inch LCD monitor. A webcam was installed on the monitor. Participants' brainwaves (EEG cap 18 Ch) and eye movements (gaze tracking device) were acquired, in addition to subjective responses to a questionnaire. We set the frame rate of the gaze-tracking device to 60 frames per second. Participants viewed four emotion-eliciting videos and responded to a questionnaire after each viewing session.

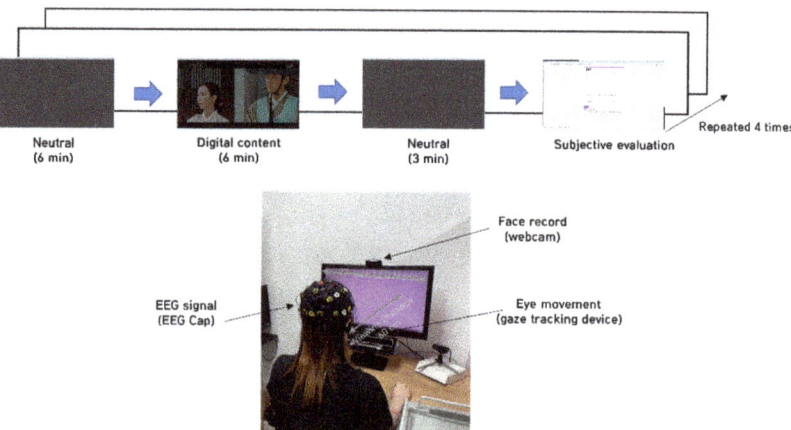

Figure 2. Experimental protocol and configuration.

3. Analysis

Our brain connectivity analysis methods were based on Jamal et al. [27], as outlined in Figure 3. The process consisted of seven stages: (1) sampled EEG signals at 500 Hz, (2) removed the noise through pre-processing, (3) conducted fast Fourier transform (FFT) at 0–30 Hz, (4) conducted band pass filter with delta (0 Hz–4 Hz), theta (4 Hz–8 Hz), alpha (8 Hz–12 Hz), and beta (12 Hz–30 Hz), (5) processed continuous wavelet transform (CWT) with complex Morlet wavelet, (6) computed the EEG frequency band-specific pairwise phase difference, and (7) determined the optimal number of states in the data using incremental k-means clustering.

We used the CWT with a complex Morlet wavelet as the basis function to analyze the transient dynamics of phase synchronization. In contrast to the discrete Fourier transform (DFT), it has a short vibration signal and an expiration date for the vibration wave. Figure 4 shows the Morlet wavelet graph. The CWT operates with a signal with scaled and shifted versions of a basic wavelet.

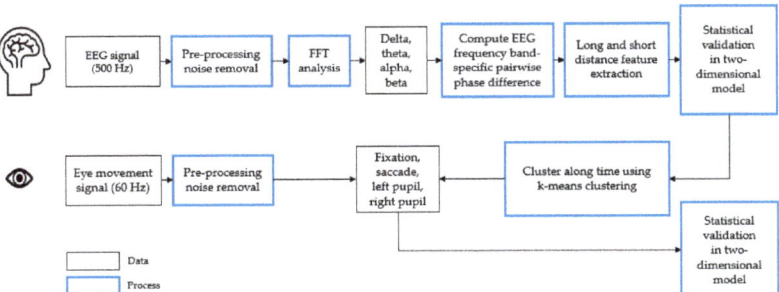

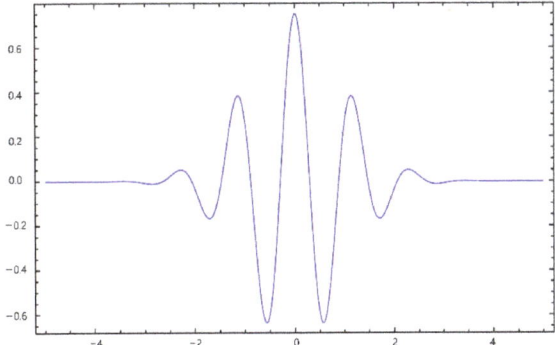

Figure 3. The process of brain connectivity analysis.

Figure 4. The Morlet wavelet graph.

Therefore, it can be expressed as the formula below in Equation (1), where a is a scale factor and b is a shift factor. Being continuous, infinite wavelets can be shifted and scaled:

$$X_w(a,b) = \frac{1}{|a|^{\frac{1}{2}}} \int_{-\infty}^{\infty} x(t)\overline{\varphi}\left(\frac{t-b}{a}\right) dt \qquad (1)$$

4. Results

We will present the results of the participants' subjective evaluation and brain connectivity analysis, followed by the results of eye movement analysis.

4.1. Subject Evaluation

We compared the subjective arousal and valence scores between the four emotion-eliciting conditions (pleasant-aroused, pleasant-relaxed, unpleasant-relaxed, and unpleasant-aroused). We conducted a series of ANOVA tests on the arousal and valence scores. Post-hoc analyses using Tukey's HSD were conducted by adjusting the alpha level to 0.0125 per test (0.05/4).

The mean arousal scores were significantly higher in the aroused conditions (pleasant-aroused, unpleasant-aroused) than in the relaxed conditions (pleasant-relaxed, unpleasant-relaxed) ($p < 0.001$), as shown in Figure 5. The pairwise comparison of the mean arousal scores indicated that the scores were significantly different from one another, as shown in Table 2. The results indicate that participants reported congruent emotional arousal with the target emotion of the stimulus.

The results indicated that the mean valence scores were significantly higher in the pleasant conditions (pleasant-aroused, pleasant-relaxed) than in the unpleasant conditions (unpleasant-aroused, unpleasant-relaxed), $p < 0.001$, as shown in Figure 6. The pairwise comparison of the mean valence scores indicated that the scores were significantly different from one another, except for two comparisons, as shown in Table 3. The results indicate that participants reported congruent emotional valence with the target emotion of the stimulus.

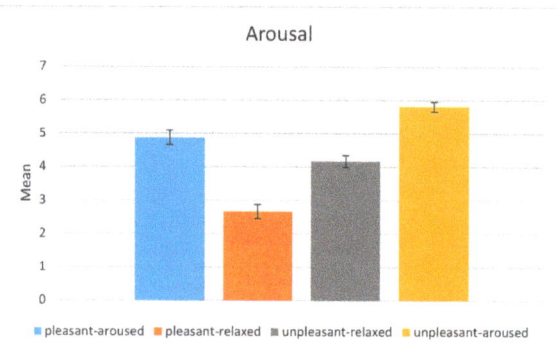

Figure 5. Analysis of the arousal values between the four emotion-eliciting conditions.

Table 2. Multiple comparisons of mean arousal scores using Tukey HSD.

Emotion Condition 1	Emotion Condition 2	Mean Difference	Lower	Upper	Reject
Pleasant-aroused	Pleasant-relaxed	−2.2083	−2.8964	−1.5202	True
Pleasant-aroused	Unpleasant-aroused	0.9375	0.2494	1.6256	True
Pleasant-aroused	Unpleasant-relaxed	−0.7083	−1.3964	−0.0202	True
Pleasant-relaxed	Unpleasant-aroused	3.1458	2.4577	3.8339	True
Pleasant-relaxed	Unpleasant-relaxed	1.5	0.8119	2.1881	True
Unpleasant-aroused	Unpleasant-relaxed	−1.6458	−2.3339	−0.9577	True

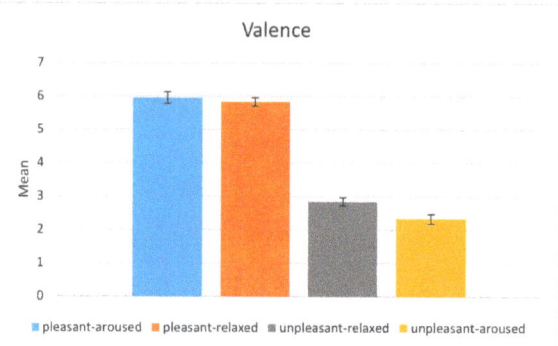

Figure 6. Analysis of the valence values between the four emotion-eliciting conditions.

Table 3. Multiple comparisons of mean valence scores using Tukey HSD.

Emotion Condition 1	Emotion Condition 2	Mean Difference	Lower	Upper	Reject
Pleasant-aroused	Pleasant-relaxed	−0.125	−0.6531	0.4031	False
Pleasant-aroused	Unpleasant-aroused	−3.625	−4.1531	−3.0969	True
Pleasant-aroused	Unpleasant-relaxed	−3.1042	−3.6322	−2.5761	True
Pleasant-relaxed	Unpleasant-aroused	−3.5	−4.0281	−2.9719	True
Pleasant-relaxed	Unpleasant-relaxed	−2.9792	−3.5072	−2.4511	True
Unpleasant-aroused	Unpleasant-relaxed	−1.6458	−2.3339	−0.9577	True

4.2. Brain Connectivity Features

We computed the EEG frequency band-specific pairwise phase differences for each emotion-eliciting condition, as shown in Figures 7–10. A total of 153 pairwise features were analyzed. If the power differences between the two brain regions are lower than the mean power value, the connectivity is relatively strong. Such cases were marked as unfilled (☐).

We further analyzed the long- and short-distance connectivity of the extracted features. The connectivity of the frontal and occipital lobes can predict the process of information transmission to the occipital lobe after emotion is generated (marked in green in Figure 11). The eigenvalue was the average (N = 47) of the connectivity sum of the two channels defined by the long-distance O-F connectivity.

The prefrontal cortex is involved in emotion regulation, recognition, judgment, and reasoning. The connectivity of the prefrontal lobe to the temporal lobe, parietal lobe, and center helps to understand the information processing process of visual-emotional stimuli (marked in yellow in Figure 11). The eigenvalue was the average (N = 47) of the connectivity sum of the two channels defined by the long-distance prefrontal connectivity.

Long- and short-range connectivity features have been extensively studied for their ability to process social emotions and interactions. Short-distance connectivity characteristics can determine the brain's different states during negative emotions, especially those related to the central-parietal lobe connectivity. We considered a distance of less than 10 cm as short connectivity (marked pink in Figure 11). The eigenvalue was the average (N = 47) of the connectivity sum of the two channels defined by the short-distance connectivity.

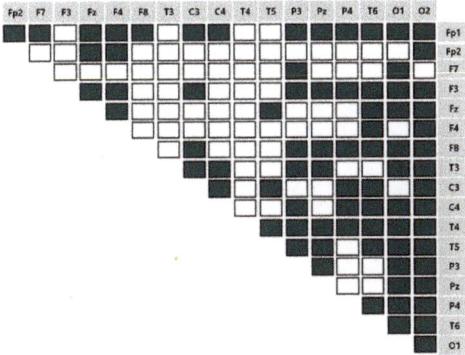

Figure 7. The brain connectivity map in the pleasant-aroused condition.

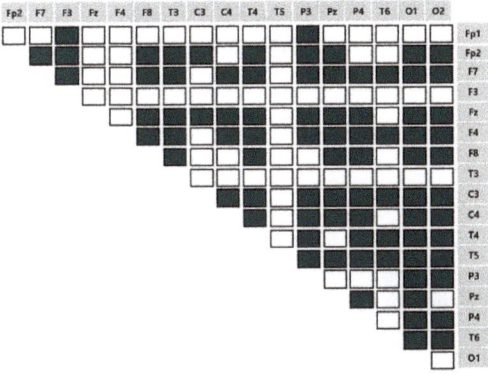

Figure 8. The brain connectivity map in the pleasant-relaxed condition.

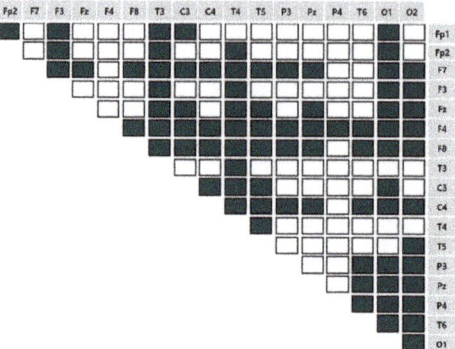

Figure 9. The brain connectivity map in the unpleasant-relaxed condition.

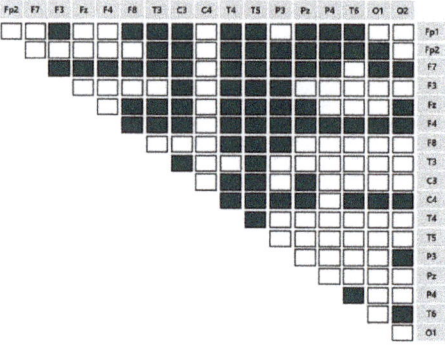

Figure 10. The brain connectivity map in the unpleasant-aroused condition.

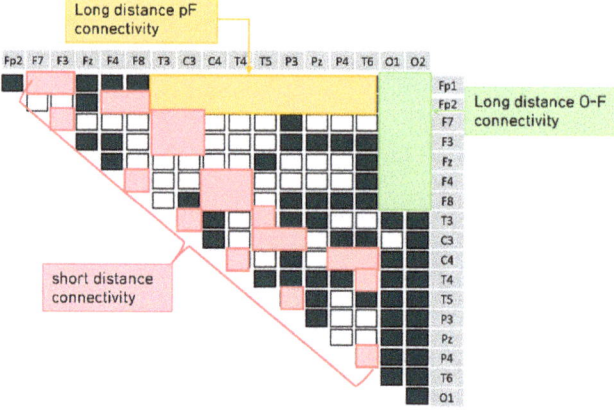

Figure 11. The three distance connectivity groups in the brain connectivity map.

4.2.1. Characteristics of Three Distance Connectivity

Figure 12 depicts the long-distance connectivity of the occipital and frontal lobes (LD_O-F connectivity) of the beta wave in the visual comparison diagram of the two-dimensional model. O-F connectivity in the unpleasant-aroused condition had the strongest connectivity. In the pleasant-relaxed condition, bi-directional connectivity was observed between the left frontal and occipital lobes. In the unpleasant-relaxed condition, bidirectional connectivity was observed from the right occipital to the frontal lobe. In the pleasant-aroused condition, cross-hemispheric connectivity was observed between the frontal and occipital lobes.

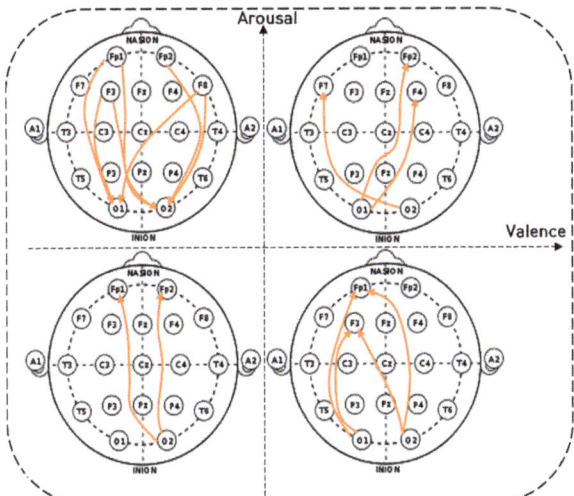

Figure 12. The long-distance connectivity of the occipital and frontal lobes (LD_O-F connectivity) of the beta wave.

Figure 13 depicts the long-distance connectivity of the prefrontal and temporal lobes, parietal lobes, and central (LD_pF connectivity) beta waves in the visual comparison diagram of the two-dimensional model. In pleasant-aroused and unpleasant-relaxed conditions, the right prefrontal lobe was strongly connected to the central, parietal, and temporal lobes of both hemispheres. In the pleasant-relaxed condition, there was strong connectivity in the left prefrontal–temporal, left prefrontal–central, and left prefrontal–parietal regions. In the unpleasant-aroused condition, the prefrontal–temporal, prefrontal–parietal, and prefrontal–central regions showed the weakest connectivity.

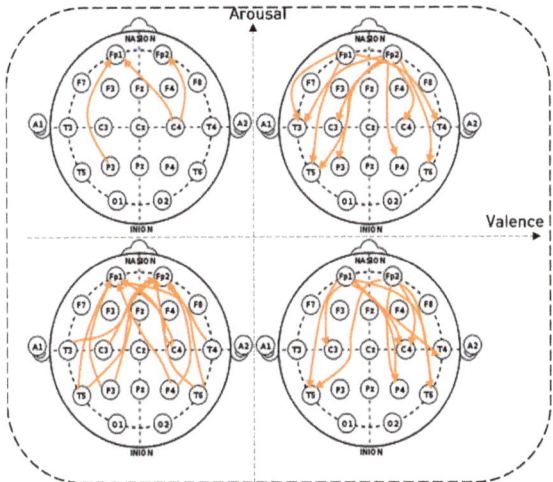

Figure 13. The long-distance connectivity of the prefrontal and temporal lobes, parietal lobes, and central (LD_pF connectivity) of the beta wave.

Figure 14 depicts the short-distance connectivity (SD connectivity) of the beta waves in the visual comparison diagram of the two-dimensional emotional model. In the aroused conditions (pleasant-aroused, unpleasant-aroused), strong frontal–temporal–central connectivity was observed. However, in the relaxed conditions (pleasant-relaxed, unpleasant-relaxed), strong central–parietal connectivity was observed.

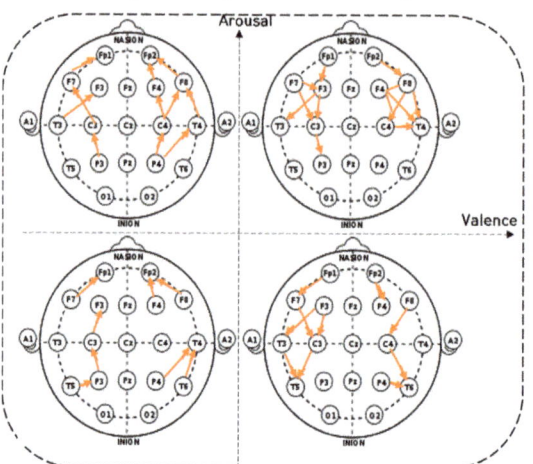

Figure 14. The short-distance connectivity of the prefrontal-temporal lobes, central-parietal lobes, and parietal-temporal lobes (SD connectivity) of the beta wave.

In summary, the analysis suggests a strong frontal activity in the unpleasant-aroused condition, indicating intense information processing and transfer involving the frontal cortex. In pleasant conditions, feedback is sent to the parietal, temporal, and central regions after the prefrontal cortex processes the information. In the unpleasant-relaxed condition, brain connectivity implies the control of the participant's eye movement.

4.2.2. Power Value Analysis in Three Distance Connectivity

To further understand the strength and directionality of brainwave connectivity, statistical analysis was performed on the power value using ANOVA, followed by post hoc analyses (see Figures 15–20).

Figure 15 depicts the eigenvalues (i.e., mean power value) of the occipital and frontal lobe connectivity. The plus-minus sign of the eigenvalue determines the causality. In the unpleasant-aroused condition, more information is processed in the frontal lobe, indicating more activity in the occipital lobe than in primary visual processing.

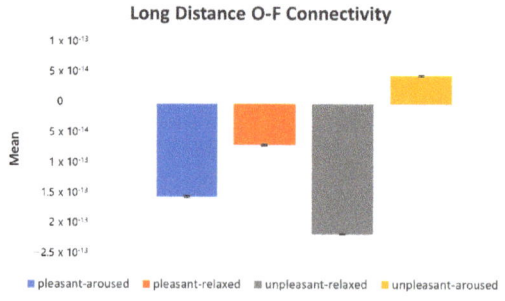

Figure 15. The eigenvalues in the long-distance O-F connectivity.

Figure 16 shows the absolute values of the mean ($|mean|$). The pleasant-relaxed and unpleasant-aroused conditions exhibited high occipital-frontal connectivity, whereas the pleasant-relaxed condition exhibited left hemisphere-frontal activation (see Figure 12).

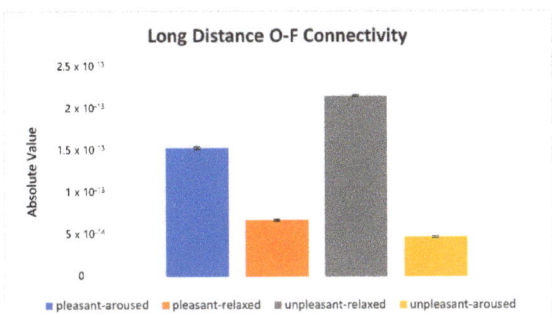

Figure 16. The absolute value in the long-distance O-F connectivity.

Figure 17 depicts the eigenvalues (i.e., the mean power value) of prefrontal connectivity. The plus-minus sign of the eigenvalue determines the causality. The results showed that activity in the prefrontal lobe in pleasant conditions (pleasant-aroused, pleasant-relaxed) was greater than that in other regions. Conversely, in the unpleasant conditions (unpleasant-aroused, unpleasant-relaxed), activity in the other regions was stronger than that in the prefrontal lobe.

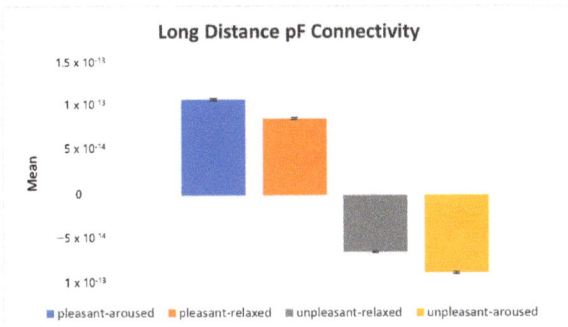

Figure 17. The eigenvalues in the long-distance prefrontal connectivity.

Figure 18 shows the absolute values of the mean ($|mean|$). The unpleasant-relaxed condition exhibited the strongest connectivity.

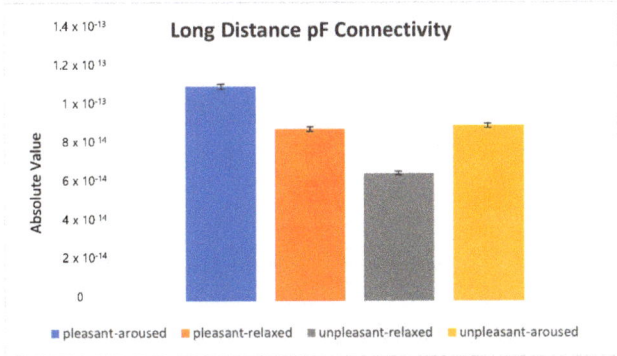

Figure 18. The absolute value in the long-distance prefrontal connectivity.

Figure 19 depicts the eigenvalues (i.e., mean power value) of the short-distance connectivity in frontal–temporal, frontal–central, and temporal–parietal connections in the four emotion-eliciting conditions. Overall, connectivity in the relaxed condition was stronger than that in the aroused condition. Specifically, central–parietal connectivity showed stronger activity than frontal–temporal and frontal–central connectivity (see Figure 14).

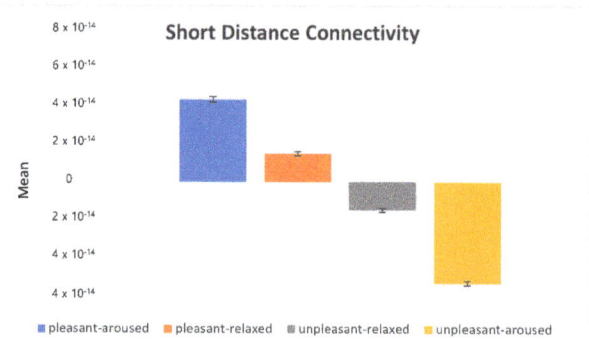

Figure 19. The eigenvalues in the short-distance connectivity.

Figure 20 shows the absolute values of mean ($|mean|$). The relaxed conditions (pleasant-relaxed and unpleasant-relaxed) showed stronger connectivity, specifically stronger P-O connectivity. Conversely, the aroused conditions (pleasant-aroused, unpleasant-aroused) showed weaker connectivity, but stronger F-T connectivity. In particular, the unpleasant-aroused, pleasant-aroused, and pleasant-relaxed conditions showed substantial premotor cortical PMDr (F7) connections associated with eye movement control. This was consistent with the saccade results.

Through statistical analysis, we found that connectivity in the pleasant-relaxed condition was the highest, while connectivity in the unpleasant-relaxed condition was higher than that in the pleasant-aroused and unpleasant-aroused conditions.

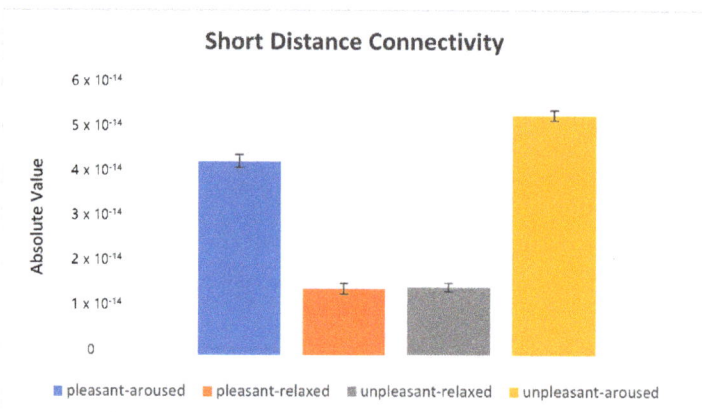

Figure 20. The absolute value in the short-distance connectivity.

By comparing the three extracted brainwave connectivity eigenvalues with subjective evaluations, we found that the long-distance prefrontal connectivity eigenvalues have similar characteristics to the valence score measures of subjective evaluations. The prefrontal cortex (PFC) makes decisions and is responsible for cognitive control. Positive valence increases the neurotransmitter dopamine, enhancing cognitive control [28–30]. This may explain prefrontal activation in pleasant conditions (see Figure 15).

In summary, in the unpleasant-aroused condition, the frontal lobe showed a stronger activation than the occipital lobe. Overall, in pleasant conditions, the prefrontal lobe showed a stronger activation than other regions. Conversely, in unpleasant conditions, the prefrontal lobe showed a weaker activation than other regions.

4.3. Clustering Eye Movement Features

The statistical results showed that the short-distance connectivity eigenvalue and subjective evaluation arousal score had similar characteristics. Connectivity in the unpleasant-relaxed condition was the strongest (Figure 16). Specifically, central-parietal connectivity showed stronger connectivity than frontal–temporal and frontal–central connectivity. Unpleasant emotions are known to activate central–parietal connectivity [31].

The three eigenvalues of the extracted EEG can be used to distinguish the four emotions in the two-dimensional emotional model. We conducted an unsupervised K-means analysis in chronological order using these three eigenvalues. We distinguished the emotional and non-emotional states of each participant while viewing the emotional video. The emotional and non-emotional states of the eye movement data were then distinguished. Figure 21 shows an instance of a participant's K-means results. Group 1 indicates the non-emotional states, whereas Group 2 indicates the emotional states. The figure implies that the participant's state changes from a non-emotional state (i.e., 0.0) to an emotional state (i.e., 1.0) as a function of time.

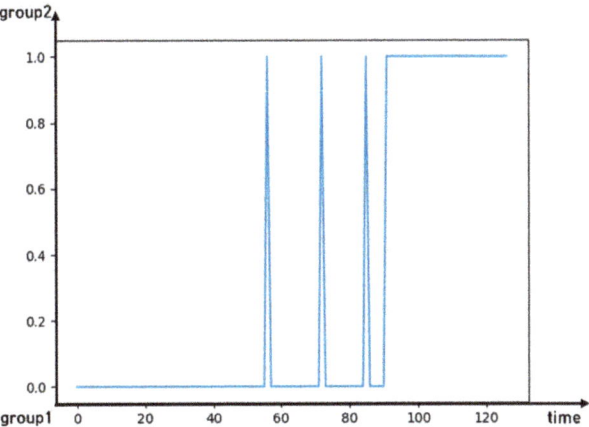

Figure 21. An instance of a participant's k-Means results.

Figures 22 and 23 depict the post-hoc analysis of the left and right pupils between the two-dimensional emotional model conditions. From the statistical results of the eye movement eigenvalues, the characteristics of the right pupil and left pupil did not change much between the four conditions; the pupil of the pleasant-aroused condition had the largest change, followed by the pleasant-relaxed and unpleasant-relaxed conditions. The least difference was observed in the unpleasant-aroused condition.

However, in relaxed conditions (pleasant-relaxed and unpleasant-relaxed), the right pupil of the unpleasant-relaxed condition was larger than the left pupil. From the first eigenvalue long-distance O-F connectivity of brain wave connectivity, we found two locations with high connectivity: the right occipital lobe and the left and right prefrontal lobes.

Figure 24 shows the results of the post hoc analysis of the fixation between the two-dimensional emotional model conditions. The fixation feature in the unpleasant-relaxed condition was larger than that in the other three conditions.

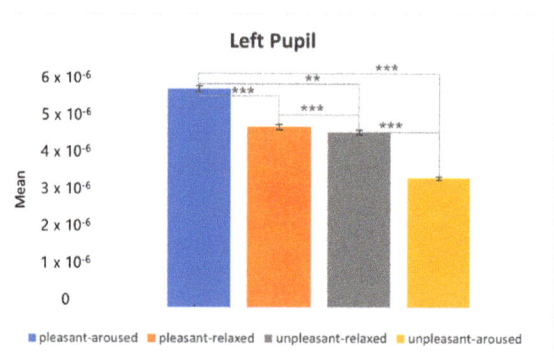

Figure 22. The post hoc analysis of the left pupil. ** $p < 0.05$. *** $p < 0.001$.

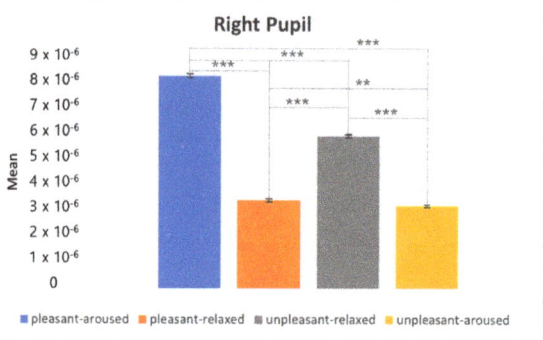

Figure 23. The post hoc analysis of the right pupil. ** $p < 0.05$. *** $p < 0.001$.

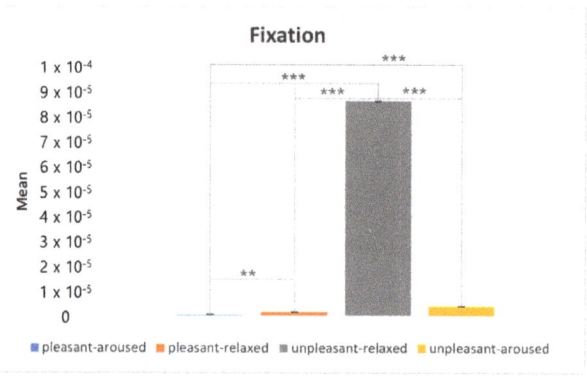

Figure 24. The post hoc analysis on the fixation. ** $p < 0.05$. *** $p < 0.001$.

Figure 25 shows the results of the post hoc analysis of the saccade between the two-dimensional emotional model conditions. The results showed the lowest change in the unpleasant-relaxed condition, and the greatest change in the pleasant-relaxed condition. The characteristics of the saccades were similar to those of the short-distance connectivity eigenvalues. Short-distance connectivity also showed weak brain connections in the unpleasant-relaxed condition (see Figure 14). After the frontal lobe makes a cognitive judgment, it gives instructions to the occipital lobe, causing saccadic eye movements.

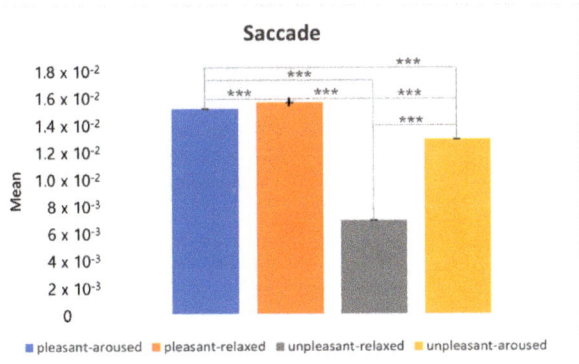

Figure 25. The post hoc analysis on the saccade. *** $p < 0.001$.

5. Conclusions and Discussion

This study aimed to understand the relationship between brain wave connectivity and eye movement characteristic values using a two-dimensional emotional model. We divided brainwave connectivity into three distinct groups: long-distance occipital–frontal connectivity, long-distance prefrontal connectivity between the prefrontal lobe and temporal lobe, parietal lobe, and central lobe, and short-distance connectivity including the characteristic relationships between the frontal lobe–temporal lobe, frontal lobe-central lobe, temporal–parietal lobe, and parietal lobe–central. Then, through unsupervised learning of these three eigenvalues, the emotional response was divided into emotional and non-emotional states in real time using K-means analysis. The two states were used to extract the feature values of the eye movements. We analyzed the relationship between eye movements and brain wave connectivity using statistical analyses.

The results revealed that the connectivity eigenvalues of the long-distance prefrontal lobe, temporal lobe, parietal lobe, and center are related to cognitive activity involving high valence. The prefrontal lobe occupies two-thirds of the human frontal cortex [32] and is responsible for recognition and decision-making, reflecting cognitive judgment from valence responses [33,34]. Specifically, the dorsolateral prefrontal cortex (dlPFC) is involved with working memory [35], decision making [36], and executive attention [37]. However, most recently, Nejati et al. [32] found that the role of dlPFC extends to the regulation of the valence of emotional experiences. Second, the saccade correlated with long-distance occipital-frontal connectivity. After making a judgment, the frontal lobe provides instructions to the occipital lobe, which moves the eye. Electrical stimulation of several areas of the cortex evokes saccadic eye movements. The prefrontal top-down control of visual appraisal and emotion-generation processes constitutes a mechanism of cognitive reappraisal in emotion regulation [38]. The short-distance connectivity results showed emotional fluctuations caused by the unconscious stimulation of audio-visual perception.

We acknowledge some limitations of the research. First, the results of our study are from one stimulus for each of the four quadrants in the two-dimensional model. Future studies may use multiple stimuli, possibly controlling the type of stimuli. Second, although pupillometry is an effective measurement for understanding brain activity changes related to arousal, attention, and salience [39], we did not find consistent and conclusive results between pupil size and brain connectivity. The size of pupils changes according to ambient light (i.e., pupillary light reflex) [40,41], which may have confounded the results. Future studies should control extraneous variables more thoroughly to find the main effect of pupil characteristics. Third, our analysis is based on participants of local university students, limiting the age range (i.e., 20 to 30 years). Age and culture may influence the results, so future studies may consider a broader range of demographic populations and conduct a cross-cultural investigation.

The study purposely analyzed brain connectivity and changes in eye movement in tandem to establish a relational basis between neural activity and eye movement features. We took the first step in unraveling such a relationship, albeit fell short in achieving a full understanding, such as the pupil size characteristics. Because the eyes' structures are connected to the brain's nerves, an exclusive analysis of eye features may lead to a comprehensive understanding of the participant's emotions. A non-contact appraisal of emotion based on eye feature analysis may be a promising method applicable to metaverse or media art.

Author Contributions: J.Z.: conceptualization, methodology, software, validation, formal analysis, investigation, resources, data curation, writing, visualization, project administration; S.P.: methodology, validation, formal analysis, investigation, writing, review, editing; A.C.: conceptualization, investigation, review, editing; M.W.: conceptualization, methodology, writing, review, supervision, funding acquisition. All authors have read and agreed to the published version of the manuscript.

Funding: This work was supported by the Electronics and Telecommunications Research Institute (ETRI) grant funded by the Korean government (22ZS1100, Core Technology Research for Self-Improving Integrated Artificial Intelligence System).

Institutional Review Board Statement: The study was conducted according to the guidelines of the Declaration of Helsinki, and approved by the Institutional Review Board of Sangmyung University (protocol code C-2021-002, approved 9 July 2021).

Informed Consent Statement: Informed consent was obtained from all subjects involved in the study. Written informed consent has been obtained from the subjects to publish this paper.

Conflicts of Interest: The authors declare no conflict of interest.

References

1. Chanel, G.; Kierkels, J.J.M.; Soleymani, M.; Pun, T. Short-term emotion assessment in a recall paradigm. *Int. Hum. J. Comput. Stud.* **2009**, *67*, 607–627. [CrossRef]
2. Friston, K.J. Functional and effective connectivity: A review. *Brain Connect.* **2011**, *1*, 13–36. [CrossRef]
3. Mauss, I.B.; Robinson, M.D. Measures of emotion: A review. *Cogn. Emot.* **2009**, *23*, 209–237. [CrossRef]
4. Kim, J.-H.; Kim, S.-L.; Cha, Y.-S.; Park, S.-I.; Hwang, M.-C. Analysis of CNS functional connectivity by relationship duration in positive emotion sharing relation. In Proceedings of the Korean Society for Emotion and Sensibility Conference, Seoul, Korea, 11 May 2012; pp. 11–12.
5. Moon, S.-E.; Jang, S.; Lee, J.-S. Convolutional neural network approach for EEG-based emotion recognition using brain connectivity and its spatial information. In Proceedings of the 2018 IEEE International Conference on Acoustics, Speech and Signal Processing (ICASSP), Calgary, AB, Canada, 15–20 April 2018; pp. 2556–2560.
6. Voorbij, H.J. Searching scientific information on the Internet: A Dutch academic user survey. *Am. J. Soc. Inf. Sci.* **1999**, *50*, 598–615. [CrossRef]
7. Thompson, K.G.; Biscoe, K.L.; Sato, T.R. Neuronal basis of covert spatial attention in the frontal eye field. *Neurosci. J.* **2005**, *25*, 9479–9487. [CrossRef]
8. Bulling, A.; Ward, J.A.; Gellersen, H.; Tröster, G. Eye movement analysis for activity recognition using electrooculography. *IEEE Trans. Pattern Anal. Mach. Intell.* **2010**, *33*, 741–753. [CrossRef]
9. Võ, M.L.H.; Jacobs, A.M.; Kuchinke, L.; Hofmann, M.; Conrad, M.; Schacht, A.; Hutzler, F. The coupling of emotion and cognition in the eye: Introducing the pupil old/new effect. *Psychophysiology* **2008**, *45*, 130–140. [CrossRef]
10. Tatler, B.W.; Brockmole, J.R.; Carpenter, R.H.S. LATEST: A model of saccadic decisions in space and time. *Psychol. Rev.* **2017**, *124*, 267. [CrossRef]
11. Carpenter, R.H.S. The neural control of looking. *Curr. Biol.* **2000**, *10*, R291–R293. [CrossRef]
12. Glimcher, P.W. The neurobiology of visual-saccadic decision making. *Annu. Rev. Neurosci.* **2003**, *26*, 133–179. [CrossRef]
13. Argyle, M.; Cook, M. *Gaze and Mutual Gaze*; Cambridge University Press: Cambridge, UK, 1976.
14. Baron-Cohen, S.; Campbell, R.; Karmiloff-Smith, A.; Grant, J.; Walker, J. Are children with autism blind to the mentalistic significance of the eyes? *Br. Dev. J. Psychol.* **1995**, *13*, 379–398. [CrossRef]
15. Emery, N.J. The eyes have it: The neuroethology, function and evolution of social gaze. *Neurosci. Biobehav. Rev.* **2000**, *24*, 581–604. [CrossRef]
16. Kleinke, C.L. Gaze and eye contact: A research review. *Psychol. Bull.* **1986**, *100*, 78. [CrossRef] [PubMed]
17. Hood, B.M.; Willen, J.D.; Driver, J. Adult's eyes trigger shifts of visual attention in human infants. *Psychol. Sci.* **1998**, *9*, 131–134. [CrossRef]

18. Pelphrey, K.A.; Sasson, N.J.; Reznick, J.S.; Paul, G.; Goldman, B.D.; Piven, J. Visual scanning of faces in autism. *J. Autism Dev. Disord.* **2002**, *32*, 249–261. [CrossRef] [PubMed]
19. Glaholt, M.G.; Reingold, E.M. Eye movement monitoring as a process tracing methodology in decision making research. *Neurosci. J. Psychol. Econ.* **2011**, *4*, 125. [CrossRef]
20. Peitek, N.; Siegmund, J.; Parnin, C.; Apel, S.; Hofmeister, J.C.; Brechmann, A. Simultaneous measurement of program comprehension with fmri and eye tracking: A case study. In Proceedings of the 12th ACM/IEEE International Symposium on Empirical Software Engineering and Measurement, Oulu, Finland, 11–12 October 2018; pp. 1–10.
21. He, Z.; Li, Z.; Yang, F.; Wang, L.; Li, J.; Zhou, C.; Pan, J. Advances in multimodal emotion recognition based on brain–computer interfaces. *Brain Sci.* **2020**, *10*, 687. [CrossRef]
22. Wu, X.; Zheng, W.-L.; Li, Z.; Lu, B.-L. Investigating EEG-based functional connectivity patterns for multimodal emotion recognition. *J. Neural Eng.* **2022**, *19*, 16012. [CrossRef]
23. Lu, Y.; Zheng, W.-L.; Li, B.; Lu, B.-L. Combining eye movements and EEG to enhance emotion recognition. In Proceedings of the 24th International Conference on Artificial Intelligence, Buenos Aires, Argentina, 25–31 July 2015; pp. 1170–1176.
24. Zheng, W.-L.; Liu, W.; Lu, Y.; Lu, B.-L.; Cichocki, A. Emotionmeter: A multimodal framework for recognizing human emotions. *IEEE Trans. Cybern.* **2018**, *49*, 1110–1122. [CrossRef]
25. Soleymani, M.; Pantic, M.; Pun, T. Multimodal emotion recognition in response to videos. *IEEE Trans. Affect. Comput.* **2011**, *3*, 211–223. [CrossRef]
26. Russell, J.A. A circumplex model of affect. *Pers. J. Soc. Psychol.* **1980**, *39*, 1161. [CrossRef]
27. Jamal, W.; Das, S.; Maharatna, K.; Pan, I.; Kuyucu, D. Brain connectivity analysis from EEG signals using stable phase-synchronized states during face perception tasks. *Phys. A Stat. Mech. Appl.* **2015**, *434*, 273–295. [CrossRef]
28. Savine, A.C.; Braver, T.S. Motivated cognitive control: Reward incentives modulate preparatory neural activity during task-switching. *Soc Neurosci.* **2010**, 10294–10305. [CrossRef] [PubMed]
29. Ashby, F.G.; Isen, A.M. A neuropsychological theory of positive affect and its influence on cognition. *Psychol. Rev.* **1999**, *106*, 529. [CrossRef]
30. Ashby, F.G.; Valentin, V.V. The effects of positive affect and arousal and working memory and executive attention: Neurobiology and computational models. In *Emotional Cognition: From Brain to Behaviour*; Moore, S.C., Oaksford, M., Eds.; John Benjamins Publishing Company: Amsterdam, The Netherlands, 2002; pp. 245–287. [CrossRef]
31. Mehdizadehfar, V.; Ghassemi, F.; Fallah, A.; Mohammad-Rezazadeh, I.; Pouretemad, H. Brain connectivity analysis in fathers of children with autism. *Cogn. Neurodyn.* **2020**, *14*, 781–793. [CrossRef] [PubMed]
32. Nejati, V.; Majdi, R.; Salehinejad, M.A.; Nitsche, M.A. The role of dorsolateral and ventromedial prefrontal cortex in the processing of emotional dimensions. *Sci. Rep.* **2021**, *11*, 1971. [CrossRef] [PubMed]
33. Stuss, D.T. New approaches to prefrontal lobe testing. In *The Human Frontal Lobes: Functions and Disorders*; Miller, B.L., Cummings, J.L., Eds.; The Guilford Press: New York, NY, USA, 2007; pp. 292–305.
34. Henri-Bhargava, A.; Stuss, D.T.; Freedman, M. Clinical assessment of prefrontal lobe functions. *Contin. Lifelong Learn. Neurol.* **2018**, *24*, 704–726. [CrossRef]
35. Barbey, A.K.; Koenigs, M.; Grafman, J. Dorsolateral prefrontal contributions to human working memory. *Cortex* **2013**, *49*, 1195–1205. [CrossRef]
36. Rahnev, D.; Nee, D.E.; Riddle, J.; Larson, A.S.; D'Esposito, M. Causal evidence for frontal cortex organization for perceptual decision making. *Proc. Natl. Acad. Sci. USA* **2016**, *113*, 6059–6064. [CrossRef]
37. Ghanavati, E.; Salehinejad, M.A.; Nejati, V.; Nitsche, M.A. Differential role of prefrontal, temporal and parietal cortices in verbal and figural fluency: Implications for the supramodal contribution of executive functions. *Sci. Rep.* **2019**, *9*, 3700. [CrossRef] [PubMed]
38. Popov, T.; Steffen, A.; Weisz, N.; Miller, G.A.; Rockstroh, B. Cross-frequency dynamics of neuromagnetic oscillatory activity: Two mechanisms of emotion regulation. *Psychophysiology* **2012**, *49*, 1545–1557. [CrossRef] [PubMed]
39. Joshi, S.; Gold, J.I. Pupil size as a window on neural substrates of cognition. *Trends Cogn. Sci.* **2020**, *24*, 466–480. [CrossRef]
40. Lowenstein, O.; Loewenfeld, I.E. Role of sympathetic and parasympathetic systems in reflex dilatation of the pupil: Pupillographic studies. *Arch. Neurol. Psychiatry* **1950**, *64*, 313–340. [CrossRef] [PubMed]
41. Toates, F.M. Accommodation function of the human eye. *Physiol. Rev.* **1972**, *52*, 828–863. [CrossRef] [PubMed]

Article

Subject-Specific Cognitive Workload Classification Using EEG-Based Functional Connectivity and Deep Learning

Anmol Gupta [1], Gourav Siddhad [1], Vishal Pandey [2], Partha Pratim Roy [1] and Byung-Gyu Kim [3,*]

[1] Department of Computer Science and Engineering, Indian Institute of Technology, Roorkee 247667, India; agupta@cs.iitr.ac.in (A.G.); g_siddhad@cs.iitr.ac.in (G.S.); partha@cs.iitr.ac.in (P.P.R.)
[2] Department of Biomedical Engineering, Institute of Nuclear Medicine and Allied Sciences, Defence Research and Development Organization, Delhi 110054, India; vishalp055@gmail.com
[3] Department of IT Engineering, Sookmyung Women's University, Seoul 04310, Korea
* Correspondence: bg.kim@sookmyung.ac.kr

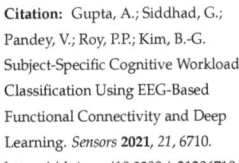

Citation: Gupta, A.; Siddhad, G.; Pandey, V.; Roy, P.P.; Kim, B.-G. Subject-Specific Cognitive Workload Classification Using EEG-Based Functional Connectivity and Deep Learning. Sensors 2021, 21, 6710. https://doi.org/10.3390/s21206710

Academic Editor: Giovanni Sparacino

Received: 19 August 2021
Accepted: 2 October 2021
Published: 9 October 2021

Publisher's Note: MDPI stays neutral with regard to jurisdictional claims in published maps and institutional affiliations.

Copyright: © 2021 by the authors. Licensee MDPI, Basel, Switzerland. This article is an open access article distributed under the terms and conditions of the Creative Commons Attribution (CC BY) license (https://creativecommons.org/licenses/by/4.0/).

Abstract: Cognitive workload is a crucial factor in tasks involving dynamic decision-making and other real-time and high-risk situations. Neuroimaging techniques have long been used for estimating cognitive workload. Given the portability, cost-effectiveness and high time-resolution of EEG as compared to fMRI and other neuroimaging modalities, an efficient method of estimating an individual's workload using EEG is of paramount importance. Multiple cognitive, psychiatric and behavioral phenotypes have already been known to be linked with "functional connectivity", i.e., correlations between different brain regions. In this work, we explored the possibility of using different model-free functional connectivity metrics along with deep learning in order to efficiently classify the cognitive workload of the participants. To this end, 64-channel EEG data of 19 participants were collected while they were doing the traditional n-back task. These data (after pre-processing) were used to extract the functional connectivity features, namely Phase Transfer Entropy (PTE), Mutual Information (MI) and Phase Locking Value (PLV). These three were chosen to do a comprehensive comparison of directed and non-directed model-free functional connectivity metrics (allows faster computations). Using these features, three deep learning classifiers, namely CNN, LSTM and Conv-LSTM were used for classifying the cognitive workload as low (1-back), medium (2-back) or high (3-back). With the high inter-subject variability in EEG and cognitive workload and recent research highlighting that EEG-based functional connectivity metrics are subject-specific, subject-specific classifiers were used. Results show the state-of-the-art multi-class classification accuracy with the combination of MI with CNN at 80.87%, followed by the combination of PLV with CNN (at 75.88%) and MI with LSTM (at 71.87%). The highest subject specific performance was achieved by the combinations of PLV with Conv-LSTM, and PLV with CNN with an accuracy of 97.92%, followed by the combination of MI with CNN (at 95.83%) and MI with Conv-LSTM (at 93.75%). The results highlight the efficacy of the combination of EEG-based model-free functional connectivity metrics and deep learning in order to classify cognitive workload. The work can further be extended to explore the possibility of classifying cognitive workload in real-time, dynamic and complex real-world scenarios.

Keywords: CNN; cognitive workload; functional connectivity analysis; LSTM; mental workload; mutual information; phase locking value; phase transfer entropy

1. Introduction

Cognitive workload is the measure of the amount of mental effort required to complete any task [1]. Working memory is required to process information for short periods of time, while long-term memory is associated with storing information for long periods of time [2]. Tasks such as arithmetic operations, reading and learning require efficient use of working memory. Cognitive workload can be defined as the amount of mental activity utilized by working memory to complete any task. Assessment of an individual's cognitive workload is an essential component in most human-machine collaboration tasks. A major application of this lies in the defense domain. Operations like driving under high-stress environmental conditions, monitoring air traffic control, piloting an aircraft or operating an unmanned vehicle are excellent examples. The optimal level of cognitive workload is pivotal in high-risk scenarios where important decisions are supposed to be made in real-time. The rate at which the information is processed determines the workload induced in any individual while performing any task. A high workload can lead to unplanned and disproportionate hazards, and too little workload can lead to being disengaged from the task. This points to the importance of maintaining optimal cognitive workload in high-risk scenarios to perform the task satisfactorily. With respect to cognitive workload, emotional intelligence and stability are regarded as essential components. An individuals' cognitive load will be affected by emotional valence as it will interfere with parallel cognitive processing. Studies show a positive relation between emotional intelligence and some cognitive tasks [3,4]. Therefore, classification of cognitive workload can be an essential indicator of emotional intelligence and stability.

Although the assessment of cognitive workload is important, it is not a trivial task. Traditional methods of the evaluation of cognitive workload included subjective measures such as interviews or questionnaire-based approaches where the participants self-reported the amount of workload caused/induced during the task. Various research groups such as Hart et al. [5] and Malekpour et al. [6] contribute towards the assessment of cognitive workload with the use of subjective methods, primarily in the form of self-assessment questionnaires, like NASA-TLX (National Aeronautics and Space Administration Task Load Index), MCH (Modified Cooper-Harper Scale) and SWAT (Subjective Workload Assessment Test). Such questionnaires generally record the various metrics involved in performing the task, such as demand (mental, physical and temporal), effort, pressure, concentration, frustration, etc., to evaluate their connection with performance during the task. These methods prove to be subjective to the individual participant, however, and can be biased and prove to be unreliable as a distinct and coherent metric for the evaluation and estimation of cognitive workload in general as they depend on the participant recalling past engagement. Another drawback of using post-task questionnaire is that it does not allow for real-time evaluation of cognitive workload.

In contrast to the subjective questionnaire based methods, the evaluation based on neuro-physiological signals present an opportunity for an objective and real time assessment of cognitive workload. However, this method of evaluation comes at the expense of limited availability of equipment, trained operators and high costs. To obtain better efficacy and efficiency, physiological measures such as Electroencephalography(EEG), Event-Related Potential (ERP), Eye Tracking (gaze entropy), and Heart Rate Variability (HRV) can be utilized [7–9]. EEG is highly accepted as a measure to assess cognitive workload in real-time [10–12]. Various EEG features including time, frequency, time-frequency, and spatial domain features extracted from raw EEG data are effective ways to gain information from EEG signals. Time domain features mainly include Event Related Potentials (ERP) [13], statistical features (mean, standard deviation, variance, etc.), higher-order crossing analysis [14], and Hjorth parameter. Frequency domain features include decomposing the frequency in multiple sub-bands such as delta, theta, alpha, beta, and gamma bands which are mainly associated with deep sleep, drowsy, relaxed, engaged, conscious, and active states, respectively [15]. Such features are commonly used for classification of workload in various machine learning experiments. Recent advancements in the application of deep learning in various domains such as emotion recognition, pattern recognition and prediction makes it an excellent choice to be used with EEG signals for classification [16–19]. EEG signals can be used to decode and classify the human cognitive state. Various studies have carried out research in the area with different combinations of EEG features and machine learning models. Bashivan et al. [20] demonstrates the use of fast Fourier transform to convert EEG data into the frequency domain and map the 3D spatial positions of electrodes to 2D, according to the distribution of the electrodes. Using theta, alpha and beta frequency bands, 3-channel spectral maps are generated and sent to CNN model for classification of mental load. Kwak et al. [21] propose a multi-level feature fusion method based on CNN to learn the spectral, spatial, as well as local and global information. Li et al. [22] reviews some deep learning models (e.g., RNN and CNN) and their applications for EEG data to decode brain activities and diagnose brain diseases.

Substantial research for estimation of cognitive workload from EEG using machine learning and deep learning is limited. Most of the studies perform binary classification of workload into high and low by extracting compute expensive EEG features from the raw data, making these non ideal to be used in real life conditions or in real time. Das et al. [23] reports an accuracy of 86.33% and 82.57% for binary and three class classification, respectively, using a BLSTM-LSTM based architecture in a subject independent study. Appriou et al. [24] performs subject specific and subject independent studies for binary classification of workload, achieving the highest mean accuracy of 72.7% and 63.7% using CNN for subject-specific and subject independent cases, respectively. In the study by Zhang et al. [25], the authors achieved an accuracy of 88.9% in binary classification using a combination of RNN and 3D CNN models with EEG topographic maps as features for classification. Using a similar technique of topographic maps in combination with a modified CNN model, highest accuracy of 91.9% in subject specific three class classification is reported [26]. However, more informative features regarding an individual's brain can be obtained from EEG data. Information acquired from signals originating from a specific brain region can be regarded to represent the brain activity of that region. This allows the study of separate brain regions in isolation when evaluating characteristics relevant to a specific cognitive state and this methodology has been adopted by various researchers. However, neuronal activity is not this straightforward as different regions of the brain contribute to the completion of a task, while different regions are still dominantly responsible for specific functions required for the completion of the task. This implores the necessity of examining the inter-regional interactions to understand the collaboration of the different brain regions. More formally, this analysis is termed as brain connectivity.

Brain Connectivity has been used to study the nature of the cerebrum in the past. Based on the attributes of connections, it can be classified into three types: structural connectivity (biophysical connections between neurons or neural elements), functional connectivity (statistical relations between anatomically un-connected cerebral regions) and effective connectivity (directional causal effects from one neural element to another) [27]. This study focuses on the exploration of functional brain connectivity as a measure to assess different levels of workload. Brain functional connectivity has been linked with cognitive deficient psycho-physiological diseases. Strong patters on connectivity in resting state EEG are evident in autism spectrum disorders as reported by [28]. Slower and less efficient connectivity is found in schizophrenia patients as reported by [29]. Another study suggested a relation between high frequency connectivity neural pattern and recurrent illness course of major depressive disorder [30]. However, few studies have investigated the links between cognitive workload and brain functional connectivity networks. Dimitrakopoulos et al. [31] is one such study that has used brain connectivity measure as a feature for classification of workload. This study uses correlation as a method of brain connectivity and achieved an accuracy of 88% for binary classification using SVM classifier. Another study by Islam et al. [32] explores the use of Mutual Information based functional connectivity for binary classification of drivers' mental workload using the SVM classifier and obtained an accuracy of 82%. There are only a limited number of studies that explore functional connectivity as a feature for classification of workload. Therefore, in this study we explore different functional brain connectivity methods as features to be used for classification of levels of cognitive workload. EEG data is known to have high inter-subject variability [33,34]. Various researchers such as Byrne et al. [35] and Pang et al. [36] study the inter-subject variability. Nentwich et al. [37] report the subject-specific nature of EEG-based functional connectivity. Given this evidence, subject specific classification of workload has been aimed at in this study. In Zhang et al. [38], the authors compared the subject-dependent and independent approach and highlighted that variations in feature distribution of EEG across subjects reduces the generalization ability of a classifier and at the same time subject-dependent approach provides a promising way to solve the problem of personalized classification. In Neto et al. [39], the authors discussed various subject specific characteristics and data splitting techniques for EEG data. A possible advantage of subject specific classification is that the classifier can learn subject-dependent features and it can be really useful in building robust and effective BCI systems [40,41].

The contributions of this paper can be summarized as follows:

- A novel method of cognitive workload estimation using EEG, functional brain connectivity and deep learning is proposed. Our pipeline included cleaning 64-channel EEG data, selecting 16 electrodes based on brodmann area, extracting a 16 × 16 connectivity matrix and using deep neural networks for classifying workload into low, medium and high classes.
- We chose model-free functional connectivity metrics (Mutual Information (MI), Phase Lag Value (PLV) and Phase Transfer Entropy (PTE)) to classify workload using simple yet effective deep learning architectures (CNN, LSTM and Conv-LSTM) in near real-time.
- The proposed method achieved state-of-the-art accuracy for three class workload classification. We achieved an average accuracy of 80.87% for three class workload classification problems using MI and CNN. PLV and PTE also perform better with CNN as compared to the other architectures with a average classification accuracy of 74.07% and 71.16%, respectively. CNN outperforms the other architectures because of the high spatial information in the input connectivity matrix.
- The efficacious results highlight the promise of using functional connectivity features of EEG for real-time workload classification.

The rest of the paper is organized as follows. Section 2 presents the materials and methods used for in the experiment. Section 3 discusses the results obtained in various experiments and Section 4 presents the implications of the reported results and the possible future directions and possible extensions of the current work.

2. Materials and Methods

2.1. Participants

A total of 19 participants (11 male and 8 females, mean age = 20.1 years, standard deviation = 1.2 years, minimum age = 19 years, maximum age = 23 years) at the Department of Biomedical Engineering, Institute of Nuclear Medicine and Allied Sciences, Delhi, India participated in this study. An institutional ethical committee approved the study at the Institute of Nuclear Medicine and Allied Sciences. Participation in the study was voluntary, and the subjects gave written consent before participating in the study. Out of 19 participants, 18 participants were right-handed, and one was left-handed. None of the participants reported neurological/psychological/mental history of any kind. All the participants hailed from a Science/Engineering/Technology/Mathematics (STEM) background. All the participants received a flat payment of INR 50, irrespective of their performance in the study.

2.2. The N-Back Task

The modern version of the n-back task [42] was designed using OpenSesame v 3.3.6 [43]. The n-back task is one of the most used psychological tests for inducing cognitive workload. In the task, the participants were required to observe a sequence of single digits separated by a small interval of time and for each letter they were required to identify whether the stimuli are a target (identical of the digit that has appeared 'n' digits back in the sequence) (see Figure 1). During a session/block the value of 'n' is kept constant. An increase in the value of 'n' induced cognitive workload according to [43]. The participants were required to interact with the appeared stimuli depending on the value of 'n'.

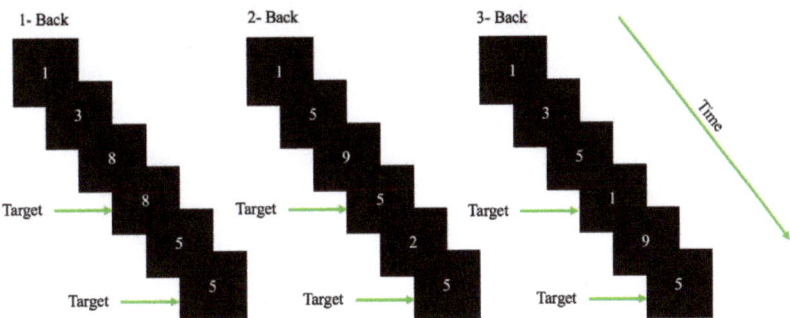

Figure 1. Schematic of the n-back task used for the cognitive workload classification. The participants were required to observe a sequence of single digits and determine whether the stimuli was a target. A target is the digit which is identical to the digit that appeared 'n' digits back in the sequence. For example, in the 2-back scenario 5 is the target as the sequence of digits were 9,5,2,5.

A total of 339 sessions were presented to each participant in a randomized manner with 113 sessions each for 1, 2 and 3 back. The sessions were initialized with an instruction set that was displayed for 5 seconds, where the participants were informed about the nature of session (type of 'n'). After the instruction block, the set of digits (1–9) appeared on the screen in sequence. The digits stayed on the screen for 500 ms, the participants were given 1500 ms to respond. The participants had to press space-bar in case the digit appeared was a target in accordance with the session. The inter-stimulus interval was 2000 ms (with 500 ms where the stimuli was displayed and 1500 ms given for response). The task was designed in accordance with standard n-back format. The n-back stimuli occurred within a visual angle of about 40° horizontally and about 4.50° vertically so the stimuli fall within the participants' visual field and for minimal eye movement. The stimuli were presented using OpenSesame [43], an open-source experiment builder. The target missed was also considered as an incorrect response in this case. The first three session of each conditions (n-back) were removed from further data analysis.

2.3. Physiological Data Acquisition and Pre-Processing

Sixty-four channel EEG were recorded through Ag/AgCl electrodes conforming with the extended 10–20 electrode system of placement. An eego™mylab amplifier (ANT Neuro, Enschede, The Netherlands) was used in the data acquisition. Electrooculogram (EOG) data was acquired from a single electrode placed below the right eye. All channels were grounded to channel CPz. Impedances were kept below 20 kΩ. The EEG data were sampled at 2048 Hz. The data were later downsampled to 256 Hz. During the recording process the participants were requested to sit in a relaxed posture to avoid potential contamination of data with movement artifacts. The data was referenced to linked mastoids in the further analyses. For pre-processing, DC offset was applied followed by band-pass with 0.1–45 Hz and finally we used ICA to get rid of the ocular and other artifacts. The data was then segmented according to the three conditions (1, 2 and 3 back) for all the 19 subjects.

2.4. Feature Extraction

Different cognitive tasks activate different specialized brain areas where the brain could dynamically coordinate the information flow to achieve the task [44]. Functional Connectivity is a method of quantifying these neuronal interactions. There exist many different algorithms for calculating these interactions using electrophysiological data. These algorithms can be divided into different domains based on the direction of the interaction among brain regions and interdependence of the signals [45]. In this study, we chose three connectivity metrics namely Mutual Information (MI), Phase Locking Value (PLV) and Phase Transfer Entropy (PTE). The reason for choosing these three metrics was to compare directed and non-directed model-free measures. One goal of the study was to build a near real-time framework for workload estimation using EEG, which is why only model-free connectivity measures were chosen. Therefore, we used only the raw (cleaned) EEG data to calculate the metrics.

Another important aspect for making the system fast was to select the dimensions of the connectivity matrix. To that end, 16 electrodes were chosen from the available 64. Choosing the 16 electrodes was done with brodmann areas in mind as functional connectivity implies interaction between different brain regions. In his article, Kaiser [46] defined a mapping between the EEG electrodes and different brodmann areas; therefore, we selected the same 16 EEG electrodes. The electrodes were Fp1, Fp2, F7, F3, F4, F8, T7, C3, C4, T8, P7, P3, P4, P8, O1 and O2. The closest associated brodmann areas with these electrodes are 10, 10, 47, 8, 8, 45, 42, 2, 1, 21, 37, 39, 39, 37, 18 and 18, respectively. This electrode placement is also supposed to be the most optimal for source localization [46]. We used the pre-processed EEG data to calculate these 16 × 16 functional connectivity metrics. Next, the different connectivity measures are discussed.

2.4.1. Mutual Information (MI)

In information theory, MI is used to quantify the interdependence between two time series [47]. For a pair of discretized random variables x and y that are recorded from time series with their respective probability distribution functions $P(x)$ and $P(y)$, and joint probability function $P(x,y)$, the MI between x and y can be defined as:

$$MI_{xy} = \sum_{x \in X, y \in Y} P(x,y) \log \frac{P(x,y)}{P(x)P(y)}. \tag{1}$$

MI was proposed as a measure to quantify the strength of functional connectivity between a pair of time series data.

2.4.2. Phase Locking Value (PLV)

Phase locking value (PLV) is a measure to quantify the synchronization of phase of different signals as acquired from separate brain areas. The analytical representations of two signals originating from brain regions, k and l, $s_k(t)$ and $s_l(t)$, are obtained by the Hilbert transform and expressed as [48,49]:

$$z_k = A_k(t) e^{j\varphi_k(t)}, \tag{2}$$

$$z_l = A_l(t) e^{j\varphi_l(t)}, \tag{3}$$

The differences in phase are then calculated at each time point by

$$\Delta \varphi_{k,l}(t) = \varphi_k(t) - \varphi_l(t). \tag{4}$$

Thereafter, by averaging over all time points (n_t being the number of time points) the PLV between the brain regions k and l is represented as:

$$PLV(k,l) = \frac{1}{n_t} \left| \sum_{t=1}^{n_t} e^{j\Delta \varphi_{k,l}(t)} \right|, \tag{5}$$

The PLV ranges between 0 (which reflects no phase synchronization) and 1 (which reflects perfect phase synchronization). After the PLV calculation is repeated for all brain regions, it is assembled to form a connectivity matrix.

2.4.3. Phase Transfer Entropy (PTE)

The flow of information between neuronal regions are quantified by the estimation of causal influence one region exercise on another. There is a plethora of methods to quantify the neuronal interactions, out of which PTE is the only measure that is phase-specific and directed in nature. For a connectivity metric to quantify the interactions amicably it should:

1. be robust to noise and linear mixing of signals [50,51]
2. computationally efficient
3. limit the number of a priori parameters
4. be able to detect transient frequency band from short data samples
5. allow the testing of statistical significance by constructing surrogate data from the experimental samples

PTE [52] is a method of quantifying directed phase interaction across trials as well as continuous data using binning methods for state-space reconstruction based on the same principle as Wiener-Granger causality [53]. In the framework of Information Theory, the Wiener-Granger causality can be re-written as: "a source signal has causal influence on the target signal, if the uncertainty of the target signal conditioned by the source signal and its own past is smaller than the uncertainty of the target signal conditioned by its own past" [54]. The instantaneous phase and amplitude of a signal $x(t)$ can be expressed by its analytic associate as expressed in Equation (1). The PTE for an analysis lag θ can be defined as:

$$PTE_{XY} = H(\varphi_y(t), \varphi_y(t')) + H(\varphi_y(t'), \varphi_x(t')) - H(\varphi_y(t')) - H(\varphi_y(t), \varphi_y(t'), \varphi_x(t')), \quad (6)$$

where $\varphi_x(t')$ and $\varphi_y(t')$ are the past states at lag θ, i.e., $\varphi_x(t') = \varphi_x(t - \theta)$ and $\varphi_y(t') = \varphi_y(t - \theta)$. The marginal and the joint entropies can then be defined as [55]:

$$H(\varphi_y(t), \varphi_y(t')) = -\sum p(\varphi_y(t), \varphi_y(t')) \log p(\varphi_y(t), \varphi_y(t')), \quad (7)$$

$$H(\varphi_y(t'), \varphi_x(t')) = -\sum p(\varphi_y(t'), \varphi_x(t')) \log p(\varphi_y(t'), \varphi_x(t')), \quad (8)$$

$$H(\varphi_y(t')) = -\sum p(\varphi_y(t')) \log p(\varphi_y(t')), \quad (9)$$

$$H(\varphi_y(t), \varphi_y(t'), \varphi_x(t')) = -\sum p(\varphi_y(t), \varphi_y(t'), \varphi_x(t')) \log p(\varphi_y(t), \varphi_y(t'), \varphi_x(t')), \quad (10)$$

where the probabilities are computed by histograms of occurrences of single, pairs or triplets of phase estimates in an epoch. The prediction delay θ and the number of bins in the histogram was set as $((L \times CH))/N_\pm$ and $e^{0.626+0.4\ln(L-\theta-1)}$ respectively, where L is the length of the epoch in sample count, CH is the number of channels and $N_\pm$ is the number of times the phase changed its sign across time and channels. The PTE values were normalized between 0 and 1 with $0.5 < PTE_{xy} < 0.5$ implying an information flow of $x \rightarrow y$, $0 < PTE_{xy} < 0.5$ implying information flow preferentially of $x \leftarrow y$ and 0.5 implying no preferential flow of information.

2.5. Classification

The classification of workload is implemented using three different variants of convolution and recurrent neural networks that provide different feature extraction and learning capabilities and a comparison of the performance is presented. The input to all the three networks were the connectivity matrices MI, PLV and PTE as described above. The shape of each of the matrix was 16×16. The networks were trained using Python 3.9 and Tensorflow 2.4 on Nvidia DGX server at Indian Institute of Technology, Roorkee. For processing the input and feeding it to the model, we used Tensorflow Datasets API and used 70,15,15 split for training, validation and testing data. As mentioned earlier, the n-back task was composed of 339 sessions, hence, we calculated a matrix corresponding to each session giving rise to 339 matrices for each participant. With the split of 70-15-15, there were 237, 51 and 51 matrices for training, validation and testing, respectively, for each of the 19 subjects. We used a batch size of 64 trained each model for 1000 epochs. During the training, early stopping [56] and learning rate scheduler [57] were used to improve the convergence time. The motivation and details of the networks used are as follows; The CNN classifier [58] was chosen based on the similarity that the input (which is a weighted square adjacency matrix) has to an image, as it's ability to extract spatial features is superior unlike the primitive ANNs. We used a Regular CNN (Table 1) (consisting of the usual 2D convolution, pooling and batchnorm layers). For all the convolution layers of the models, stride of 1, 'same' padding, and ReLU [59] as activation was used. The last dense layer consisted of 3 units and softmax activation [60] for classifying the three levels of workload. Similarly, in LSTM (Table 2), the input was flattened and all LSTM layers make use of ReLU activation. In Conv-LSTM (Table 3), all Conv2D layers have ReLU activation. After reshaping the output, they are followed by LSTM layers, followed by 2 dense layers and a softmax layer same as the above models. The overview of the classification framework can be visualized as shown in Figure 2. Additionally, Figure 3 shows the architecture of the CNN, LSTM and the Conv-LSTM models used.

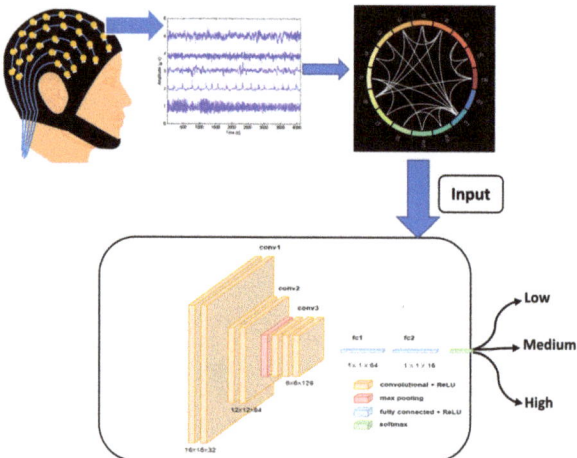

Figure 2. Overview of the classification workflow using EEG signals.

Table 1. Configuration of CNN Architectures used for the ablation study. C-A, C-B and C-C refers to the three variations of CNN Networks. The bottom half of the table is common to all the three variations.

C-A	C-B	C-C
Input [16, 16, 1]	Input [16, 16, 1]	Input [16, 16, 1]
Conv2D (32, 5 × 5)	Conv2D (32, 5 × 5)	Conv2D (32, 5 × 5)
Conv2D (64, 3 × 3)	Conv2D (64, 3 × 3)	Conv2D (64, 5 × 5)
MaxPooling (2 × 2)	MaxPooling (2 × 2)	MaxPooling (2 × 2)
Conv2D (128, 5 × 5)	Conv2D (128, 5 × 5)	Conv2D (128, 3 × 3)
	Conv2D (128, 5 × 5)	
Flatten		
Dense (64)		
Dropout (0.25)		
Dense (16)		
Dense (3)		

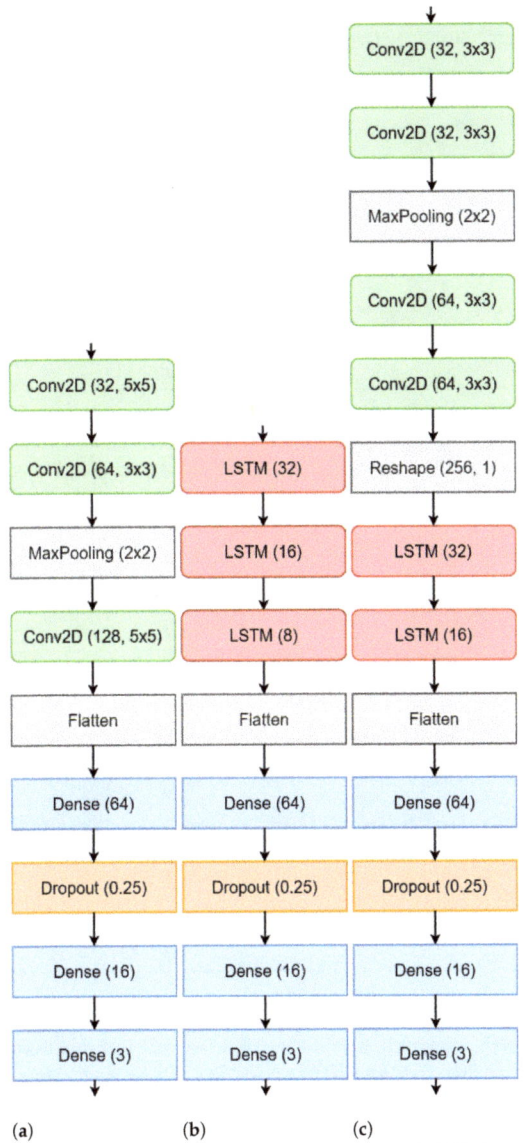

Figure 3. Model architectures for (**a**) CNN C-A (**b**) LSTM L-A (**c**) Conv-LSTM CL-A.

Table 2. Configurations of LSTM Architectures used for the ablation study. L-A, L-B and L-C refers to the three variations of LSTM Networks. The bottom half of the table is common to all three variations.

L-A	L-B	L-C
Input [256, 1]	Input [256, 1]	Input [256, 1]
	LSTM (64)	LSTM (64)
LSTM (32)	LSTM (32)	LSTM (32)
LSTM (16)	LSTM (16)	LSTM (16)
LSTM (8)	LSTM (8)	LSTM (16)
Flatten		
Dense (64)		
Dropout (0.25)		
Dense (16)		
Dense (3)		

Table 3. Configuration of Conv-LSTM Architectures used for the ablation study. CL-A, CL-B and CL-C refers to the three variations of Conv-LSTM Networks. The bottom half of the table is common to all the three variations.

CL-A	CL-B	CL-C
Input [16, 16, 1]	Input [16, 16, 1]	Input [16, 16, 1]
Conv2D (32, 3 × 3)	Conv2D (16, 3 × 3)	Conv2D (32, 3 × 3)
Conv2D (32, 3 × 3)	Conv2D (16, 3 × 3)	Conv2D (32, 3 × 3)
MaxPooling (2 × 2)	MaxPooling (2 × 2)	MaxPooling (2 × 2)
Conv2D (64, 3 × 3)	Conv2D (64, 3 × 3)	Conv2D (64, 3 × 3)
Conv2D (64, 3 × 3)	Conv2D (64, 3 × 3)	Conv2D (64, 3 × 3)
Reshape (256, 1)	Reshape (256, 1)	Reshape (256, 1)
LSTM (32)	LSTM (64)	LSTM (64)
LSTM (16)	LSTM (16)	LSTM (32)
		LSTM (16)
Flatten		
Dense (64)		
Dropout (0.25)		
Dense (16)		
Dense (3)		

3. Results and Discussion

In this research, the efficacy of three different functional brain connectivity analysis methods (MI, PLV and PTE) to classify cognitive workload into high, medium and low using three different deep learning architectures (CNN, LSTM and Conv-LSTM) was investigated. Nineteen participants executed the the modern version of the n-back task on a computer screen with three levels of cognitive workload, high, medium and low.

The input to the deep learning networks was 16 × 16 connectivity metrics. Sixteen brain regions were chosen from the brodmann atlas [61] to cover the different brain regions and at the same time keep the computations as fast as possible. Figure 4 shows the differences (for a random participant) between low, medium and high workloads of MI, PTE and PLV, respectively. Although the differences among the three connectivity metrics are visible, there are no explicit and visible differences among the three workload conditions, i.e., low, medium and high.

However, in the statistical analysis, significant differences were found among the three conditions. The mean accuracy (in percentage) for the three n-back condition was- 75.42 (SD = 16.10), 62.27 (SD = 15.64), 37.84 (SD = 14.18) for 1-back, 2-back and 3-back, respectively. There were significant differences among the groups ($F(2, 75) = 40.22$, $p < 0.01$, $\eta^2 = 0.56$). Similarly we found significant differences in the reaction time as well (1-back =

492.58 (SD = 91.1), 2-back = 673.58 (SD = 150.57), 3-back = 824.84 (SD = 147.32), ANOVA = $F(2, 75) = 40.98$, $p < 0.01$, $\eta^2 = 0.48$). Differences between all possible combinations (1 vs. 2, 1 vs. 3, 2 vs. 3) across both mean accuracy (in percentage) and mean reaction time (in ms) were also found to be significant ($p < 0.01$).

Based on the statistical results, we hypothesized that there will be differences in the brain connectivity matrices (although not visible to the naked eye) in the three workload settings and the deep learning classifiers will be able to utilize these differences for successful classification. It was expected that PTE would perform best in terms of connectivity metric, with it being directed and phase-specific.

Several experiments (ablation study) were performed to find best hyperparameter settings for the three deep learning architectures. The results of the ablation study are compiled in Table 4. As shown in Table 4, for MI, a mean accuracy of 80.87% was achieved with CNN, 71.87% was achieved with LSTM and 71.16% was achieved with Conv-LSTM. Similarly, for PLV a mean accuracy of 75.88% was achieved with CNN, 71.82% was achieved with LSTM and 69.68% was achieved with Conv-LSTM. Lastly, for PTE a mean accuracy of 71.16% was achieved with CNN, 69.63% was achieved with LSTM and 69.74% was achieved with Conv-LSTM. The highest accuracy (among all subjects) was achieved with the combination of PLV with Conv-LSTM and CNN at 97.92%. This is followed by MI with CNN at 95.83%. Besides the accuracy, Precision, Recall and F1-score of the classifiers are also reported in Table 5. Figure 5 shows the box-plot containing the accuracy and statistical results (standard error, quartiles, and outliers) of all the classifiers in combination with different functional connectivity methods. The combination of CNN and MI indicates the best classification performance. The achieved accuracy outperforms the state-of-the-art in multi-class classification in the context of workload classification in the n-back task with various EEG features and machine-learning algorithms. The comparison of the proposed method with others is given in Table 6. Since, the number of trials for the three workload settings were balanced, accuracy was indicative of the performance of the classifiers. Nevertheless, we reinforced the results with the analysis of the confusion matrices and ROC curves. Figure 6 shows the confusion matrix and Figure 7 shows the ROC curves for all combinations of the classifiers and the connectivity metrics of the best subject. From these figures, it can be substantiated that the classification performance of the models is high for the multiclass-classification problem as the true positive rate is high. The high value class-wise area under the curve shows that the classifier is able to learn and classify each class separately with high accuracy.

Figure 8 shows the features learned by the CNN when MI was given as an input. MI was chosen as it gave the highest accuracy and similarly, input image of medium workload was chosen since the recall of medium workload was highest. It is visible that the filters are actually learning similar activation as in the input image indicating that the classifier was successful. Overall, given the consistent performance of the classifiers across all the metrics and the significant differences found in the statistical tests, it can be concluded that the classifier was successful.

Although state-of-the-art results were obtained, the study had some limitations. One important limitation of the study is the hypothesis itself. We hypothesized that there will be differences in the connectivity matrices in the three workload conditions. However, the study was limited to calculating the connectivity using raw(cleaned) EEG data. This was done to test whether all inclusive connectivity (not band limited) would yield conceivable differentiation in workload or not. This would have implications in making the entire framework close to real-time since band-limiting the signals would have increased the computational complexity. In the future we will consider doing a comparison with our approach and investigations in connectivity with different frequency bands to make a comprehensive and exhaustive hypothesis. Another limitation was the subject-dependent classification. The subject-dependent classifiers can extract subject-dependent features and can effectively tackle the issue of accuracy and generalization encountered in subject-independent EEG classifiers. However, it also gives rise to the issues of long collaboration sessions and collection of large quantities of data [38,39]. Lastly, the choice of 16 brain regions for computing the connectivity matrices. The choice of the brain regions could have been empirical instead of hypothesis and use-case driven. Exhaustive search and feature selection algorithms could be used in the future for validating the selection of brain regions empirically.

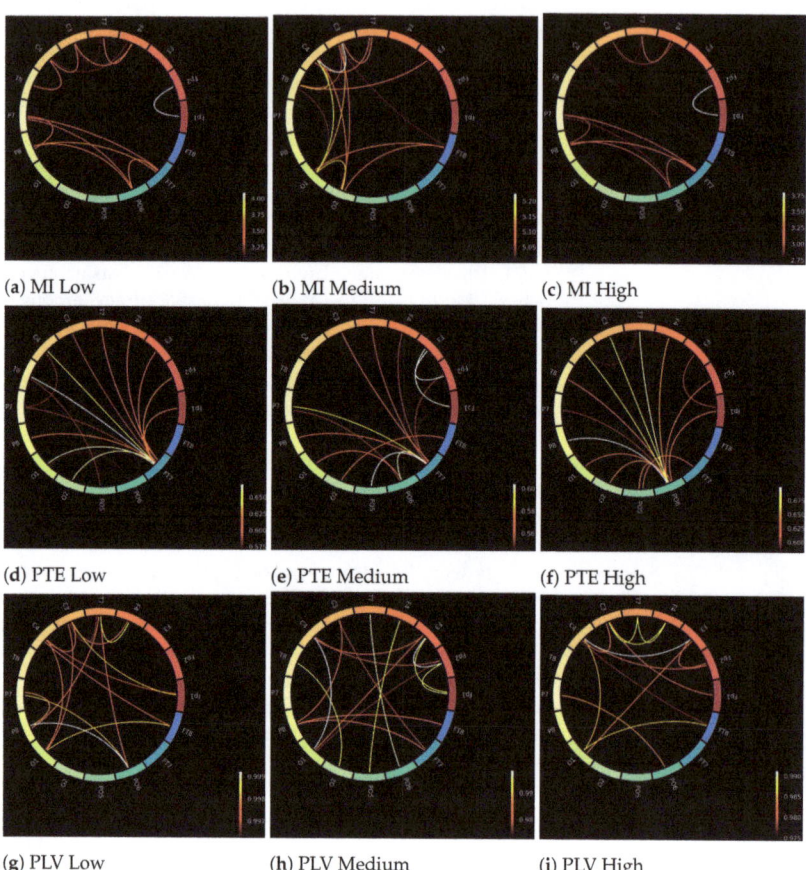

Figure 4. Brain connectivity maps of a random subject obtained through MI, PTE, and PLV for different workload states (low, medium, and high) using Brodmann atlas [61].

Table 4. Ablation Study of different variations of the hyper-parameter combinations for used classifiers as described in Tables 1–3.

Methods	Best Subject			Average Accuracy ± Std. Dev.		
	MI	PLV	PTE	MI	PLV	PTE
CNN						
C-A	93.75	89.58	85.42	80.87 ± 10.24	74.07 ± 08.28	71.16 ± 06.38
C-B	91.67	89.58	83.33	80.87 ± 10.29	71.49 ± 10.85	71.05 ± 10.85
C-C	95.83	97.92	79.17	80.21 ± 11.26	75.88 ± 11.01	70.72 ± 05.34
LSTM						
L-A	87.50	91.67	79.17	71.87 ± 06.56	71.82 ± 08.15	69.63 ± 05.66
L-B	85.42	79.17	81.25	69.52 ± 07.77	65.24 ± 07.79	67.00 ± 08.47
L-C	87.50	89.58	79.17	70.29 ± 07.30	69.41 ± 08.30	67.76 ± 06.80
Conv-LSTM						
CL-A	93.75	97.92	81.25	71.16 ± 10.03	69.68 ± 10.46	67.32 ± 05.05
CL-B	87.50	87.50	79.17	70.61 ± 08.27	68.64 ± 07.23	68.09 ± 04.73
CL-C	91.67	89.58	79.17	67.49 ± 07.12	67.87 ± 07.50	69.74 ± 05.54

Table 5. Precision, recall and F1-score for the different architectures used in the ablation study as described in Tables 1–3.

Methods	Precision			Recall			F1-Score		
	MI	PLV	PTE	MI	PLV	PTE	MI	PLV	PTE
CNN									
C-A	94.31	88.79	86.93	94.23	88.46	84.62	94.22	88.44	84.07
C-B	92.39	89.74	81.44	92.31	88.46	80.77	92.19	88.35	80.45
C-C	96.54	98.18	79.33	96.15	98.08	78.85	96.13	98.08	78.74
LSTM									
L-A	87.09	92.63	77.35	86.54	92.31	76.92	86.40	92.27	76.54
L-B	84.51	80.33	80.00	84.62	80.33	83.00	84.36	80.33	83.00
L-C	91.35	90.33	80.00	88.46	90.33	78.66	88.05	90.33	78.33
Conv-LSTM									
CL-A	95.05	98.18	81.44	94.23	98.08	80.77	94.17	98.07	80.45
CL-B	88.46	87.21	80.12	88.46	86.54	78.85	88.46	86.47	78.50
CL-C	90.48	90.44	78.85	90.38	90.38	78.85	90.38	90.26	78.81

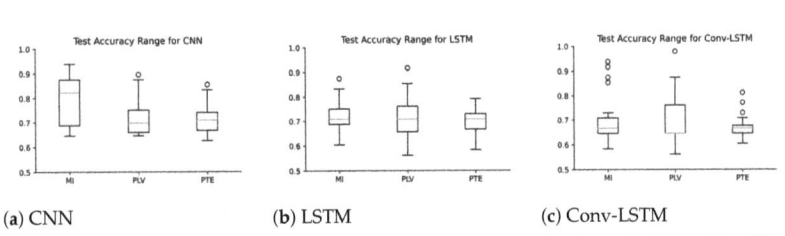

(a) CNN (b) LSTM (c) Conv-LSTM

Figure 5. Box Plots representing the range of accuracy (with standard error) achieved by different subjects with deep learning architectures used (a) CNN (b) LSTM and (c) Conv-LSTM.

Table 6. Comparison of the proposed work with state-of-the-art results. The comparison includes different features and classifiers used for EEG-based cognitive workload classification in the n-back task. The proposed work achieves the highest accuracy in multi-class classification.

Paper	Feature	Classifier	Accuracy	Subject Dependency	Number of Classes
Appriou et al. [24]	Preprocessed EEG	CNN	72.7% 63.7%	Subject Specific Subject Independent	2 Classes
Dimitrakopoulous et al. [31]	Functional Connectivity (Pearson Correlation)	SVM classifier (RBF kernel and Least Squares Learning Method)	88%	Subject Independent	2 Classes
Zhang et al. [25]	Topographic Maps	RNN and 3D CNN structures (R3DCNN)	88.9%	Subject Independent	2 Classes
Zhang et al. [26]	Topographic Maps	Modified CNN	91.9%	Subject Specific	3 Classes
Proposed	Functional Connectivity (PLV)	Conv-LSTM, CNN	97.92%	Subject Specific	3 Classses

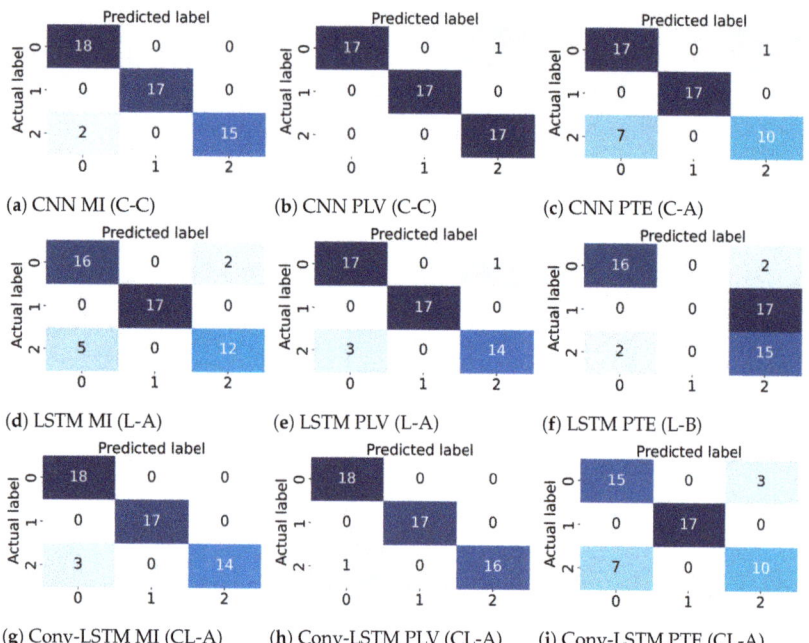

Figure 6. Confusion Matrix for the best performing subject for different combinations of the deep learning architectures (CNN, LSTM, and Conv-LSTM) and the functional connectivity metrics (MI, PLV and PTE).

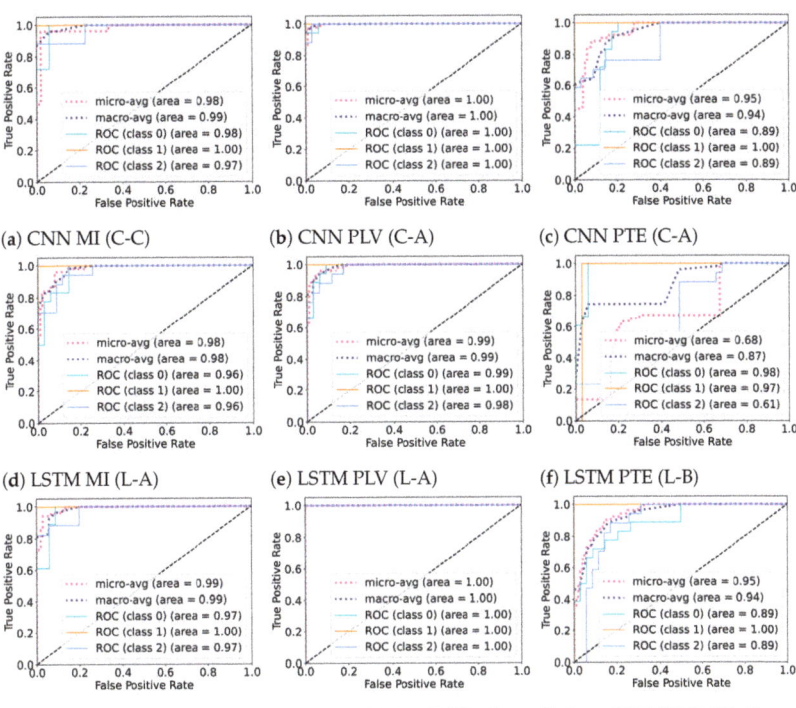

Figure 7. ROC (Receiver Operating Characteristics) curves for the best performing subject for different combinations of the deep learning architectures (CNN, LSTM, and Conv-LSTM) and functional connectivity metrics (MI, PLV and PTE).

(a) Medium Workload MI matrix   (b) 64 Filters of the 2nd Conv2D layer.

Figure 8. (a) Input given to the CNN network (b) Visualization of feature maps of the convolution layer in the CNN network.

4. Conclusions

Workload Classification can be used as an indicator of the Emotional Intelligence and stability. The aim of the study was to build a fast and accurate workload classifier which can be extended to real-time workload classification. Real-time workload classification is an important and very useful cognitive construct for the development of robust BCI systems [62] and useful in several other domains like Virtual Reality [63] and Human-Machine Teaming [64]. In this research, EEG was chosen as the neuroimaging modality with its advantages of being cheap, portable and having high time resolution [65]. Model-free functional connectivity was chosen for the feature extraction with the concomitant advantages of being fast and associated with cognitive control in the context of mental workload [66]. Also, it has been shown that there are subject-specific differences in EEG-based functional connectivity measures [37].

Thereby, a combination of various directed/non-directed model-free brain functional connectivity algorithms and state-of-the-art deep learning algorithms were utilized for efficient subject-specific classification of cognitive workload into three levels, high, medium and low. Three functional brain connectivity algorithms (Mutual Information, Phase Transfer Entropy and Phase Locking Value) were used to generate the functional connectivity networks, which represents the neuronal interactions between the different regions of the brain. These connectivity networks are used as inputs to the classification models to classify different levels of workload. We employed three different deep learning architectures (CNN, LSTM and Conv-LSTM) for classification of cognitive workload. Intra-subject method of classification was applied on the data of 19 participants. The best classification performance was obtained with CNN in combination of each of the three connectivity networks over LSTM and Conv-LSTM. CNN outperforms the other two deep learning architectures because of the spatial information provided by the connectivity analysis in the form of input data upon which the classification is being performed. With CNN, MI produces the best classification results with an accuracy of 80.87%, followed by CNN with PLV with an accuracy of 75.88% and LSTM with MI with an accuracy of 71.87%.

We achieved state-of-the-art accuracy for multi-class workload classification using EEG and functional connectivity. From the results, it can be concluded that indeed EEG-based model-free functional connectivity metrics, when combined with deep-learning, provides an accurate, reliable and fast method of classifying cognitive workload. Although there is not much literature available on this, it was hypothesized that the connectivity method PTE will outperform MI and PLV as PTE is the only connectivity measure that is phase-specific and directed in nature. However, in our experiments MI outperformed PTE in the classification performance. This can be due to the fact that this study had lesser number of participants' and the choice of brain regions. Therefore, no significant conclusions can be made about which model-free connectivity measure is the best. A future study can be performed with higher number of participants and different permutations and combinations of brain regions to make better and clear conclusions regarding the comparative analysis of the different connectivity measures.

Since these brain connectivity methods enable extremely rapid (specially MI) and accurate connectivity matrix generation from raw EEG data, the proposed architecture (a combination of MI/PLV/PTE and state-of-the-art CNN) can be used for effective and efficient cognitive state monitoring and other BCI applications. In addition to that, brain connectivity coupled with hybrid deep learning architectures can be used to classify higher-order cognitive processes like executive functioning and complex decision-making in the future. The subject-specific classification also sanctions the analysis and extraction of subject-specific features. Together, this could enable BCIs to become more reliable and efficient exponents of effective state monitoring in complex real world scenarios.

Author Contributions: Conceptualization, A.G., G.S. and V.P.; methodology, G.S.; software, A.G.; validation, G.S. and V.P.; formal analysis, A.G.; investigation, G.S. and V.P.; resources, P.P.R.; writing—original draft preparation, G.S. and V.P.; writing—review and editing, A.G., P.P.R. and B.-G.K.; visualization, G.S. and V.P.; supervision, P.P.R. and B.-G.K.; project administration, A.G. All authors have read and agreed to the published version of the manuscript.

Funding: This research received no specific grant from any funding agency in the public, commercial, or not-for-profit sectors.

Institutional Review Board Statement: The study was conducted according to the guidelines of the Declaration of Helsinki, and approved by the Institutional Review Board (IRB) at the Institute of Nuclear Medicine and Allied Sciences (INMAS), Defence R & D Organization, Delhi.

Informed Consent Statement: All the participants had provided written informed consent for taking part in the study.

Data Availability Statement: The raw data will be made available on request by the authors, without undue reservation.

Acknowledgments: Our sincere thanks to Sushil Chandra, Head of Department of Biomedical Engineering, Institute of Nuclear Medicine and Allied Sciences, Defence Research and Development Organization for his invaluable guidance and support. We extend our gratitude to him for sharing the data so this research study could be conducted.

Conflicts of Interest: The authors declare no conflict of interest.

References

1. Paas, F.; Renkl, A.; Sweller, J. Cognitive load theory and instructional design: Recent developments. *Educ. Psychol.* **2003**, *38*, 1–4. [CrossRef]
2. Burgess, N.; Hitch, G. Computational models of working memory: Putting long-term memory into context. *Trends Cogn. Sci.* **2005**, *9*, 535–541. [CrossRef] [PubMed]
3. Ojha, A.; Ervas, F.; Gola, E. Emotions as Intrinsic Cognitive Load: An Eye Movement Analysis of High and Low Intelligent Individuals. In Proceedings of the 3rd IEEE International Conference on Cybernetics, Exeter, UK, 21–23 June 2017 pp. 1–6. [CrossRef]
4. José, G.C.M.; Rosario, C.; Pablo, F.B. The Relationship between Emotional Intelligence and Cool and Hot Cognitive Processes: A Systematic Review. *Front. Behav. Neurosci.* **2016**, *10*, 101–114 [CrossRef]
5. Hart, S.G. NASA-task load index (NASA-TLX); 20 years later. In *Proceedings of the Human Factors and Ergonomics Society Annual Meeting*; Sage Publications: Los Angeles, CA, USA, 2006; Volume 50, pp. 904–908.
6. Malekpour, F.; Mohammadian, Y.; Malekpour, A.; Mohammadpour, Y.; Sheikh Ahmadi, A.; Shakarami, A. Assessment of mental workload in nursing by using NASA-TLX. *Nurs. Midwifery J.* **2014**, *11*. Available online: http://unmf.umsu.ac.ir/article-1-1699-en.html (accessed on 1 October 2021).
7. Jaquess, K.J.; Gentili, R.J.; Lo, L.C.; Oh, H.; Zhang, J.; Rietschel, J.C.; Miller, M.W.; Tan, Y.Y.; Hatfield, B.D. Empirical evidence for the relationship between cognitive workload and attentional reserve. *Int. J. Psychophysiol.* **2017**, *121*, 46–55. [CrossRef]
8. Causse, M.; Fabre, E.; Giraudet, L.; Gonzalez, M.; Peysakhovich, V. EEG/ERP as a measure of mental workload in a simple piloting task. *Procedia Manuf.* **2015**, *3*, 5230–5236. [CrossRef]
9. Mansikka, H.P. Fighter Pilots' Mental Workload and Performance: A Comparison of Simulated Instrument Approaches and Air Combat. Ph.D. Thesis, Coventry University, Coventry, UK, 2016. Available online: https://pureportal.coventry.ac.uk/en/studentTheses/fighter-pilots-performance-and-mental-workload (accessed on 1 October 2021).
10. Matthews, G.; Reinerman-Jones, L.E.; Barber, D.J.; Abich, J., IV. The psychometrics of mental workload: Multiple measures are sensitive but divergent. *Hum. Factors* **2015**, *57*, 125–143. [CrossRef]
11. Hopstaken, J.F.; Van Der Linden, D.; Bakker, A.B.; Kompier, M.A. A multifaceted investigation of the link between mental fatigue and task disengagement. *Psychophysiology* **2015**, *52*, 305–315. [CrossRef]
12. Wascher, E.; Rasch, B.; Sänger, J.; Hoffmann, S.; Schneider, D.; Rinkenauer, G.; Heuer, H.; Gutberlet, I. Frontal theta activity reflects distinct aspects of mental fatigue. *Biol. Psychol.* **2014**, *96*, 57–65. [CrossRef]
13. Monitoring Pilot's Mental Workload Using ERPs and Spectral Power with a Six-Dry-Electrode EEG System in Real Flight Conditions. *Sensors* **2019**, *19*, 1324. [CrossRef]
14. Zhang, L.; Wade, J.; Bian, D.; Fan, J.; Swanson, A.; Weitlauf, A.; Warren, Z.; Sarkar, N. Cognitive Load Measurement in a Virtual Reality-Based Driving System for Autism Intervention. *IEEE Trans. Affect. Comput.* **2017**, *8*, 176–189. [CrossRef]
15. Kabbara, A. Brain Network Estimation from Dense EEG Signals: Application to Neurological Disorders. Ph.D. Thesis, Université Rennes 1, Rennes, France, 2018. Available online: https://tel.archives-ouvertes.fr/tel-01943768/ (accessed on 1 October 2021).
16. Choi, Y.J.; Lee, Y.W.; Kim, B.G. Residual-based Graph Convolutional Network (RGCN) for Emotion Recognition in Conversation (ERC) for Smart IoT. *Big Data* **2021**, *9*, 279–288. [CrossRef]
17. Chhetri, M.; Kumar, S.; Roy, P.P.; Kim, B.G. Deep BLSTM-GRU Model for Monthly Rainfall Prediction: A Case Study of Simtokha, Bhutan. *Remote Sens.* **2020**, *12*, 3174. [CrossRef]
18. Jeong, D.; Kim, B.G. Suh-Yeon Dong, Deep Joint Spatiotemporal Network (DJSTN) for Efficient Facial Expression Recognition. *Sensors* **2020**, *20*, 1936. [CrossRef]
19. Ji-Hae, K.; Byung-GYU, K.; Pratim, P.; Roy, D.M. Efficient Facial Expression Recognition Algorithm Based on Hierarchical Deep Neural Network Structure. *IEEE Access* **2019**, *7*, 41273–41285.
20. Bashivan, P.; Rish, I.; Yeasin, M.; Codella, N. Learning Representations from EEG with Deep Recurrent-Convolutional Neural Networks. *arXiv* **2015**, arXiv:1511.06448.

21. Kwak, Y.; Kong, K.; Song, W.J.; Min, B.-G.K.; Kim, S.E. Multilevel Feature Fusion with 3D Convolutional Neural Network for EEG Based Workload Estimation. *IEEE Access* **2020**, *8*, 16009–16021. [CrossRef]
22. Li, G.; Lee, C.; Jung, J.; Youn, Y.; Camacho, D. Deep learning for EEG data analytics: A survey. *Concurr. Comput. Pract. Exp.* **2020**, *e519*, e5199. [CrossRef]
23. Das Chakladar, D.; Dey, S.; Roy, P.P.; Dogra, D.P. EEG-based mental workload estimation using deep BLSTM-LSTM network and evolutionary algorithm. *Biomed. Signal Process. Control.* **2020**, *60*, 101989. [CrossRef]
24. Appriou, A.; Cichocki, A.; Lotte, F. Towards robust neuroadaptive HCI: Exploring modern machine learning methods to estimate mental workload from EEG signals. In Proceedings of the Extended Abstracts of the 2018 CHI Conference on Human Factors in Computing Systems, Montreal, ON, Canada, 21–26 April 2018; pp. 1–6.
25. Zhang, P.; Wang, X.; Zhang, W.; Chen, J. Learning Spatial-Spectral-Temporal EEG Features With Recurrent 3D Convolutional Neural Networks for Cross-Task Mental Workload Assessment. *IEEE Trans. Neural Syst. Rehabil. Eng.* **2019**, *27*, 31–42. [CrossRef] [PubMed]
26. Zhang, P.; Wang, X.; Chen, J.; You, W.; Zhang, W. Spectral and temporal feature learning with two-stream neural networks for mental workload assessment. *IEEE Trans. Neural Syst. Rehabil. Eng.* **2019**, *27*, 1149–1159. [CrossRef]
27. Rubinov, M.; Sporns, O. Complex network measures of brain connectivity: Uses and interpretations. *Neuroimage* **2010**, *52*, 1059–1069. [CrossRef] [PubMed]
28. Murias, M.; Webb, S.J.; Greenson, J.; Dawson, G. Resting state cortical connectivity reflected in EEG coherence in individuals with autism. *Biol. Psychiatry* **2007**, *62*, 270–273. [CrossRef] [PubMed]
29. Yin, Z.; Li, J.; Zhang, Y.; Ren, A.; Von Meneen, K.M.; Huang, L. Functional brain network analysis of schizophrenic patients with positive and negative syndrome based on mutual information of EEG time series. *Biomed. Signal Process. Control.* **2017**, *31*, 331–338. [CrossRef]
30. Whitton, A.E.; Deccy, S.; Ironside, M.L.; Kumar, P.; Beltzer, M.; Pizzagalli, D.A. EEG source functional connectivity reveals abnormal high-frequency communication among large-scale functional networks in depression. *Biol. Psychiatry. Cogn. Neurosci. Neuroimaging* **2018**, *3*, 50. [PubMed]
31. Dimitrakopoulos, G.N.; Kakkos, I.; Dai, Z.; Lim, J.; deSouza, J.J.; Bezerianos, A.; Sun, Y. Task-independent mental workload classification based upon common multiband EEG cortical connectivity. *IEEE Trans. Neural Syst. Rehabil. Eng.* **2017**, *25*, 1940–1949. [CrossRef]
32. Islam, M.; Barua, S.; Ahmed, M.; Begum, S.; Aricò, P.; Borghini, G.; Di Flumeri, G. A Novel Mutual Information Based Feature Set for Drivers' Mental Workload Evaluation Using Machine Learning. *Brain Sci.* **2020**, *10*, 551. [CrossRef]
33. Saha, S.; Baumert, M. Intra-and inter-subject variability in EEG-based sensorimotor brain computer interface: A review. *Front. Comput. Neurosci.* **2020**, *13*, 87. [CrossRef]
34. Croce, P.; Quercia, A.; Costa, S.; Zappasodi, F. EEG microstates associated with intra-and inter-subject alpha variability. *Sci. Rep.* **2020**, *10*, 1–11. [CrossRef]
35. Byrne, A.; Murphy, A.; McIntyre, O.; Tweed, N. The relationship between experience and mental workload in anaesthetic practice: An observational study. *Anaesthesia* **2013**, *68*, 1266–1272. [CrossRef]
36. Pang, L.; Guo, L.; Zhang, J.; Wanyan, X.; Qu, H.; Wang, X. Subject-specific mental workload classification using EEG and stochastic configuration network (SCN). *Biomed. Signal Process. Control.* **2021**, *68*, 102711. [CrossRef]
37. Nentwich, M.; Ai, L.; Madsen, J.; Telesford, Q.K.; Haufe, S.; Milham, M.P.; Parra, L.C. Functional connectivity of EEG is subject-specific, associated with phenotype, and different from fMRI. *NeuroImage* **2020**, *218*, 117001. [CrossRef] [PubMed]
38. Zhang, K.; Xu, G.; Chen, L.; Tian, P.; Han, C.; Zhang, S.; Duan, N. Instance transfer subject-dependent strategy for motor imagery signal classification using deep convolutional neural networks. *Comput. Math. Methods Med.* **2020**, *2020*, 1683013. [CrossRef] [PubMed]
39. Neto, E.C.; Pratap, A.; Perumal, T.M.; Tummalacherla, M.; Snyder, P.; Bot, B.M.; Trister, A.D.; Friend, S.H.; Mangravite, L.; Omberg, L. Detecting the impact of subject characteristics on machine learning-based diagnostic applications. *NPJ Digit. Med.* **2019**, *2*, 1–6.
40. Thomas, K.P.; Robinson, N.; Vinod, A.P. Utilizing Subject-Specific Discriminative EEG Features for Classification of Motor Imagery Directions. In Proceedings of the 2019 IEEE 10th International Conference on Awareness Science and Technology (iCAST), Morioka, Japan, 23–25 October 2019; pp. 1–5.
41. Nijboer, F.; Morin, F.O.; Carmien, S.P.; Koene, R.A.; Leon, E.; Hoffmann, U. Affective brain-computer interfaces: Psychophysiological markers of emotion in healthy persons and in persons with amyotrophic lateral sclerosis. In Proceedings of the 2009 3rd International Conference on Affective Computing and Intelligent Interaction and Workshops, Amsterdam, The Netherland, 10–12 September 2009; pp. 1–11.
42. Kane, M.J.; Conway, A.R.; Miura, T.K.; Colflesh, G.J. Working memory, attention control, and the N-back task: A question of construct validity. *J. Exp. Psychol. Learn. Mem. Cogn.* **2007**, *33*, 615. [CrossRef]
43. Mathôt, S.; Schreij, D.; Theeuwes, J. OpenSesame: An open-source, graphical experiment builder for the social sciences. *Behav. Res. Methods* **2012**, *44*, 314–324. [CrossRef]
44. Wang, X.J. Neurophysiological and computational principles of cortical rhythms in cognition. *Physiol. Rev.* **2010**, *90*, 1195–1268. [CrossRef]

45. Bastos, A.M.; Schoffelen, J.M. A tutorial review of functional connectivity analysis methods and their interpretational pitfalls. *Front. Syst. Neurosci.* **2016**, *9*, 175. [CrossRef]
46. Kaiser, D.A. Cortical cartography. *Biofeedback* **2010**, *38*, 9–12. [CrossRef]
47. Ince, R.; Giordano, B.; Kayser, C.; Rousselet, G.; Gross, J.; Schyns, P. A Statistical Framework for Neuroimaging Data Analysis Based on Mutual Information Estimated via a Gaussian Copula. *Hum. Brain Mapp.* **2017**, *38*, 1541–1573. [CrossRef] [PubMed]
48. Celka, P. Statistical analysis of the phase-locking value. *IEEE Signal Process. Lett.* **2007**, *14*, 577–580. [CrossRef]
49. Aydore, S.; Pantazis, D.; Leahy, R.M. A note on the phase locking value and its properties. *Neuroimage* **2013**, *74*, 231–244. [CrossRef] [PubMed]
50. Palva, S.; Palva, J. Discovering Oscillatory Interaction Networks with M/EEG: Challenges and Breakthroughs. *Trends Cogn. Sci* **2012**, *16*, 219–230. [CrossRef] [PubMed]
51. Haufe, S.; Nikulin, V.V.; Müller, K.R.; Nolte, G. A critical assessment of connectivity measures for EEG data: A simulation study. *Neuroimage* **2013**, *64*, 120–133. [CrossRef] [PubMed]
52. Lobier, M.; Siebenhühner, F.; Palva, S.; Palva, J.M. Phase Transfer Entropy: A Novel Phase-Based Measure for Directed Connectivity in Networks Coupled by Oscillatory Interactions. *NeuroImage* **2014**, *85*, 853–872. [CrossRef]
53. Granger, C. Investigating Causal Relations by Econometric Models and Cross-Spectral Methods. *Econometrica* **1969**, *37*, 424–438. [CrossRef]
54. Hlaváčková-Schindler, K.; Paluš, M.; Vejmelka, M.; Bhattacharya, J. Causality detection based on information-theoretic approaches in time series analysis. *Phys. Rep.* **2007**, *441*, 1–46. [CrossRef]
55. Di Flumeri, G.; Aricò, P.; Borghini, G.; Sciaraffa, N.; Ronca, V.; Vozzi, A.; Storti, S.F.; Menegaz, G.; Fiorini, P.; Babiloni, F. EEG-based workload index as a taxonomic tool to evaluate the similarity of different robot-assisted surgery systems. In Proceedings of the International Symposium on Human Mental Workload: Models and Applications, Rome, Italy, 14–15 November 2019; Springer: Cham, Switzerland, 2019; pp. 105–117.
56. Song, H.; Kim, M.; Park, D.; Lee, J.G. How does Early Stopping Help Generalization against Label Noise? *arXiv* **2019**, arXiv:1911.08059.
57. Berger, L.; Hyde, E.; Pavithran, N.; Mumtaz, F.; Bragman, F.; Cardoso, M.J.; Ourselin, S. How to control the learning rate of adaptive sampling schemes. In Proceedings of the Medical Imaging with Deep Learning, Amsterdam, The Netherlands, 4–6 July 2018.
58. Lecun, Y.; Bengio, Y.; Hinton, G. Deep Learning. *Nature* **2015**, *521*, 436–444. [CrossRef]
59. Ide, H.; Kurita, T. Improvement of Learning for CNN with ReLU Activation by Sparse Regularization. In Proceedings of the International Joint Conference on Neural Networks, Anchorage, AK, USA, 14–19 May 2017; pp. 2684–2691.
60. Dunne, R.A.; Campbell, N.A. On the Pairing of the Softmax Activation and Cross-Entropy Penalty Functions and the Derivation of the Softmax Activation Function. Available online: https://citeseerx.ist.psu.edu/viewdoc/summary?doi=10.1.1.49.6403 (accessed on 1 October 2021).
61. Yao, Z.; Hu, B.; Xie, Y.; Moore, P.; Zheng, J. A review of structural and functional brain networks: Small world and atlas. *Brain Inform.* **2015**, *2*, 45–52. [CrossRef]
62. Aricò, P.; Borghini, G.; Di Flumeri, G.; Sciaraffa, N.; Babiloni, F. Passive BCI beyond the Lab: Current Trends and Future Directions. *Physiol. Meas.* **2018**, *39*, 08TR02. [CrossRef] [PubMed]
63. Luong, T.; Martin, N.; Raison, A.; Argelaguet, F.; Diverrez, J.M.; Lécuyer, A. Towards Real-Time Recognition of Users Mental Workload Using Integrated Physiological Sensors Into a VR HMD. In Proceedings of the 2020 IEEE International Symposium on Mixed and Augmented Reality (ISMAR), Porto de Galinhas, Brazil, 9–13 November 2020; pp. 425–437.
64. Knisely, B.M.; Joyner, J.S.; Vaughn-Cooke, M. Cognitive task analysis and workload classification. *MethodsX* **2021**, *8*, 101235. [CrossRef] [PubMed]
65. Michel, C.M.; Murray, M.M. Towards the utilization of EEG as a brain imaging tool. *Neuroimage* **2012**, *61*, 371–385. [CrossRef] [PubMed]
66. Dimitriadis, S.I.; Sun, Y.; Kwok, K.; Laskaris, N.A.; Bezerianos, A. A tensorial approach to access cognitive workload related to mental arithmetic from EEG functional connectivity estimates. In Proceedings of the 2013 35th Annual International Conference of the IEEE Engineering in Medicine and Biology Society (EMBC), Osaka, Japan, 3–7 July 2013; pp. 2940–2943.

EEG Emotion Classification Network Based on Attention Fusion of Multi-Channel Band Features

Xiaoliang Zhu, Wenting Rong, Liang Zhao *, Zili He, Qiaolai Yang, Junyi Sun and Gendong Liu

National Engineering Research Center of Educational Big Data, Central China Normal University, Wuhan 430079, China; zhuxl@ccnu.edu.cn (X.Z.); rwt_0706@mails.ccnu.edu.cn (W.R.); hzlzero@mails.ccnu.edu.cn (Z.H.); yql2020113547@mails.ccnu.edu.cn (Q.Y.); sunjunyi@mails.ccnu.edu.cn (J.S.); gendong@mails.ccnu.edu.cn (G.L.)
* Correspondence: liang.zhao@ccnu.edu.cn

Abstract: Understanding learners' emotions can help optimize instruction sand further conduct effective learning interventions. Most existing studies on student emotion recognition are based on multiple manifestations of external behavior, which do not fully use physiological signals. In this context, on the one hand, a learning emotion EEG dataset (LE-EEG) is constructed, which captures physiological signals reflecting the emotions of boredom, neutrality, and engagement during learning; on the other hand, an EEG emotion classification network based on attention fusion (ECN-AF) is proposed. To be specific, on the basis of key frequency bands and channels selection, multi-channel band features are first extracted (using a multi-channel backbone network) and then fused (using attention units). In order to verify the performance, the proposed model is tested on an open-access dataset SEED ($N = 15$) and the self-collected dataset LE-EEG ($N = 45$), respectively. The experimental results using five-fold cross validation show the following: (i) on the SEED dataset, the highest accuracy of 96.45% is achieved by the proposed model, demonstrating a slight increase of 1.37% compared to the baseline models; and (ii) on the LE-EEG dataset, the highest accuracy of 95.87% is achieved, demonstrating a 21.49% increase compared to the baseline models.

Keywords: EEG; learning emotions; emotion recognition; attention; convolutional neural network; multi-channel band features

1. Introduction

As a high-level psychological state, emotion is composed of many kinds of feelings, thoughts, and other factors, and has been broadly used in the medical, educational, and other related fields because of its capability to reflect people's real psychological reactions to different things. With the rapid development of artificial intelligence, emotion recognition research has become a hotspot. Generally speaking, the existing research in the field of emotion recognition is carried out from one of the two following aspects. The first type of research is a variety of manifestations (e.g., voice, text, and images) based on external behavior, which is acquired through non-contact methods. For example, in 2005, Burkhardt et al. established a speech dataset, called the Berlin database, which contained seven emotions [1]. In 2016, Lim et al. converted the original speech signal in this dataset into a spectrogram by time–frequency analysis and proposed a shallow convolutional neural network (CNN) and long short-term memory (LSTM) fusion network to identify the seven emotions [2]. Socher et al. built a text dataset containing the five emotions of very positive, positive, neutral, negative, and very negative [3], while Kim et al. used CNN to learn sentence feature vectors from this dataset and identify the emotions [4]. Anderson et al. proposed that facial muscle movements can represent emotional states, in which the support vector machine (SVM) was used to identify six basic emotions commonly associated with facial expressions [5]. The second type of research is based on the neurophysiological state, that is, the acquisition of various physiological signals [6–10], such as electrocardiogram (ECG), photoplethysmography (PPG), and electroencephalogram (EEG), among many others. Although this type of research requires subjects to wear certain appropriate physiological signal acquisition equipment, compared with the former external behavioral research, focusing on neurophysiological states is a more objective method of representing emotions. The collected physiological signals address better the problems associated with facial expression deception, and among them, the EEG signal is a focus of great concern [11]. A number of researchers previously constructed their own EEG signal datasets to study the basic emotions (i.e., anger, disgust, fear, happiness, sadness, and surprise) proposed by Ekman et al. [12]. For example, Petrantonakis et al. developed an EEG dataset in an attempt to distinguish the six basic emotional states proposed by Ekman et al. [13]. Schaaff et al. developed an EEG dataset in an attempt to distinguish three emotions (including pleasant, neutral, and unpleasant) [14]. Duan et al. created the SEED dataset to distinguish between negative, neutral, and positive emotions in subjects [15]. Koelstra et al. created the DEAP dataset, which measures two types of emotional states obtained from potentiation and arousal [16]. D'Mello et al. pointed out that, although the six basic emotions proposed by Ekman et al. [12] are common in our daily life, most of them do not exist for the study time of 30 min to 2 h; hence, six learning emotions (i.e., boredom, engagement, confusion, frustration, delight, and surprise) are defined and further ranked in an ascending order of persistence on a time scale: (delight = surprise) < (confusion = frustration) < (boredom = engagement) [17]. Meanwhile, Graesser et al. proposed that, for college students, the main emotions centered on learning include confusion, frustration, boredom, engagement, curiosity, anxiety, delight, and surprise [18].

Distinguishing the learners' emotions in an intelligent educational environment is very important; thus, in recent years, research on learning emotions has gradually attracted the attention of scientists. For instance, Tonguc et al. recorded the facial expressions of students during their speech process and recognized seven different types of learning emotions [19]. Sharma et al. studied students' engagement states in conjunction with their eye, head, and facial muscle movements in an online learning scenario [20]. Actually, in a real learning scenario, students mostly showed their normal emotions, i.e., it is quite difficult to capture the facial expressions at that moment, due to the fact that the facial muscles possessed small amplitudes and short durations. In addition, facial expressions showed defects (such as falsifiability) that cannot truly reflect emotions, bringing challenges to learning emotion recognition. Therefore, the present study attempts to explore the learning emotion classification algorithm based on EEG signals. Although EEG causes a lot of inconveniences due to contact measurement, its ability to capture and represent real learning emotions for students is quite helpful. In our preliminary research, the six learning emotions proposed in [17] were taken into account initially; however, considering the time scale and the probability of emotion occurrence, it was found that the chances of recognizing confusion, delight, and curiosity are small. Therefore, in this study, a learning emotion EEG dataset (LE-EEG) is constructed, which only focuses on three emotions (i.e., boredom, neutrality, and engagement) that can last for a longer time. The main contributions of this study are as follows:

(1) An EEG emotional classification network based on the attentional fusion (ECN-AF) of multi-channel band features is proposed, focusing on the relationship among the frequency bands, channels, and time series features.
(2) An induction experiment of an online learning scenario is designed, resulting in the self-collected LE-EEG dataset with relatively large sample size ($N = 45$).
(3) The cross-dataset validation demonstrates that the proposed ECN-AF model outperforms the baseline models, showing not only a good performance on the public data SEED, but also significant advantages on the self-collected LE-EEG dataset.

The remainder of this paper is organized as follows: Section 2 introduces the commonly used emotion classification algorithms; Section 3 presents the framework of the proposed ECN-AF model; Section 4 discusses the experimental design; Section 5 analyzes the experimental results; and Section 6 makes a summary and lists the future research directions.

2. Related Works

To realize emotion classification, the key methods of feature extraction based on EEG signals tend to be developed around the three aspects of time, frequency, and time–frequency domains [21]. First, the time domain methods focus on the EEG signals' temporal information, including the typical features of Hjorth parameters, fractal dimensional features, and higher-order crossover features. Second, the frequency domain methods often convert the collected EEG signals (0–50 Hz) into five sub-bands (i.e., delta (1–4 Hz), theta (4–7 Hz), alpha (8–13 Hz), beta (13–30 Hz), and gamma (31–50 Hz)) [22] and extract features, such as power spectral density, differential entropy and asymmetry, and rational asymmetry in different frequency bands [15]. Meanwhile, the time–frequency domain method combines the characteristics of both time and frequency domains, converting the EEG signals into sub-bands and using the windowing method for emotion classification.

Typical EEG emotion recognition methods tend to extract features and adopt machine learning, such as Support vector machines (SVM), k-nearest neighbor (KNN), and other algorithms for classification and recognition [23–25]. For example, Arnau-Gonzalez et al. conducted emotion classification experiments on the DEAP dataset, where frequency domain features (e.g., PSD) and mutual information in each frequency band of the channel were extracted, and a final classification accuracy of 66.7% for valence and 69.6% for arousal was obtained using the SVM [23]. Li et al. conducted experiments on the SEED dataset by extracting features (such as peak-to-peak average, alignment entropy, and Hjorth parameters), and their average classification accuracy using the SVM reached 83.3% [24]. Algumaei et al. used linear discriminant analysis (LDA), achieving an average accuracy of 90.93% on the SEED data set [25].

Compared with traditional machine learning models, deep neural networks show a more efficient performance [26–29]. They can not only automatically extract effective features, but also mark key frequency bands and brain regions. Therefore, more and more researchers use deep learning models to study EEG-based emotion classification. For example, on the SEED dataset, Zheng et al. proposed an emotion classification model using SVM and deep belief networks (DBN), and investigated the effect of the combinations of different frequency bands on emotion classification accuracy. Their final experimental results showed that the accuracy under the 12-channel combination could surpass that under the 62-channel combination. In addition, the direct concatenation of the DE features of five frequency bands under the DBN network led to an average classification accuracy of 86.08% [30]. Many researchers have improved the emotion recognition accuracy by developing advanced convolutional networks, such as the self-organizing graph neural network (SOGNN) [31] and dynamic graph convolutional neural network (DGCNN) [32], which respectively achieved 86.81% and 90.4% classification accuracy. To be specific, Li et al. proposed SOGNN, which constructs inter-channel correlations from self-organizing graphs, and explores the aggregation of these inter-channel connections and time–frequency features in frequency bands. The final experimental average accuracy (ACC) and the standard deviation (STD) were 86.81% and 5.79%, respectively [31]. Song et al. proposed DGCNN, which uses a graph to model the multi-channel EEG features and dynamically learn the intrinsic relationship between different EEG channels. As a result, they achieved 90.4% highest accuracy and 8.49% STD [32].

By contrast, studying emotion classification by exploring frequency bands and their correlation has made fruitful achievements. Yang et al. did not distinguish between the sub-bands on the SEED dataset to study the channel combination, but proposed the usage of directional RNNs to extract independent features of left and right brain regions. Consequently, they acquired 93.12% ACC and 6.06% STD [33]. Wang et al. improved the bidirectional long- and short-term memory network by proposing a similarity-learning network, achieving a classification accuracy of 94.62% on the SEED dataset [34]. Shen et al. proposed a four-dimensional convolutional recurrent neural network (4D_CRNN) that converted full EEG channels into a two-dimensional picture. They superimposed all sub-bands to convert the features into three dimensions and finally extracted the channel and band features using 2DCNN, as well as the temporal features using LSTM. They acquired 94.08% ACC and 2.55% STD [35].

The attention mechanism [36,37] was successfully introduced into neural networks, which greatly improved the performance of classification models. Researchers in the field of EEG emotion recognition found that the attention mechanism is like the idea of focusing on emotion-related brain regions and started to try using this in the field of EEG emotion recognition to improve the model performance. For instance, Li et al. proposed the transferable attention neural network (TANN) with 93.34% ACC and 6.64% STD, which used two directed RNN modules to extract features from whole brain regions and global attention layer fusion features to highlight the key brain regions for emotion classification [38].

In summary, existing research faces the following problems: (1) the exploration of multiple channel combinations for emotion classification fails to combine well the five sub-band features; and (2) exploring band correlations to synthesize all-channel studies is a mainstream method; however, not all brain regions of EEG signals contain valid emotion information, and thus this approach fails to focus on capturing the important emotion channels. To address these problems, in this study, ECN-AF is proposed, focusing on specific channels and some frequency bands for the fusion of attention units.

3. Methodology

3.1. Model Framework

Figure 1 depicts the framework of the proposed ECN-AF model consisting of the following three main modules:

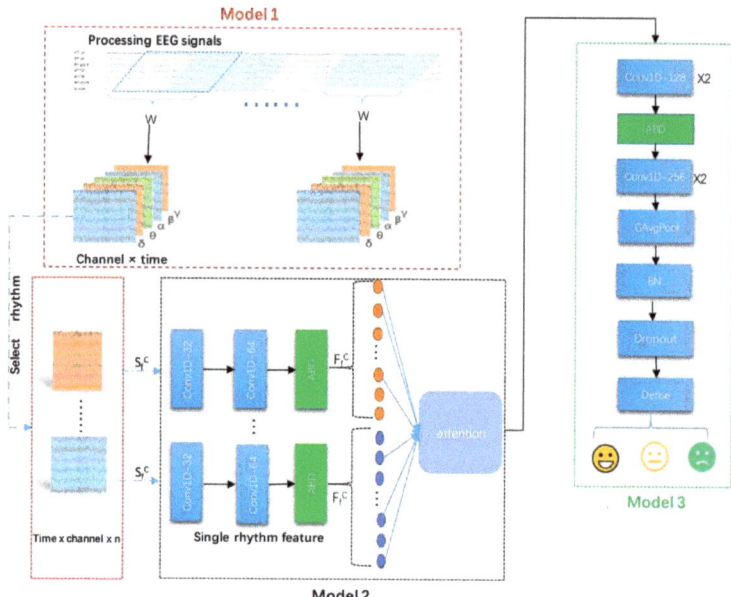

Figure 1. ECN-AF framework diagram.

(1) Module 1: frequency band division and channel selection module. In this module, first, the acquired EEG signal were divided into raw segments by a sliding window with a window size 10 s and a step size 2 s; second, five different frequency bands were extracted by passing the raw segments through bandpass filters; third, the final segments were generated, which were the optimal combinations of EEG channels obtained by multi-channel filtering operation.

(2) Module 2: frequency band attention feature extraction module. This module comprised a multi-channel convolutional backbone network with a frequency band attention fusion unit. First, the EEG sequences output from Module 1 were put into the multi-channel convolutional backbone network, which extracted not only the channel and time series features but also the features in different frequency bands. Second, the features extracted from different frequency bands were further put into a frequency band attention fusion unit, which performed the fusion of the channels and time series features across different frequency bands.

(3) Module 3: feature fusion and classification module. In this module, the combined features obtained from the fusion unit were taken as the input to the classification network; subsequently, the fused features were extracted using the depth network and then input to the fully connected layer, giving the final classification results.

3.2. Module 1: Frequency Band Division and Channel Selection Module

After data cleaning, the SEED dataset contained 62 channels of EEG signals from 15 subjects with a sampling rate of 200 Hz [15]. The LE-EEG dataset contained 32 channels of EEG signals from 45 subjects with a sampling rate of 128 Hz. Both the SEED and LE-EEG datasets were divided using a window

$$W = T \times C \tag{1}$$

In Equation (1), W is the segment size, T is the time duration after splitting, and C is the number of channels. The datasets were all segmented using a sliding window with a window length of 10 s and a step size 2 s. In the SEED and LE-EEG datasets, W values are 2000×62 and 1280×32, respectively.

$$S = \{W_1, W_2, W_3, \ldots W_i, \ldots W_{n-1}, W_n\} \tag{2}$$

$$Y = \{Y_1, Y_2, Y_3, \ldots, Y_i, \ldots, Y_{n-1}, Y_n\}, \ Y_i \in \{-1, 0, 1\} \tag{3}$$

In Equations (2) and (3), S denotes a subject's dataset, W_i denotes the sequential segment data, n denotes the total number of samples, Y denotes a subject's sentiment label set, and Y_i denotes the label of the ith segment data.

Finally, a sample size of 4896 for each subject and a total sample size of 73,440 for all the 15 subjects were collected in the SEED dataset. Meanwhile, a sample size of one subject ranging from 1082 to 1650 and a total sample size of 60,376 for all the 45 subjects were collected in the LE-EEG dataset.

$$|H(w)|^2 = \frac{1}{1 + \left(\frac{W}{W_{f_1 \sim f_2}}\right)^{2N_f}} \tag{4}$$

$$H(S) = \begin{cases} S_\delta, \ w \in (1, 4) \\ S_\theta, \ w \in (4, 7) \\ S_\alpha, \ w \in (8, 13) \\ S_\beta, \ w \in (13, 30) \\ S_\gamma, \ w \in (31, 50) \end{cases} \tag{5}$$

In Equations (4) and (5), a fourth-order Butterworth bandpass filter was used to filter the EEG signal into five wave sub-bands [39–42]. N_f is the order of the filter, i.e., $N_f = 4$. W is the frequency; $W_{f_1 \sim f_2}$ is the normalized frequency band; and the range of frequencies f_1 to f_2 is the passband interval of the bandpass filter. $H(S)$ is the EEG signal filtered by the fourth-order Butterworth bandpass filter, w is the frequency band, and $\delta, \theta, \alpha, \beta$, and γ denote the data of the five different frequency bands.

$$S_f = \frac{H(S) - AVG(H(S))}{STD(H(S))}, \ f \in \{\delta, \theta, \alpha, \beta, \gamma\} \tag{6}$$

In Equation (6), S_f is the normalized EEG segment data; f is one of the five sub-bands; H denotes the five different frequency band EEG signals of one subject; AVG is the average value; STD is the standard deviation.

Previous studies have found that, a combination of frequency channels can improve the recognition performance. For example, Zheng et al. used six channel combinations of "FT7," "FT8," "T7," "T8," "TP7," and "TP8" for emotion classification [43]. Zheng et al. designed four different electrode placement patterns based on the peak characteristics of the weight distribution and the asymmetry of the emotion processing, finally "FT7," "T7," "TP7," "P7," "C5," "CP5," "FT8," "T8," "TP8," "P8," "C6," and "CP6" were used, achieving the best result of 86.65% classification accuracy. This confirmed that it is possible to achieve better experimental results with fewer channel combinations than full-channel recognition [30]. Combining the abovementioned studies, we obtain the following setting:

$$X_f^C = \begin{cases} S_f^{C1} \\ S_f^{C2} \end{cases} f \in \{\delta, \theta, \alpha, \beta, \gamma\} \tag{7}$$

In Equation (7), X_f^C is the EEG signal at f frequency under the Cth channel combination; C is the channel combination method; and in our study, C1 and C2 are taken as C1 = {"FT7," "FT8," "T7," "T8," "TP7," "TP8"} and C2 = {"FT7," "T7," "TP7," "P7," "C5," "CP5," "FT8," "T8," "TP8," "P8," "C6," "CP6"}, respectively.

3.3. Module 2: Frequency Band Attention Feature Extraction Module

This section presents the combination of two sub-modules, a multi-channel convolutional backbone network and a band attention fusion unit.

3.3.1. Multi-Channel Convolutional Backbone Network

The backbone network was built using two layers of CNN, AvgPool1D, BatchNormalization, and SpatialDropout1D, with the parameters shown in Table 1. We used the X_f^C in Module 1 input to the multichannel convolutional backbone network to extract channel and time features.

$$F_f^C = \text{ReLU}\left((f * g)_{\times 2}\left(X_f^C\right)\right), f \in \{\delta, \theta, \alpha, \beta, \gamma\} \tag{8}$$

$$F^C = \left\{F_f^C\right\}, f \in \{\delta, \theta, \alpha, \beta, \gamma\} \tag{9}$$

Table 1. Multi-channel convolutional backbone network construction.

Stage	Stage Setting	Output
Conv-1	32, strides = 2, activation = "relu"	(1000,32)
Conv-2	64, strides = 2, activation = "relu"	(498,64)
Pool_1	2, AvgPool	(249,64)
Batch_norm1	BatchNormalization	(249,64)
Drop_1	Dropout1D	(249,64)

In Equations (8) and (9), F_f^C is the feature of the output of the convolutional network in the f-band under the Cth channel combination, and F^C is the set of different band features extracted by the convolutional backbone network under the Cth channel combination.

3.3.2. Frequency Band Attention Fusion Unit

The feature F^C was used as the input of the band attention fusion unit. First, the bands were selected from the feature F^C for combination. Next, the attention weights were generated by the sigmoid function using the feature vector. Finally, the weights were attached to the corresponding features to finally obtain the channel, time, and band fusion features. This three-step process is expressed as follows, also see Figure 2:

$$\text{Weight}_k = \text{Sigmoid}\left(q^T \text{Mult}\left(\text{Select}\left(F^C\right)_{\times n}\right)\right) \tag{10}$$

$$F' = \text{Mult}\left(\text{Select}\left(F^C\right)_{\times n}\right) \times \text{Weight}_k \tag{11}$$

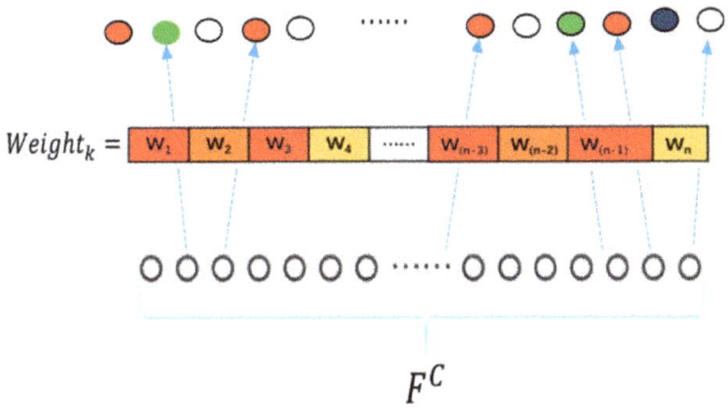

Figure 2. Band attention fusion unit.

3.4. Module 3: Feature Fusion and Classification Module

After the band attention feature extraction module, we input the fused features F' into the classification network built by CNN, AvgPool1D, BatchNormalization, Spatial-Dropout1D, GlobalAvgPool1D, Dropout, and Dense. Table 2 lists the specific parameters. We used convolution to extract the depth features in the upper layers of the classification network. The fully connected layer output the triple classification results. We set the BatchNormalization behind the convolutional network to normalize the segment data and transform the features in a state with zero mean and a variance of 1. It not only sped up the convergence speed but also effectively prevented gradient explosion and disappearance.

Table 2. Classification network construction.

Stage	Stage Setting	Output
Conv-1	128, strides = 2, activation = "relu"	(245,128)
Conv-2	128, strides = 2, activation = "relu"	(245,128)
Pool_1	2, AvgPool	(122,128)
Batch_norm1	BatchNormalization	(122,128)
Drop_1	Dropout	(122,128)
Conv-3	256, strides = 2, activation = "relu"	(118,256)
Conv-4	256, strides = 2, activation = "relu"	(118,256)
Pool_2	GlobalAvgPool	(256)
Drop_2	Dropout	(256)
Dense	Activation = "softmax"	(3)

4. Experiments

4.1. Experimental Materials

We want to control the following variables: take a graduate student majoring in big data artificial intelligence as the subject's educational background; ensure that the video duration is not much different; and select popular courses and the knowledge points of the selected courses which cover multiple disciplines.

4.1.1. Sources of Emotional Materials

At this stage, no standardized learning emotion induction course video is available in China. Hence, we used the well-known domestic learning websites https://www.icourse163.org/ (accessed on 21 March 2021) (Chinese University MOOC Network) and https://www.bilibili.com/ (accessed on 21 March 2021) (Learning section in Bilibili). The lessons were selected from these two sites according to the learners' comments about engagement and boredom-related vocabulary. With computer-related courses as the academic background, 50 learning videos of computer majors and science-, literature-, history-, and philosophy-related learning courses were finally selected to induce learning clips with focused and boring emotional labels. Note that the China University MOOC is the largest online classroom in China. Its course categories are classified according to the students' professional background (e.g., computer, foreign language, and science). Bilibili.com is a popular video platform used by young people in China to learn knowledge, exchange ideas, and spread culture. The website contains many excellent user-uploaded learning resources.

4.1.2. Emotional Material Clipping

Fifty videos were collected through the abovementioned means, among which 18 videos were marked as engaging, 17 videos were marked as boring; and 15 videos were marked as neutral. To clip a knowledge point in the videos, all acquired course videos were edited using Cut Screening for Windows Professional, which ensured that the content of the clip was complete, and the video length was not excessively long. The clipped video clips were edited into MP4 format video files, with a resolution of 1920 × 1080 px (30 fps). The clipping video duration was 76–293 s, with an average of 166 s. The emotion-inducing materials mainly consisted of Chinese materials and explanations. A few of them were English clips with Chinese subtitles.

4.1.3. Evaluation of Emotional Materials

In this study, 49 graduate students were recruited as subjects for the emotional material assessment experiment. The participants were 23 male students and 26 female students aged 20–25 years, with an average of (22 ± 1.19) years. All subjects were physically healthy, right-handed, and free of significant emotional problems and mental illness. Forty-nine subjects were taking majors in computer and science technology, electronic information, educational information technology, and educational technology. To avoid the subjects' prior knowledge from interfering with the emotion induction results, those who previously participated in rating the emotion material did not participated in the current data collection experiment.

For the experiment, all subjects were given a "Self-assessment of Learning Status" questionnaire. After each video clip was shown, the subjects were asked to report their actual feelings and score the questionnaire. Each question was scored using a 5-point scale:

- 0: really boring, I don't want to listen at all;
- 1: a little boring;
- 2: average;
- 3: not boring, can keep up with the teacher's rhythm;
- 4: not boring, very focused.

According to careless/insufficient effort (C/IE) detection (see Appendix A), finally 44 valid questionnaires were collected in this study. All data were imported into SPSS 27.0 statistical software according to the required SPSS format. The data were statistically analyzed by descriptive statistics, correlation analysis, reliability analysis, group analysis, and analysis of variance.

Figure 3 shows the 5-point scoring of 22 video clips marked as boredom and engagement by 44 subjects. The X-axis depicts 22 target videos. The Y-axis represents the ratings of the 44 subjects for each target video. The set of red dots indicates the rating of the 14 engaging emotional clips, while the set of green dots implies the rating of eight boring clips. Lighter scatters represent fewer subjects giving a score with the y-axis value, and darker scatters represent more subjects giving a score with the y-axis value. Figure 4 represents the mean scores of 44 subjects after the 5-point scoring for the 28 selected target video clips. The X-axis shows 28 target videos. The Y-axis is the mean score of 44 subjects for each target video. The blue bars indicate the mean scores of the 14 engaging emotion clips, while the red bars illustrate the mean scores of six neutral clips. The orange bars show the mean scores of eight boring emotion clips.

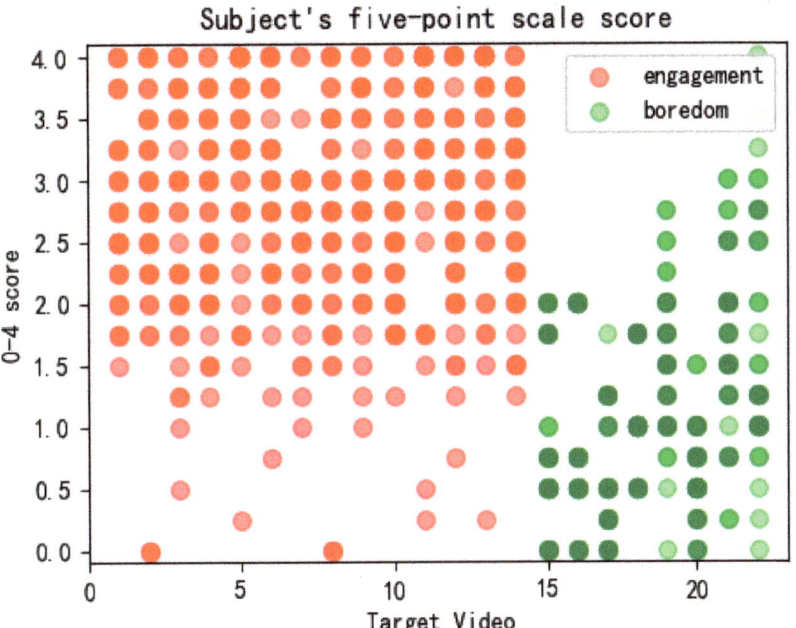

Figure 3. 5-point scale score of the subjects.

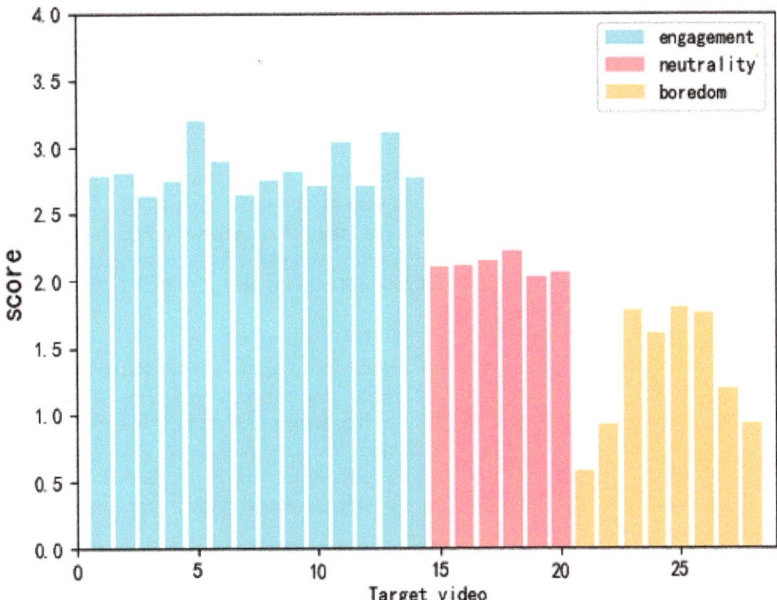

Figure 4. Description statistics of the 28 target videos, with 0–4 ratings.

Gross et al. pointed out that the indicators for judging the success of emotion induction include the intensity and discreteness of emotion induction [44]. Intensity refers to the average score of different emotional segments. The greater the intensity of the emotional response, the higher the average score. The discreteness was judged by the hit rate (hit rate = the type of video discriminated by the subjects/the number of all emotions discriminated). The higher the hit rate, the better the singleness of the emotions induced by the emotional video clips. Figures 3 and 4 depict the dispersion and the intensity of the subject's response induced by the target video clip. According to the discrete scoring points in Figure 3, the hit rate of the engaging emotion was $79.48 \pm 4.54\%$, while that of the boredom emotion was $81.73 \pm 16.03\%$, proving that the singleness induced by the two emotions was good. In Figure 4, the average score of the input emotion was 2.873, while those of the boredom emotion and the neutral segments were 1.256 and 2.036, respectively. These results proved that the intensity of the induced emotional response was high. Finally, according to 44 valid questionnaires, 28 videos were effectively distinguished from the three emotions. We had 14 engaging segments, 8 boring emotional segments, and 6 neutral segments.

4.2. Experimental Procedure and Signal Pre-Processing

4.2.1. Experimental Procedure

In the experiment, we selected seven each of the engagement and boredom clips and six neutral videos as the target emotions from the 28 induced emotion materials. After each video clip was shown, all subjects were asked to answer the questionnaire, report their actual feelings, and rate the questionnaire. The questionnaire consisted of nine questions, each of which was scored on a 5-point (0–4) scale, except for the first two questions. The more intense the subject's concentration, the closer the question score was to 4. The more intense the boredom, the closer the question score was to 0.

We used a pseudo-randomized approach to play the induction video to prevent the boredom caused by the subjects watching the same emotional video for a long time. After the researcher played a video clip, the subjects were given 1 min to fill out the questionnaire and take a short break. The process was repeated for 20 times, with a 10 min break until all video clips had been studied.

The hardware device used to collect the data in this experiment was the EPOC Flex Saline Sensor Kit. The software device was EmotivPRO v2.0. During the experimental acquisition, we asked the subjects to keep their limbs still and try to avoid continuous blinking to minimize the presence of artifacts. The final experiment collected 940 segments of EEG data and 940 assessment questionnaires, of which 777 questionnaires were identified as valid data based on the subjects' completion and the researcher's screening. All valid questionnaires were labeled as boredom, neutrality, and engagement. The EEG data collected for the sentiment classification contained 745 segments because of the equipment acquisition failures and other reasons.

4.2.2. Signal Pre-Processing

The pre-processing and removal of artifacts from the EEG signals are a demanding step in the EEG processing process. In Figure 5, the LE-EEG dataset was preprocessed using MATLAB R2020b, eeglab toolbox [45], ICLab [46–49], and adjusted [50] for bandpass filtering and automatic artifact processing of EEG signals. After the artifacts were processed using the automatic toolkit, some of the bad data were manually removed by visual inspection to finally obtain relatively clean EEG data.

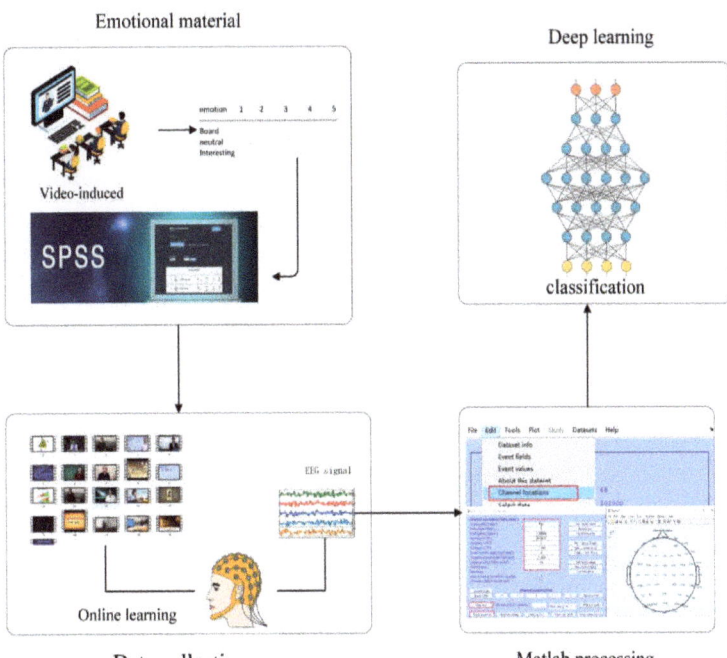

Figure 5. Experimental flow of the LE-EEG dataset.

5. Results and Analysis

We trained the model on an NVIDIA GTX 1080 GPU. The model learning rate was set to 0.001. The learning rate decay was set to 0.00001. The optimization function was set to Adam optimization. The loss function was set to categorical_crossentropy. The number of multi-channel convolutional backbone network settings depended on the number of band combinations. We conducted experiments on the SEED and LE-EEG dataset separately. The ACC and the STD were used as the evaluation criteria for all subjects in the dataset, dividing the data into training and test sets in a ratio of 8:2 in each fold of cross validation. On the SEED dataset, we performed the subject-dependent experiments, we performed a comparison with several baseline models using cross-validation to assess the model performance. On the LE-EEG data, we cited the paper containing the code for comparison with the model in this paper. In contrast to the approach to the SEED dataset prediction, we fused all subject data for data partitioning.

5.1. Ablation Study

We conducted two sets of ablation study experiments on the SEED dataset to validate the effectiveness of the combined band and attention fusion units in the model for sentiment classification. One experiment explored the effects of split-band prediction and combined band prediction on emotion classification to validate the importance of integrating the band features. Another experiment discussed multiple fusion approaches to validate the need for attentional fusion units.

5.1.1. Sub-Band Prediction and Combined Band Prediction

In our experiments, we compared the emotional classification accuracy in two cases: one uses a single-channel backbone network to extract the sub-band features, while the other uses a multi-channel backbone network combination to extract the sub-band features. Table 3 shows the experimental results on the two datasets. First, on the SEED dataset, C1 and C2 are different channel combination methods, as described in Section 3.2. We recall that C1 represents the combination of "FT7," "FT8," "T7," "T8," "TP7," and "TP8," and C2 represents the combination of "FT7," "T7," "TP7," "P7," "C5," "CP5," "FT8," "T8," "TP8," "P8," "C6," and "CP6." Second, on the LE-EEG dataset, All_band indicates that all available EEG channels are used instead of C1 and C2. This is because the number of available EEG channels from the two datasets are not consistent, which are 64 and 32 for the SEED and LE-EEG datasets, respectively. Furthermore, in Table 3, in order to ensure the consistency of the algorithm migration benchmark and further make a fair comparison, C3 was proposed as the combination of "T7," "P7," "CP5," "T8," "P8"and "CP6," as shown in Figure 6. In Figure 6a, the scatter points shown are all 62 electrode points used in the seed data set, of which the blue scatter points are C1 combined electrodes; In Figure 6b, the scatter points shown are the electrical poles used in the LE-EEG data set, and the blue scatter points are C3 combined electrodes. Notably, the channels involved in C3 (see the blue points in Figure 6b) aimed to match the locations of the channels involved in C2 (see the blue points in Figure 6a) as closely as possible.

Table 3. Accuracy comparison (i.e., ACC/STD) of different frequency bands (average 5-fold cross validation results).

Frequency Band	SEED			LE-EEG
	C1	C2	C3	All_Band
δ	83.18/2.42	84.23/2.85	93.69/0.40	95.22/0.49
θ	67.05/7.71	69.88/7.52	93.06/0.45	94.64/1.15
α	77.55/6.82	82.68/5.58	93.09/1.11	94.64/0.63
β	81.46/7.27	87.09/4.17	93.56/0.44	94.97/0.51
γ	83.60/4.91	90.90/4.38	93.83/0.48	95.52/0.62
$\beta + \gamma$	84.14/6.12	92.10/4.02	-	-
$\beta \times \gamma$	91.30/4.56	93.39/2.42	-	-
Attention (β, γ)	90.03/3.40	94.20/2.38	-	-

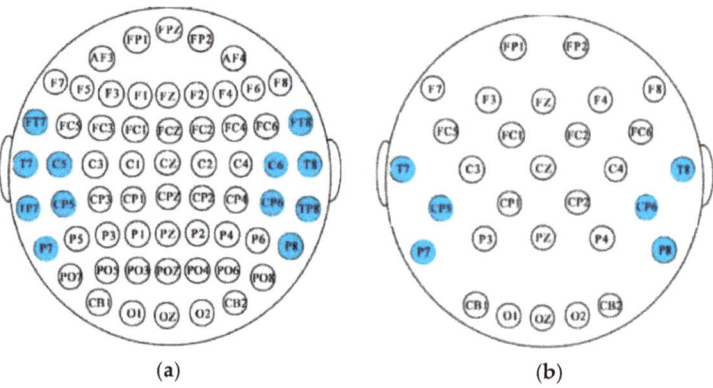

Figure 6. Channel selection maps: (a) C2 on the SEED dataset; (b) C3 on the LE-EEG dataset.

Table 3 shows the classification accuracy of the five sub-bands (i.e., δ, θ, α, β, and γ) in the SEED. $\beta+\gamma$ means the add fusion method. $\beta \times \gamma$ means the multiply fusion method. These two operations have been widely used in deep learning network design. Specifically, the add fusion method is described as having the corresponding elements of the feature matrix (which outputs from the multi-channel convolutional network) for each sub-band be added together. Similarly, the multiplicative fusion method is described as having the corresponding elements of the feature matrix for each sub-band be multiplied. Attention (β, γ) indicates that the attention fusion unit is used for the feature-level fusion. Take C2 (see the third column of Table 3) as an example. Based on the experimental results of the single-channel network, on the SEED dataset, we found that the β and γ bands performed a better prediction than the other bands, the accuracy of these two bands were 87.09% and 90.90%, respectively. Therefore, we combined the β and γ frequency bands, input them to the multi-channel backbone network to extract features, and adopted three feature-level fusion methods for emotion prediction. The final experimental results showed that the fusion of the frequency band information (i.e., Attention (β, γ)) could improve the model accuracy; the resulting accuracy was 94.20%.

Furthermore, on the LE-EEG dataset, the emotion classification accuracy in each sub-band was high. We believe that the possible reasons for this phenomenon include (i) compared with the SEED dataset (N = 15), the LE-EEG dataset had relatively larger sample size (N = 45); (ii) after data fusion, the training samples (of the LE-EEG dataset) became even larger, which results in better model performance after the training. In addition, from the comparison between the last two columns in Table 3, we can see that the performance of All_band has higher classification accuracy than the C3 combination of channels in each sub-band, so the channel selection does not yield better classification results. We believe that the reason for this phenomenon is that the types of emotions on the two datasets were different. To be specific, the SEED data were designed to explore three basic emotions containing negative, neutral, and positive, while the LE-EEG dataset explored three learned emotions of engagement, neutrality, and boredom. Therefore, the relevant channels for studying basic emotions may not be applicable to the study of learning emotions, and at this stage, there is no past reference literature regarding learning emotion channel studies, so in future work, learning emotion-related channel exploration should be the research focus. In this paper, the optimal combination of channels for learning emotions will not be discussed for the time being.

5.1.2. Comparison of the Results of Fusion Methods

In this subsection, we verified the effectiveness of combining frequency band features to improve the model performance. This subsection focuses on analyzing the impact of multiple fusion methods on the model accuracy and verifying the necessity of attention fusion units. We compared three fusion methods, namely feature summation fusion, feature multiplication fusion, and attention weight fusion, which are denoted as *Add*, *Mult*, and *Attention* in Table 4, respectively. Table 4 shows the classification accuracy of the five sub-bands (i.e., δ, θ, α, β, and γ) in the SEED dataset after inputting different frequency band combinations into the multi-channel backbone network to extract features.

Table 4. Accuracy comparison (i.e., ACC/STD) of various fusion methods validated on SEED dataset (average 5-fold cross validation results).

Method	C1			C2		
	Add	Mult	Attention	Add	Mult	Attention
α, β	72.34/10.70	72.54/11.50	**72.75/7.54**	83.16/4.84	87.63/7.67	**89.80/4.13**
α, γ	69.48/12.10	78.84/10.22	**79.26/7.10**	80.56/8.80	**95.04/3.80**	90.77/4.59
δ, β	**94.81/2.20**	77.62/11.56	93.77/2.27	94.68/3.45	**95.36/3.96**	87.40/4.41
δ, γ	95.03/2.45	82.41/8.30	**95.63/1.92**	92.00/2.26	95.60/2.75	**95.70/3.67**
β, γ	84.14/6.12	**91.30/4.56**	90.03/3.40	92.10/4.02	93.39/2.42	**94.20/2.38**
δ, α, β	94.79/3.22	**95.11/3.60**	94.95/2.73	94.24/3.32	**96.09/3.00**	95.87/4.17
θ, β, γ	**94.10/4.50**	92.23/4.99	92.46/6.92	95.44/2.35	**95.77/3.90**	94.89/4.06
α, β, γ	92.70/5.52	**95.17/4.27**	93.84/3.63	95.31/3.21	94.66/5.43	**96.02/5.54**
δ, β, γ	95.17/2.17	95.13/3.67	**95.32/3.53**	95.78/3.45	96.15/2.13	**96.45/3.56**
δ, α, β, γ	**94.28/5.46**	87.07/12.96	77.0/16.81	**94.68/2.72**	80.99/14.82	86.49/17.90

Notably, Add means to directly add and fuse the features; Mult means that the features are multiplied and fused; Attention means that the attention fusion unit is used for feature-level fusion, and Bold indicates the best accuracy achieved using different fusion methods (for a given channel combination, C1 or C2).

Our experiments revealed that first, the proposed attention fusion unit pair model has a better performance on more frequency band combinations in general; however, more frequency band combinations cannot always guarantee a higher performance of emotion classification. For example, compared with the sub-band combinations shown in the other rows of Table 4, in the case of the sub-band (δ, α, β, γ) shown in the last row of Table 4, (i) the model performance using the fusion mode of *Add* decreased (see the 2nd and 5th columns of the last row in Table 4), but remained relatively stable; (ii) the model performance using fusion mode of either *Mult* or *Attention* (see the 3rd and 6th columns or the 4rd and 7th columns of the last row in Table 4) was seriously degraded. The reason for this might include that when the model was trained, the fusion method of *Mult* and *Attention* made the model training parameters exponentially increase, resulting in severe overfitting caused by model overtraining.

Second, we can see that, the best performance obtained by C2 (see the 5th–7th columns of Table 4) was always higher than that of C1 (see the 2nd–4th columns of Table 4). For clarification, let us take the sub-band (δ, γ) as an example. From the 4th row in Table 4, we can see that, (i) regarding C1, the best performance with 95.63% was achieved using the fusion method of *Attention*; (ii) regarding C2, the best performance with 95.70% was achieved again using the fusion method of *Attention*, i.e., compared with C1, 0.07% accuracy improvement was achieved by C2.

Third, regarding C2, the top two performances were achieved by the sub-bands (α, β, γ) and (δ, β, γ) using the fusion method of *Attention*, which were 96.02% and 96.45%, respectively (see the 2nd and 3nd last rows of the last column in Table 4). Take the sub-band (δ, β, γ) as an example. Compared with *Add* and *Mult*, 0.67% and 0.30% accuracy improvements were obtained by the fusion method of *Attention*. This demonstrated that the classification performance can be improved using the fusion method of *Attention*, due to those more important features were assigned by attention weights.

5.2. Comparison

Based on above experiments, we take δ, β, and γ bands and attention fusion to complete comparison. On the SEED dataset, the model herein was compared with the baseline models. Table 5 presents the results. Compared with that of the optimal baseline model (see the row of "DCCA [39]" in Table 5), the performance of our model was improved by 1.37%.

Table 5. Accuracy comparison (i.e., ACC/STD) versus baseline models (average 5-fold cross validation results).

Method	SEED	LE-EEG
SVM [24]	83.30/—	—
DBN [30]	86.08/—	—
SOGNN [31]	86.81/5.79	74.38/1.50
LDA [25]	90.93/—	—
DGCNN [32]	90.40/8.48	—
BiHDM [33]	93.12/6.06	—
TANN [38]	93.34/6.64	—
3DCNN-BiLSTM [27]	93.38/2.66	—
4D_CRNN [35]	94.08/2.55	67.48/0.39
RGNN [51]	94.24/5.95	—
DE-CNN-BiLSTM [26]	94.82/—	—
DCCA [39]	95.08/6.42	—
ECN-AF (C1)	95.32/3.53	—
ECN-AF (C2)	**96.45/3.56**	—
ECN-AF (C3)	—	94.80/0.57
ECN-AF (All_band)	95.7/4.71	**95.87/0.38**

Dotted line (i.e., "—") indicates that data was not provided; and bold indicates the best accuracy achieved for a given dataset.

Referring to the baseline models on the SEED dataset, two baseline models 4D_CRNN [35] and SOGNN [31] that can be reproduced with the shared code were selected for comparison when validating on the LE-EEG dataset. Table 5 presents the comparison with the baseline models. Compared with that of these two baseline models, the performance of our model was improved by 28.39% and 21.49% (see the 3rd column of the rows of "4D_CRNN [35]," "SOGNN [31]," and "ECN-AF(All_band)" in Table 5), confirming that the network was robust across datasets. Figure 7 shows the validation set accuracy of the three different models during the training process. We still find that the ECN-AF model yields a better performance.

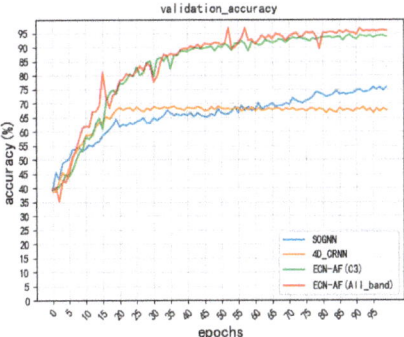

Figure 7. Accuracy of the model's validation set.

6. Conclusions

In this study, we collected the EEG signals of 45 subjects while they were watching learning materials. We established the LE-EEG dataset and tried to use the EEG signals to recognize learning emotions. The proposed ECN-AF first extracted the frequency band features through a multi-channel backbone network, and then fused the frequency band features with attention, which could effectively improve the model performance. Using the complementarity of the frequency band combination effectively improved the model's accuracy and robustness and yielded better results compared to a single sub-band. This is a conclusion similar to that of previous studies [30,31]. The ablation experiments performed herein also demonstrated the necessity of multi-channel backbone blocks and attention blocks. The experiments on the SEED and LE-EEG datasets showed that the proposed model outperforms baseline models with a better cross-dataset performance.

Our future work will focus on the expansion of the LE-EEG dataset and on the construction of a physiological signal dataset for multimodal learning emotion recognition. At the same time, the learning of emotion-related frequency bands and related brain regions and channels must be continuously explored and optimized, e.g., to further improve the performance by exploring the optimal combination of EEG channels on the LE-EEG dataset. The accuracy of the proposed model still needs improvement in across-participant research. The generalization ability and robustness of the algorithm must also be further improved.

Author Contributions: Conceptualization, X.Z.; methodology, X.Z., Z.H. and L.Z.; software, W.R.; Data collection, W.R., Q.Y., J.S. and G.L.; validation, W.R.; investigation, Z.H.; writing—original draft preparation, W.R.; writing—review and editing, X.Z. and L.Z.; supervision, X.Z. and L.Z.; project administration, X.Z. All authors have read and agreed to the published version of the manuscript.

Funding: The authors would like to thank support from the National Key R&D Program of China (2020AAA0108804), National Natural Science Foundation of China (61937001) and the National Natural Science Foundation of Hubei Province (2021CFB157).

Institutional Review Board Statement: Our Institutional Review Board approved the study.

Informed Consent Statement: Informed consent was obtained from all subjects involved in the study.

Data Availability Statement: The open access dataset SEED is used in our study. Its links is as follows, https://bcmi.sjtu.edu.cn/~seed/seed.html (granted on 7 May 2020; accessed on 25 April 2022).

Conflicts of Interest: The authors declare that they have no conflict of interest to report regarding the present study.

Appendix A

Referring to [52–57], a questionnaire is taken as invalid if one or more than one of the six factors in Table A1 is/are involved.

Table A1. Summary of methods of careless/insufficient effort (C/IE) detection.

Index	Method	Type	Description
1	bogus or infrequency [52–55]	check items	Odd items placed in scale to solicit particular responses.
2	long-string analysis [52–55]	invariance	Length of longest sequential string of the same response
3	self-report data [52–55]	self-report	Items which ask the participant how much effort they applied or how they judge the quality of their data
4	semantic antonyms/synonyms [52–55]	consistency	Within-person correlations on sets of semantically matched pairs of items with opposite or similar meaning
5	instructional manipulation checks [52–55]	check items	Items with extended instructions which include instructing participant to answer in unique manner
6	polytomous guttman errors [52]	consistency	Count of the number of instances where a respondent broke the pattern of monotonically increasing response on the set of survey items ordered by difficulty.

References

1. Burkhardt, F.; Paeschke, A.; Rolfes, M.; Sendlmeier, W.F.; Weiss, B. A database of German emotional speech. In Proceedings of the 9th European Conference on Speech Communication and Technology (INTERSPEECH2005), Lisbon, Portugal, 4–8 September 2005; pp. 1517–1520.
2. Lim, W.; Jang, D.; Lee, T. Speech emotion recognition using convolutional and recurrent neural networks. In Proceedings of the Asia-Pacific Signal and Information Processing Association Annual Summit and Conference (APSIPA2016), Jeju, Korea, 13–16 December 2016; pp. 1–4.
3. Socher, R.; Perelygin, A.; Wu, J.; Chuang, J.; Manning, C.D.; Ng, A.Y.; Potts, C. Recursive deep models for semantic compositionality over a sentiment treebank. In Proceedings of the 2013 Conference on Empirical Methods in Natural Language Processing (EMNLP2013), Seattle, WA, USA, 18–21 October 2013; pp. 1631–1642.
4. Kim, Y. Convolutional neural networks for sentence classification. In Proceedings of the 2014 Conference on Empirical Methods in Natural Language Processing (EMNLP2014), Doha, Qatar, 25–29 October 2014; pp. 1746–1751.
5. Anderson, K.; Mcowan, P.W. A real-time automated system for the recognition of human facial expressions. *IEEE Trans. Syst. Man. Cybern. B Cybern.* **2006**, *36*, 96–105. [CrossRef] [PubMed]
6. Kim, K.H.; Bang, S.W.; Kim, S.R. Emotion recognition system using short-term monitoring of physiological signals. *Med. Biol. Eng. Comput.* **2004**, *42*, 419–427. [CrossRef]
7. Bulagang, A.F.; Weng, N.G.; Mountstephens, J.; Teo, J. A review of recent approaches for emotion classification using electrocardiography and electrodermography signals. *Inform. Med. Unlocked* **2020**, *20*, 100363. [CrossRef]
8. Suzuki, K.; Laohakangvalvit, T.; Matsubara, R.; Sugaya, M. Constructing an emotion estimation model based on eeg/hrv indexes using feature extraction and feature selection algorithms. *Sensors* **2021**, *21*, 2910. [CrossRef] [PubMed]
9. Fujii, A.; Murao, K.; Matsuhisa, N. disp2ppg: Pulse wave generation to PPG sensor using display. In Proceedings of the ACM International Symposium on Wearable Computers (ISWC2021), Virtual Event, 21–26 September 2021; pp. 119–123.
10. Tong, Z.; Chen, X.X.; He, Z.; Kai, T.; Wang, X. Emotion Recognition Based on Photoplethysmogram and Electroencephalogram. In Proceedings of the IEEE 42nd Annual Computer Software and Applications Conference (COMPSAC2018), Tokyo, Japan, 23–27 July 2018; pp. 402–407.
11. Coan, J.A.; Allen, J.J. Frontal EEG asymmetry as a moderator and mediator of emotion. *Biol. Psychol.* **2004**, *67*, 7–49. [CrossRef] [PubMed]
12. Ekman, P. Expression and the nature of emotion. *Approaches Emot.* **1984**, *3*, 319–344.
13. Petrantonakis, P.C.; Hadjileontiadis, L.J. Emotion recognition from brain signals using hybrid adaptive filtering and higher order crossings analysis. *IEEE Trans. Affect. Comput.* **2010**, *1*, 81–97. [CrossRef]
14. Schaaff, K.; Schultz, T. Towards emotion recognition from electroencephalographic signals. In Proceedings of the Third International Conference and Workshops on Affective Computing and Intelligent Interaction(ACII2009), Amsterdam, The Netherlands, 10–12 September 2009; pp. 1–6.

15. Duan, R.N.; Zhu, J.Y.; Lu, B.L. Differential entropy feature for EEG-based emotion classification. In Proceedings of the 6th International IEEE/EMBS Conference on the Neural Engineering (NER2013), San Diego, CA, USA, 6–8 November 2013; pp. 81–84.
16. Koelstra, S.; Muhl, C.; Soleymani, M.; Lee, J.; Yazdani, A.; Ebrahimi, T.; Pun, T.; Nijholt, A.; Patras, I. DEAP: A database for emotion analysis using physiological signals. *IEEE Trans. Affect. Comput.* **2012**, *3*, 18–31. [CrossRef]
17. D'mello, S.; Graesser, A. Emotions during learning with AutoTutor. In *Adaptive Technologies for Training and Education*; Cambridge University Press: Cambridge, UK, 2012; pp. 117–139.
18. Graesser, A.C.; D'mello, S. Emotions during the learning of difficult material. *Psychol. Learn Motiv.* **2012**, *57*, 183–225.
19. Tonguc, G.; Ozkara, B.O. Automatic recognition of student emotions from facial expressions during a lecture. *Comput. Educ.* **2020**, *148*, 103797. [CrossRef]
20. Sharma, P.; Joshi, S.; Gautam, S.; Maharjan, S.; Filipe, V.; Reis, M.J. Student engagement detection using emotion analysis, eye tracking and head movement with machine learning. *arXiv* **2019**, arXiv:1909.12913.
21. Jenke, R.; Peer, A.; Buss, M. Feature extraction and selection for emotion recognition from EEG. *IEEE Trans. Affect. Comput.* **2017**, *5*, 327–339. [CrossRef]
22. Davidson, R.J. What does the prefrontal cortex "do" in affect: Perspectives on frontal EEG asymmetry research. *Biol. Psychol.* **2004**, *67*, 219–233. [CrossRef] [PubMed]
23. Arnau-González, P.; Arevalillo-Herráez, M.; Ramzan, N. Fusing highly dimensional energy and connectivity features to identify affective states from EEG signals. *Neurocomputing* **2017**, *244*, 81–89. [CrossRef]
24. Li, X.; Song, D.; Zhang, P.; Zhang, Y.; Hou, Y.; Hu, B. Exploring EEG features in cross-subject emotion recognition. *Front. Neurosci.* **2018**, *12*, 162. [CrossRef]
25. Algumaei, M.; Hettiarachchi, I.T.; Veerabhadrappa, R.; Bhatti, A. Wavelet packet energy features for eeg-based emotion recognition. In Proceedings of the IEEE International Conference on Systems, Man, and Cybernetics (SMC2021), Melbourne, Australia, 17–20 October 2021; pp. 1935–1940.
26. Cui, F.; Wang, R.; Ding, W.; Chen, Y.; Huang, L. A Novel DE-CNN-BiLSTM Multi-Fusion Model for EEG Emotion Recognition. *Mathematics* **2022**, *10*, 582. [CrossRef]
27. Xing, M.; Hu, S.; Wei, B.; Lv, Z. Spatial-Frequency-Temporal Convolutional Recurrent Network for Olfactory-enhanced EEG Emotion Recognition. *J. Neurosci. Methods* **2022**, *376*, 109624. [CrossRef]
28. Li, J.; Wu, X.; Zhang, Y.; Yang, H.; Wu, X. DRS-Net: A spatial–temporal affective computing model based on multichannel EEG data. *Biomed. Signal Process. Control.* **2022**, *76*, 103660. [CrossRef]
29. Toraman, S.; Dursun, Ö.O. GameEmo-CapsNet: Emotion Recognition from Single-Channel EEG Signals Using the 1D Capsule Networks. *Traitement Signal* **2021**, *38*, 1689–1698. [CrossRef]
30. Zheng, W.L.; Lu, B.L. Investigating critical frequency bands and channels for EEG-based emotion recognition with deep neural networks. *IEEE Trans. Auton. Ment. Dev.* **2015**, *7*, 162–175. [CrossRef]
31. Li, J.; Li, S.; Pan, J.; Wang, F. Cross-subject EEG emotion recognition with self-organized graph neural network. *Front. Neurosci.* **2021**, *15*, 611653. [CrossRef] [PubMed]
32. Song, T.Z.W.; Song, P.; Cui, Z. EEG emotion recognition using dynamical graph convolutional neural networks. *IEEE Trans. Affect. Comput.* **2020**, *3*, 532–541. [CrossRef]
33. Li, Y.; Wang, L.; Zheng, W.; Zong, Y.; Qi, L.; Cui, Z.; Zhang, T.; Song, T. A novel bi-hemispheric discrepancy model for EEG emotion recognition. *IEEE Trans. Cogn. Dev. Syst.* **2020**, *13*, 354–367. [CrossRef]
34. Wang, Y.; Qiu, S.; Li, J.; Ma, X.; Liang, Z.; Li, H.; He, H. EEG-based emotion recognition with similarity learning network. In Proceedings of the 41st Annual International Conference of the IEEE Engineering in Medicine & Biology Society (EMBC2019), Berlin, Germany, 23–27 July 2019; pp. 1209–1212.
35. Shen, F.; Dai, G.; Lin, G.; Zhang, J.; Kong, W.; Zeng, H. EEG-based emotion recognition using 4D convolutional recurrent neural network. *Cogn. Neurodyn.* **2020**, *14*, 815–828. [CrossRef] [PubMed]
36. Hu, J.; Shen, L.; Albanie, S.; Sun, G.; Wu, E. Squeeze-and-excitation networks. *IEEE Trans. Pattern. Anal. Mach. Intell.* **2020**, *42*, 2011–2023. [CrossRef]
37. Woo, S.; Park, J.; Lee, J.Y.; Kweon, I.S. CBAM: Convolutional block attention module. In Proceedings of the 15th European Conference on Computer Vision (ECCV2018), Munich, Germany, 8–14 September 2018; Springer: Cham, Switzerland, 2018; Volume VII, pp. 3–19.
38. Li, Y.; Fu, B.; Li, F.; Shi, G.; Zheng, W. A novel transferability attention neural network model for EEG emotion recognition. *Neurocomputing* **2021**, *447*, 92–101. [CrossRef]
39. Wu, X.; Zheng, W.L.; Li, Z.; Lu, B.L. Investigating EEG-based functional connectivity patterns for multimodal emotion recognition. *J. Neural Eng.* **2022**, *19*, 016012. [CrossRef]
40. Keelawat, P.; Thammasan, N.; Numao, M.; Kijsirikul, B. A comparative study of window size and channel arrangement on EEG-emotion recognition using deep CNN. *Sensors* **2021**, *21*, 1678. [CrossRef]
41. Garg, N.; Garg, R.; Parrivesh, N.S.; Anand, A.; Abhinav, V.A.S.; Baths, V. Decoding the neural signatures of valence and arousal from portable EEG headset. *bioRxiv* **2021**. [CrossRef]
42. Kasim, Ö.; Tosun, M. Effective removal of eye-blink artifacts in EEG signals with semantic segmentation. *Signal Image Video Processing* **2022**, *16*, 1289–1295. [CrossRef]

43. Zheng, W.L.; Liu, W.; Lu, Y.; Lu, B.L.; Cichocki, A. EmotionMeter: A multimodal framework for recognizing human emotions. *IEEE Trans. Cybern.* **2019**, *49*, 1110–1122. [CrossRef] [PubMed]
44. Gross, J.J.; Levenson, R.W. Emotion elicitation using films. *Cogn. Emot.* **1995**, *9*, 87–108. [CrossRef]
45. Delorme, A.; Makeig, S. EEGLAB: An open source toolbox for analysis of single-trial EEG dynamics including independent component analysis. *J. Neurosci. Methods* **2004**, *134*, 9–21. [CrossRef]
46. Pion-Tonachini, L.; Kreutz-Delgado, K.; Makeig, S. ICLabel: An automated electroencephalographic independent component classifier, dataset, and website. *NeuroImage* **2019**, *198*, 181–197. [CrossRef] [PubMed]
47. Zhang, H.; Zhao, M.; Wei, C.; Mantini, D.; Li, Z.; Liu, Q. Eegdenoisenet: A benchmark dataset for deep learning solutions of eeg denoising. *J. Neural Eng.* **2021**, *18*, 056057. [CrossRef]
48. Klug, M.; Gramann, K. Identifying key factors for improving ICA-based decomposition of EEG data in mobile and stationary experiments. *Eur. J. Neurosci.* **2021**, *54*, 8406–8420. [CrossRef]
49. Plechawska-Wójcik, M.; Tokovarov, M.; Kaczorowska, M.; Zapała, D. A three-class classification of cognitive workload based on EEG spectral data. *Appl. Sci.* **2019**, *9*, 5340. [CrossRef]
50. Leach, S.C.; Morales, S.; Bowers, M.E.; Buzzell, G.A.; Debnath, R.; Beall, D.; Fox, N.A. Adjusting ADJUST: Optimizing the ADJUST algorithm for pediatric data using geodesic nets. *Psychophysiology* **2020**, *57*, e13768. [CrossRef]
51. Zhong, P.; Wang, D.; Miao, C. EEG-based emotion recognition using regularized graph neural networks. *IEEE. Trans. Affect. Comput.* **2020**. [CrossRef]
52. Curran, P.G. Methods for the detection of carelessly invalid responses in survey data. *J. Exp. Soc. Psychol.* **2016**, *66*, 4–19. [CrossRef]
53. DeSimone, J.A.; Harms, P.D.; DeSimone, A.J. Best practice recommendations for data screening. *J. Organ. Behav.* **2015**, *36*, 171–181. [CrossRef]
54. DeSimone, J.A.; Harms, P.D. Dirty data: The effects of screening respondents who provide low-quality data in survey research. *J. Bus. Psychol.* **2018**, *33*, 559–577. [CrossRef]
55. Murana, S.; Rahimin, R. Application of SPSS software in statistical learning to improve student learning outcomes. *Indo-MathEdu Intellect. J.* **2021**, *2*, 12–23. [CrossRef]
56. Maison, M.; Kurniawan, D.A.; Anggraini, L. Perception, attitude, and student awareness in working on online tasks during the covid-19 pandemic. *J. Pendidik. Sains Indones.* **2021**, *9*, 108–118. [CrossRef]
57. Chen, C. Research on teaching effect and course evaluation based on spss and analysis of influencing factors. In Proceedings of the 2021 4th International Conference on E-Business, Information Management and Computer Science, Hong Kong, China, 29–31 December 2021; pp. 229–234.

Article

Fear Detection in Multimodal Affective Computing: Physiological Signals versus Catecholamine Concentration

Laura Gutiérrez-Martín [1,2], Elena Romero-Perales [1,2], Clara Sainz de Baranda Andújar [1,3], Manuel F. Canabal-Benito [1,2], Gema Esther Rodríguez-Ramos [1], Rafael Toro-Flores [4], Susana López-Ongil [4] and Celia López-Ongil [1,2,*]

[1] UC3M4Safety Team, Universidad Carlos III de Madrid, c/Butarque, 15, 28911 Madrid, Spain; lagutier@ing.uc3m.es (L.G.-M.); eleromer@ing.uc3m.es (E.R.-P.); cbaranda@hum.uc3m.es (C.S.d.B.A.); mcanabal@ing.uc3m.es (M.F.C.-B.); gerodrig@pa.uc3m.es (G.E.R.-R.)
[2] Departamento de Tecnología Electrónica, c/Butarque, 15, 28911 Madrid, Spain
[3] Departamento de Comunicación, c/Madrid, 126, 28903 Madrid, Spain
[4] Fundación para la Investigación Biomédica del Hospital Universitario Príncipe de Asturias, Ctra. Alcalá-Meco s/n, 28805 Madrid, Spain; rafael.toro@uah.es (R.T.-F.); slorgil@salud.madrid.org (S.L.-O.)
* Correspondence: celia@ing.uc3m.es

Citation: Gutiérrez-Martín, L.; Romero-Perales, E.; de Baranda Andújar, C.S.; F. Canabal-Benito, M.; Rodríguez-Ramos, G.E.; Toro-Flores, R.; López-Ongil, S.; López-Ongil, C. Fear Detection in Multimodal Affective Computing: Physiological Signals versus Catecholamine Concentration. *Sensors* **2022**, *22*, 4023. https://doi.org/10.3390/s22114023

Academic Editor: Mincheol Whang

Received: 30 April 2022
Accepted: 18 May 2022
Published: 26 May 2022

Publisher's Note: MDPI stays neutral with regard to jurisdictional claims in published maps and institutional affiliations.

Copyright: © 2022 by the authors. Licensee MDPI, Basel, Switzerland. This article is an open access article distributed under the terms and conditions of the Creative Commons Attribution (CC BY) license (https://creativecommons.org/licenses/by/4.0/).

Abstract: Affective computing through physiological signals monitoring is currently a hot topic in the scientific literature, but also in the industry. Many wearable devices are being developed for health or wellness tracking during daily life or sports activity. Likewise, other applications are being proposed for the early detection of risk situations involving sexual or violent aggressions, with the identification of panic or fear emotions. The use of other sources of information, such as video or audio signals will make multimodal affective computing a more powerful tool for emotion classification, improving the detection capability. There are other biological elements that have not been explored yet and that could provide additional information to better disentangle negative emotions, such as fear or panic. Catecholamines are hormones produced by the adrenal glands, two small glands located above the kidneys. These hormones are released in the body in response to physical or emotional stress. The main catecholamines, namely adrenaline, noradrenaline and dopamine have been analysed, as well as four physiological variables: skin temperature, electrodermal activity, blood volume pulse (to calculate heart rate activity. i.e., beats per minute) and respiration rate. This work presents a comparison of the results provided by the analysis of physiological signals in reference to catecholamine, from an experimental task with 21 female volunteers receiving audiovisual stimuli through an immersive environment in virtual reality. Artificial intelligence algorithms for fear classification with physiological variables and plasma catecholamine concentration levels have been proposed and tested. The best results have been obtained with the features extracted from the physiological variables. Adding catecholamine's maximum variation during the five minutes after the video clip visualization, as well as adding the five measurements (1-min interval) of these levels, are not providing better performance in the classifiers.

Keywords: multimodal affective computing; catecholamines; emotion classification; wearable devices

1. Introduction

Affective computing, the study, analysis, and interpretation of human emotional reactions by means of artificial intelligence [1], has become a hot topic in the scientific community. Possible applications include accurate neuromarketing techniques, more efficient human-machine interfaces and new wellness and/or healthcare practices, with innovative therapies for phobias and mental illnesses [2–6]. Recently, the prevention of violent attacks on vulnerable people by means of the early detection of fear or panic emotional reactions is under research in this area [7].

In affective computing, many research areas merge to provide efficient and accurate systems capable of classifying the emotion felt by a person. Apart from psychology, neuroscience and physiology, other disciplines are required to automate the emotion detection process as well as to allow in-depth data analysis and useful feedback.

Human emotions are the consequence of biochemical reactions in the brain. External stimuli are processed in certain brain regions such as the amygdala, insula and prefrontal cortex [8–10]. These areas activate the autonomic nervous system, which triggers physiological changes as an emotional response. From the global emotional response, we can distinguish conscious and unconscious processes. The cognitive component in the emotion obtains a high degree of consciousness and can feedback the physiological reactions chain.

The measuring and processing of these physiological reactions allow automatizing the emotion detection and classification process, known as affective computing. If this detection involves several sources of information, it is known as multimodal affective computing. Validity and corroboration issues have made physiological variables the most attractive to researchers. Multimodal recordings commonly used are Galvanic Skin Response (GSR), ElectroMyoGraphy (EMG) (frequency of muscle tension), Heart Rate (HR), Respiration Rate (RR), ElectroEncephaloGraphy (EEG), functional Magnetic Resonance Imaging (fMRI), and Positron Emission Tomography (PET) [11], even though behavioural measurements such as facial expressions, voice, movement, and subjective self-reporting can also be useful for experimental purposes.

In this sense, some authors have related non-external physiological variables with emotional reactions [12]. For example, the levels of neurotransmitters in the brain or circulating catecholamines vary depending on a person's emotional state, affecting activity of physiological variables. Although their measures are very invasive, the relation between physiological variable changes and the concentration of these molecules makes them interesting in some applications of affective computing. For example, in risk situations, this early detection of fear or panic emotions would trigger a protection response for the person in danger. To date, there is no study using catecholamine concentration in blood plasma for emotion detection that includes an experimental sample in humans, just theoretical studies.

The concentration of catecholamines is usually measured in urine to diagnose or rule out the presence of certain tumours such as pheochromocytoma or neuroblastoma because these tumours raise the levels significantly. However, in basal conditions, the levels are low and can be detected in blood by high-performance liquid chromatography (HLPC) techniques.

Continuous and autonomous measurement of these molecules is not available currently, but if they prove useful, wearable analysis devices could be designed and developed, similar to insulin micropumps [13].

In this work, a methodology and protocol are proposed to connect the elicitation of human emotions with the variation of plasma catecholamine concentration. For this first test, fear is chosen as the target emotion for two main reasons. On the one hand, the relationship between neurotransmitters and stress or fear is well documented in the literature, as they are responsible for the activation of the body's fight or flight mechanisms. On the other hand, the protection of women against gender-based violence has been chosen as a target application. For this purpose, the objective is to be able to detect fear automatically so that an alarm is triggered to protect women in danger. Although there is already work in this area, so far only physiological variables have been used. In order to validate if the inclusion of catecholamine plasma concentration improves the results, an immersive virtual reality environment has been arranged to provoke realistic situations where the volunteer could have intense emotional reactions. Continuous monitoring of physiological variables, with a research toolkit system (for the sake of comparison with other affective computing research works), is connected with the virtual environment, as well as to an interface for the classification of the emotions elicited. The detection of emotions in humans through the plasma concentration of catecholamines has been analysed and compared with externally measured physiological variables, such as SKT, HR and

EDA. The main obtained results are very positive with regard to physiological variables while they are not conclusive for the levels of catecholamine concentration in blood plasma.

The main contributions of this work can be summarized as:

- The design of a methodology for plasma catecholamine concentration measurement along with physiological variables under audiovisual stimuli for automatic fear detection.
- An experimental test involving 21 volunteers where dopamine, adrenaline and noradrenaline are measured along with blood volume pulse, skin temperature, galvanic skin response, respiration rate, and electromyography.
- An analysis of the data collected, including both physiological variables and catecholamine concentration separately and also combined.
- An implementation and comparison of three artificial intelligence methods for fear detection using the measurements collected in the experimental test in order to validate the convenience of including plasma catecholamine concentration in fear detection systems.

The rest of this paper is organized as follows: Section 2 provides a review of the state of the art regarding emotion theory, automatic emotion detection, and physiological response related to catecholamines and emotion. As result, we can formulate the hypothesis of this work. Section 3 describes the methodology used in this work for the experimental setup, including the sample description, the design of the study, the stimuli used, the labelling method, and the collected measurements. Section 4 presents the experimental results (for labelling, physiological variables and catecholamine concentration). Additionally, we present an artificial intelligence algorithm analysis in order to validate the hypothesis formulated previously. The discussion is presented in Section 5, and finally, Section 6 concludes the work.

2. State of the Art: Emotions, Physiological Response and Affective Computing

2.1. Emotions

Emotions are fundamental for human beings since they play an important role in individual and social behaviour and mental processes, such as decision making, perception, memory, attention, etc. [14]. However, they have been partially ignored in the past, generally due to the difficulties they intrigue for experimental methodology.

The identification and classification of emotions for improving people's lives have gained interest in recent years as several fields can take advantage of the results in this area [15–17]. such as mental health, human-machine interfaces, learning and teaching methods, video games or neuromarketing. In psychology, emotions are described as "psychological states that include three components: subjective personal experience, associated physiological response, and behaviours" [18,19].

Within the literature and the state of the art in emotion identification and classification, there are two trends: (1) the classification of emotions as discrete elements, and (2) their inclusion in a continuous vector space. Within the first option, different classifications have been proposed. The first classification was presented by Ekman [20] using six basic emotions (happiness, sadness, disgust, fear, surprise, and anger). Since then, other classifications have been presented, adding emotions, or changing some of them [21,22]. Within the second option, we find the representation in the affective space. This consists of the multidimensional representation (usually within two or three axes) of the emotion so that the affective space becomes a continuous space in which every emotional state is represented by two or three coordinates. The most lately used space [23] proposes three dimensions (valence, arousal, and dominance). In this space, valence-pleasure (P) indicates positive or negative emotions; arousal (A) ranges from calm to high excitement levels; and finally, dominance (D) denotes the ability to control the emotion [24]. Several studies [25] of emotion classification use only a 2-dimensional space (PA space) using the valence and arousal axes previously described. That generates four quadrants in the space for locating emotions (Q1, Q2, Q3, and Q4). Some authors [26,27] have tried to place the discrete emotions in the quadrants according to the valence and arousal presumably experienced by each of them (see Figure 1a). Adding the third dimension (D) allows for differentiating discrete emotions sharing similar values in the PA space, such as fear and anger in Q2.

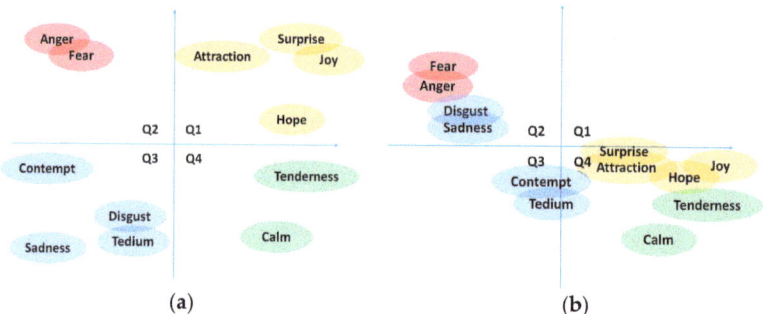

Figure 1. (a) Discrete emotion mapping in PA space in the literature. (b) Results extracted from Spanish study [28].

Both emotion classification systems present difficulties when applied to the automatic identification of emotions and their experimental validation. On the one hand, the use of discrete emotions is considerably biased by the sociocultural environment of the person [28], especially the background and the country of origin. In addition, there is reasonable dependence on the correct understanding of the description of the emotion or its nuances when identifying it [29]. In an attempt to address this, several emotions have been added to the list making it longer, but this also leads to problems for automatic emotion classification methods (as they add subtle differences in the responses). On the other hand, PAD affective space systems are often also related to the difficulty in understanding the three classification axes.

2.2. Emotion Detection

Affective computing has emerged to shed light on the gap where technology and emotions converge. One of the goals of this field is trying to model emotional response to a wide variety of stimuli by evaluating emotional states. These states become measurable regarding subjective self-reports, physiological variables and behaviour.

The main elements involved in affective computing systems are the emotions theory [30] which connects human affective reactions to external stimuli, attending to intrinsic and extrinsic factors, with externally measurable physical and physiological changes; collecting data with smart sensors, first through emotion elicitation experiments in the lab and secondly through live in-the-wild monitoring; and the generation, training and integration of artificial intelligence algorithms in autonomous systems [3].

In affective computing, those changes are objectively measured in the person to determine the emotion felt. External (behavioural) aspects, such as facial expression, voice, movement, etc., are voluntary and biased through culture and society, making them difficult to apply to user-independent emotion detection. On the other hand, physiological changes (involuntary reactions) with an external effect (it is possible to measure them in a non-invasive way), have been preferred [31]. Typical variables used in affective computing include galvanic skin response, which increases linearly with a person's level of arousal [32,33] electromyography (frequency of muscle tension), which is correlated with emotions of negative valence [34]; heart rate, which increases with negative valence emotions like fear [35,36]; respiration rate(how deep and fast the breath is), which becomes irregular with more aroused emotions like anger [37]; electroencephalography [38,39] and functional magnetic resonance imaging [40].

All these variables differ in many aspects, some of them are ease of measurement, which is related to how internal or external the target signal is; consciousness, because some variables can be consciously controlled and altered by the individual; and invasiveness, which means that some variables can be measured with low/high invasiveness for the individual. Many affective computing systems combine several variables in order to increase the performance of the application integrating solutions known as multimodal affective computing [41–43]. This allows combining several features from different sources making the automatic detection usually more complex but also with higher accuracy.

Intelligent algorithms should be trained with these measured physiological variables together with subjective perceived emotion during stimuli application. Among the different available options, we can feature according to the literature [44] those used in constrained devices as: Support Vector Machine (SVM) [45], K-Nearest Neighbours (KNN) [46] and Ensemble Methods (ENS) [47]. For training and research purposes, there are different databases compiling all these data for helping in the generation of affective computing systems [48,49].

The measurement of these physiological variables with wearable devices during daily life is associated with a high amount of noise due to interferences and users' movements [50]. There are several works proposing solutions to eliminate or reduce this noise, through filters, algorithms, and even, fuzzy logic [51], but these techniques are expensive in terms of power consumption, the time required, and computation effort.

In order to try to overcome this problem, other variables could be tested in order to validate its inclusion pertinence. Among them, catecholamines' presence in blood plasma, saliva or sweat could be an interesting option, even if its measurement is more invasive, as they could be more robust against artifacts.

2.3. Chatecolamines in Emotion Detection

Since the first half of the 20th century, explanatory theories emerged to explain the physiological changes caused by stressful stimuli that altered the body's homeostasis. These theories somehow evolved from the "stress non-specificity" approach to the "stress specificity" approach [52]. This means that the first theories of stress regarded this response relatively independent of the type of threat. Whether it was exposure to cold, haemorrhage or distressing emotional encounters, the stress response would be essentially the same [53]. However, recent data and observations indicate the probable existence of a variety of stressors with different targets and different effects on homeostasis [54]. These theories tend to explain the stress response by considering that it has a primitive type of specificity, with differential responses of the sympathetic nervous and adrenomedullary hormonal systems, depending on the type and intensity of the stressor perceived by the organism and interpreted in the light of experience [55]. The activation of the adrenomedullary hormonal system has been linked to glucoprivation and emotional distress such as fear. There is some evidence to confirm an accumulated association between noradrenaline and active escape, avoidance or attack, and a link between adrenaline and passive, immobile fear [56].

Catecholamines are hormones made in nerve tissue, the brain, and the adrenal glands. If they are found in the synapses of the nervous system, they are classified as neurotransmitters, and if they are found in the bloodstream, they are classified as hormones. The adrenal glands produce large amounts of catecholamines in response to acute stress or elevated arousal [57]. The main catecholamines are adrenaline (epinephrine), noradrenaline (norepinephrine) and dopamine. Catecholamines help the body to respond to stress or fear and prepare the body for "fight or flight" reactions [58]. This reaction to states of threat or high arousal results in a general discharge of catecholamines from three peripheral systems: the sympathetic branch of the autonomic nervous system, the adrenomedullary hormonal system and the autocrine/paracrine dopaminergic system. The activation of these systems favours the secretion of catecholamines into the bloodstream, where they trigger a cascade of physiological changes in peripheral tissues after binding to their receptors. Catecholamines increase heart rate, blood pressure, respiratory rate, muscle strength, and alertness. They also reduce the amount of blood going to the skin and intestines and increase blood going to major organs, such as the brain, heart, and kidneys [59].

Theoretical studies such as [12] propose that there is a direct relationship between neurotransmitter levels (dopamine, noradrenaline, and serotonin) and emotions. In this model, for example, fear is related to a combination of a low level of serotonin, a low level of noradrenaline and a high level of dopamine, (see Figure 2).

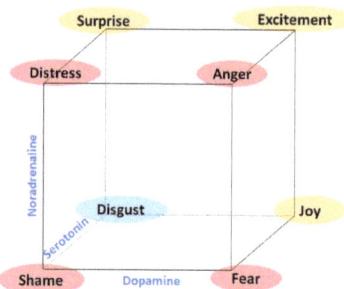

Figure 2. Loveheim cube showing correspondence among catecholamines and emotions (based on [12]).

Loveheim's study describes a theoretical framework that, if measurable, could improve multimodal affective computing systems for the automatic identification and classification of emotions. In fact, the study proposes to continue this research with a further experimental test that allows validating his proposal. Walker also proposes a theoretical framework that includes cortisol (a hormone produced in the adrenal gland) as an indicator related to fear and stress [60]. Again, this work suggested validating this framework with experimental tests. There are no results for catecholamines and human emotions experiments, although some previous tests have been performed in cats [61]. Directly measuring the presence of neurotransmitters is very invasive and nearly impossible on a day-to-day basis, so measuring catecholamines' presence in blood plasma in an experimental setup in order to confirm whether there is a relationship between this presence associated with different emotional states is a good starting point for future developments in affective computing research.

2.4. Hypotheses

Once the state of the art is reviewed, it can be stated that there is a lack of experimental studies that validate the relationship and convenience of using the concentration of plasma catecholamine in affective computing. So, in this work, the authors propose that:

- The emotional states of fear and no-fear can be discriminated through the plasma catecholamine concentration levels
- Using catecholamine concentration level improves the results for fear detection provided by the use of solely physiological variables.

If this hypothesis is proved correct, an automatic system for early detection of emotional states of fear can be implemented, reducing the effect of interferences and noise in the measured signals. Better protection for people in dangerous situations will be provided through the activation of early protective responses.

3. Material and Methods

In this section, we present the proposed methodology for data collection of both physiological variables and catecholamines in an immersive environment for emotion elicitation. Since the design of this experiment involves the extraction of blood samples for the analysis of catecholamines in blood plasma, and the number of samples cannot be high, fear has been chosen as the target emotion, since, as discussed in Section 2, it is highly related to the release of catecholamines.

In addition, some considerations have to be taken into account. As stated before, one of the objectives of the authors is to apply multimodal affective computing to the protection of women victims of gender-based violence. For this reason, the sample of this study is entirely composed of women, and the proposed final application also influences the choice of one of the audio-visual stimuli, which is directly related to gender violence.

3.1. Sample of the Study

The study population consisted of 21 volunteers, all of them apparently healthy women. All of them were Spanish women, and healthcare workers. Study subjects were not allowed to perform strenuous exercise, smoke, eat some foods, or take drugs or some medicines (Table 1) at least 24 h before analysis, to avoid interference with catecholamines measurement.

Table 1. Foods, drinks, and drugs can interfere with the analysis of catecholamines.

Food	Drinks	Drugs	Medicines
Cocoa	Coffee	Amphetamine	Paracetamol
Citric Fruits	Tea	Caffeine	Phenoxybenzamine, phenothiazine
Walnuts	Chocolate	Nicotinic Acid	Levodopa
Beans	Beer	Cocaine	Monoamine oxidase inhibitors
Avocado, Banana	Red wine		Reserpine
Vanilla			Pseudoephedrine

Main data of female volunteers are registered in Table 2. The mean age of the volunteers is 36. Only 5 of them had one child, and 13 volunteers were single. With regard to Body Mass Index (BMI), only 4 volunteers presented values between 25 and 30, overweight indicative. Finally, 4 volunteers are in their menopause. Some volunteers (6) were taking treatments for chronic illnesses (hypertension, chronic pain, heart failure, ulcerative colitis, anaemia, and diabetes).

Table 2. Characteristics of women volunteers.

Parameter	Mean ± Std Deviation (SD)/Nb.
Age (year)	36.19 ± 13.43
Weight (kg)	61.20 ± 8.68
Height (cm)	164.29 ± 5.09
BMI (kg/m^2)	22.75 ± 3.56
Food, drinks, drugs	Citric fruits (3), coffee (11), tea (2) and alcohol (1)
Medicines reported	Analgesic (5), chronic illness treatment (3), contraceptives (1), and vitamin (1)
Stress situation	5
Intense exercise	2

The study conforms to the ethical principles outlined in the Declaration of Helsinki. Design of the study was approved by the Research Ethics Committee (REC) of Principe de Asturias Hospital with protocol number: CLO (LIB 10/2019). All participants received a detailed description of the purpose and design of the study and signed informed consent approved by the REC.

3.2. Design of the Study

The study consisted in measuring the physiological variables of a set of volunteers while they were watching a set of 4 emotion-related videos in an immersive virtual reality environment. Additionally, several blood extractions were performed after the visualization of three of these videos to analyse the plasma catecholamine levels (dopamine, adrenaline, and nor-adrenaline). Besides, after every video watching, the volunteer labelled the emotions elicited during the visualization.

Each participant fasted at least twelve hours before the experiment. Previously to the experiment, the participant filled in a form providing information such as personality traits, sex, age group, recent physical activity, or medication (which could alter the participant's physiological response), self-identified emotional loads, and mood bias (fears, phobias, or traumatic experiences), summarized in Table 2. This information could be relevant and informative to the emotional reactions of the participants during the experiment, affecting their cognition, appraisal, and attention.

The experiment was designed to last globally 2 h. In Figure 3, the schedule of the experiment is shown. After the interview, filling in the questionnaire, and signing the informed consent, the test schedule and protocol were explained to every volunteer and some demo was performed in relation to the virtual reality environment. Then, the sensors for measuring the physiological variables were located. The BioSignalPlux® research toolkit system was used to register the physiological variables evolution throughout the study, such as forearm skin temperature, galvanic skin response, finger blood volume pulse (BVP), trapezoidal electromyogram, and chest respiration. The system is placed in different locations in the volunteer's body (arm, hand, chest, and finger), (Figure 4). These physiological signals were selected because they could be easily implemented in an inconspicuous and comfortable wearable device, avoiding any disadvantage to the user. There are smartwatches that already integrate BVP, GSR, and SKT sensors. Respiration and EMG could be integrated into a patch or band. This characteristic is mandatory for this type of application.

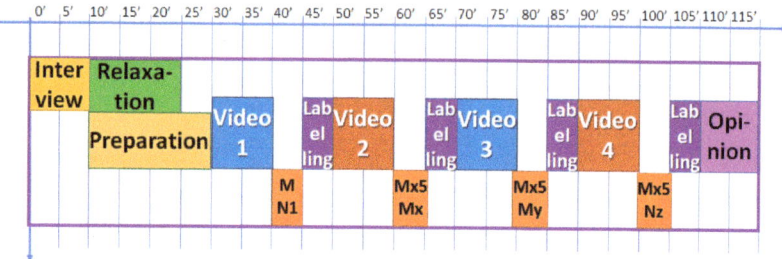

Figure 3. Schedule of the experiment for each volunteer.

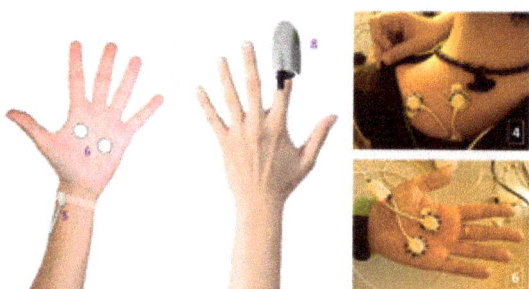

Figure 4. Electrodes and sensors position for experiment.

Once explained how to handle the equipment to label each video, the nurse proceeded to put a via in the antecubital vein to extract blood samples at different time points of the study, at the beginning (basal point) and after each video (5 samples). Each subject watched four unexpected videos related to different emotions that had to be labelled according to what she was feeling at that moment. Just after finishing each video a blood sample was taken. After videos 2, 3 and 4, five samples were collected, separated 1 min each, to monitor the changes in catecholamine levels, (Figure 5).

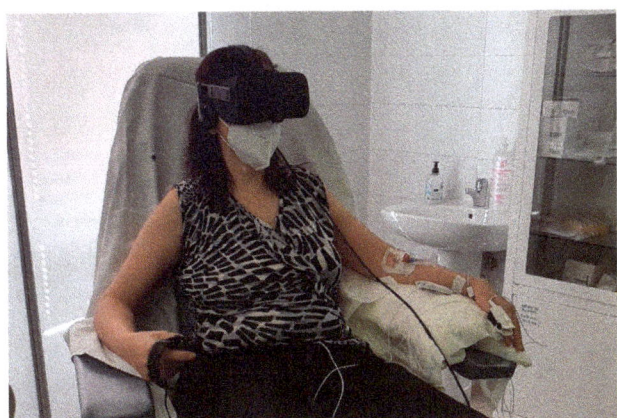

Figure 5. Volunteer ready to start the experiment.

3.3. *Audiovisual Stimulus*

Every subject watched four videos, two of them related to the emotion of fear, one related to calm and the other one related to joy. The schedule is Calm Fear Joy Fear. The order of fear-related videos is randomly set for each volunteer.

The video clips used for the experiment were selected from the UC3M4Safety Database of audiovisual stimuli aimed to elicit different emotional reactions through an immersive virtual reality environment [62] (see Figure 6). Most of the clips were 360-degree scenes providing more realistic experiences.

Figure 6. Screenshots for fear and calm video visualization.

The Oculus™ Rift S Headset was used under an application built on Unity™ that connects the video clips projection to the physiological monitoring system and records the emotion labelling. The whole data recording system was initiated by the virtual reality environment that manages both video stimuli and sensor measurement. A TCP/IP port connection was created at the beginning of the trial to communicate with the OpenSignals application. The information storage was divided by scenes, meaning each file contained the information collected between two timestamps (start and end of each screen) set by the environment, thus enabling synchronization.

The four video clips were V1, V2, V3, and V4, aimed to provoke calm, fear (gender-based violence related), joy and fear, respectively.

- V1: "Nature"—calm
- V2: "Refugiado"—fear related to gender-based violence

- V3: "Don't stop me now"—joy
- V4: "Inside chamber of horrors"—general fear

These videos obtained a very good unanimity in discrete emotion, higher in the case of women for the fear and joy clips while the mean and standard deviations in the PAD affective space dimension are also closer than expected for fear clips and for women, (Table 3). In this table, the discrete emotion labelled for every video is shown for the experiment detailed in [28], as well as the three dimensions of the PAD affective space. As it could be seen, V2 has a very high unanimity in the discrete emotion of fear in women, and also V4. Regarding PAD variables, the dispersion and the mean are complying with the expected ranges.

Table 3. Emotional Labelling of the video clips used in the experiment [28].

Video Clip	Target Emotion	Duration	Unanimity (Discrete) Men	Women	PAD (Mean/SD) Men	Women
V1	Calm	60 s	78%	74,4%	V: 7.3 (1.7) A: 2.1 (1.1) D: 6.8 (1.8)	V: 7.7 (1.7) A: 2.0 (1.7) D: 6.6 (2.4)
V2	Fear Gender-based violence	93 s	62.1%	93.2%	V: 2.5 (1.8) A: 7.1 (1.2) D: 4.2 (1.7)	V: 1.7 (0.7) A: 7.7 (0.9) D: 3.4 (1.6)
V3	Joy	101 s	71.9%	83.3%	V: 7.3 (1.6) A: 4.6 (2.1) D: 6.6 (2.0)	V: 7.8 (1.3) A: 4.5 (2.2) D: 7.2 (1.9)
V4	Fear	119 s	75.0%	84.2%	V: 2.9 (1.7) A: 6.6 (1.7) D: 4.3 (2.3)	V: 2.7 (1.6) A: 6.9 (1.7) D: 4.3 (2.2)

3.4. Labelling

In order to try to overcome the problems related to labelling method mentioned above, in this work, we have decided to include both a discrete classification of emotions (joy, hope, surprise, attraction, tenderness, calm, tedium, contempt, sadness, fear, disgust, and anger), plus an indicator of emotional intensity to be able to detect more nuances, and the classification in the PAD affective space using the SAM methodology [63] (see Figure 7). As depicted in Figure 3, the labelling is carried out just after the blood sample collection.

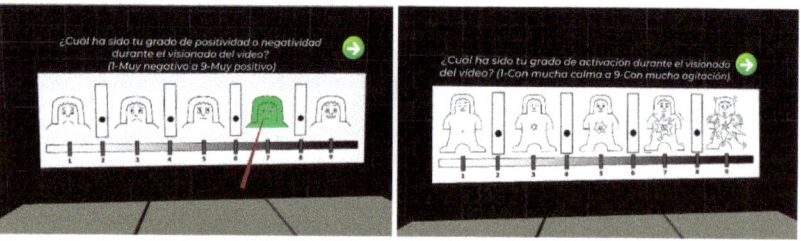

Figure 7. Labelling screen used in the experiment.

3.5. Measurement of Dopamine, Adrenaline and Noradrenaline

We have carried out the determination of catecholamines in 3 mL of plasma by high-performance liquid chromatography (HPLC). Blood samples were collected in pre-chilled EDTA-treated tubes, in the morning after a 12-h overnight fast and resting period. As several samples had to be taken every few times after watching each video, a via was placed to assist sample collection from each point of the study. Plasmas were immediately separated, to prevent catecholamines degradation, by centrifugation at 2000× g for 15 min at 4 °C. After that, the plasma was collected in clean and pre-chilled tubes and then stored at −80 °C until measured. All plasmas were properly submitted to Reference Laboratory S.A. (L'Hospitalet de Llobregat, Barcelona, Spain) to measure by HPLC the adrenaline, noradrenaline and dopamine in each sample.

Measurement of serotonin requires serum instead of plasma, needing the extraction of additional 5 mL blood samples from each volunteer. Apart from the extra cost, equivalent to measuring the other three catecholamines, the large number of samples required has prevented the authors from analysing the evolution of serotonin concentrations during the study.

4. Experimental Results

The experiments were performed from December 2020 to January 2021, on 12 and 9 volunteers, respectively.

4.1. Emotion Labeling

As it was already mentioned, emotional labelling is a complex task, not only because sometimes the target emotions are not the ones that are elicited to the volunteers, but also because of the terminology.

For that reason, at first, it is important to analyse the distribution of the labels reported during the experiment and study how well the clips have been eliciting their target emotions.

Taking into consideration discrete classification, (Figure 8), the clip targeting general fear emotion (V4) is the one with the highest agreement among the volunteers, 95% of them labelled it as fear. In the case of the clips of calm (V1) and joy (V3), a unique emotion does not obtain a clear majority; however, if the quadrants of PAD space are analysed, these videos show 76% and 90% of agreement, respectively.

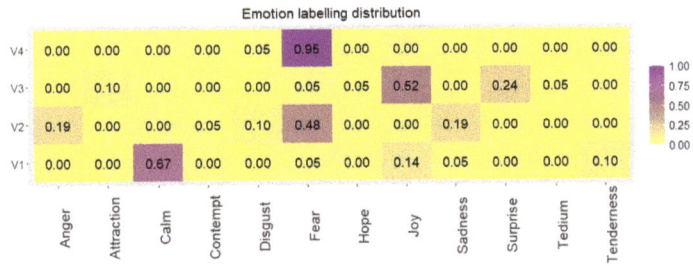

Figure 8. Emotion labelling distribution (0.00–1.00) between emotions reported by the volunteers w.r.t. each video clip visualized.

On the other hand, V2 shows the highest dispersion, although fear is the most used label (48%), anger (19%), and sadness (19%) represent approximately 40% of the reported classifications. This scattering is mainly due to the scenes presented in the clip. As we have already found in previous works [28], gender-based violence videos elicit this variety of emotions depending on the volunteer's perspective (first person or external).

As regards continuous labelling, independently from the dispersion found in discrete labelling, both fear clips are represented in their theoretical ideal position in the PAD space, low-valence, low-dominance and high-arousal corner.

The same occurs with the calm and joy clips which are placed at spots of high-valence, medium-high dominance, and medium-low arousal, with the joy clip being slightly above in terms of arousal.

Looking at previous results, and to observe the intercorrelation between volunteers when classifying all the clips, the correlation coefficient is computed considering all continuous reported labels. As result, a high positive relationship is obtained between all the volunteers, except for V002 and V005, who barely correlate with the rest, Figure 9. These results allow us to check that the emotions elicited are not only close to the original target (at least in the quadrant) but also inter-volunteer.

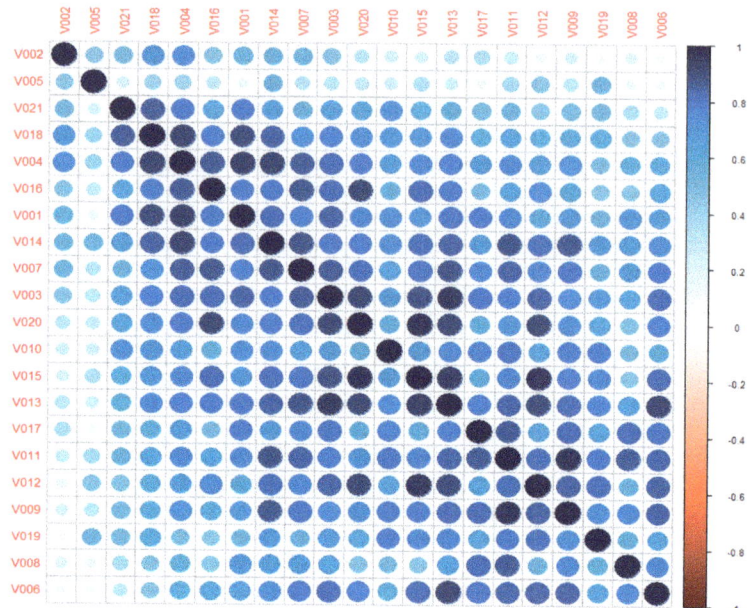

Figure 9. Correlation matrix between volunteers considering continuous reporting labelling.

4.2. Physiological Variables

From the physiological variables measured, the authors extracted features from the forearm skin temperature, skin conductance (GSR), finger blood volume pulse (BVP), and respiration. These variables have been measured throughout the whole experiment for every volunteer. First, a global analysis of the whole group of volunteers was carried out, for every video clip watched and, consequently, for every emotion. Later, temporal evolution of every physiological variable was also performed to find patterns of evolution during the visualization of the different emotion-related video clips.

4.2.1. Median and Quartile Distribution of Extracted Features per Video Clip

This analysis has been performed on the measurements from all the volunteers, considering the target labels of emotion, normalizing every volunteer with respect to their own values.

Although Clip 2 (V2) and Clip 4 (V4) have the same fear label, V2 includes gender-based violence and the emotional reactions are very different from the reactions on V4, as it has been detailed in the previous section.

The extracted features from the physiological variables are Inter-Bit-Interval (IBI) and Heart Rate Variability (HRV) extracted from BVP, which are very related to the degree of arousal, and the phasic peaks of GSR and the mean of GSR, which have been identified with the variables that work better for artificial intelligent algorithms in affective computing. These features are computed in 60 s windows.

As it can be observed in the Figure 10, the median and quartile distribution (box plots) IBI (a) and HRV (d) are the physiological features that better differentiate fear-related emotions, while the mean (c) and peaks (b) of GSR are clearly different for fear emotions (V4). Even, gender-based violence (V2) reactions are not distinguishable from calm or joy in terms of median values.

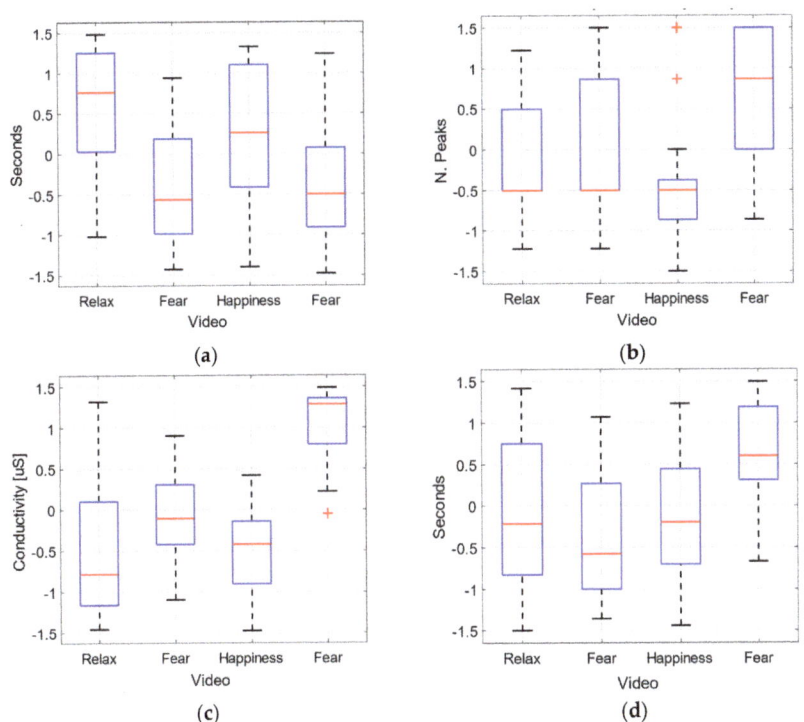

Figure 10. Normalized physiological features per video. (**a**) IBI. (**b**) number of phasic GSR peaks. (**c**) mean of GSR. (**d**) HRV rmssd.

The statistical analysis ANOVA on the features extracted from the physiological variables has provided some differences in the effect of different emotions elicited. In Table 4, the *p*-values for the comparison between videos are shown. We have observed significant values for the comparison between the effect of video clip V1 (calm) and video clips V2 and V4, for the mean of GSR. Additionally, there are significant differences in the effect of V1 and V4 for the IBI, and V3 and V4 for the number of peaks of GSR.

Table 4. *p*-values results from Kruskal-Wallis one-way ANOVA test for physiological data grouped by video clip.

Group A	Group B	GSR_mean	GSR_npeaks	HRV	IBI
V1	V2	0.22291	0.99970	0.63272	** 0.00155
V1	V3	1	0.75762	0.97703	0.66276
V1	V4	** 1.82×10^{-7}	0.03096	0.07931	** 0.00119
V2	V3	0.22578	0.70152	0.86245	0.06034
V2	V4	* 0.00163	0.04035	** 0.00196	0.99989
V3	V4	** 1.89×10^{-7}	** 0.00111	0.02655	0.05052

NOTE: Significant codes: '**' 0.001 '*' 0.01 ' ' 0.05.

4.2.2. Temporal Evolution of Physiological Variables

Temporal evolution analysis provides information about the evolution of the emotional state during the video. It should be noted that videos are labelled according to the prevailing emotion, but the same video could elicit more than one emotion, and the intensity could be non-homogeneous. This is a limitation of this type of experiment where continuous labelling is not possible. The result is dispersion/noise in the data, hindering their classification and modelling. Figure 11 shows the mean evolution of the four features used in the previous section.

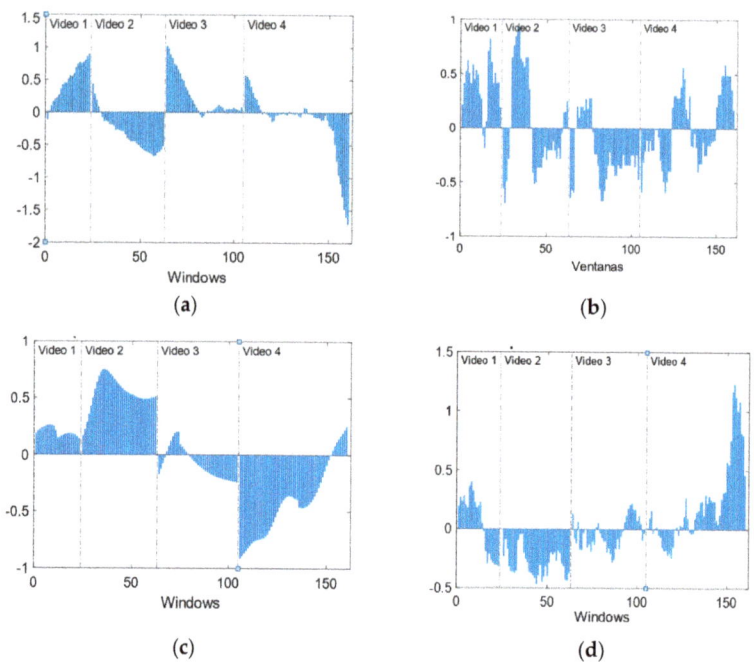

Figure 11. Temporal evolution of normalized features. (**a**) IBI. (**b**) Number of phasic GSR peaks. (**c**) Mean GSR. (**d**) HRV.

The four videos present a high variation of the selected features, especially V4. These variations correlate with scenes in the videos. In Figures 12 and 13, details on the scenes of both videos, V2 and V4, related to the fear emotion, are provided. As it could be seen, the most intense period of stress-fear in V2 is between seconds 32 and 58 when the boy is trying to open the bathroom's door. In Figure 11, features extracted from physiological variables present a very different behaviour in this period of time that, in some cases, it is maintained untill the end of the video due to the empathizing effect with the escaping mother and boy. Until they discover the aggressor is not in the lift, second 90, the climax is maintained.

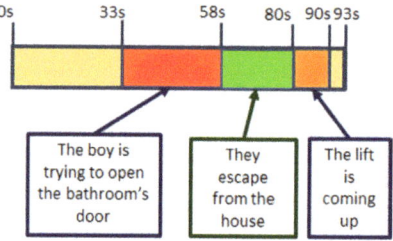

Figure 12. V02 main stressful events. "Refugiado" Diego Lerma 2014. Available at [62].

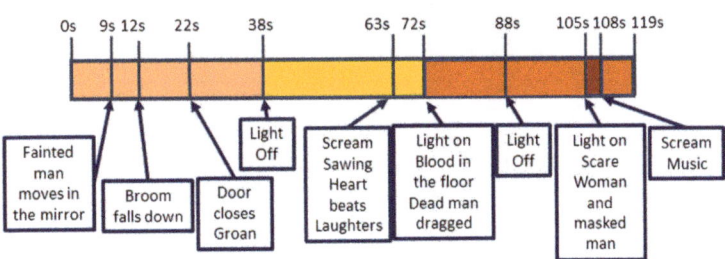

Figure 13. V04 main stressful events. "Chamber of horrors" Inside 360 VR Prod 2018. Available at [62].

With regard to V4, all the scenes are stressful but peak instants are when lights go off (seconds 38 and 88) and there are screams or sudden hits/blows (seconds 12, 22, 63, and 105). The worst moment is when two people appear suddenly in front of the viewer, no-faced, with loud music and screams (105); all features show a change of behaviour around this final scare that has been under preparation right from second 63.

4.3. Catecholamine Concentration

The concentration of adrenaline, dopamine and nor-adrenaline catecholamines, has been measured as detailed in Section 3, with the HPLC technique. In Table 5 the concentration values for these catecholamines are detailed per volunteer. A global analysis of these values has been performed to determine the relationship between the emotional reaction and these concentrations. First, the box plots of mean and quartile for every video clip were obtained, Figure 14. Second, to analyse the temporal evolution of these concentrations, temporal graphs were plotted, in Figures 15 and 16.

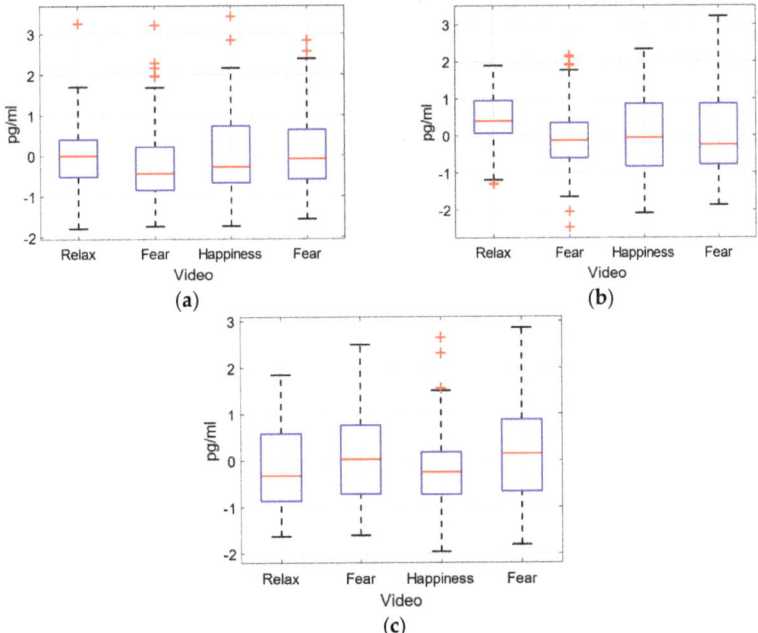

Figure 14. Normalized concentrations for dopamine, adrenaline and nor-adrenaline (a–c) for every video clip.

Table 5. Plasma catecholamine concentration levels for every volunteer for every sample (pg/mL), for adrenaline (A), dopamine (DA) and noradrenaline (NA).

		Volunteer 1			Volunteer 2			Volunteer 3			Volunteer 4			Volunteer 5			Volunteer 6			Volunteer 7		
	Sample	A	DA	NA	A	DA	NA	A	DA	NA	A	DA	NA	A	DA	NA	A	DA	NA	A	DA	NA
Video 1 - Basal	1	12	11	274	15	12	503	16	9	292	45	13	309	31	13	338	32	29	566	41	52	331
Video 2: Refugee (Fear GBV)	2	47	11	492	13	25	480	59	9	434	42	37	346	16	13	270	29	10	454	30	24	538
	3	48	12	379	23	16	614	32	11	455	44	22	371	26	11	336	22	11	579	37	34	591
	4	29	10	287	22	17	456	29	10	500	45	21	411	32	15	249	26	12	467	40	27	642
	5	30	11	360	23	11	604	32	8	520	32	21	310	29	11	294	31	19	500	25	20	601
	6	23	29	232	32	16	547	43	13	434	42	28	424	29	17	231	40	10	435	32	16	491
Video 3: Queen (Joy)	7	17	31	335	19	25	445	21	9	373	26	30	362	33	11	247	23	10	415	25	31	267
	8	21	20	302	24	10	569	38	9	396	46	29	368	24	22	234	12	11	451	49	28	376
	9	22	23	344	14	11	633	23	12	375	40	32	415	30	13	238	37	8	402	21	32	337
	10	37	11	300	28	14	542	50	9	363	30	17	313	38	18	237	12	10	446	43	49	371
	11	22	13	302	33	18	469	21	9	351	40	26	410	38	22	201	19	9	333	13	13	376
Video 4: Inside de chamber of horror (Fear)	12	10	27	284	27	11	492	58	9	289	46	26	413	28	16	300	14	13	474	21	47	279
	13	11	14	374	20	16	520	37	9	402	48	17	442	30	14	298	41	8	451	28	40	414
	14	32	17	410	29	15	558	28	11	330	46	21	415	36	13	273	21	12	446	45	34	343
	15	42	11	280	25	14	595	30	9	426	39	30	397	27	14	264	52	9	338	27	41	267
	16	20	20	368	19	15	623	17	13	450	31	21	361	26	15	271	29	11	478	14	37	293

		Volunteer 8			Volunteer 9			Volunteer 10			Volunteer 11			Volunteer 12			Volunteer 13			Volunteer 14		
	Sample	A	DA	NA	A	DA	NA	A	DA	NA	A	DA	NA	A	DA	NA	A	DA	NA	A	DA	NA
Video 1 - Basal	1	19	15	363	27	16	144	28	14	475	23	13	233	27	12	233	39	31	225	24	14	315
Video 2: Refugee (Fear GBV)	2	26	9	437	17	11	129	20	9	406	17	13	238	23	17	229	16	16	242	14	12	370
	3	18	9	475	13	11	114	34	16	289	28	19	239	21	18	212	41	34	268	12	14	387
	4	20	18	492	37	16	137	22	13	576	28	11	212	27	9	253	20	24	278	13	15	449
	5	17	10	481	11	12	108	27	13	521	33	16	270	25	8	183	17	16	256	14	9	280
	6	13	9	642	21	17	95	16	8	419	28	18	299	14	9	239	14	16	244	16	10	279
Video 3: Queen (Joy)	7	16	8	311	35	17	125	42	11	421	22	16	319	31	10	210	35	20	241	33	11	239
	8	15	10	375	29	33	107	25	20	370	16	16	267	29	22	239	22	47	468	19	13	458
	9	12	9	375	23	13	119	13	14	619	30	22	277	20	9	235	19	33	240	11	9	328
	10	19	15	233	23	20	108	11	10	615	13	9	250	26	12	197	26	42	348	14	31	420
	11	20	12	243	12	13	100	22	11	148	33	30	256	35	19	226	25	43	303	19	9	416
Video 4: Inside de chamber of horror (Fear)	12	22	9	380	21	12	114	14	11	160	19	15	228	35	18	178	15	46	452	18	9	675
	13	13	12	370	17	18	121	13	21	255	12	12	247	20	9	253	23	38	429	13	12	423
	14	21	13	338	18	20	141	35	11	296	17	12	229	23	14	280	16	45	453	11	15	530
	15	14	11	246	26	41	212	44	17	476	11	11	217	23	13	293	17	27	333	17	13	554
	16	43	14	322	27	13	171	41	16	295	31	14	251	30	16	238	16	18	457	31	11	643

		Volunteer 15			Volunteer 16			Volunteer 17			Volunteer 18			Volunteer 19			Volunteer 20			Volunteer 21		
	Sample	A	DA	NA	A	DA	NA	A	DA	NA	A	DA	NA	A	DA	NA	A	DA	NA	A	DA	NA
Video 1 - Basal	1	49	13	153	38	20	447	31	15	288	29	15	333	33	16	627	15	16	359	23	18	312
Video 2: Refugee (Fear GBV)	2	33	10	147	31	17	609	15	18	138	24	10	297	28	12	710	22	20	332	17	11	324
	3	43	15	187	24	14	539	26	14	150	26	10	286	26	25	704	12	11	438	24	13	302
	4	38	19	186	34	17	481	20	12	160	21	16	388	26	12	630	25	10	285	24	13	407
	5	29	13	171	39	12	586	29	17	143	35	30	259	23	14	462	12	17	300	16	14	335
	6	33	11	159	31	12	519	39	18	143	17	22	284	27	11	552	19	14	278	16	11	318
Video 3: Queen (Joy)	7	33	13	186	43	15	516	11	21	167	31	52	288	16	15	411	18	16	391	12	18	325
	8	48	15	204	40	10	498	14	13	223	21	19	375	13	15	583	44	14	331	13	16	347
	9	37	10	228	46	19	496	15	16	243	25	33	578	30	11	606	27	11	220	15	19	247
	10	27	14	211	33	14	643	30	15	147	19	29	383	25	11	575	36	14	233	14	16	270
	11	39	11	191	37	11	624	35	12	166	19	16	323	36	12	516	17	13	310	14	11	301
Video 4: Inside de chamber of horror (Fear)	12	44	14	255	31	17	508	12	17	239	30	18	306	27	15	604	14	19	395	41	20	367
	13	45	33	199	39	19	433	17	10	285	22	11	444	22	13	483	35	16	387	22	11	345
	14	47	28	282	30	14	594	11	11	192	19	11	418	36	10	603	15	18	240	29	29	298
	15	23	40	234	27	14	387	13	12	203	20	19	280	41	14	600	33	18	187	38	26	312
	16	35	20	259	26	14	368	21	14	222	19	11	435	30	11	496	34	15	282	21	15	401

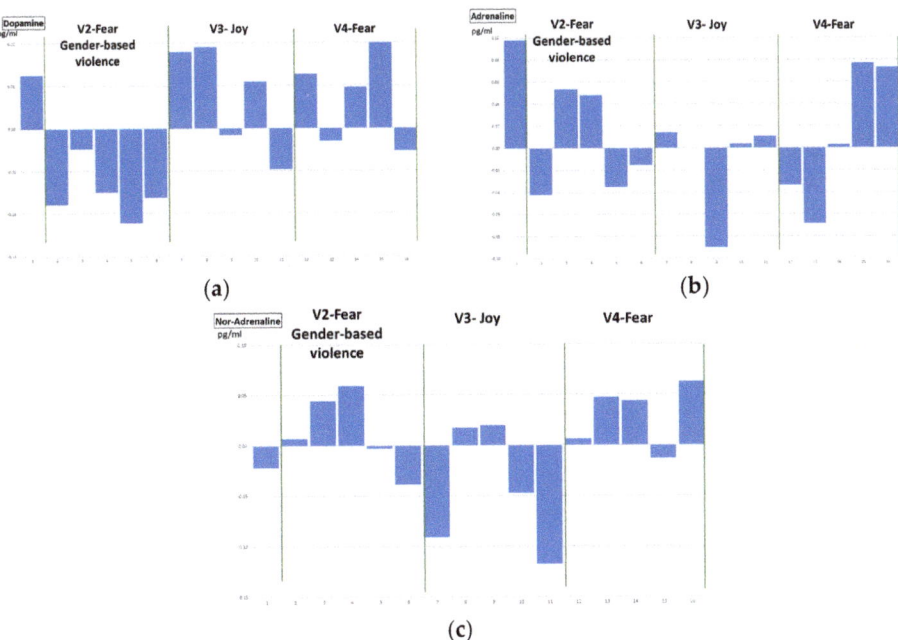

Figure 15. Temporal evolution of normalized concentrations for dopamine, adrenaline and noradrenaline (**a**–**c**) for every video clip, mean for all volunteers.

4.3.1. Catecholamine Concentration and Quartile Distribution

Data was collected per video clip, normalized per volunteer, and mean values were calculated for all the volunteers.

The obtained values do not show differences in catecholamine concentrations for different emotion-related video clips, especially for adrenaline and dopamine. Furthermore, for these catecholamines (A and DA), the gender-based violence fear video clip (V2) presents very dispersed values, while the fear video clip (V4) provides higher dispersion just for dopamine, Figure 14.

The statistical analysis ANOVA of the plasma concentration level has not provided a clear difference between the effects of different emotions elicited for the three catecholamines measured. In Table 6 the *p*-values for the comparison between the videos are shown. No significant values have been obtained for any pair compared.

Table 6. *p*-values results from Kruskal-Wallis one-way ANOVA test for catecholamine concentration data grouped by video clip.

Group A	Group B	Adrenaline	Noradrenaline	Dopamine
V1	V2	0.82591	0.90859	0.62776
V1	V3	0.65790	0.99983	0.97443
V1	V4	0.76604	0.95005	0.99913
V2	V3	0.95784	0.56652	0.53611
V2	V4	0.99743	0.99573	0.25316
V3	V4	0.98951	0.71117	0.95883

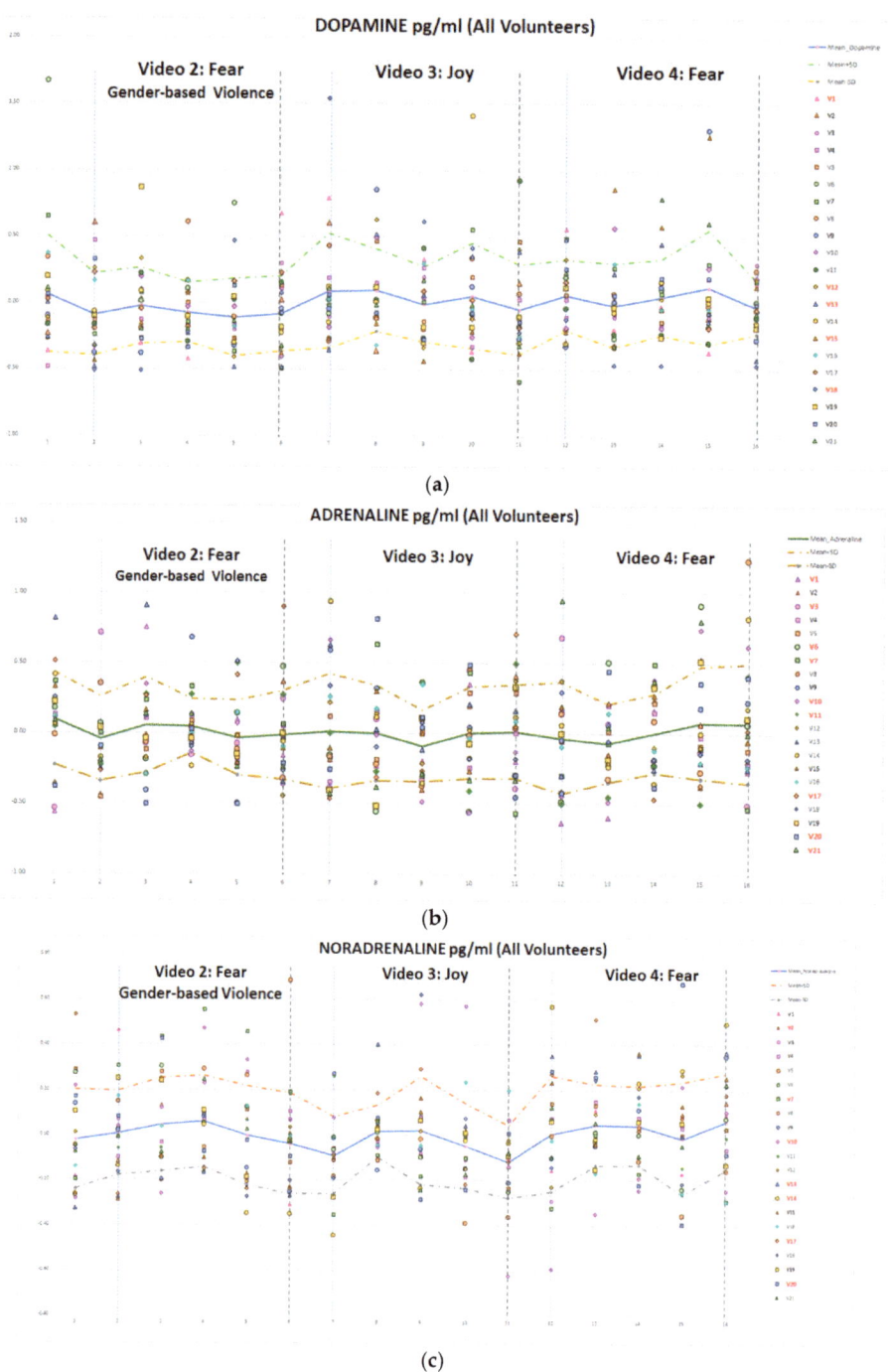

Figure 16. Temporal evolution of normalized catecholamine concentration for video clips 2, 3 and 4 for DA (**a**), A (**b**) and NA (**c**).

4.3.2. Temporal Evolution of Catecholamines after Video Clip Watching

Figure 15 shows the temporal evolution of dopamine (a), adrenaline (b) and noradrenaline (c) for video clips V2, V3 and V4, related to fear (gender-based violence related), joy and fear, respectively. The graphs represent the concentration of catecholamines, per sample (five per video per volunteer), as well as the mean value (continuous line) and the mean plus/minus standard deviation (dashed lines) for all the volunteers. Catecholamine concentration values have been normalized with respect to the mean value of every volunteer. For the sake of clarity, and for comparison with respect to the behaviour of physiological variables, in Figure 15 the temporal evolution of the mean value (for all volunteers) has been plotted for the three catecholamines. Dopamine concentrations show a slightly different evolution after watching the video clips related to fear with gender-based violence than in those related to joy or fear, where a final drop can be appreciated, (Figure 15a). Adrenaline concentration shows a continuous rising tendency for the fear-related clip (V4) while for joy (V3), a stabilization is observed in the final samples (Figure 15b). In the gender-based violence clip (V2), the stressful/relieving situation may provoke a rise and a drop in the adrenaline's concentration. Finally, in the noradrenaline's concentration (Figure 15c), a similar evolution can be observed in V2 and V3 (fear with gender-based violence and joy) with a final drop in the normalized value, while V4 (intense fear) is not presenting the final drop, since the stressful situation continues to get even more stressful until the end of the clip.

4.4. Artificial Intelligent Algorithms

Considering our goal, which is to study the improvement that catecholamines measurements can bring to our fear/not-fear detection model and compare the results with physiological models, the data were normalized, reorganized, and grouped by clip for both data types to generate supervised techniques and evaluate performance metrics individually and together.

In this work the standardization selected is a modified version of self-dependent z-score; it consists of subtracting the mean value and dividing by the standard deviation of the complete experiment for each volunteer independently.

The algorithms tested to classify the data were support vector machine (SVM), k-nearest neighbour (KNN), and ensemble (ENS). This selection was based on the target application, a wearable device with memory and computation power constraints. In addition, these methods are the most common ones used in the literature [44].

Each model's hyper-parameters were tuned using Bayesian optimization to minimize the misclassification rate over iterations and supported by 5 k-fold cross-validation strategy. Specifically, the selected technique is a sequential model-based optimization, which has shown substantial improvements over combinational space approaches [64]. Besides, this training and validation scheme was based on previous works and results in [7]. The performance values presented were the mean validation results of 10 iterations. No testing was carried out due to the lack of data.

Table 7 shows the characteristics of the different models used to generate classifiers regarding the information source, number of features, and windowing. A detailed explanation is provided in the next subsections. Videos V02, V03, and V04 were considered in all cases.

The metrics selected to evaluate the classifiers' performance are geometrical mean (Gmean) between Sensitivity (true positive rate, TPR) and Specificity (true negative rate, TNR) according to Equation (1). The TPR is the ratio between true positive (TP) and the sum of true positive and false negative (FN). The TNR is the ratio between true negative (TN) and the sum of true negative and false positive (FP).

$$Gmean = \sqrt{(Sensitivity * Specificity)} = \sqrt{\left(\frac{TP}{TP+FN}\right) * \left(\frac{TN}{TN+FP}\right)} \qquad (1)$$

Table 7. Characteristics of each configuration.

Nb. Configs	Physio	Cat.	Observations	Features	Window Size	Overlap
1	✓	-	63	47	60 s	-
2	✓	-	315	47	20 s	10 s
3	-	✓	63	15	-	-
4	-	✓	63	3	-	-
5	✓	✓	63	48	60 s	-
6	✓	✓	315	48	20 s	10 s

4.4.1. Physiological Supervised Models

The classification of physiological data with supervised machine learning techniques is a common approach in affective computing due to the complex relationships that implies. The models presented in this work are user-independent because there is not enough data for user-dependent solutions.

Two configurations were tested with the same number of features but with a different window size and overlapping. The features used are 22 for BVP, 7 for GSR, 6 for SKT, and 12 for respiration. The segmentation and windowing were applied following two strategies. Firstly, the configuration 1 used a 60 s window per video clip aiming to reduce data dispersion in the video. The second one has five windows per video, 20 s with 10 s overlap. This strategy helped algorithm training by providing more data and more temporal resolution; however, this could also lead to information redundancy.

The results in Table 8 showed that it is possible to classify the data between fear and no fear generally (Gmean above 0.5). The best performance was achieved by ENS (Adaboost) with the first model.

Table 8. Performance metrics for physiological configurations.

Nb. Config	Algorithm	G. Mean	TPR	TNR
1	SVM	0.59	0.83	0.51
1	KNN	0.74	0.83	0.67
1	ENS	0,91	0.83	1.00
2	SVM	0.56	0.86	0.45
2	KNN	0.64	0.83	0.50
2	ENS	0.74	0.83	0.66

4.4.2. Catecholamines Supervised Models

As in the physiological section, three algorithms KNN, SVM, and ENS (RandomForest) were applied (Table 9).

Table 9. Performance metrics for catecholamines models.

Nb. Config	Algorithm	G. Mean	TPR	TNR
3	SVM	0.49	0.47	0.55
3	KNN	0.53	0.51	0.58
3	ENS	0.45	0.47	0.50
4	SVM	0.33	0.29	0.73
4	KNN	0.37	0.25	0.64
4	ENS	0.44	0.42	0.53

Firstly, each observation was associated with a clip and each feature to a sample of that clip, resulting in a data matrix of 63 rows (21 volunteers × 3 clips) and 15 columns (5 samples per clip × 3 catecholamines).

After achieving in almost all cases overfitted models or poor-quality metrics, a transformation of the data was applied to compute the maximum in-video variations, considering the sign positive if this variation was increasing (minimum previous maximum) or negative if it was decreasing (maximum previous minimum). This variable was obtained and then normalized for each catecholamine, resulting in a data matrix of 63 rows (21 volunteers × 3 clips) and 3 columns (1 maximum variation per clip × 3 catecholamines).

As in previous models and mainly due to the lack of enough data and an imbalanced configuration, overfitted models were achieved and performance results worsened (Gmean values between 0.33 and 0.44) and showed the model would work randomly, such as flipping a coin.

4.4.3. Fusion Models

The data fusion applied followed two strategies based on physiological configurations. The first configuration was merged with the variation in plasma catecholamine concentration levels, per video clip, as explained previously (Model 5) and the physiological variables in a unique 60 s window. The second one used the plasma catecholamine concentration level directly, five samples per video clip. Each sample was paired with a 20 s physio window.

Table 10 shows the performance metrics obtained with the fusion models. The results were slightly worse than physiological models alone, i.e., the model was not learning from this data.

Table 10. Performance metrics for merged models.

Nb. Config	Model	G. Mean	TPR	TNR
	SVM	0.57	0.88	0.46
5	KNN	0.72	0.81	0.65
	ENS	0.90	0.81	1.00
	SVM	0.52	0.88	0.41
6	KNN	0.64	0.82	0.52
	ENS	0.74	0.82	0.67

5. Discussion

The study conducted in this work presents four main results. First, a methodology and protocol have been defined to connect the elicitation of human emotions with the variation of plasma catecholamine concentration. An immersive virtual reality environment has been arranged to provoke realistic situations where the volunteer could have intense emotional reactions. A continuous monitoring of physiological variables, with a research toolkit system (for the sake of comparison with other affective computing research works), is connected with the virtual environment, as well as a labelling procedure for discrete emotions and continuous PAD affective space dimensions. These three elements have been presented in previous works by the authors [65]. The novelty added to this method is to determine whether a person's emotions can be reliably recorded, assessing the differences or similarities between recording different physiological variables and measuring plasma catecholamine levels. The blood extraction must be performed after the video clip visualization to not interfere in the emotion elicitation but as soon as possible to detect the concentration peaks and valleys due to the emotion processed in the brain, which provokes a change in plasma catecholamine concentration. A pattern in the concentration variation has been looked for, as well as different classifiers, typical in affective computing, to determine the feasibility of using catecholamines for detecting fear emotions in a person.

Second, the emotion labels obtained during the study guaranteed the elicitation of the target emotions. The video clips selected were those with the best scores in terms of unanimity, in discrete and continuous emotions classifications, from the UC3M4Safety database [62]. The video clips' durations were between 60 s and 119 s. The 21 volunteers labelled the emotion felt during the video clip visualization in a very close way to the target emotion, especially for video clips V04 (fear) and V01 (calm), while for the other clips, at least the PA quadrant is maintained, (Figure 8). Every video clip provoked the target emotions, and, except for two volunteers, every volunteer labelling process matched with the rest of them, (Figure 9). Therefore, the variation in the measures of physiological variables and plasma catecholamine concentration per video clip, whatever they were, can be associated with a specific emotion.

Third, the physiological variables measured during the study, and the features extracted from them (IBI, GSR number of peaks, GSR mean and HRV) present similar behaviour as in previous works [7,65]. Statistically representative differences between fear-related video clip V04 and joy and calm clips (V03 and V01) were found for the GSR mean, as well as between V01 (calm), V02 (fear related to Gender-based violence) and V04 (fear) for IBI. The classifiers applied to generate an artificial intelligence algorithm to detect fear emotional reactions present good results for windows of 20 s and 60 s, although the results were better for wider windows, and ENS model, with a True Negative Rate of 1 and a True Positive Rate of 0.83, (Table 8).

It should be noted that the amount of data compiled during the experiment was large due to the sampling frequency (200 Hz), making easier the training and testing processes for affective computing tasks.

Finally, the plasma catecholamine concentration measurements provided data with apparently no connection with the emotion elicited. The ANOVA analysis provided no significant differences between the levels of catecholamines in blood plasma after visualizing the video clips of the different emotions. Besides, the clustering analysis (fear/no-fear emotions) on the data obtained from the 21 volunteers did not produce a valid result. Moreover, the classifiers selected as artificial intelligence algorithms to detect fear emotional reactions present poor-quality metrics, mainly due to the lack of enough data for training, testing and generalizing.

This problem of insufficient data on plasma catecholamine concentration (only five samples per video, i.e., per emotion) is difficult to solve. Even in an experimental study, the ethical research advises to not make volunteers suffer unnecessarily. Sixteen blood samples per session per volunteer, although taken through a via, while visualizing emotional intensive video clips within a virtual reality environment, are a fairly good number to test the hypothesis of the research work. In the literature, up to our knowledge, there is no similar study, with most of the proposals being theoretical hypotheses and/or based on analysing previous experimental results for other purposes.

However, the data obtained should have provided some patterns of responses to different target emotions and, although in the temporal evolution of the concentration levels of adrenaline and nor-adrenaline a similar behaviour can be observed after both V02 and V04 fear-related clips, neither statistically significant relations have been found nor affective computing classifiers provided good results.

It is true, that the plasma catecholamine levels are altered by the effect of some foods, drinks, and medicines or drugs, as well as by strong physical exercise and/or recent intense stressful episodes. Amines found in banana, avocado, walnuts, beans, cheese, beer and red wine can modify the concentration of these hormones in the blood. Additionally, foods/drinks with cocoa, coffee, tea, chocolate, liquorice, or vanilla, as well as drugs (nicotine, cocaine and ethanol) and medicines (aspirin, tricyclic antidepressants, tetracycline, theophylline, blood pressure control agents, and nitro-glycerine) have similar effects.

Besides, the emotional response is altered by prior experiences during a lifetime, and so does the emotional response to stress and the conditioned response to fear. Traumatic stress-induced fear memories may affect the physiological response and plasma catecholamine levels. There is strong evidence supporting that central catecholamines are involved in the regulation of fear memory, by activation of the sympathetic nervous system with elevated basal catecholamine levels are common in patients suffering from post-traumatic stress disorder (PTSD).

In the study presented, attention is paid to the activity of the volunteers before the experiment, as well as the different substances taken and, also, previous traumatic stressful experiences.

Although we previously informed about the recommendations, the volunteers reported the following data. With regard to medicines as regular treatment, six volunteers reported five chronic diseases: diabetes mellitus (1), hypertension (2), cardiac failure (1), ulcerative colitis (1), anaemia (1), and chronic pain (1). Additionally, one volunteer was taking contraceptives. On the other hand, four volunteers were taking ibuprofen or another type of anti-inflammatory drugs for the two days prior to the experiment. Respect to avoiding stimulants in food, drinks and drugs in the 24 h prior to the experiment, 13 volunteers took coffee or tea in that period of time, and one volunteer drank alcohol. Additionally, three of them ate citric fruits in that period.

Only four volunteers (v06, v11, v13, v19) exactly complied with the recommendations with regard to avoiding stimulant foods, drinks and drugs; and did not take any medication. They were young women with ages 23, 30, 29, and 23, respectively. Likewise, three volunteers (v01, v04, and v17) only had a coffee, complying with the rest of the recommendations, and did not take any medication either. Their ages were 21, 55, and 24 respectively. There are seven volunteers that only took a coffee and medicaments not presenting differences in the levels of catecholamine concentrations (v02, v05, v09, v12, v14, v15, and v20). In summary, we can consider that 14 volunteers were fully compliant and 7 could have some objection with respect to regular catecholamine activity.

Regarding prior stressful experiences, or specific fears, seven volunteers reported some previous traumas that activate themselves in situations like video clips V02 and V04, (v01, v03, v04, v12, v15, v16, and v20). Two of them identified as gender-based violence victims. However, the evolution of their plasma catecholamine concentration levels were not different from the other volunteers', (Figures 15 and 16).

Apart from the extrinsic and intrinsic factors that can be affecting the results of the study, the authors wish to highlight the low levels of the concentration of these catecholamines present in the blood plasma. We tested the technique ELISA that produced worse results in terms of sensitivity of these catecholamines. Nine women volunteers followed a similar experimental study, and 15 blood samples per volunteer were analysed with ELISA kits.

With respect to the hypothesis stated in this work, the measurement of the levels of dopamine, noradrenaline and adrenaline concentration in blood plasma is neither providing better classifications nor a more accurate differentiation of fear-emotion reactions in women.

6. Conclusions

In this work, a methodology and a protocol have been proposed to connect the elicitation of human emotions with the variation of plasma catecholamine concentration. For them, an immersive virtual reality environment has been arranged to provoke realistic situations where the volunteer could have intense emotional reactions. A continuous monitoring of physiological variables, with a research toolkit system (for the sake of comparison with other affective computing research works) was connected to the virtual environment, as well as a labelling procedure for discrete emotions and continuous PAD affective space dimensions.

Using this methodology, an experimental study with 21 volunteers has been conducted, using fear as a target emotion, thus provoking fear and non-fear while measuring physiological variables and extracting blood samples after the visualization of every video stimulus. In this first study, 16 blood samples have been extracted per volunteer; 1 for basal measure and 5 after the three emotion-related video clips (fear (gender-based violence related), joy and fear). These samples have been extracted in 1-min intervals after the visualization of the video clip. Along with the blood sample for catecholamine plasma analysis, physiological variables have been measured during the visualization of the video clips. Skin temperature, galvanic skin response, blood volume pulse, respiration, and Trapezoidal Electromyogram were the selected variables, measured with a commercial research toolkit.

Additionally, the emotion labelling for every video clip by all the volunteers has been analysed and there is a high degree of agreement in the discrete emotion, which was even better in the PAD affective space dimensions, especially for fear-related video V04. Therefore, we can affirm that the selected video clips are meaningful for the experiment.

The results for the evolution of the features extracted from the physiological variables, as well as an ANOVA statistical analysis, are in accordance with previous works. Differences between features measured during fear-related and during calm and joy-related video clips have been found for the mean of GSR (60 s windows). Additionally, differences have been found between calm-related and fear/gender-based-violence fear-related video clips for the IBI (for heart rate,). Furthermore, the temporal evolution of these features has been analysed and correlated with the fear-related video clips, identifying precise moments where the features' behaviour can be associated with the scene development.

We can conclude that there are no significant p-values (ANOVA statistical analysis performed) that allow differentiating the emotion elicited using only the evolution of the plasma catecholamine concentration levels as a variable. Additionally, the temporal evolution of these levels has been analysed, not identifying precise patterns for fear-related video clips different from the joy-related video clip.

Finally, artificial intelligence algorithms for fear classification with physiological variables and plasma catecholamine concentration levels (separately and together) have been tested. The best results have been obtained with the features extracted from the physiological variables. Adding the maximum variation of catecholamines during the five minutes after the video clip visualization, as well as adding the five measurements (1-min interval) of these levels, do not provide better performance in the classifiers.

The small number of samples together with the low concentration of catecholamines in blood plasma make it not possible to use these data for machine learning techniques for fear classification in this experiment.

Finally, we can state that research on this topic should continue considering the following future actions:

1. Although it is true that the results of this study show that the measurement of catecholamine concentration does not improve the detection and identification of emotions, it would be desirable to have a larger sample of volunteers in order to detect patterns of variation in this concentration that validate this conclusion.
2. Following Lovehëim's theory work, adding the measurement of blood serotonin concentration would be recommendable since it could allow us to improve the classification of fear from joy, which are both emotions with a high theoretical degree of activation. For this study, although its inclusion was considered, adding the serotonin measurement entailed the use of another analysis technique, which meant extracting twice as many samples from each volunteer, which was not recommended from an ethical point of view.
3. In the search for non-invasive emotion detection systems, it would be interesting to analyse the effect of the concentration of catecholamine in sweat (cortisol) or in saliva (alpha-amylase). If significant differences were found, it would be possible to include these variables in automatic emotion detection systems design.
4. However, in the search for any other extra information, instead of clustering fear and not-fear emotions, a behaviour pattern for each volunteer was examined according to Khrone [66] which suggests that there are two main strategies in stress reaction: vigilance and avoidance. From an unsupervised standpoint and after applying k-

means algorithms four clear groups were observed, two of them being a symmetrical representation of the other two. In two of the groups, the third clip contains a negative variation, which is below the other two clips. On the other hand, the other two groups have a peak in the third clip (V3) which is above the values representing the other two videos.

Author Contributions: Conceptualisation, L.G.-M., E.R.-P., C.S.d.B.A., M.F.C.-B., G.E.R.-R., R.T.-F., S.L.-O. and C.L.-O.; methodology L.G.-M., E.R.-P., C.S.d.B.A., M.F.C.-B., G.E.R.-R., R.T.-F., S.L.-O. and C.L.-O.; software L.G.-M. and M.F.C.-B.; formal analysis L.G.-M., E.R.-P., C.S.d.B.A., M.F.C.-B., G.E.R.-R., R.T.-F., S.L.-O. and C.L.-O.; data curation, L.G.-M., E.R.-P., M.F.C.-B., G.E.R.-R., S.L.-O. and C.L.-O.; writing—original draft preparation, L.G.-M., E.R.-P., M.F.C.-B., G.E.R.-R., S.L.-O. and C.L.-O.; writing—review and editing, L.G.-M., E.R.-P., C.S.d.B.A., M.F.C.-B., G.E.R.-R., R.T.-F., S.L.-O. and C.L.-O.; visualization, L.G.-M., M.F.C.-B. and C.L.-O.; supervision, E.R.-P., C.S.d.B.A., S.L.-O. and C.L.-O.; project administration, E.R.-P., C.S.d.B.A. and C.L.-O.; funding acquisition, E.R.-P., C.S.d.B.A. and C.L.-O. All authors have read and agreed to the published version of the manuscript.

Funding: This research has been supported by the Madrid Governement (Comunidad de Madrid, Spain) under the ARTEMISA-UC3M-CM research project (reference 2020/00048/001), the EMPATIA-CM research project (reference Y2018/TCS-5046) and the Multiannual Agreement with UC3M in the line of Excellence of University Professors (EPUC3M26), and in the context of the V PRICIT (Regional Programme of Research and Technological Innovation).

Institutional Review Board Statement: Not applicable.

Informed Consent Statement: Not applicable.

Data Availability Statement: Not applicable.

Acknowledgments: The authors acknowledge the technical help given by José Ángel Miranda.

Conflicts of Interest: The authors declare no conflict of interest.

References

1. Picard, R.W. *Affective Computing*; MIT Press: Cambridge, MA, USA, 1997.
2. Tao, J.; Tan, T. Affective computing: A Review. In Proceedings of the 1st International Conference on Affective Computing and Intelligent Interaction, ACII 2005, Beijing, China, 22–24 October 2005; pp. 981–995.
3. Picard, R.W.; Vyzas, E.; Healey, J. Toward machine emotional intelligence: Analysis of affective physiological state. *IEEE Trans. Pattern Anal. Mach. Intell.* **2001**, *23*, 1175–1191. [CrossRef]
4. Pagán, J.; De Orbe, M.I.; Gago, A.; Sobrado, M.; Risco-Martín, J.L.; Mora, J.V.; Moya, J.M.; Ayala, J.L. Robust and Accurate Modeling Approaches for Migraine Per-Patient Prediction from Ambulatory Data. *Sensors* **2015**, *15*, 15419–15442. [CrossRef] [PubMed]
5. Poh, M.Z.; Loddenkemper, T.; Reinsberger, C.; Swenson, N.C.; Goyal, S.; Sabtala, M.C.; Madsen, J.R.; Picard, R.W. Convulsive seizure detection using a wrist-worn electrodermal activity and accelerometry biosensor. *Epilepsia* **2012**, *53*, e93–e97. [CrossRef] [PubMed]
6. Hickey, B.A.; Chalmers, T.; Newton, P.; Lin, C.-T.; Sibbritt, D.; McLachlan, C.S.; Clifton-Bligh, R.; Morley, J.; Lal, S. Smart Devices and Wearable Technologies to Detect and Monitor Mental Health Conditions and Stress: A Systematic Review. *Sensors* **2021**, *21*, 3461. [CrossRef]
7. Miranda, J.A.; Canabal, M.F.; Gutiérrez-Martín, L.; Lanza-Gutierrez, J.M.; Portela-García, M.; López-Ongil, C. Fear Recognition for Women Using a Reduced Set of Physiological Signals. *Sensors* **2021**, *21*, 1587. [CrossRef] [PubMed]
8. Andreassi, J.L. *Psychophysiology: Human Behavior and Physiological Response*, 5th ed.; Psychology Press: Hove, UK, 2006. [CrossRef]
9. Kreibig, S.D. Autonomic nervous system activity in emotion: A review. *Biol. Psychol.* **2010**, *84*, 394–421. [CrossRef] [PubMed]
10. Best, B. Anatomical Basis of Mind. In *The Amygdala and the Emotions*; 2009; Chapter 9. Available online: https://www.benbest.com/science/anatmind/anatmind.html (accessed on 17 May 2022).
11. Picard, R.W. *Affective Computing*; Technical Report N 321; MIT Media Laboratory, Perceptual Computing Section: Cambridge, MA, USA, 1995.
12. Lövheim, H. A new three-dimensional model for emotions and monoamine neurotransmitters. *Med. Hypotheses* **2012**, *78*, 341–348. [CrossRef] [PubMed]
13. Yan, B.; An, D.; Wang, X.; DeLong, B.J.; Kiourti, A.; Dungan, K.; Volakis, J.L. Battery-free implantable insulin micropump operating at transcutaneously radio frequency-transmittable power. *Med. Devices Sens.* **2019**, *2*, e10055. [CrossRef]
14. Damasio, A.R. *Descartes' Error: Emotion, Reason, and the Human Brain*; Harper Perennial: New York, NY, USA, 1995.

15. Imani, M.; Montazer, G.A. A survey of emotion recognition methods with emphasis on E-Learning environments. *J. Netw. Comput. Appl.* **2019**, *147*, 102423. [CrossRef]
16. Filipovic, F.; Despotovic-Zrakic, M.; Radenkovic, B.; Jovanic, B.; Živojinovic, L. An Application of Artificial Intelligence for Detecting Emotions in Neuromarketing. In Proceedings of the International Conference on Artificial Intelligence: Applications and Innovations (IC-AIAI), Belgrade, Serbia, 30 September–4 October 2019. [CrossRef]
17. Alakus, T.B.; Gonen, M.; Turkoglu, I. Database for an emotion recognition system based on EEG signals and various computer games–GAMEEMO. *Biomed. Signal Process. Control* **2020**, *60*, 101951. [CrossRef]
18. Mauss, I.B.; Robinson, M.D. Measures of emotion: A review. *Cogn. Emot.* **2009**, *23*, 209–237. [CrossRef]
19. Lang, P.J. Fear reduction and fear behavior: Problems in treating a construct. In *Research in Psychotherapy Conference*; Shlien, J.M., Ed.; American Psychological Association: Washington, DC, USA, 1968; Volume 3, pp. 90–102.
20. Ekman, P. Are there basic emotions? *Psycol. Rev.* **1992**, *99*, 550–553. [CrossRef]
21. Plutchik, R. Emotions: A general psychoevolutionary theory. *Approaches Emot.* **1984**, *1984*, 197–219.
22. Frijda, N.H. *The Emotions*; Cambridge University Press: Cambridge, UK, 1986.
23. Fontaine, J.R.J.; Scherer, K.R.; Roesch, E.B.; Ellsworth, P.C. The World of Emotions is not Two-Dimensional. *Psychol. Sci.* **2007**, *18*, 1050–1057. [CrossRef] [PubMed]
24. Verma, G.K.; Tiwary, U.S. Affect representation and recognition in 3D continuous valence–arousal–dominance space. *Multimed. Tools Appl.* **2017**, *76*, 2159–2183. [CrossRef]
25. Russell, J. A circumplex model of affect. *J. Personal. Soc. Psychol.* **1980**, *39*, 1161–1178. [CrossRef]
26. Hamann, S. Mapping discrete and dimensional emotions onto the brain: Controversies and consensus. *Trends Cogn. Sci.* **2012**, *16*, 458–466. [CrossRef] [PubMed]
27. Hoffman, H.; Scheck, A.; Schuster, T.; Walter, S.; Limbrecht, K.; Traue, H.; Kessler, H. Mapping discrete emotions into the dimensional space: An empirical approach. In Proceedings of the IEEE International Conference on Systems, Man, and Cybernetics, Seoul, Korea, 14–17 October 2012.
28. Blanco-Ruiz, M.; Sainz-de-Baranda, C.; Gutiérrez-Martín, L.; Romero-Perales, E.; López-Ongil, C. Emotion Elicitation Under Audiovisual Stimuli Reception: Should Artificial Intelligence Consider the Gender Perspective? *Int. J. Environ. Res. Public Health* **2020**, *17*, 8534. [CrossRef]
29. Robinson, D.L. Brain function, emotional experience and personality. *Neth. J. Psychol.* **2008**, *64*, 152–168. [CrossRef]
30. Bakker, I.; van der Voordt, T.; Vink, P.; de Boon, J. Pleasure, arousal, dominance: Mehrabian and russell revisited. *Curr. Psychol. Res. Rev.* **2014**, *33*, 405–421. [CrossRef]
31. Dzedzickis, A.; Kaklauskas, A.; Bucinskas, V. Human Emotion Recognition: Review of Sensors and Methods. *Sensors* **2020**, *20*, 592. [CrossRef] [PubMed]
32. Vijaya, P.A.; Shivakumar, G. Galvanic Skin Response: A Physiological Sensor System for Affective Computing. *Int. J. Mach. Learn. Comput.* **2013**, *3*, 31. [CrossRef]
33. Posada-Quintero, H.F.; Chon, K.H. Innovations in electrodermal activity data collection and signal processing: A systematic review. *Sensors* **2020**, *20*, 479. [CrossRef]
34. Gruebler, A.; Suzuki, K. Design of a Wearable Device for Reading Positive Expressions from Facial EMG Signals. *IEEE Trans. Affect. Comput.* **2014**, *5*, 227–237. [CrossRef]
35. Hayashi, N.; Someya, N.; Maruyama, T.; Hirooka, Y.; Endo, M.Y.; Fukuba, Y. Vascular responses to fear-induced stress in humans. *Physiol. Behav.* **2009**, *98*, 441–446. [CrossRef] [PubMed]
36. Nardelli, M.; Valenza, G.; Greco, A.; Lanata, A.; Scilingo, E.P. Recognizing Emotions Induced by Affective Sounds through Heart Rate Variability. *IEEE Trans. Affect. Comput.* **2015**, *6*, 385–394. [CrossRef]
37. Kolodyazhniy, V.; Kreibig, S.D.; Gross, J.J.; Roth, W.T.; Wilhelm, F.H. An affective computing approach to physiological emotion specificity: Toward subject-independent and stimulus-independent classification of film-induced emotions. *Psychophysiology* **2011**, *48*, 908–922. [CrossRef] [PubMed]
38. Qing, C.; Qiao, R.; Xu, X.; Cheng, Y. Interpretable Emotion Recognition Using EEG Signals. *IEEE Access* **2019**, *7*, 94160–94170. [CrossRef]
39. Krishna, N.M.; Sekaran, K.; Vamsi, A.V.N.; Ghantasala, G.S.P.; Chandana, P.; Kadry, S.; Blazauskas, T.; Damasevicius, R.; Kaushik, S. An Efficient Mixture Model Approach in Brain-Machine Interface Systems for Extracting the Psychological Status of Mentally Impaired Persons Using EEG Signals. *IEEE Access* **2019**, *7*, 77905–77914. [CrossRef]
40. Han, J.; Ji, X.; Hu, X.; Guo, L.; Liu, T. Arousal Recognition Using Audio-Visual Features and FMRI-Based Brain Response. *IEEE Trans. Affect. Comput.* **2015**, *6*, 337–347. [CrossRef]
41. Zhang, J.; Yin, Z.; Chen, P.; Nichele, S. Emotion recognition using multi-modal data and machine learning techniques: A tutorial and review. *Inf. Fusion* **2020**, *59*, 103–126. [CrossRef]
42. Tzirakis, P.; Trigeorgis, G.; Nicolaou, M.A.; Schuller, B.W.; Zafeiriou, S. End-to-end multimodal emotion recognition using deep neural networks. *IEEE J. Sel. Top. Signal Process.* **2017**, *11*, 1301–1309. [CrossRef]
43. Šalkevicius, J.; Damaševičius, R.; Maskeliunas, R.; Laukienė, I. Anxiety Level Recognition for Virtual Reality Therapy System Using Physiological Signals. *Electronics* **2019**, *8*, 1039. [CrossRef]
44. Schmidt, P.; Reiss, A.; Dürichen, R.; Laerhoven, K.V. Wearable-Based Affect Recognition—A Review. *Sensors* **2019**, *19*, 4079. [CrossRef] [PubMed]

45. Chauhan, V.K.; Dahiya, K.; Sharma, A. Problem formulations and solvers in linear svm: A review. *Artif. Intell. Rev.* **2018**, *52*, 803–855. [CrossRef]
46. Sen, P.C.; Hajra, M.; Ghosh, M. Supervised classification algorithms in machine learning: A survey and review. In *Emerging Technology in Modelling and Graphics*; Springer: Singapore, 2020; pp. 99–111.
47. Wang, W.; Sun, D. The improved adaboost algorithms for imbalanced data classification. *Inf. Sci.* **2021**, *563*, 358–374. [CrossRef]
48. Lichtenauer, J.; Soleymani, M. MAHNOB-hci-Tagging Database. 2011. Available online: https://mahnob-db.eu/ (accessed on 24 April 2022).
49. Koelstra, S.; Mühl, C.; Soleymani, M.; Lee, J.-S.; Yazdani, A.; Ebrahimi, T.; Pun, T.; Nijholt, A.; Patras, I. DEAP: A Database for Emotion Analysis: Using Physiological Signals. *IEEE Trans. Affect. Comput.* **2011**, *3*, 18–31. [CrossRef]
50. Larradet, F.; Niewiadomski, R.; Barresi, G.; Caldwell, D.G.; Mattos, L.S. Toward emotion recognition from physiological signals in the wild: Approaching the methodological issues in real-life data collection. *Front. Psychol.* **2020**, *11*, 1111. [CrossRef]
51. Alam, S.; Gupta, R.; Sharma, K.D. On-board signal quality assessment guided compression of photoplethysmogram for personal health monitoring. *IEEE Trans. Instrum. Meas.* **2021**, *70*, 1–9. [CrossRef]
52. Kvetnansky, R.; Sabban, E.L.; Palkovits, M. Catecholaminergic systems in stress: Structural and molecular genetic approaches. *Physiol. Rev.* **2009**, *89*, 535–606. [CrossRef]
53. Cannon, W.B. *The Wisdom of the Body*; WW Norton and Company: New York, NY, USA, 1939.
54. Selye, H. *Stress without Distress*; New American Library: New York, NY, USA, 1974.
55. Goldstein, D.S. *Stress, Catecholamines, and Cardiovascular Disease*; Oxford University Press: New York, NY, USA, 1995.
56. Goldstein, D.S. Catecholamines and stress. *Endocr. Regul.* **2003**, *37*, 69–80. [PubMed]
57. Laverty, R. Catecholamines: Role in Health and Disease. *Drugs* **1978**, *16*, 418–440. [CrossRef] [PubMed]
58. Cannon, W.B. *Bodily Changes in Pain, Hunger, Fear and Rage*; D. Appleton & Co.: New York, NY, USA, 1929.
59. Tank, A.W.; Wong, D.L. Peripheral and Central Effects of Circulating Catecholamines. *Compr. Physiol.* **2015**, *5*, 1–15. [CrossRef]
60. Walker, F.R.; Pfingst, K.; Carnevali, L.; Sgoifo, A.; Nalivaiko, E. In the search for integrative biomarker of resilience to psychological stress. *Neurosci. Biobehav. Rev.* **2017**, *74*, 310–320. [CrossRef] [PubMed]
61. Kojima, K.; Maki, S.; Hirata, K.; Higuchi, S.; Akazawa, K.; Tashiro, N. Relation of emotional behaviors to urine catecholamines and cortisol. *Physiol. Behav.* **1995**, *57*, 445–449. [CrossRef]
62. Blanco Ruiz, M.Á.; Gutiérrez Martín, L.; Miranda Calero, J.Á.; Canabal Benito, M.F.; Rituerto González, E.; Luis Mingueza, C.; Robredo García, J.C.; Morán González, B.; Páez Montoro, A.; Ramírez Bárcenas, A.; et al. UC3M4Safety Database—List of Audiovisual Stimuli (Video). v1. 2021; e-cienciaDatos. [CrossRef]
63. Lang, P.J. The emotion probe: Studies of motivation and attention. *Am. Psychol.* **1995**, *50*, 372–385. [CrossRef] [PubMed]
64. Hutter, F.; Hoos, H.H.; Leyton-Brown, K. Sequential model-based optimization for general algorithm configuration. In *International Conference on Learning and Intelligent Optimization*; Springer: Heidelberg/Berlin, Germany, 2011; pp. 507–523.
65. Miranda, J.A.; Rituerto-González, E.; Gutiérrez-Martín, L.; Luis-Mingueza, C.; Canabal, M.F.; Ramírez Bárcenas, A.; Lanza-Gutiérrez, J.M.; Peláez-Moreno, C.; López-Ongil, C. WEMAC: Women and Emotion Multi-modal affective computing DATASET. *arXiv* **2022**, arXiv:2203.00456.
66. Krohne, H.W. The concept of coping modes: Relating cognitive person variables to actual coping behavior. *Adv. Behav. Res. Ther.* **1989**, *11*, 235–248. [CrossRef]

Article

Non-Contact Measurement of Empathy Based on Micro-Movement Synchronization

Ayoung Cho [1], Sung Park [2], Hyunwoo Lee [1] and Mincheol Whang [3],*

1. Department of Emotion Engineering, University of Sangmyung, Seoul 03016, Korea; joa6391@gmail.com (A.C.); lhw4846@naver.com (H.L.)
2. School of Design, Savannah College of Art and Design, Savannah, GA 31401, USA; spica7601@gmail.com
3. Department of Human Centered Artificial Intelligence, Sangmyung University, Seoul 03016, Korea
* Correspondence: whang@smu.ac.kr; Tel.: +82-2-2287-5293

Abstract: Tracking consumer empathy is one of the biggest challenges for advertisers. Although numerous studies have shown that consumers' empathy affects purchasing, there are few quantitative and unobtrusive methods for assessing whether the viewer is sharing congruent emotions with the advertisement. This study suggested a non-contact method for measuring empathy by evaluating the synchronization of micro-movements between consumers and people within the media. Thirty participants viewed 24 advertisements classified as either empathy or non-empathy advertisements. For each viewing, we recorded the facial data and subjective empathy scores. We recorded the facial micro-movements, which reflect the ballistocardiography (BCG) motion, through the carotid artery remotely using a camera without any sensory attachment to the participant. Synchronization in cardiovascular measures (e.g., heart rate) is known to indicate higher levels of empathy. We found that through cross-entropy analysis, the more similar the micro-movements between the participant and the person in the advertisement, the higher the participant's empathy scores for the advertisement. The study suggests that non-contact BCG methods can be utilized in cases where sensor attachment is ineffective (e.g., measuring empathy between the viewer and the media content) and can be a complementary method to subjective empathy scales.

Keywords: video content empathy; micro-movement synchronization; non-contact empathy measurement; empathic advertisement

1. Introduction

Empathy, a crucial factor in successful digital content marketing [1], is generally conceptualized as a multidimensional construct that includes both cognitive and affective responses to others in dyadic interactions [2–4]. However, empathy for digital content involves the emotional engagement of a viewer with a character in a causal and probable narrative [5]. For example, eliciting a consumer's emotions congruent to content emotions may maximize an advertisement's effect. Viewers empathizing with content tend to better understand the story and have more positive attitudes. They are more attentive and engaged [6–8], feel favorably toward products and brands [9,10], and are less likely to skip an advertisement [11,12]. Moreover, heightened empathy promotes the consumption of content in addition to attitudinal acceptance [13,14]. Such behavioral acceptance implies that viewer empathy is a critical predictor of the success of media content.

Empathy has been measured to predict the success of commercials. Escalas and Stern developed a battery of scale items to measure empathy toward advertisements, which has been widely used in consumer research [15]. Other prominent subjective measures include Schlinger's Viewer Response Profile [16–18], the Balanced Emotional Empathy Scale [19], the Empathy Quotient [20], the Toronto Empathy Questionnaire [21], the Interpersonal Reactivity Index [22,23], the Basic Empathy Scale [24], and the Hogan Empathy Scale [25].

However, such subjective evaluations cannot measure the dynamics of empathy over time. Empathic questionnaires are limited to assessing dispositional empathy, which refers to an individual's capability (i.e., personality trait) to empathize with others.

The dynamics of empathy when consuming digital content require a novel measurement that can capture the fluctuation of emotions over time. The ever-changing interplay between the viewer's emotions and the content emotions demands a more direct, sensitive, and real-time measurement, such as physiological measures, to properly assess the degree of empathy. The unconscious level of empathy that is not verbally reportable (i.e., subjective evaluation) can be acquired through more direct physiological measures.

1.1. Psychophysiological Basis of Empathy

Empathy includes motor mimicry and emotional contagion associated with autonomically activated neural mechanisms of the other's feelings [26–29]. It also includes mirroring responses between people, in which explicit and implicit physiology become synchronized [30–32]. Explicit responses from empathy involve the synchronization of faces, gestures, and body movements. Changes in body motion synchronization are associated with the degree of empathy during face-to-face communication [33,34]. Greater synchronization of head motion was observed when a listener empathized with a speaker in a lecture [35]. In addition, body synchronization was reported between counselors and clients when they shared empathy [36,37].

Such observable synchronized behavior is a result of an implicit empathic response. The implicit process constitutes the synchronization of physiological activities between individuals [38], which can be measured through electroencephalography (EEG) [39,40], electrocardiography (ECG) [41–44], and skin conductance [45,46]. For example, the synchronization of electrodermal activities (i.e., skin response) between a therapist and a patient correlates to the patient's perceived empathy toward the therapist [47,48].

Neuroscientific bases have been identified for the synchronization of brain activity among participants during empathic communication [49–51]. Empathy researches using EEG have been mainly focused on understanding the sharing of painful experiences. Several asymmetries or activations in the pain-related brain areas have been reported, which were elicited by empathy. The left frontal asymmetry has been related to the suffering of the other, and the right frontal asymmetry has been associated with the pain and sorrow of the other [52]. Moreover, empathy-related activation in fronto-insula and anterior cingulate cortices was reported, which have been related to pain [53]. Peng et al. have shown that brain-to-brain synchronization could be triggered by sharing painful experiences and could strengthen social bonds [54].

1.2. Cardiovascular Measures of Empathy

Measures of cardiovascular activity reflect both attentional and affective states [55]. Cardiovascular measures can be achieved using a piezoelectric transducer, ECG, or analysis of facial micromovements. Cardiovascular activity in empathy research has been understudied compared to other physiological measures [56], but recent advances in vision technology have shed light on novel and innovative methodologies such as remote ballistocardiography (rBCG).

Kodama et al. [57] examined a psychotherapy session between a counselor and a client and found synchronization in heart rate, suggesting a promising indicator that leads to the building of rapport and empathy. Salminen et al. [58] found that higher synchrony in respiration rates, which has a positive relationship with heart rate, is associated with higher empathy. The synchronization of the heart rate can also enhance closeness [59] and intimacy [60].

However, the measurement of synchronization between the cardiovascular activities of viewers and people in media content has been less studied, mainly due to multiple technical issues.

First, viewers need a sensor attachment to capture physiological measurements, which is a significant barrier to general adoption. Second, to evaluate empathy, measuring dyadic synchronization is paramount. The cardiovascular information of both the viewer and the person in the media content must be obtained and analyzed. Obviously, acquiring the latter is impossible with sensor attachment because it is digital content.

However, advances in vision technology for cardiac measurements, such as remote photoplethysmography (rPPG) and rBCG, suggest promising methods for overcoming these challenges. The rPPG evolved to detect changes in blood volume remotely without direct contact between the photosensor (i.e., PPG) and the skin [61]. Non-contact data acquisition is possible through various means, including infrared [62], thermal [63], and RGB [64] cameras. The rPPG uses band-pass filters to eliminate motion components in images [65] but has less effect on cardiovascular activities that include the motion itself, referred to as ballistocardiography motion [66]. The rBCG is a measurement of ballistocardiographic head movements through remote means using a camera and vision-based analysis. These vision technologies have improved considerably in recent years, enabling the estimation of the heartbeat signals of both the viewer and the person in the digital content without needing skin contact.

Specifically, BCG motion causes microscopic vibration (i.e., micro-movement), which appears in the face through the carotid artery [67]. Micro-movement implies the subtle movement of a face that the human eye cannot easily see. This is caused by regular vibrations from the heart that are transmitted to the face. Micro-movement can be obtained by filtering the frequency corresponding to the regular heart rate band from the frontal facial video capture [68–71]. Analyzing the similarity of micro-movement between viewers and digital content (e.g., advertisements) may provide insights into whether the viewer is empathizing with the content. We intended to analyze the similarity of micro-movements through cross-entropy analysis and compare it to the participants' subjective empathy through a questionnaire. To our knowledge, no study has investigated the relationship of micro-movements through an rBCG method for a participant and a person in real-world media content, such as an advertisement.

2. Materials and Methods

2.1. Research Hypothesis

This study sought to verify the following hypothesis:

Hypothesis 1 (H1). *The more similar the micro-movements between the participant and the person in the advertisement, the higher the participant's empathy scores for the advertisement.*

The following section explains our operational definition of micro-movement signals, how the signals were measured from the participant and the advertisement, and how the participant's subjective assessment of empathy was acquired.

2.2. Experimental Design

The main experiment was a one-factor design (empathy factor) with two levels (empathy and non-empathy). Each participant viewed two empathy conditions (i.e., within-subject design), manifested in an empathy or non-empathy advertisement, and responded to an empathy questionnaire. The design of the stimuli (i.e., advertisement) and the questionnaire are explained in Section 2.3.

The dependent measurements involved the similarity of micro-movements between the participant and the stimulus, specifically, the similarity between the micro-movement signals extracted from the participant and those from the person in the advertisement. Cross-entropy was used as a similarity metric. Cross-entropy is suitable for the comparison of periodic distributions. The more similar the two distributions, the closer the cross-entropy is to zero [72]. This study extracted the micro-movement signals by filtering the power spectrum between 0.75 Hz and 2.5 Hz corresponding to 45~150 bpm when static.

However, this filtering range may vary according to the context, situation, and use cases. The details of the analysis are explained in Section 3.

2.3. Participants

Thirty participants (15 males and 15 females) voluntarily participated in the experiment. The mean age of participants was 22 (±2) years. None of the participants had a medical history of cardiovascular disease. The participants had an uncorrected or corrected visual acuity of 0.6 or better and were able to wear soft contact lenses but not glasses. Written informed consent was obtained from all the participants prior to the experiment. All participants were compensated for their participation.

Empathy varies with demographic characteristics, such as age [28], race [73], education [74], and gender [75]. Researchers have suggested an inverse-U-shaped pattern as a function of age, with middle-aged adults showing higher empathy than young adults [28]. Meta-analyses of gender differences in empathy support that women have more empathy than men [28,75,76]. One study reported a decline in empathy among undergraduate nursing students as they advanced through training [74]. The empathic neural response is increased for members of the same race, but not for other races [73]. Due to such demographic variance, the most recent (2021) massive survey (n = 3486) on the experience of empathy [77] quota sampled to reflect the U.S. population on demographic parameters. However, all empirical lab studies on empathy, including ours, have limitations when generalizing. We balanced the N of gender (15) and confirmed that gender did not have an effect on the dependent measures and ensured that the ethnicity of the participants (i.e., Korean) was consistent with the characters in the video stimuli. However, we acknowledge the limitation for generalizing the findings, such that the results may only apply to younger adults. Further studies are needed to confirm this hypothesis.

2.4. Procedures and Materials

The experimental procedure is shown in Figure 1. The participants stared at the blank screen for four minutes to stabilize their physiological state. For each stimulus, participants viewed an advertisement video and responded to a self-report questionnaire. Each condition (empathy and non-empathy) had 12 stimuli, so participants viewed 24 advertisements in total. The stimuli were presented in random order.

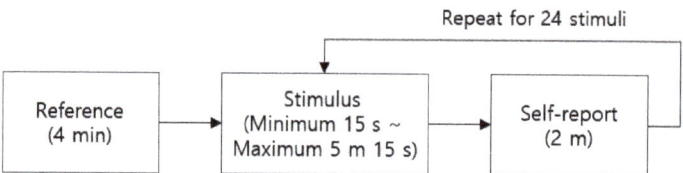

Figure 1. Experimental procedure.

Participants' frontal views, which were necessary for extracting the micro-movement signals, were recorded at 30 fps, 1920 × 1080 pixels, using a web camera installed on the monitor while they viewed the stimuli, as shown in Figure 2.

2.4.1. Video Stimuli (Advertisements)

Marketing researchers have explored empathy as a construct for estimating advertising effects. Escalas and Stern suggested that well-developed stories elicit higher levels of empathy than poorly developed ones [15]. Classical drama advertisements that have clear causality have been better able to hook viewers into commercials than vignettes. Emotionally driven advertisements have a positive impact on consumers' engagement and empathy [8,78,79]. In short, advertisements that elicit viewers' empathy tend to provide a clear context behind the story, in addition to an emotional appeal [14,79,80]. As a result, we chose three criteria for selecting the video stimuli: (1) causality of the storyline, (2) advertising appeal type, and (3) the degree of empathy.

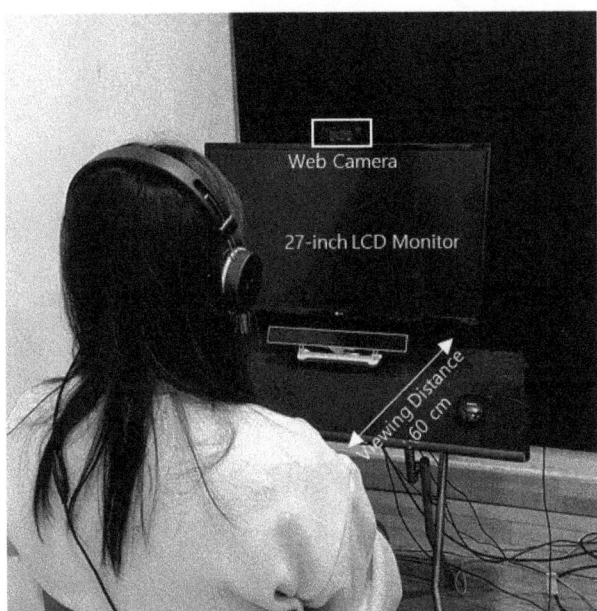

Figure 2. Experimental environment.

Nine emotion researchers viewed and assessed 50 candidate advertisements. The candidates were limited to those targeting the younger generation in their 20 s and 30 s, consistent with the participant pool. For each criterion related to the candidate, the researchers responded from −3 to +3 on a six-point Likert scale. Per criteria 1, researchers scored from −3 (ambiguous causality) to +3 (clear causality) for the story of the advertisement. Per criteria 2, they scored from −3 (rational appeal) to +3 (emotional appeal) for the advertising appeal type. Finally, according to criteria 3, they scored from −3 (not empathetic) to +3 (empathetic).

We classified the candidates into empathy advertisements if the average score for the evaluators was above zero for all three criteria. Conversely, we classified them into non-empathy advertisements if the score was below zero. For each advertisement group (empathy and non-empathy), we sorted the advertisements into four product advertisements (energy boosters, snacks, computer peripheral devices (e.g., printer)) and selected the three best advertisements for each product group. That is, we selected 12 advertisements for each condition (empathy and non-empathy).

Empathy advertisements tend to be longer than non-empathy advertisements because the viewer requires some time for the narrative to "sink in". In contrast, non-empathy advertisements focus on the presentation of prominent models and products. For example, an energy booster's empathy advertisement has a story involving a student exhausted from studying being revitalized after drinking an energy drink. The non-empathy advertisement, however, featured a character dancing with an energy drink and did not have a particular narrative.

2.4.2. Subjective Evaluations

As empathy is a multifaceted construct that includes both cognitive and affective processes, we adopted a comprehensive and empirically validated questionnaire with the participants' ethnicity (i.e., Korean). We used the Consumer Empathic Response to

Advertising Scale [81,82], which consists of 11 items, as shown in Table 1. The factor loading exceeded 0.4 and Cronbach's alpha exceeded 0.8. The questionnaire included three empathy factors: cognitive empathy, affective empathy, and identification empathy. The dependent variable for analysis was the sum of all 11 items.

Table 1. Questionnaire about Empathy to Video Contents.

	Questionnaire	Empathy Factor
1	I understood the characters' needs.	Cognitive empathy
2	I understood how the characters were feeling.	
3	I understood the situation of the video.	
4	I understood the motives behind the characters' behavior.	
5	I felt as if the events in the video were happening to me.	Affective empathy
6	I felt as if I was in the middle of the situation.	
7	I felt as if I was one of the characters.	
8	I experienced many of the same feelings that the characters portrayed.	Identification empathy
9	I felt the characters' needs were similar to mine.	
10	The events in the video were similar to my experience.	
11	I felt as if the events in the video could happen to me.	

All questions were rated on a seven-point Likert scale. We asked for the degree of agreement with each empathy statement, with the lowest scale labeled "strongly disagree" and the highest scale labeled "strongly agree". The survey was collected through a web survey rather than a paper questionnaire.

3. Analysis

This study aimed to analyze whether the similarity of micro-movement signals between participants and advertisements differs according to the user's perceived empathy (i.e., subjective evaluation) with the advertisement video. The signal processing to filter only the micro-movements caused by the heartbeat is described in detail in Section 3.1. In addition, a method for calculating the cross-entropy, an indicator of similarity between the two signals, is described. Section 3.2 describes the statistical difference in the similarity between the participant and advertisement measured by cross-entropy according to the empathy score.

3.1. Signal Processing

The micro-movement signals were measured from the participant's facial videos, as shown in Figure 3. Ballistocardiographic changes are reflected to the face and can be measured at a distance, as validated by Balakrishnan [68]. The face was detected from the facial video using the Viola-Jones face detector and was defined as a region of interest (ROI). As the forehead and nose were more robust to facial expressions than other facial regions, the ROI was divided into multiple ROIs by cropping to the middle 60% of the width and top 12% of the height (i.e., forehead region) and the middle 10% of the width and middle 30% of the height (i.e., nose region).

Determining the feature point within multiple ROIs was necessary to measure the movements induced by the BCG. Although several studies on remote BCG employed the good-feature-to-track (GFTT) algorithm [83,84], their feature point numbers were not fixed because the algorithm determined the feature points based on the solid edge components. It was difficult to employ the GFTT algorithm in this study because the feature points needed to be re-determined quickly owing to the frequent change of the screen and the face movement.

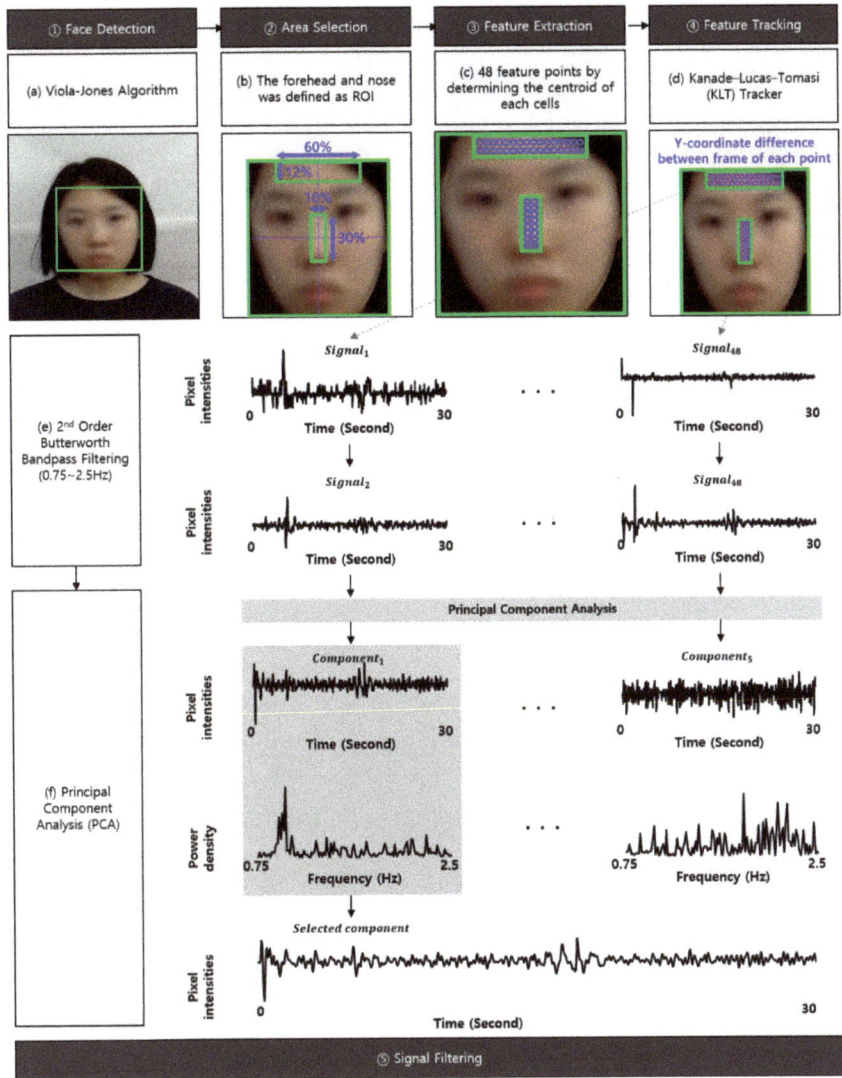

Figure 3. Signal processing of the micro-movements [71]. (**a**) Face detection using Viola-Jones algorithm; (**b**) Area selection using the forehead and nose defined as ROIs; (**c**) Feature extraction using the GFTT algorithm; (**d**) Feature tracking using the KLT tracker; (**e**) Bandpass filtering for signals in 30 s window buffer using the second order Butterworth filter; (**f**) Decomposition of noise using PCA.

Thus, the ROIs of the forehead and nose regions were divided into cells using 16 × 2 and 2 × 8 grids, respectively. This study employed 48 feature points by determining the centroid of each cell as a feature point. The movements were measured by tracking the y-coordinate difference between frames of each feature point using the Kanade-Lucas-Tomasi (KLT) tracker because the BCG movements were generated up and down by the heartbeat [85–87].

The movements measured from the face are a combination of facial expressions, voluntary head movements, and micro-movements. Therefore, it is essential to remove motion artifacts due to facial expressions and voluntary head movements from the measured movements. First, the movements were filtered by a second order Butterworth bandpass filter with a cut-off 0.75–2.5 Hz corresponding to 45–150 bpm. Then, the movements were normalized from their mean value (i.e., μ) and standard deviation (σ) by z-score. If the movements exceeded the $\mu + -2\sigma$, they were determined to be noise, due to the subtle movements, and their mean value (i.e., μ) was corrected. Finally, principal component analysis (PCA) was performed to estimate the micro-movement from the mixed movements by decomposing the noise from facial expressions and voluntary head movements. This study extracted five components using PCA and then selected one component with the highest peak in their power spectrum converted using a fast Fourier transform. The selected component was finally determined to be micro-movements.

3.2. Statistical Analysis

As empathy is an individualized experience, the manner in which each stimulus affects each participant varies. Individualized response is affected by factors, such as the individual's empathy capability, predisposed tendency, and past experience (for an extensive review of empathy as a concept, see [88]). The observer's (i.e., the person who empathizes) mood and personality are also an important modulating factor [89]. Such individual differences mean that, in our study, the empathy stimuli selected by the emotion experts do not necessarily elicit empathy from the participants. Therefore, we applied an inclusion criterion to the participants' subjective empathy scores to select response sets from certain stimuli for analysis. We selected data obtained from stimuli that scored, on average (i.e., the mean of all 30 participants), on or higher than four for the empathy condition. In the seven-point Likert scale, four was the middle point, labeled as "Neutral". Conversely, we selected data obtained from stimuli that scored less than four on average for the non-empathy condition. This selection process yielded response sets from four out of the original 12 stimuli in the empathy condition and six out of the original 12 stimuli in the non-empathy condition.

In short, we analyzed 60 samples (30 participants in two empathy conditions) consisting of subjective empathy scores and cross-entropy data. A paired t-test was used to test this hypothesis.

4. Results

The study analyzed differences in the micro-movement similarity between empathy and non-empathy conditions using a t-test. The results showed that there was a significant difference in the subjective empathy score between empathy and non-empathy conditions induced by advertisements ($t(29) = -11.754$, $p < 0.001$), as shown in Figure 4. The subjective empathy score was significantly higher when watching empathy advertisements ($\mu = 5.149$, $\sigma = 0.564$) than non-empathy advertisements ($\mu = 3.341$, $\sigma = 0.759$).

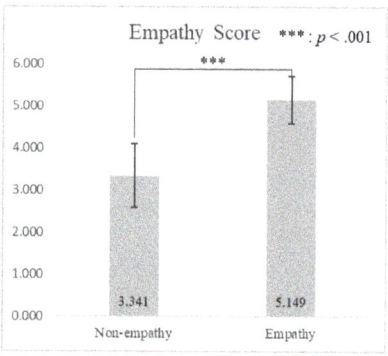

Figure 4. A comparison of empathy scores for non-empathy and empathy advertisements by paired t-test.

There was a significant statistical difference in cross-entropy between empathy and non-empathy advertisements ($t(29) = 61.019$, $p < 0.001$), as shown in Figure 5. As predicted, cross-entropy was significantly lower when watching empathy advertisements ($\mu = 0.00317$, $\sigma = 0.00005$) than non-empathy advertisements ($\mu = 0.00392$, $\sigma = 0.00005$). This supported hypothesis H_1, which stated that the more similar the micro-movements (i.e., the lower the cross-entropy) between the participant and person in the advertisement, the higher the participant's empathy scores for the advertisement (i.e., empathy advertisements).

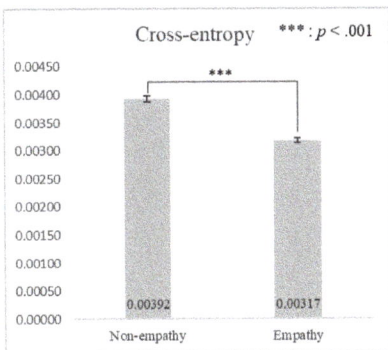

Figure 5. A comparison of cross-entropy between non-empathy and empathy advertisements by paired t-test.

The Pearson correlation indicated that cross-entropy was also significantly associated with empathy score ($r = -0.796$, $p < 0.001$), indicating an inverse relationship between cross-entropy and the empathy scores. That is, the lesser cross-entropy, the higher the empathy scores.

5. Discussion

In summary, our study invited participants to view advertisements classified as empathy or non-empathy advertisements by experts. During each viewing of the advertisement, we recorded their facial data and obtained their subjective empathy scores after each viewing. We analyzed the cross-entropy between the participant's and the person's facial data and found that it was significantly lower when viewing empathy advertisements than when viewing non-empathy advertisements.

To the best of our knowledge, this is the first study to apply remote BCG methods to understand empathy-based micro-movement synchronization in a real-world use case (i.e., viewing an advertisement). Our research confirmed that the higher the similarity of micro-movement between the participants and the advertisements, the higher the subjective empathy. The results validate the remote BCG methods with the accompanying analysis process (e.g., cross-entropy analysis), suggesting an alternative or complementary method to the subjective empathy scales.

Our findings also provide implications for understanding the empathic interactions of human dyads. In human communication, information is shared through natural language (i.e., explicit channels), whereas empathy is mainly shared through embodied synchrony (i.e., implicit channels). The latter synchronization is widely observed in human communication and is reflected in the harmonization of the heart rhythm. In other words, the heartbeat tends to follow the rhythm of someone who empathizes. Such mutual entrainment has been defined as two interacting nonlinear oscillating systems with different periods becoming a common period [90]. Although challenging, advances in technology enable us to tap into heartbeat traces through the carotid artery, reflected in the facial micro-movement. Our study confirmed that microscopic vibration is a valid indicator of dyadic empathy synchronization in an ecologically valid scenario.

In previous studies that measured empathy based on unconscious physiological responses [41,43], it was also verified that the correlation between the heartbeat patterns of two people was higher in the empathy condition than in the non-empathy condition. They measured heart rate patterns by attaching an ECG sensor to the participant's skin. The task of eliciting empathy was overly simplified, such as facing each other, and only momentary emotions were of concern, resulting in limitations to generalization. Although they can effectively elicit a definite empathic response, the emotion dynamics were not considered.

In addition, there were fewer applications measuring empathy for digital content because of the challenge in solving the barrier of obtrusive measurement and consideration of the dynamic nature of empathy. This study suggested a practical method for measuring empathy that complements the issue of contact-based empathy measurement that obstructs users' immersion in the content.

The hypothesis of the present study was tested under experimental conditions by manipulating product advertisements. This study acknowledges that there were large differences among the durations of the stimuli, and the stimuli were only focused on product commercials. However, the differences in time duration among stimuli did not affect the similarity because the similarity between the two signals was analyzed in the frequency domain. That is, because the similarity of the periodicity of the two signals was analyzed, the time length of the signal did not have a significant effect. Even if there was an effect, the empathy stimuli, which had a long duration, were difficult to make similar to the non-empathy stimuli, which had a short duration, because they had to vibrate at a similar frequency for a longer period of time.

This study suggests an application framework for evaluating empathy in interaction (e.g., viewing) with digital content. As our suggested method is non-contact and unobtrusive to real-life behavior (e.g., consuming media), future research agendas seem promising. Specifically, future studies may investigate content in other media domains (e.g., movies, TV shows, video games).

However, we acknowledge that a larger N would be needed to achieve the appropriate power to completely rule out false positives. We acknowledge that our N is small (30) and, as such, we conducted a post hoc power analysis with the program G*Power [91] with power set at 0.8 and $\alpha = 0.05$, d = 0.5, two-tailed. The results suggest that an N value of approximately 34 would be needed to achieve appropriate statistical power.

Empathy is a multifaceted social psychological construct that is affected by many factors, such as the relationship and history between the observer (i.e., empathizer) and the observed. Such social relationships are also shaped by intimacy, while favorability also comes into play. As empathy is dependent on context and task [89,92], our study has an inherent limitation in generalization.

We also acknowledge that empathic expression is a result of a combination of many nonverbal modalities (e.g., voice, facial expression, posture). We focused on a singular modality, the facial movements captured from the involuntary heartbeat, because such measures could also be confounded by noise. Moreover, there can be a gap between the actual emotion the actor felt and the physiological measurement we acquired. Such a gap can be measured through a combination of expressive measures (facial muscle movement, gestures) and implicit measures (heart rate, GSR). Future studies may investigate multimodal recognition of empathy, in addition to facial micro-movements.

We strived to filter out the signals that represent empathy from the signal spectrum as closely as possible to the target population by guiding the participant not to move and to refrain from exaggerating facial expressions. We did not include any participants who may have made significant movements that would confound our results, such as participants with Tourette syndrome or a person with bruxism.

Privacy issues that may arise from identifying individuals can be crucial in research that considers prosocial behaviors. However, the suggested method of recognizing empathy can enhance privacy by not saving personal identification data (i.e., original record video) in the database. Only the processed secondary data (i.e., micro-movement signals) can be saved in the database by analyzing video frames in real-time without recording the face images. Then, the synchronization data can be analyzed if only a key can match (i.e., random number) an advertisement and a viewer. The analyzed micro-movement features are hardly restored to the original facial image, so it is impossible to identify its data.

Author Contributions: A.C. designed the study with an investigation of previous studies and performed the experiments and original draft preparation; S.P. wrote the review and edited with an investigation of previous studies; H.L. analyzed raw data and developed micromovement algorithm; M.W. conceived the study and was in charge of overall direction and planning. All authors have read and agreed to the published version of the manuscript.

Funding: "This work was supported by the National Research Foundation of Korea (NRF) grant funded by the Korea government (MSIT) (NRF-2020R1A2B5B02002770)" and "This work was supported by Basic Science Research Program through the National Research Foundation of Korea (NRF) funded by the Ministry of Education (NRF-2021R1I1A1A01045641)".

Institutional Review Board Statement: The study was conducted according to the guidelines of the Declaration of Helsinki, and approved by the Institutional Review Board of the Sangmyung University, Seoul, Korea (SMUIRB C-2019-015).

Informed Consent Statement: Informed consent was obtained from all subjects involved in the study.

Conflicts of Interest: The authors declare no conflict of interest.

References

1. Zillmann, D. Empathy: Affective reactivity to others' emotional experiences. *Psychol. Entertain.* **2006**, 151–181.
2. Hoffman, M.L. Interaction of affect and cognition in empathy. *Emot. Cogn. Behav.* **1984**, 103–131.
3. Kerem, E.; Fishman, N.; Josselson, R. The Experience of Empathy in Everyday Relationships: Cognitive and Affective Elements. *J. Soc. Pers. Relatsh.* **2001**, *18*, 709–729. [CrossRef]
4. Davis, M.H. Empathy. In *Handbook of the Sociology of Emotions*; Springer: Boston, MA, USA, 2006; pp. 443–466.
5. Busselle, R.; Bilandzic, H. Measuring Narrative Engagement. *Media Psychol.* **2009**, *12*, 321–347. [CrossRef]
6. Eisenberg, N. Empathy and Sympathy: A Brief Review of the Concepts and Empirical Literature. *Anthrozoös* **1988**, *2*, 15–17. [CrossRef]
7. Galimberti, C.; Gaggioli, A.; Brivio, E.; Caroli, F.; Chirico, A.; Rampinini, L.; Trognon, A.; Vergine, I. Transformative Conversations. Questioning collaboration in digitally mediated interactions. *Ann. Rev. Cyberther. Telemed.* **2020**, *18*, 77–80.
8. Teixeira, T.; Wedel, M.; Pieters, R. Emotion-Induced Engagement in Internet Video Advertisements. *J. Mark. Res.* **2012**, *49*, 144–159. [CrossRef]
9. Batra, R.; Ray, M.L. Affective Responses Mediating Acceptance of Advertising. *J. Consum. Res.* **1986**, *13*, 234–249. [CrossRef]
10. Howard, D.J.; Gengler, C. Emotional Contagion Effects on Product Attitudes: Figure 1. *J. Consum. Res.* **2001**, *28*, 189–201. [CrossRef]
11. Belanche, D.; Flavián, C.; Rueda, A.P. Understanding Interactive Online Advertising: Congruence and Product Involvement in Highly and Lowly Arousing, Skippable Video Ads. *J. Interact. Mark.* **2017**, *37*, 75–88. [CrossRef]
12. Jeon, Y.A. Skip or Not to Skip: Impact of Empathy and Ad Length on Viewers' Ad-Skipping Behaviors on the Internet. In *International Conference on Human-Computer Interaction*; Springer: Cham, Switzerland, 2018; pp. 261–265.
13. Adelaar, T.; Chang, S.; Lancendorfer, K.M.; Lee, B.; Morimoto, M. Effects of Media Formats on Emotions and Impulse Buying Intent. *J. Inf. Technol.* **2003**, *18*, 247–266. [CrossRef]
14. Deighton, J.; Romer, D.; McQueen, J. Using Drama to Persuade. *J. Consum. Res.* **1989**, *16*, 335–343. [CrossRef]
15. Escalas, J.E.; Stern, B.B. Sympathy and Empathy: Emotional Responses to Advertising Dramas. *J. Consum. Res.* **2003**, *29*, 566–578. [CrossRef]
16. Steyn, P.; Pitt, L.; Chakrabarti, R. Financial services ads and viewer response profiles: Psychometric properties of a shortened scale. *J. Financ. Serv. Mark.* **2011**, *16*, 210–219. [CrossRef]
17. Stout, P.A.; Rust, R.T. Emotional Feelings and Evaluative Dimensions of Advertising: Are They Related? *J. Advert.* **1993**, *22*, 61–71. [CrossRef]
18. Hyun, S.S.; Kim, W.; Lee, M.J. The impact of advertising on patrons' emotional responses, perceived value, and behavioral intentions in the chain restaurant industry: The moderating role of advertising-induced arousal. *Int. J. Hosp. Manag.* **2011**, *30*, 689–700. [CrossRef]
19. Balconi, M.; Canavesio, Y. Emotional contagion and trait empathy in prosocial behavior in young people: The contribution of autonomic (facial feedback) and Balanced Emotional Empathy Scale (BEES) measures. *J. Clin. Exp. Neuropsychol.* **2013**, *35*, 41–48. [CrossRef] [PubMed]

20. Lawrence, E.J.; Shaw, P.; Baker, D.; Baron-Cohen, S.; David, A.S. Measuring empathy: Reliability and validity of the Empathy Quotient. *Psychol. Med.* **2004**, *34*, 911–920. [CrossRef]
21. Spreng, R.N.; McKinnon, M.C.; Mar, R.A.; Levine, B. The Toronto Empathy Questionnaire: Scale Development and Initial Validation of a Factor-Analytic Solution to Multiple Empathy Measures. *J. Pers. Assess.* **2009**, *91*, 62–71. [CrossRef]
22. Jabbi, M.; Swart, M.; Keysers, C. Empathy for positive and negative emotions in the gustatory cortex. *NeuroImage* **2007**, *34*, 1744–1753. [CrossRef]
23. De Corte, K.; Buysse, A.; Verhofstadt, L.L.; Roeyers, H.; Ponnet, K.; Davis, M.H. Measuring Empathic Tendencies: Reliability And Validity of the Dutch Version of the Interpersonal Reactivity Index. *Psychol. Belg.* **2007**, *47*, 235–260. [CrossRef]
24. Albiero, P.; Matricardi, G.; Speltri, D.; Toso, D. The assessment of empathy in adolescence: A contribution to the Italian validation of the "Basic Empathy Scale. *J. Adolesc.* **2009**, *32*, 393–408. [CrossRef] [PubMed]
25. Hogan, R. Development of an empathy scale. *J. Consult. Clin. Psychol.* **1969**, *33*, 307–316. [CrossRef]
26. Singer, T.; Klimecki, O.M. Empathy and compassion. *Curr. Biol.* **2014**, *24*, R875–R878. [CrossRef]
27. Pfeifer, J.H.; Iacoboni, M.; Mazziotta, J.C.; Dapretto, M. Mirroring others' emotions relates to empathy and interpersonal compe-tence in children. *Neuroimage* **2008**, *39*, 2076–2085. [CrossRef]
28. O'Brien, E.; Konrath, S.; Grühn, D.; Hagen, L. Empathic Concern and Perspective Taking: Linear and Quadratic Effects of Age Across the Adult Life Span. *J. Gerontol. Ser. B* **2013**, *68*, 168–175. [CrossRef] [PubMed]
29. Rizzolatti, G.; Craighero, L. The mirror-neuron system. *Annu. Rev. Neurosci.* **2004**, *27*, 169–192. [CrossRef] [PubMed]
30. Barsade, S.G. The Ripple Effect: Emotional Contagion and its Influence on Group Behavior. *Adm. Sci. Q.* **2002**, *47*, 644–675. [CrossRef]
31. Husserl, E. *Phenomenology and the Foundations of the Sciences*; Springer Science & Business Media: Berlin/Heidelberg, Germany, 2001.
32. Husserl, E. *Cartesian Meditations: An Introduction to Phenomenology*; Springer Science & Business Media: Berlin/Heidelberg, Germany, 2013.
33. Duffy, K.A.; Chartrand, T.L. Mimicry: Causes and consequences. *Curr. Opin. Behav. Sci.* **2015**, *3*, 112–116. [CrossRef]
34. Bavelas, J.B.; Black, A.; Lemery, C.R.; Mullett, J. I show how you feel": Motor mimicry as a communicative act. *J. Pers. Soc. Psychol.* **1986**, *50*, 322. [CrossRef]
35. Yokozuka, T.; Ono, E.; Inoue, Y.; Ogawa, K.-I.; Miyake, Y. The Relationship between Head Motion Synchronization and Empathy in Unidirectional Face-to-Face Communication. *Front. Psychol.* **2018**, *9*. [CrossRef]
36. Komori, M.; Nagaoka, C. The relationship between body movements of clients and counsellors in psychother-apeutic counselling: A study using the video-based quantification method. *Jpn. J. Cogn. Psychol.* **2011**, *8*, 1–9.
37. Nagaoka, C.; Komori, M. Body Movement Synchrony in Psychotherapeutic Counseling: A Study Using the Video-Based Quantification Method. *IEICE Trans. Inf. Syst.* **2008**, *E91-D*, 1634–1640. [CrossRef]
38. Palumbo, R.V.; Marraccini, M.E.; Weyandt, L.L.; Wilder-Smith, O.; McGee, H.A.; Liu, S.; Goodwin, M.S. Interpersonal Autonomic Physiology: A Systematic Review of the Literature. *Pers. Soc. Psychol. Rev.* **2016**, *21*, 99–141. [CrossRef] [PubMed]
39. Stratford, T.; Lal, S.; Meara, A. Neurophysiology of therapeutic alliance. *Gestalt. J. Aust.* **2009**, *5*, 19–47.
40. Stratford, T.; Lal, S.; Meara, A. A Neuroanalysis of therapeutic alliance in the symptomatically anxious: The physiological con-nection revealed between therapist and client. *Am. J. Psychother.* **2012**, *66*, 1–21. [CrossRef]
41. Park, S.; Choi, S.J.; Mun, S.; Whang, M. Measurement of emotional contagion using synchronization of heart rhythm pattern between two persons: Application to sales managers and sales force synchronization. *Physiol. Behav.* **2019**, *200*, 148–158. [CrossRef]
42. Ferrer, E.; Helm, J.L. Dynamical systems modeling of physiological coregulation in dyadic interactions. *Int. J. Psychophysiol.* **2013**, *88*, 296–308. [CrossRef]
43. Chatel-Goldman, J.; Congedo, M.; Jutten, C.; Schwartz, J.-L. Touch increases autonomic coupling between romantic partners. *Front. Behav. Neurosci.* **2014**, *8*, 95. [CrossRef] [PubMed]
44. Reed, R.G.; Randall, A.K.; Post, J.H.; Butler, E.A. Partner influence and in-phase versus anti-phase physiological linkage in romantic couples. *Int. J. Psychophysiol.* **2013**, *88*, 309–316. [CrossRef]
45. Feijt, M.A.; de Kort, Y.A.; Westerink, J.H.; Okel, S.; IJsselsteijn, W.A. The effect of simulated feedback about psychophysiological synchronization on perceived empathy and connectedness. *Annu. Rev. Cyberther. Telemed.* **2020**, *117*.
46. Marci, C.D.; Ham, J.; Moran, E.; Orr, S.P. Physiologic Correlates of Perceived Therapist Empathy and Social-Emotional Process during Psychotherapy. *J. Nerv. Ment. Dis.* **2007**, *195*, 103–111. [CrossRef] [PubMed]
47. Koole, S.L.; Tschacher, W. Synchrony in Psychotherapy: A Review and an Integrative Framework for the Therapeutic Alliance. *Front. Psychol.* **2016**, *7*, 862. [CrossRef] [PubMed]
48. Ramseyer, F.; Tschacher, W. Nonverbal synchrony of head-and body-movement in psychotherapy: Different sig-nals have different associations with outcome. *Front. Psychol.* **2014**, *5*, 979. [CrossRef] [PubMed]
49. Stephens, G.; Silbert, L.J.; Hasson, U. Speaker-listener neural coupling underlies successful communication. *Proc. Natl. Acad. Sci. USA* **2010**, *107*, 14425–14430. [CrossRef]
50. Yun, K.; Watanabe, K.; Shimojo, S. Interpersonal body and neural synchronization as a marker of implicit social interaction. *Sci. Rep.* **2012**, *2*, srep00959. [CrossRef]
51. Zaki, J.; Ochsner, K.N. The neuroscience of empathy: Progress, pitfalls and promise. *Nat. Neurosci.* **2012**, *15*, 675–680. [CrossRef]

52. Tullett, A.M.; Harmon-Jones, E.; Inzlicht, M. Right frontal cortical asymmetry predicts empathic reactions: Support for a link between withdrawal motivation and empathy. *Psychophysiology* **2012**, *49*, 1145–1153. [CrossRef] [PubMed]
53. Singer, T.; Seymour, B.; O'Doherty, J.P.; Stephan, K.E.; Dolan, R.J.; Frith, C.D. Empathic neural responses are modulated by the perceived fairness of others. *Nature* **2006**, *439*, 466–469. [CrossRef]
54. Peng, W.; Lou, W.; Huang, X.; Ye, Q.; Tong, R.K.-Y.; Cui, F. Suffer together, bond together: Brain-to-brain synchronization and mutual affective empathy when sharing painful experiences. *NeuroImage* **2021**, *238*, 118249. [CrossRef]
55. McGuigan, F.J.; Andreassi, J.L. Psychophysiology—Human Behavior and Physiological Response. *Am. J. Psychol.* **1981**, *94*, 359. [CrossRef]
56. Neumann, D.L.; Westbury, H.R. The psychophysiological measurement of empathy. In *Psychology of Empathy*; Griffith University: Queenland, Australia, 2011; pp. 119–142.
57. Kodama, K.; Tanaka, S.; Shimizu, D.; Hori, K.; Matsui, H. Heart Rate Synchrony in Psychological Counseling: A Case Study. *Psychology* **2018**, *09*, 1858–1874. [CrossRef]
58. Salminen, M.; Jarvela, S.; Ruonala, A.; Harjunen, V.; Jacucci, G.; Hamari, J.; Ravaja, N. Evoking Physiological Synchrony and Empathy Using Social VR with Biofeedback. *IEEE Trans. Affect. Comput.* **2019**, *3045*, 1. [CrossRef]
59. Werner, J.; Hornecker, E. United-Pulse: Feeling Your Partner's Pulse. In Proceedings of the MobileHCI, Amsterdam, The Netherlands, 2–5 September 2008; ACM: New York, NY, USA, 2008; pp. 535–538.
60. Janssen, J.H.; Bailenson, J.N.; Ijsselsteijn, W.; Westerink, J.H.D.M. Intimate Heartbeats: Opportunities for Affective Communication Technology. *IEEE Trans. Affect. Comput.* **2010**, *1*, 72–80. [CrossRef]
61. Rouast, P.; Adam, M.; Chiong, R.; Cornforth, D.; Lux, E. Remote heart rate measurement using low-cost RGB face video: A technical literature review. *Front. Comput. Sci.* **2018**, *12*, 858–872. [CrossRef]
62. Van Gastel, M.M.; Stuijk, S.; De Haan, G.G. Motion Robust Remote-PPG in Infrared. *IEEE Trans. Biomed. Eng.* **2015**, *62*, 1425–1433. [CrossRef]
63. Blanik, N.; Abbas, A.K.; Venema, B.; Blazek, V.; Leonhardt, S. Hybrid optical imaging technology for long-term remote monitoring of skin perfusion and temperature behavior. *J. Biomed. Opt.* **2014**, *19*, 016012. [CrossRef]
64. Verkruysse, W.; Svaasand, L.O.; Nelson, J.S. Remote plethysmographic imaging using ambient light. *Opt. Express* **2008**, *16*, 21434–21445. [CrossRef]
65. Kamshilin, A.A.; Miridonov, S.; Teplov, V.; Saarenheimo, R.; Nippolainen, E. Photoplethysmographic imaging of high spatial resolution. *Biomed. Opt. Express* **2011**, *2*, 996–1006. [CrossRef] [PubMed]
66. Moco, A.V.; Stuijk, S.; De Haan, G. Ballistocardiographic artifacts in PPG imaging. *IEEE Trans. Biomed. Eng.* **2015**, *63*, 1804–1811. [CrossRef] [PubMed]
67. Starr, I.; Rawson, A.J.; Schroeder, H.A.; Joseph, N.R. Studies on the Estimation of Cardiac Output in Man, and of Abnormalities in Cardiac Function, from the heart's Recoil and the blood's Impacts; the Ballistocardiogram. *Am. J. Physiol. Leg. Content* **1939**, *127*, 1–28. [CrossRef]
68. Balakrishnan, G.; Durand, F.; Guttag, J. Detecting Pulse from Head Motions in Video. In Proceedings of the 2013 IEEE Conference on Computer Vision and Pattern Recognition, Portland, OR, USA, 23–28 June 2013; pp. 3430–3437.
69. Shan, L.; Yu, M. Video-based heart rate measurement using head motion tracking and ICA. In Proceedings of the 2013 6th International Congress on Image and Signal Processing (CISP), Melbourne, Australia, 15–18 September 2013; Volume 1, pp. 160–164.
70. Hassan, M.A.; Malik, A.S.; Fofi, D.; Saad, N.M.; Ali, Y.S.; Meriaudeau, F. Video-Based Heartbeat Rate Measuring Method Using Ballistocardiography. *IEEE Sens. J.* **2017**, *17*, 4544–4557. [CrossRef]
71. Lee, S.; Cho, A.; Whang, M. Vision-Based Measurement of Heart Rate from Ballistocardiographic Head Movements Using Unsupervised Clustering. *Sensors* **2019**, *19*, 3263. [CrossRef] [PubMed]
72. Liu, W.; Pokharel, P.P.; Principe, J.C. Correntropy: A localized similarity measure. In Proceedings of the 2006 IEEE international Joint Conference on Neural Network Proceedings, Vancouver, BC, Canada, 16 July 2006; pp. 4919–4924.
73. Chiao, J.Y.; Mathur, V.A. Intergroup Empathy: How Does Race Affect Empathic Neural Responses? *Curr. Biol.* **2010**, *20*, R478–R480. [CrossRef] [PubMed]
74. Ward, J.; Cody, J.; Schaal, M.; Hojat, M. The Empathy Enigma: An Empirical Study of Decline in Empathy among Undergraduate Nursing Students. *J. Prof. Nurs.* **2012**, *28*, 34–40. [CrossRef]
75. Christov-Moore, L.; Simpson, E.A.; Coudé, G.; Grigaityte, K.; Iacoboni, M.; Ferrari, P.F. Empathy: Gender effects in brain and be-havior. *Neurosci. Biobehav. Rev.* **2014**, *46*, 604–627. [CrossRef]
76. Thompson, A.; Voyer, D. Sex differences in the ability to recognise non-verbal displays of emotion: A meta-analysis. *Cogn. Emot.* **2014**, *28*, 1164–1195. [CrossRef]
77. Depow, G.J.; Francis, Z.; Inzlicht, M. The Experience of Empathy in Everyday Life. *Psychol. Sci.* **2021**, *32*, 1198–1213. [CrossRef]
78. Bagozzi, R.P.; Moore, D.J. Public service advertisements: Emotions and empathy guide prosocial behavior. *J. Mark.* **1994**, *58*, 56–70. [CrossRef]
79. Rawal, M.; Saavedra Torres, J.L. Empathy for Emotional Advertisements on Social Networking Sites: The Role of Social Identity. 2017. Available online: http://www.mmaglobal.org/publications/MMJ/MMJ-Issues/2017-Fall/MMJ-2017-Fall-Vol27-Issue2-Rawal-SaavedraTorres-pp88-102.pdf (accessed on 10 October 2021).
80. Green, M.C.; Brock, T.C. *In the Mind's Eye: Transportation-Imagery Model of Narrative Persuasion*; Psychology Press: New York, NY, USA, 2002.

81. Soh, H. Measuring Consumer Empathic Response to Advertising Drama. *J. Korea Contents Assoc.* **2014**, *14*, 133–142. [CrossRef]
82. Yoo, S.; Whang, M. Vagal Tone Differences in Empathy Level Elicited by Different Emotions and a Co-Viewer. *Sensors* **2020**, *20*, 3136. [CrossRef]
83. Haque, M.A.; Irani, R.; Nasrollahi, K.; Moeslund, T.B. Facial video-based detection of physical fatigue for maximal muscle activity. *IET Comput. Vis.* **2016**, *10*, 323–330. [CrossRef]
84. Tommasini, T.; Fusiello, A.; Trucco, E.; Roberto, V. Making good features track better. In Proceedings of the 1998 IEEE Computer Society Conference on Computer Vision and Pattern Recognition (Cat. No.98CB36231), Santa Barbara, CA, USA, 6 August 2002; pp. 178–183.
85. Bouguet, J.Y. Pyramidal implementation of the affine lucas kanade feature tracker description of the algorithm. *Intel. Corp.* **2001**, *5*, 4.
86. Yang, S.; Luo, P.; Loy, C.C.; Tang, X. WIDER FACE: A Face Detection Benchmark. In Proceedings of the 2016 IEEE Conference on Computer Vision and Pattern Recognition (CVPR), Las Vegas, NV, USA, 27–30 June 2016; pp. 5525–5533.
87. Tomasi, C.; Kanade, T.T. Detection and Tracking of Point Features, Carnegie Mellon University Technical Report CMU-CS-91-132, April 1991. Available online: https://cecas.clemson.edu/~{}stb/klt/tomasi-kanade-techreport-1991.pdf (accessed on 10 October 2021).
88. Cuff, B.; Brown, S.; Taylor, L.; Howat, D.J. Empathy: A Review of the Concept. *Emot. Rev.* **2016**, *8*, 144–153. [CrossRef]
89. De Vignemont, F.; Singer, T. The empathic brain: How, when and why? *Trends Cogn. Sci.* **2006**, *10*, 435–441. [CrossRef]
90. Schimansky-Geier, L.; Kuramoto, Y. *Chemical Oscillations, Waves, and Turbulence*; Springer Series in Synergetics 19; Springer: Berlin/Heidelberg, Germany; New York, NY, USA; Tokyo, Japan, 1984; ISBN 3-540-13322-4.
91. Erdfelder, E.; Faul, F.; Buchner, A. GPOWER: A general power analysis program. *Behav. Res. Methods Instrum. Comput.* **1996**, *28*, 1–11. [CrossRef]
92. Davis, M.H. *Empathy: A Social Psychological Approach*; Routledge: London, UK, 2018.

Article

Multimodal Data Collection System for Driver Emotion Recognition Based on Self-Reporting in Real-World Driving

Geesung Oh [1], Euiseok Jeong [1], Rak Chul Kim [1], Ji Hyun Yang [2], Sungwook Hwang [3], Sangho Lee [3] and Sejoon Lim [2,*]

[1] Graduate School of Automotive Engineering, Kookmin University, Seoul 02707, Korea; gsethan17@kookmin.ac.kr (G.O.); euiseok_jeong@kookmin.ac.kr (E.J.); valance95@kookmin.ac.kr (R.C.K.)
[2] Department of Automobile and IT Convergence, Kookmin University, Seoul 02707, Korea; yangjh@kookmin.ac.kr
[3] Chassis System Control Research Lab, Hyundai Motor Group, Hwaseong 18280, Korea; gazz@hyundai.com (S.H.); imprince@hyundai.com (S.L.)
* Correspondence: lim@kookmin.ac.kr; Tel.: +82-2-910-5469

Abstract: As vehicles provide various services to drivers, research on driver emotion recognition has been expanding. However, current driver emotion datasets are limited by inconsistencies in collected data and inferred emotional state annotations by others. To overcome this limitation, we propose a data collection system that collects multimodal datasets during real-world driving. The proposed system includes a self-reportable HMI application into which a driver directly inputs their current emotion state. Data collection was completed without any accidents for over 122 h of real-world driving using the system, which also considers the minimization of behavioral and cognitive disturbances. To demonstrate the validity of our collected dataset, we also provide case studies for statistical analysis, driver face detection, and personalized driver emotion recognition. The proposed data collection system enables the construction of reliable large-scale datasets on real-world driving and facilitates research on driver emotion recognition. The proposed system is avaliable on GitHub.

Keywords: driver emotion recognition; multimodal; self-report; real-world driving

1. Introduction

In recent decades, the use of data-driven state-of-the-art techniques such as deep learning has increased interest in and performance of human affect recognition [1]. This has increased interest in the development of driver emotion recognition systems. Since driving is significantly affected by the driver's emotions [2–4], driver emotion recognition studies have been conducted for various purposes such as driving safety, adjusting vehicle dynamics, and emotion elicitation of drivers [4–6]. All studies are affected by the quality and quantity of data. Therefore, research on quantitative and qualitative datasets for driver emotion recognition is being actively conducted [7–14].

Although large-scale and high-quality datasets are collected through various studies, the collection conditions vary significantly. First, the experimental environment is largely divided into simulation and real-world driving. Second, the modalities of collected signals are also diverse. When broadly classified, there are video, audio, bio-physiological, and controller area network (CAN) data. In detail, the position of cameras and microphones differ, and the collection list of biophysiological or CAN data is not unified. Lastly, the annotation of emotional states is various, which is critical for emotion recognition. The simplest way to classify a driver's emotional state is by driving experiments (e.g., assume that heavy traffic on the urban is high stress, and light traffic on the highway is low stress) [7–9]. There is also an approach in which external annotators judge a driver's emotional state based on the collected information about the driver. However, this approach has limitations in that it has a high-cost and requires others to report their emotional states [10,11]. In the self-reporting approach, drivers report their emotional states, but this should not interfere with the main task of driving. Hence, it is restricted to experiments through simulation or they have to report their emotional states after the completion of the experiments [12–14]. As previously stated, since data collection environments, measured data types, and annotation methods very, Zepf et al. have argued that a consistent dataset is needed to facilitate research on driver emotion recognition [15].

In this paper, we propose a data collection system that can be used for a variety of driver emotion recognition studies. The proposed system collects multimodal datasets such as videos from various views, audio, biophysiological, CAN data, and drivers' emotional states, which are data representatively used for driver emotion recognition. A driver's emotional state is collected by a driver self-reporting their emotional state while driving through a human–machine interaction (HMI) application. To realize a universal dataset, the collection experiment should be conducted in the real world environment, not through a simulator. To conduct a real-world driving experiment, it is necessary to prevent the behavioral and cognitive disturbances of drivers in advance to avoid potential traffic accidents. To prevent behavioral disturbance, the proposed system collects biophysiological data using wearable sensors, instead of biometric sensors attached to the body. The self-reporting application for minimizing cognitive disturbances comprises a haptic, acoustic response, and graphical user interface (GUI) based on user experience (UX). In addition, there are concerns about the reflection of strong bias during self-reporting due to false memories or the desire to impress others [15]. To address these concerns, we focused on making the self-reporting interaction occur periodically. All considerations for reliable data are detailed in Section 3. The data collection system is installed on a vehicle, and data collection is performed under real-world driving conditions. Figure 1 shows the data collection vehicle driving during real-world driving.

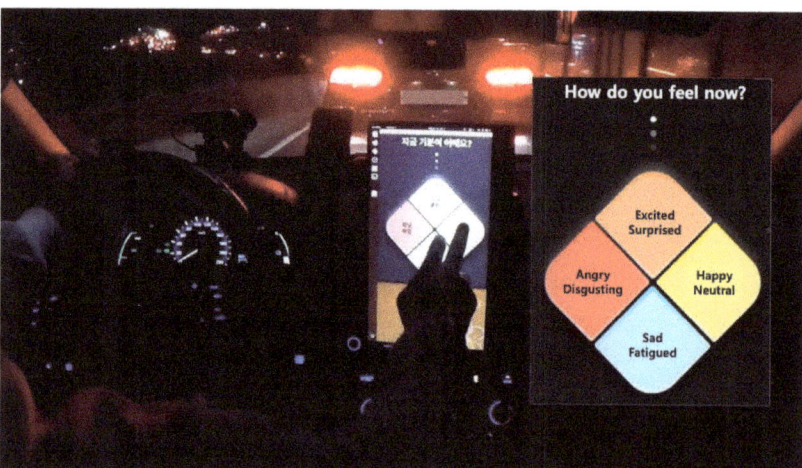

Figure 1. A scene in which a driver's emotional state data is being collected during real-world driving using the proposed data collection system. The driver is self-reporting their emotional state by touching the HMI application mounted on the vehicle center fascia. The screenshot on the right is the English translation of the GUI of the HMI application implemented in Korean.

According to the real-world data collection experiment results using the proposed system, the experiment was completed without any accidents over four months. A large-scale dataset of over 122 h, 4446 km, and 787 GB was collected, along with 6356 self-reporting data points of drivers while driving. Through the statistical analysis of the collected data, the imbalance of self-reported emotion labels and the need for personalized driver emotion recognition were confirmed. In addition, case studies of driver face detection and personalized single and multimodel driver emotion recognitions are presented, and comprehensive understanding is provided.

Our main contributions can be described as follows:

- We proposed a data collection system that can collect the multimodal data of drivers during real-world driving tasks. The proposed system is capable of collecting real-world driving big data for driver emotion recognition while considering the minimization of behavioral disturbances.
- The proposed system comprises an HMI application through which drivers can report their emotional states. This application is designed to collect selected emotional states from the driver without cognitive disturbance during real-world driving by utilizing the haptic, acoustic response, and GUI, and eliminating the bias problem that may occur with the self-reporting by setting the interaction period.
- We deployed the proposed system on a vehicle and collected high-quality multimodal sensor data without any accidents during real-world driving experiments for over 122 h. To demonstrate the validity of our collected dataset, we provided various case studies such as statistical analysis, driver face detection, and personalized single and multimodal driver emotion recognition.

The rest of this paper is organized as follows. Section 2 introduces related works on the data collection system for driver emotion recognition. Section 3 discusses the proposed data collection system in real-world driving. Section 4 provides data collection experiments, analysis of collected data, and case studies using the collected data. Section 5 concludes this work and describes further work. Appendix A describes details of terminologies and variables used in this paper.

2. Related Works

Driver state recognition research is being conducted from various viewpoints, from the recognition of inattention [16], distraction [17], stress [5], and behavior [18] for safety to readiness [19] for autonomous driving. This has resulted in research on driver emotion recognition, along with the improvement of data-based human emotion recognition performance [20–22]. Data used for driver emotion recognition is classified into video [11], audio [10], biophysiological [12], and CAN data [15]. In most cases, these data are not used alone but are fused to recognize driver's emotional states [6–9]. However, real-world driving data resources that account for data types do not exist. Ma et al. [11] only collected the video of a driver's face, and CIAIR [23] and DriveDB [7] collected video, audio, and biophysiological data, excluding CAN data. UTDrive DB collected various CAN data, along with video and audio but did not collect bio-physiological data [8]. In this study, we propose the various multimodal data collection system in real-world driving.

Emotion annotation data are as important as sensor data in driver emotion recognition. To annotate a driver's emotional state, three major methods are employed: experimental context, external annotators, and self-reports. The experimental context is the simplest way to annotate an emotional state by estimating the driver's emotional state with the driving situation or environment, e.g., annotate the driver's stress level by road type or congestion level [7–9]. Since this approach presupposes strong assumptions, there are limitations in annotating an accurate emotional state. Although using external annotators requires additional manpower and cost, it enables objective annotation. Jones and Jonsson recorded a driver's speech while driving using a simulator, and an external annotator annotated the driver's emotional state by listening to the recorded speech for driver emotion recognition [10]. Ma et al. developed an annotation tool to allow external annotators to annotate two emotion categories at five levels each based on driver face images collected during real-world driving [11]. This approach also has limitations in that experienced and trained annotators are required. Because self-reporting is an approach to self-report how drivers feel while driving, it can overcome the limitations of other approaches. However, driving is a task that requires considerable concentration, and drivers' self-reporting while driving affects the experiment. Hence, most self-reporting is performed immediately after the driving experiments. Taib et al. [13] and Ihme et al. [14] conducted a driving simulation experiment for driver frustration and asked participants who drove for self-reporting information after the experiment. Taib et al. used a 9-point Likert scale and Ihme et al. used a self-assessment manikin (SAM) [24] for self-reporting. Kato et al. proposed a self-report application that can visualize data and selected the driver's emotional state while driving [12]. The proposed application enables a driver's self-reporting to be performed in real time while driving, not after the experiment. This application was only used in a simulation experiment, and to use it in real-world driving experiments, additional safety considerations are required. In addition, concerns about subjective biases that may be included in self-reports are another challenge to overcome [15]. In this study, we propose an HMI application that allows drivers to safely report their emotional states while real-world driving.

3. Proposed Work

In this section, a system that enables the simultaneous collection of videos, audio, biophysiology, and CAN data during real-world driving is described. The system also includes an HMI application that interacts with the driver and collects the driver's emotional state. In other words, this section demonstrates methods for developing hardware and software systems for a multimodal dataset based on self-reporting in real-world driving for driver emotion recognition. All systems are built into the vehicle, as the data collection is performed under driving conditions. We used an IONIQ 1.6 Hybrid vehicle (Hyundai, Seoul, KR, https://www.hyundai.com/, accessed on 31 March 2022) shown in Figure 2a as the base environment for building the proposed system. Figure 3 shows the flowchart of the entire system. When the system starts, the first thing to check is whether the vehicle is ignited. The system is designed to start after the vehicle is ignited because the surge voltage generated when the vehicle is ignited can reduce the quality of data collected using electronic sensors. In addition, for safety reasons, whether the vehicle is stopped before starting and ending the system is checked (blue rhombus in Figure 3). This prevents the driver from operating the system while driving. After confirming that data collection is possible, two types of metadata are requested before the main data collection. One is the name of the driver, which must be input by the driver manually. The other is the current odometer, which can be obtained automatically via vehicle CAN data. After obtaining the current odometer and treating it as the starting odometer, the main data collection process starts. The main data collection process uses multiprocessing to efficiently collect different multimodel data (orange rectangle in Figure 3). When a suitable end request is input into the system by the driver, the main data collection process is terminated, and if the vehicle is stopped, the vehicle odometer is obtained once more and treated as the ending odometer. Finally, all data, metadata, and collected data (green box in Figure 3) are integrated into one dataset (red rectangle in Figure 3), and the entire system is shut down. All processes in the proposed system are performed using a computer, shown as Figure 2d. The proposed system is released as an open source repository on GitHub (https://github.com/KMUIMLAB/DMS, accessed on 27 May 2022) and the details of each data type for multimodal data collection are discussed in the following sections.

3.1. Video

We use two RealSense D435i cameras (Intel, Santa Clara, CA, USA, https://www.intel.com/, accessed on 31 March 2022) to collect video data composed of various modalities. The RealSense camera provides a maximum of three video modalities: red, green, and blue (RGB), infrared (IR), and depth. In addition to the RGB image, the IR image, which is robust to environment changes, such as illumination changes, is essential in real-world driving. One camera is installed on the dashboard to capture the driver's face, as shown in Figure 2b, and the other is installed on the top of the passenger seat window to capture the driver's posture, as shown in Figure 2c. Since the sample rate of the camera can be set, we set it as R_v Hz. Alternatively, each camera sequentially captures R_v individual images per second.

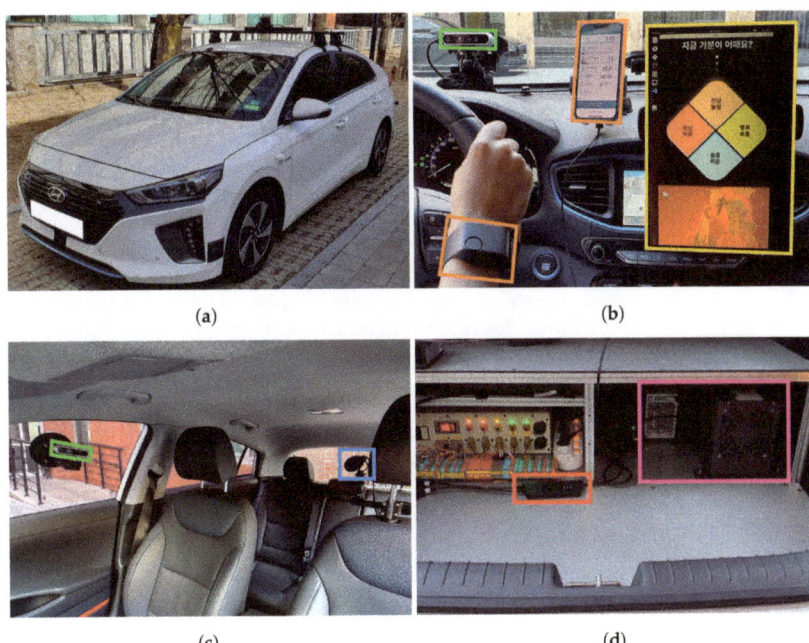

Figure 2. Figures of the dataset collection system hardware interface build in the vehicle. (**a**) Vehicle exterior; (**b**) Inside view of the vehicle center fascia; (**c**) Inside view of the vehicle passenger seat; (**d**) Vehicle trunk. Two cameras are installed to collect the image data of a driver's face and posture (green). A microphone is installed on the right side of the driver seat's headrest to collect audio data in the cabin (blue). Wristband-type wearable sensor is worn on the driver's wrist to collect the driver's bio-physiological data, and the collecting status can be monitored through a smartphone (orange). The CAN interface device supports the collection of vehicle CAN data (red). The monitor installed on the center fascia is a touch screen for interaction with the driver (yellow). The computer installed in the trunk of the vehicle integrates the collected data (magenta).

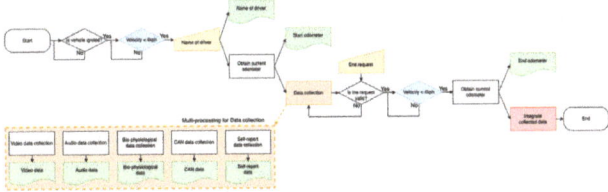

Figure 3. Flow chart of the proposed data collection system during real-world driving.

3.2. Audio

The CVM-VM10 II microphone (CoMica Technology, Shenzhen, Guangdong, CN, https://www.comica-audio.com/, accessed on 31 March 2022) was used to collect audio information in the cabin while driving. To collect data with audio information similar to what the driver hears, the cardioid condenser microphone was selected and placed close to the driver's ear. To minimize noise and vibrations that occur during real-world driving, the microphone was installed on the right side of the driver's seat headrest, along with the shock mount and wind muff, as shown in Figure 2c. The audio data collection system collects R_a audio data samples per second until the system stops according to the sample rate, R_a Hz.

3.3. Biophysiological

To collect biophysiological data of the driver, the biometric sensor must be in contact with the driver's body. The attached sensor may cause behavioral disturbances, resulting in potential accidents. For safe biophysiological data collection during real-world driving, it is necessary to prevent behavioral disturbances in advance, and we used an E4 wristband (Empatica, Boston, MA, USA, https://www.empatica.com/, accessed on 31 March 2022) as a solution. The E4 wristband (E4) is a wearable biometric sensor and is used as an alternative sensor while exhibiting similar data quality 85% of the time compared to the clinician standard device [25]. As a result of comparing the E4 and laboratory biometric sensor data in terms of emotion recognition performance, similar accuracy was realized [26]. Hence, we used the E4 for biophysiological data collection during real-world driving. E4 provides skin temperature, electrodermal activity (EDA), photoplethysmography (PPG), and 3-axis acceleration of the band, along with interbeat interval (IBI) and heart rate (HR) through postprocessing. As shown in Figure 2b, biophysiological data collection is possible by simply wearing E4 on the wrist while driving, and real-time monitoring is also possible using a mobile device through the application provided by E4. Unlike video or audio data, E4 collects each data at an optimized sampling rate, so no separate setting is required. Each sample rate is shown in Table 1.

3.4. CAN

The method of mounting additional sensors or collecting on-board diagnostics (OBD) signals can also be used to access vehicle signals; however, since we can access vehicle CAN, we can collect vehicle signals with the CAN interface device. CAN is a message-based protocol designed to allow vehicle controllers to communicate with each other. The USBcan Pro 2xHS v2 (KVASER, Mission Viejo, CA, USA, https://www.kvaser.com/, accessed on 31 March 2022) is a CAN interface device used to access vehicle CAN signals to collect vehicle data. As shown in Figure 2d, the device is located in the trunk of the vehicle and connects the vehicle CAN line to the computer. Among the many signals on CAN, we select key signals closely related to the driver. Since the selected key signals are updated according to the set cycle time, the sample rate of CAN data, R_c, is set according to the cycle time. The collected key data and the sample rate are presented in Table 1.

3.5. HMI

Drivers' emotion annotation is essential in datasets for driver's emotion recognition. Although external annotators or the experimental context can be employed to estimate and annotate drivers' emotional states, we focused on annotating the driver's emotional state using reports from the driver rather than via estimation. This method is called self-report and will be performed in real-world driving experiments. It must be designed with an emphasis on safety. Requiring drivers to report driving conditions may cause cognitive disturbances, probably leading to severe traffic accidents on the road.

To minimize cognitive disturbances, we proposed the HMI application that periodically interacts with the driver through haptic and acoustic response and receives the emotional state response from the driver. We used a TFX133T DEX monitor (HANSUNG, Seoul, KR, https://www.monsterlabs.co.kr/, accessed on 31 March 2022), and the touch screen has a built-in speaker to realize haptic and acoustic responses. The screen was installed on the center fascia of the vehicle, as shown in Figure 2b. When data collection starts, the HMI application requests that the driver report their emotional state with a sound announcement as follows: "Please enter your current state". If there is no response from the driver for I_{rr} seconds from the request, the application requests once more with the same sound announcement. If there is no response from the driver within I_s seconds from the first request, not to disturb the driver, it is treated as a nonresponse with a sound announcement as follows: "The input is delayed, so it enters in a nonresponse state". This skipping process is essential as frequent response requests can interfere with safe driving. The driver can input an answer by only touching the screen, and when the input is completed, the input emotional state is displayed on the screen in large fonts; and at the same time, a sound announcement is provided as follows: "Your input is complete". This feedback minimizes confusion for the driver.

In addition to cognitive disturbances, self-reported emotion labels have limitations in that they reflect strong bias because of false memories or the desire to impress other people [15]. Repeated sampling in real-time is necessary to minimize this bias [27]. That is, the self-reporting requests should be continuously made at periodic intervals. Hence, the proposed HMI application continuously requests the response at an interval, I_r, from when driving starts to when it ends. The interval between response requests, I_r, is tuned through test driving. Moreover, our system allows the driver to report their emotional states at any time by touching the screen even between response intervals. This feature enables logging drivers' rapidly changing emotional changes in real-world varying driving conditions.

The proposed HMI application can apply any representative emotional states as long as they are discretely expressed states. However, since the driver has to choose the most similar to their current emotional state among them, cognitive disturbances can occur if there is difficulty in choosing an emotion no matter how well the interaction with the driver is completed. Therefore, the discrete representative emotional states should be simple, not numerous, and suitable for the driving situation.

3.6. GUI

We propose a GUI design to reduce drivers' cognitive disturbance in self-reporting through HMI while driving. To propose UX-based GUI of the HMI application, the following four representative driver emotional states by referring to the emotions that can be induced in a driving situation [28] are suggested.

- Happy | Neutral;
- Excited | Surprised;
- Angry | Disgusting;
- Sad | Fatigued.

The proposed GUI designs are shown in Figure 4. There are two factors to consider in the GUI design: the layout and color of the emotional states. The layout of the emotional states refers to the valence–arousal plane, a popular concept used in emotional representation [29]. Based on the division of the x-axis into pleasure and misery in the valence–arousal plane, we placed "Happy | Neutral" and "Angry | Disgusting" on the right and left of the screen: "Happy | Neutral" is on the right and "Angry | Disgusting" is on the left. Based on the division of the y-axis into arousal and sleepiness in the valence–arousal plane, we placed "Excited | Surprised" and "Sad | Fatigued" on the top and bottom of the screen: "Excited | Surprised" is on the top and "Sad | Fatigued" is on the bottom. The overall layout of the emotional states is in the form of a rhombus, as shown in Figure 4. In the GUI shown in Figure 4, each emotional state is expressed in different colors. The correlation between basic colors and human psychological state was identified, and states that can be felt by humans were classified according to color characteristics [30]. Based on this, appropriate colors were used for each emotional state. The GUI design provides not only a default GUI, as shown in Figure 4a, but also a touch GUI, as shown in Figure 4b. Therefore, when the driver inputs the current emotional state by touching the screen, it provides visual feedback, as shown in Figure 4c, along with the sound announcement. The UX-based GUI of the HMI application gives the driver more accurate intuition about the proposed representative emotional states.

(a)   (b)   (c)

Figure 4. GUI of HMI application for self-reporting of driver emotional state. (**a**) GUI in default; (**b**) GUI in touch; (**c**) GUI example where "Angry | Disgusting" state is touched.

4. Experiments

This section presents the details of the data collection experiment conducted on the basis of the proposed data collection system and some case studies based on the collected data from the experiment.

4.1. Data Collection Experiment

Motivated by the need for a dataset in real-world driving, the data collection experiment with the proposed system described in Section 3 was conducted on the road. During real-world driving, the cameras are used to capture RGB and IR modalities at the sample rate, R_v, of 15 Hz, and audio data are collected at the sample rate, R_a, of 44,100 Hz. Biophysiological data are collected, as described in Section 3.3. The following CAN data signals are collected: accelerator pedal position, brake pedal position, steering wheel angle, yaw rate, longitudinal acceleration, and lateral acceleration. All CAN data are collected at the sample rate, R_c, of 100 Hz. The self-reportable application collected the driver's emotional state in five states involving four representative emotional states mentioned in Section 3.5 and nonresponse. The response request time interval, I_r, is set to 60 s, and then the sample rate of self-reported emotion label, R_s, is $\frac{1}{60}$ Hz. Because the driver is encouraged to self-report whenever there is a change in their emotional state even without that response request, the self-reported emotional state annotation includes information on the driver's emotional change for unexpected or urgent events. The rerequest time interval, I_{rr}, and the skip time interval, I_s, are set to 10 and 20 s, respectively. All interval times have been adjusted through several test drives in real-world driving, so that there is no safely issue. Details, including save format and unit for all data collected through the experiment, are described in Table 1.

To address the lack of long-term datasets, the experiment was conducted with a few people who could participate continuously for a long period. Four males participated in the experiment for four months from July 2021 to October 2021. The detailed information of these participants is described in Table 2.

During these four months, a large-scale dataset was collected by the participants' driving in wild, uncontrolled conditions. The weather conditions were divided into four categories, and the proportions are as follows: Sunny: 20.4%, Cloudy: 40.6%, Overcast: 11.8%, Rainy: 27.3%. Because safety is considered in the proposed data collection system, no accidents occurred during this period, and according to the data collection experiment results, the total experiment time was 122 h 15 min, the total driving mileage was 4446 km, the total number of self-reported emotion labels was 6356, and 787 GB data were collected.

Table 1. Details of data collected by experiment.

Data		Sample Rate (Hz)	Format	Unit
Video	RGB-front	15	.avi	-
	RGB-side	15	.avi	-
	IR-front	15	.avi	-
	IR-side	15	.avi	-
Audio	-	44,100	.wav	-
Bio-physiological	Skin temperature	4	.csv	°C
	EDA	4	.csv	µS
	PPG	64	.csv	nW
	IBI	-	.csv	s
	HR	1	.csv	bpm
	3-axis acceleration	32	.csv	$\frac{1}{64}g$
CAN	Accelerator pedal position	100	.csv	%
	Brake pedal position	100	.csv	%
	Steering wheel angle	100	.csv	°
	Yaw rate	100	.csv	rad/s
	Longitudinal acceleration	100	.csv	m/s^2
	Lateral acceleration	100	.csv	m/s^2
Self-reported emotions	Emotional state	no less than $\frac{1}{60}$	.csv	-

Table 2. Detailed information of participated drivers.

	Gender	Age (Year)	Driving Experience (Year)	Experiment Time (h)	Driving Mileage (km)
Driver A	Male	27	more than 15	38	1375
Driver B	Male	32	between 11–15	43	1449
Driver C	Male	26	between 6–10	21	852
Driver D	Male	28	less than 5	20	770

4.2. Case Studies

This section presents some case studies using the collected multimodal dataset for driver emotion recognition. Section 4.2.1 discusses the detailed analysis of the dataset collected in real-world driving. Sections 4.2.2 and 4.2.3 present case studies of driver emotion recognition using single-modal or multimodal inputs.

4.2.1. Statistical Analysis

In this section, we discuss the detailed analysis results for the collected dataset in the real-world driving experiment. Figure 5 depicts the self-report proportion for each driver as a pie chart. The emotion with the highest proportion was "Happy | Neutral". More than 50% of the drivers' self-reported emotion labels are "Happy | Neutral", and they often account for up to approximately 82%. The proportion of the other three emotions varies by the driver, but it accounts for a small proportion compared to the "Happy | Neutral".

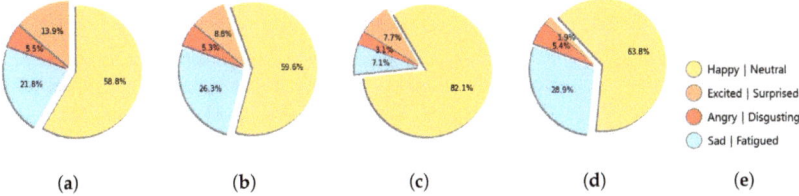

Figure 5. Pie charts for self-reported emotion label proportion by driver. (**a**) Driver A; (**b**) Driver B; (**c**) Driver C; (**d**) Driver D; (**e**) Legend of the pie charts.

To confirm the self-reported emotion label tendency of each emotion, the distribution of self-reports and vehicle speed by emotion for all drivers is depicted in Figures 6 and 7.

In Figure 6, the start and end of all individuals driving were normalized from 0 to 100 steps and divided into 50 sections. The number of self-reported emotion labels for each section is displayed as a histogram and kernel density estimate plot to evaluate the distribution by emotion. "Happy | Neutral" had several distributions at the start and end of the driving, and had an even distribution throughout the driving process, as shown in Figure 6a. Overall, "Excited | Surprised" and "Angry | Disgusting" had an irregular distribution. "Excited | Surprised" seemed to have a greater variance than "Angry | Disgusting", as shown in Figure 6b,c, and it is judged that "Excited | Surprised" was more maintained when the emotion was induced than "Angry | Disgusting". As shown in Figure 6d, the distribution of "Sad | Fatigued" emotion increases toward the middle and late stages of driving. Figure 7 shows the number of self-reported emotion labels at that vehicle speed with a histogram and kernel density estimate plot to evaluate the distribution of vehicle speed by self-reported emotion labels. "Happy | Neutral" had high distributions from 0 to about 15 kph, and an even distribution throughout the driving process, as shown in Figure 7a. In Figure 7b,c, the fact that the vehicle speed had a relatively irregular distribution compared to "Happy | Neutral" and "Sad | Fatigued" in "Excited | Surprised" and "Angry | Disgusting" is a common feature with the distribution of self-reported emotion labels in Figure 6. As shown in Figure 7d, the distribution of the "Sad | Fatigued" emotion had a particularly high distribution from 0 to about 30 kph. Based on the distribution of self-reports and vehicle speed by emotion (especially in Figure 6a), "Happy | Neutral" was the default emotion and the others were induced while driving.

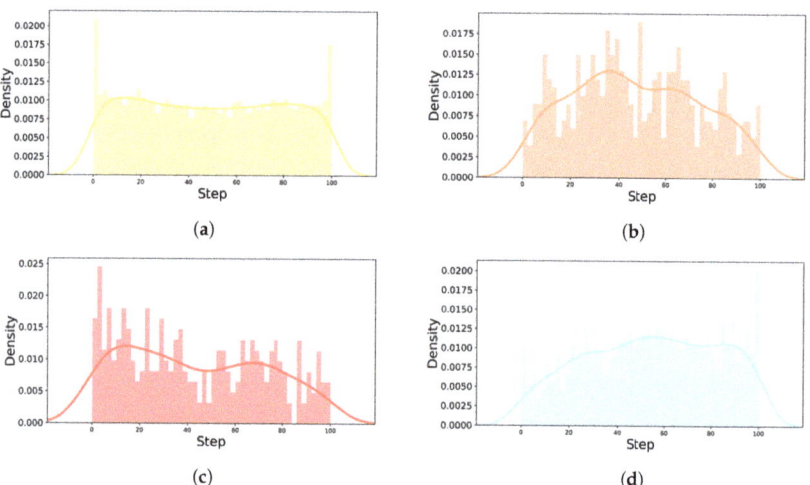

Figure 6. Distribution of self-reported emotion labels in real-world driving. (a) Happy | Neutral; (b) Excited | Surprised; (c) Angry | Disgusting; (d) Sad | Fatigued.

In addition to self-reported emotion label data, we used the statistical hypothesis test to analyze the significance of the collected sensor data. We built the null hypothesis (H_0) that the structured data collected did not differ according to the self-reported emotion label and confirmed the difference by the emotion of each structured data through a Kruskal–Wallis H test [31,32]. According to the Kruskal–Wallis H test results, if the significance probability expressed as the p-value is less than the significance level, 0.05, the null hypothesis (H_0) can be rejected and the alternative hypothesis (H_1) can be accepted as true. The statistical significance by self-reported emotion label of each data is described using the p-value and which hypothesis was accepted as true in Table 3. If the statistical significance between the four self-reported emotion labels is confirmed by the Kruskal–Wallis H test, it is also necessary to confirm how many of the pairs show statistical significance through the post-hoc test. We confirmed the statistical significance of a total of six self-reported emotion label pairs through the Mann–Whitney U test [33,34], a nonparametric statistical hypothesis test, and the total number of the null hypothesis (H_0) rejection pairs is also listed in Table 3. As shown in Table 3, all collected structured data had statistically different distributions for self-reported emotion labels, and three or more pairs out of six pairs were statistically significant.

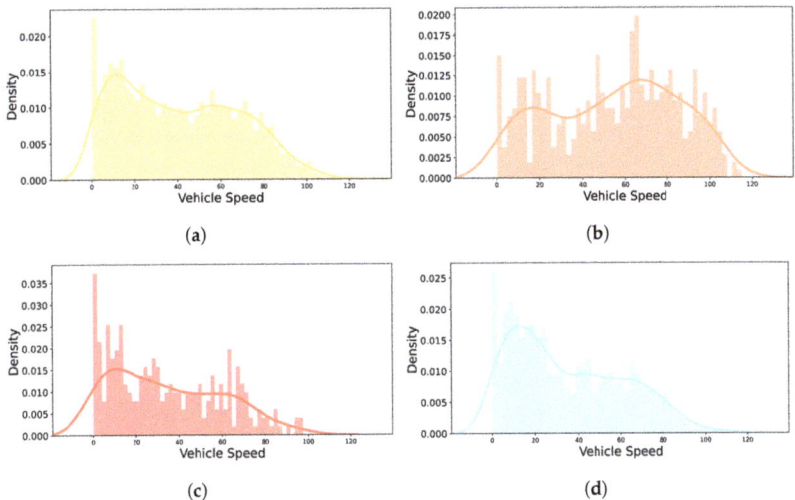

Figure 7. Distribution of vehicle speed by self-reported emotion labels in real-world driving. (a) Happy | Neutral; (b) Excited | Surprised; (c) Angry | Disgusting; (d) Sad | Fatigued.

Table 3. Statistical hypothesis test results of structured data by self-reported emotion label.

Data		Statistical Hypothesis Test	Post-Hoc Test
		Reject H_0	Number of Reject H_0 Pairs (Total Number of Pairs is 6)
Bio-physiological	Skin temperature	Yes	6
	EDA	Yes	5
	PPG	Yes	3
	HR	Yes	4
CAN	Accelerator pedal position	Yes	5
	Brake pedal position	Yes	6
	Steering wheel angle	Yes	6
	Yaw rate	Yes	3
	Longitudinal acceleration	Yes	6
	Lateral acceleration	Yes	5

Although the statistical hypothesis test results can explain the significance of the emotion recognition of the collected sensor data, another aspect that requires analysis is whether there is a significant distribution difference according to the driver. Therefore, the same statistical hypothesis test as above was repeated by separating the data for each driver, and the results are shown in Table 4. EDA and steering wheel angle are the only structured data with the same results for all drivers. Not only were the post-hoc results different, but also the results of determining whether to reject the null hypothesis were different for each driver. That means the collected data significantly vary from driver to driver. This may be because each driver has a different way of expressing their emotions while driving. Therefore, different data will be required to recognize each driver's emotion. In other words, emotion recognition research requires personalization.

Table 4. Statistical hypothesis test results of structured data by self-reported emotion label according to driver.

Data		Statistical Hypothesis Test				Post-Hoc Test			
		Reject H_0				Number of Reject H_0 Pairs (Total Number of Pairs is 6)			
		Driver A	Driver B	Driver C	Driver D	Driver A	Driver B	Driver C	Driver D
Bio-physiological	Skin temperature	Yes	Yes	Yes	Yes	5	6	6	6
	EDA	Yes	Yes	Yes	Yes	6	6	6	6
	PPG	No	Yes	No	Yes	-	1	-	3
	HR	Yes	Yes	Yes	Yes	4	5	5	2
CAN	Accelerator pedal position	Yes	Yes	Yes	Yes	5	6	6	6
	Brake pedal position	Yes	Yes	Yes	Yes	6	5	6	6
	Steering wheel angle	Yes	Yes	Yes	Yes	6	6	6	6
	Yaw rate	Yes	Yes	Yes	Yes	6	6	4	5
	Longitudinal acceleration	Yes	Yes	Yes	Yes	6	5	3	6
	Lateral acceleration	Yes	Yes	Yes	Yes	5	6	6	6

4.2.2. Driver Face Detection

One of the most common approaches to recognizing a driver's emotional state is using face images. Studies adopting this approach generally use a well-known face detector to crop only the face image from the driver's frontal image and use it as input data. The most popular face detectors have proven their performance only on in-the-wild datasets such as FDDB [35] or WIDER FACE [36]. Thus, we evaluate the performance of five popular face detectors, Haar [37], Dlib [38], OpenCV [39], MMOD [40], and MTCNN [40], on detecting the driver's front image in the collected real-world driving dataset. First, the detection results of the five detectors for the collected IR-front images were output and qualitatively compared. Figure 8 is an example of the detection results of the five detectors. According to the results, Haar has a high false positive rate, i.e., nonfaces are detected, and Dlib has a high false negative rate, i.e., faces are not detected. In contrast to Haar and Dlib, other detectors are capable of detecting the driver's face to a similar degree.

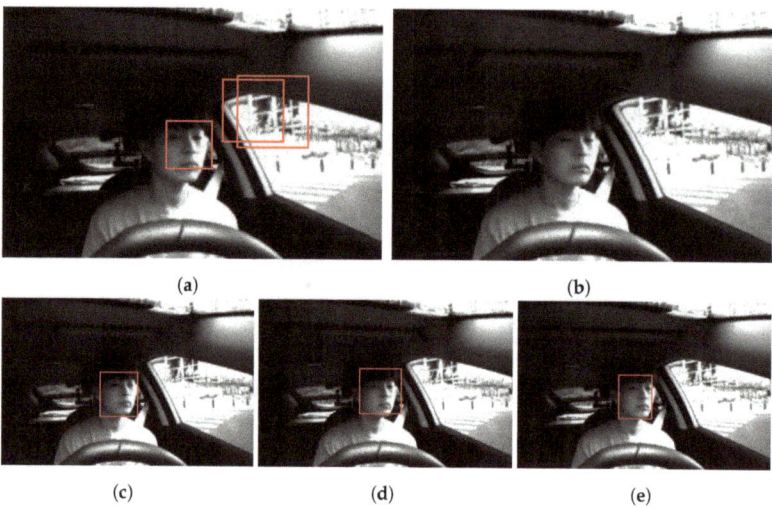

Figure 8. Example of the detection results of five face detectors. The bounding boxes (red) are face detection results. (**a**) Haar; (**b**) Dlib; (**c**) OpenCV; (**d**) MMOD; (**e**) MTCNN.

For accurate performance comparison of the similar three face detectors, we selected 200 different images and labeled face bounding boxes. If the intersection over union (IoU) value between the labeled bounding box and the detection bounding box is greater than or equal to the threshold, it is considered true positive (TP); if the IoU value is less than the threshold, it is considered false positive (FP). Figure 9 shows the precision–recall (PR) curve drawn using the considered TP and FP. Quantitative performance comparison of face detectors can be made with the average precision (AP) value calculated by the area under the PR curve. Depending on whether the threshold is 0.5, 0.75, or 0.95, AP performance is expressed as AP50, AP75, or AP95, respectively. Refer to Table 5 for detailed comparison results. Since the inference speed of the face detector is as important as detection accuracy, Table 5 describes the inference speed and the GPU specifications.

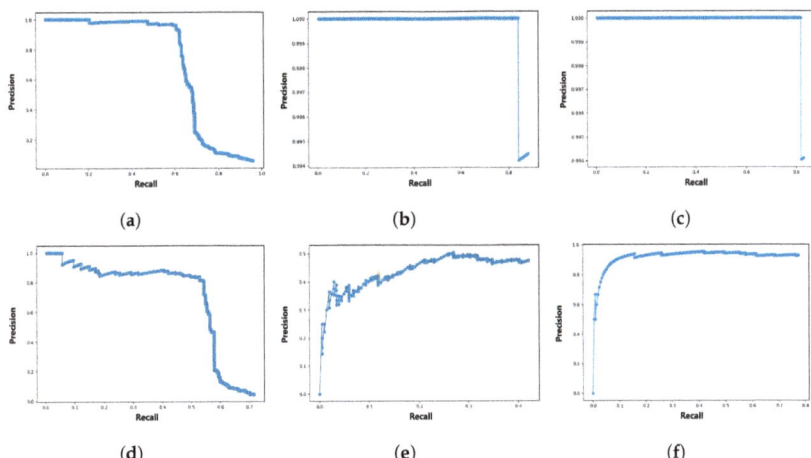

Figure 9. PR curve for face detectors capable of detecting the driver's face. The thresholds are 0.5 and 0.75. (a) OpenCV, threshold is 0.5; (b) MMOD, threshold is 0.5; (c) MTCNN, threshold is 0.5; (d) OpenCV, threshold is 0.75; (e) MMOD, threshold is 0.75; (f) MTCNN, threshold is 0.75.

Table 5. Driver's face detection performance comparison of face detectors.

	AP50	AP75	AP95	Speed	GPU
OpenCV	68.4	51.4	0.0	400 FPS	Nvidia GTX 3080
MMOD	83.8	18.1	0.0	260 FPS	Nvidia GTX 3080
MTCNN	81.4	72.0	0.0	4 FPS	Nvidia GTX 3080

OpenCV has the fastest inference speed, but its detection performance is low. For MMOD and MTCNN, AP50 is at a similar level, but at AP75, the detection performance of MMOD decreases rapidly. Although the AP75 performance of MTCNN is inferior to AP50, it is insignificant. Conversely, in the case of inference speed performance, MMOD significantly outperforms MTCNN. Since the inference speed of MTCNN is also insufficient, it seems appropriate to use a suitable face detector as the driver face detector depending on the purpose or computational sources. In terms of AP95, the performance of all detectors is 0.0. This is due to the small area occupied by the driver's face in the driver's front image, and the IoU value may not exceed the threshold value of 0.95 due to differences in determining whether only the eyes and nose are included, or including the forehead or chin when the bounding box is labeled. Figure 10 shows an example image of the detected and labeled driver face bounding boxes with an IoU value of 0.68, it detects the driver's facial expression sufficiently. In face detection for driver emotion recognition, the threshold should not be as high as 0.5 or 0.95. Therefore, we crop the face image using the MMOD face detector, which achieved the highest detection performance in AP50 for driver emotion recognition, as discussed in Section 4.2.3.

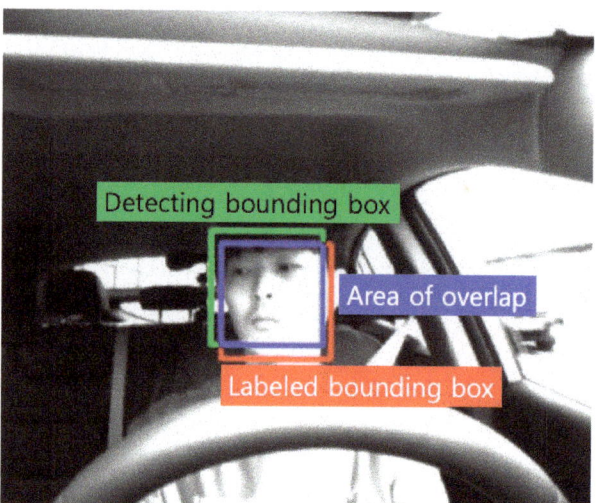

Figure 10. Example image with IoU of 0.68. Area of union (green and red) is 7441, and area of overlap (blue) is 5040.

4.2.3. Personalized Driver Emotion Recognition

This section discusses the results of personalized driver emotion recognition utilizing single or multimodal data. Since individual driver data are required for personalized driver emotion recognition training, the data required to complete the training should be as small as possible, and the performance of the trained recognition model should be preserved for as long as possible. Therefore, the collected data are sorted in ascending order of mileage, and the mileage for completing the collection of training data, K, is determined. The data collected during K km driving from the initial mileage for each individual are used as training data, and the data from thereafter to the last data are used as test data. We set the completing mileage for the training data, K, to 500 km, and to obtain more test data than training data, we experimented with drivers A and B, who collected data over 1000 km.

We proposed a personalized driver emotion recognition model based on deep learning networks that recognize a driver's emotional state using four multimodal inputs: front and side image, biophysiological, and CAN data. The proposed model is trained and verified using only individual data, and, as shown in Figure 11, each multimodal input performs single-modal emotion recognition and multimodal emotion recognition through an ensemble model. Each single-modal model and multimodal recognition model are described as follows.

- Single-modal of front image (S_f): The single-modal recognition model of the front image uses front IR images for 2 s from 4 s to 2 s before the driver's self-reporting. Because RGB images are vulnerable to changes in illuminance, IR images that can always capture a stable image are used as input. From 2 s before self-reporting, it shows uniform motion for self-reporting, so it is excluded from the input data. The input images are evenly time-divided into six equal parts and input to a face detector; the MMOD-based face detector outputs one cropped face image with the highest confidence value for each input. The cropped images are resized to the input shape of the feature extractor and sequentially fed into a feature extractor and a classifier based on CAPNet [41]. Because the classification form is different from that of CAPNet, only the number of units in the top layer of the classifier is modified to the number of representative driver emotional states. The last activation function is softmax and outputs the probability of each representative driver emotional state.

- Single-modal of side image (S_s): The single-modal recognition model of the side image uses the side IR image captured 2 s before self-reporting. The reason for using the image from 2 s ago is the same as that for using the front image. The input image is fed into a feature extractor based on AlphaPose [42]. The feature extractor consists of layers up to just before outputting feature points in the form of histograms in AlphaPose. The classifier consists of a global max pooling layer and fully connected layers. The top layer of the classifier is the same as other classifiers to output the probability of each representative driver emotional states.
- Single-modal of biophysiological (S_b): The single-modal recognition model of biophysiological data uses the PPG and EDA data for 10 s before the driver's self-reporting. Since PPG and EDA have different sample rates, up-sampling using linear interpolation is applied to the EDA data to match the input shape. The biophysiological input is directly fed into the classifier without a feature extractor to output the probability of each representative driver emotional state. The classifier is composed of the fully connected and batch normalization layers.
- Single-modal of CAN (S_c): The single-modal recognition model of CAN data uses all collected signals for 10 s before the driver's self-reporting. The input data are down-sampled by a tenth before being fed into the feature extractor. The feature extractor is an encoder of long short-term memory-based autoencoder that extracts the feature vector for driving propensity. The classifier consists of fully connected layers and a dropout and outputs the probability of each representative driver emotional states by receiving the feature vector.
- Multimodal (M): The multimodal recognition model uses the input vectors of each classifier of single-modal as input vectors. The model is a deep learning-based ensemble model that outputs the probability of each representative driver emotional states by fusing all input vectors. The feature vectors of the front image, CAN, and side image are flattened using flatten and pooling layers. The flattened vectors are concatenated using the concatenate layer. The concatenated vector undergoes the normalization, fully connected layers, and softmax activation function to become the final output. The input modalities to fuse can be chosen, and the modals are denoted by a subscript, e.g., M_{fb} is the ensemble model that fuses the front image and biophysiological data. We evaluated three or more input modal combinations for multimodal models.

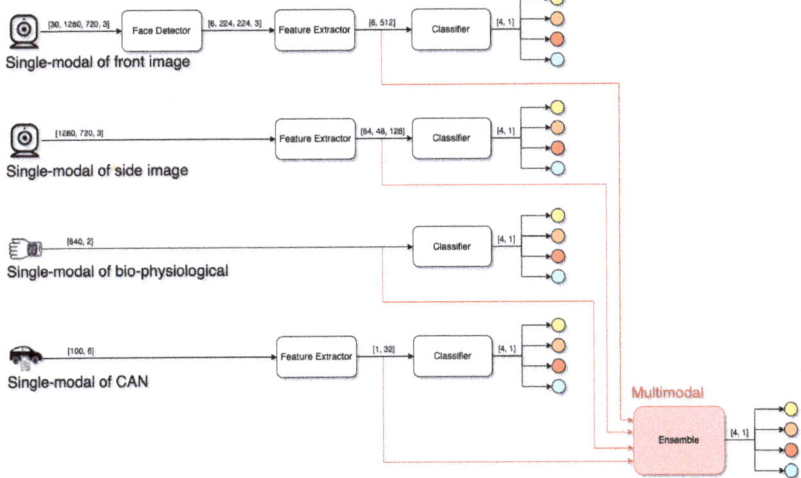

Figure 11. Deep learning-based personalized driver emotion recognition model.

It is necessary to define a loss function when training the proposed models. Because the self-reported emotion label has data imbalance, as described in Section 4.2.1, high performance cannot be expected if a typical loss function is used such as cross entropy. We overcome the data imbalance problem by making the precision and recall differentiable by computing the likelihood values of TP, FP, and false negative (FN) using probabilities. The loss function we used is shown as follows:

$$L(\mathbf{y}, \hat{\mathbf{y}}) = 1 - \frac{1}{N} \left(\frac{p_1^{TP}}{p_1^{TP} + p_1^{FP} + \epsilon} + \sum_{i=2}^{N} \frac{p_i^{TP}}{p_i^{TP} + p_i^{FN} + \epsilon} \right) \quad (1)$$

$$\mathbf{p}^{TP} = \mathbf{y} \circ \hat{\mathbf{y}} \quad (2)$$

$$\mathbf{p}^{FP} = (\begin{bmatrix} 1. \\ 1. \\ 1. \\ 1. \\ 1. \end{bmatrix} - \mathbf{y}) \circ \hat{\mathbf{y}} \quad (3)$$

$$\mathbf{p}^{FN} = \mathbf{y} \circ (\begin{bmatrix} 1. \\ 1. \\ 1. \\ 1. \\ 1. \end{bmatrix} - \hat{\mathbf{y}}) \quad (4)$$

where $\mathbf{y}$ and $\hat{\mathbf{y}}$ represent a one-hot vector of the self-reported emotion and predicted emotion, respectively, where the first element of each vector represents the default emotion, "Happy | Neutral". $\mathbf{p}^{TP}$, $\mathbf{p}^{FP}$, and $\mathbf{p}^{FN}$ are the likelihood values of TP, FP, and FN, respectively, where $\circ$ is an element-wise product.

Equation (1) is a loss function for increasing the precision of default emotion and for increasing the recall of induced emotions, where N represents the total number of representative emotions, and ϵ represents a very small value that prevents the precision or recall values from going to infinity. This loss function, $L(\mathbf{y}, \hat{\mathbf{y}})$, can be used for backpropagation by probabilistically expressing the precision and recall for each prediction class. It increases precision for the majority class, the default emotional state, and increases recall for minority class, inducible emotional states.

The evaluation results with test data are in terms of F1 score, precision, and recall, and are described for each driver. As mentioned in Section 4.2.1, since the representative driver emotional states are divided into default and inducible emotions, the recognition performance of inducible emotions is evaluated first. Tables 6 and 7 summarize the performance of inducible emotion recognition between default and inducible emotions for each driver. The highest recognition performance is the F1 score 0.698 of S_s for Driver A and 0.667 of M_{sbc} for Driver B. As expected in Section 4.2.1, the input modals with the best performance for each driver differed. Driver A achieved the best performance in a single front image, and Driver B achieved the best in a side image, biophysiological, CAN data combination. However, their performance was similar. Driver B had similar performance between all evaluated models from 0.562 to 0.667. For Driver A, models without CAN data had a similar performance from 0.613 to 0.696, but models with CAN data such as S_c, M_{fsc}, M_{fbc}, M_{sbc}, and M_{fsbc} had a significantly lower performance from 0.228 to 0.469. Driver B can interpret that when inducible emotions are induced while driving, emotions are expressed overall in the front and side images and biophysiological, and CAN data, whereas driver A can interpret that the induction of emotion is not expressed in CAN data. These results may support the fact that driver emotion recognition necessitates personalization.

Table 6. Performance of inducible emotion recognition of Driver A.

	S_f	S_s	S_b	S_c	M_{fsb}	M_{fsc}	M_{fbc}	M_{sbc}	M_{fsbc}
F1	0.696	**0.698**	0.619	0.355	0.613	0.430	0.469	0.469	0.228
Precision	0.541	0.537	0.478	0.248	0.446	0.280	0.311	0.314	0.231
Recall	0.975	0.998	0.879	0.630	0.982	0.923	0.950	0.927	0.225

Table 7. Performance of inducible emotion recognition of Driver B.

	S_f	S_s	S_b	S_c	M_{fsb}	M_{fsc}	M_{fbc}	M_{sbc}	M_{fsbc}
F1	0.584	0.613	0.593	0.536	0.562	0.646	0.661	**0.667**	0.615
Precision	0.419	0.442	0.475	0.492	0.420	0.539	0.522	0.500	0.468
Recall	0.963	1.000	0.790	0.589	0.852	0.805	0.900	1.000	0.900

The performance of driver emotion recognition among the inducible emotions for each driver is also summarized. The recognition performance for each of the three inducible emotions and the average of three F1 scores are described in Tables 8 and 9. Comparing the recognition performance using the F1 scores of each emotion and average value, none of the input models with the best performance matched among the drivers. The common results, regardless of the driver, were that "Sad I Fatigued" emotion had the best recognition performance and "Angry I Disgusting" emotion had the worst recognition performance. "Sad I Fatigued" emotion recognition performance was 0.835 and 0.859 and "Excited I Surprised" emotion recognition performance was 0.653 and 0.583 for Drivers A and B, respectively. Both of which are similar performances. However, in the case of "Angry I Disgusting" emotion, recognition performance differed, 0.571 and 0.373 for each driver. Notably, there was very little performance difference between all evaluated models. The difference between the highest and lowest average F1 score was 0.163 and 0.061 for Drivers A and B, respectively. This can be a fail-safe method of the driver emotion recognition model, and each input modal will ensure each other's redundancy.

Table 8. Performance of driver emotion recognition among inducible emotions of Driver A.

		S_f	S_s	S_b	S_c	M_{fsb}	M_{fsc}	M_{fbc}	M_{sbc}	M_{fsbc}
Average F1		0.496	0.444	0.447	0.561	0.456	0.500	**0.607**	0.557	0.483
Excited	F1	0.359	0.301	0.362	**0.653**	0.344	0.487	0.444	0.465	0.417
I	Precision	0.591	1.000	0.563	0.593	1.000	0.950	0.800	0.909	1.000
Surprised	Recall	0.258	0.177	0.267	0.727	0.208	0.328	0.308	0.313	0.263
Angry	F1	0.293	0.196	0.147	0.263	0.216	0.280	**0.571**	0.400	0.200
I	Precision	0.579	1.000	1.000	0.500	1.000	0.875	0.667	1.000	0.667
Disgusting	Recall	0.196	0.109	0.080	0.179	0.121	0.167	0.500	0.250	0.118
Sad	F1	**0.835**	0.833	0.830	0.768	0.808	0.733	0.807	0.806	0.831
I	Precision	1.000	1.000	1.000	0.977	0.995	1.000	0.926	1.000	1.000
Fatigued	Recall	0.717	0.714	0.710	0.632	0.680	0.578	0.714	0.675	0.711

Table 9. Performance of driver emotion recognition among inducible emotions of Driver B.

		S_f	S_s	S_b	S_c	M_{fsb}	M_{fsc}	M_{fbc}	M_{sbc}	M_{fsbc}
Average F1		0.488	0.472	0.481	0.450	0.491	0.468	0.491	0.501	**0.511**
Excited	F1	0.450	0.403	0.333	0.286	0.511	0.417	0.537	0.511	**0.583**
|	Precision	0.636	1.000	0.452	1.000	1.000	1.000	0.846	0.923	0.539
Surprised	Recall	0.348	0.252	0.264	0.167	0.344	0.263	0.393	0.353	0.636
Angry	F1	0.270	0.270	**0.373**	0.204	0.321	0.194	0.227	0.273	0.233
|	Precision	1.000	1.000	0.452	1.000	0.907	0.429	1.000	1.000	1.000
Disgusting	Recall	0.156	0.156	0.264	0.114	0.195	0.125	0.128	0.158	0.132
Sad	F1	0.744	0.743	0.736	**0.859**	0.641	0.794	0.710	0.719	0.717
|	Precision	1.000	1.000	1.000	1.000	1.000	1.000	1.000	0.958	0.864
Fatigued	Recall	0.593	0.592	0.582	0.753	0.472	0.658	0.550	0.575	0.613

5. Conclusions

Although real-world datasets for driver emotion recognition are diverse, to overcome the limitation of the lack of consistency in collected data, we proposed a data collection system capable of collecting multimodal datasets during real-world driving. The proposed system was installed in a vehicle and collected the following multimodal data while driving on the real road: videos captured from two viewpoints, audio inside the cabin, driver's biophysiological data, and vehicle sensor signals via CAN. We designed a self-reportable HMI application to annotate driver emotional states, used as labels for driver emotion recognition. This application allows the driver to select the emotion most similar to their current emotional state among representative emotions. Thus, emotion labels are collected as self-reported emotion labels and no longer inferred by others. In addition, continuous and repeated report requests were made over a long-term period, making the driver's bias not be reflected in the self-reported emotion label. Since safety is the most important factor in real-world driving, we focused on minimizing drivers' behavioral and cognitive disturbances in all processes, including sensor selection, flow, and GUI design while designing the data collection system.

According to the results of the data collection experiment in real-world driving, more than 122 h, 4446 km of driving, and 787 GB of data were collected without any accidents. Through statistical analysis of the collected data, the imbalance and report characteristics of self-reported emotion labels were identified, and default and inducible emotions were distinguished. Based on the statistical hypothesis test, the null hypothesis (H_0) that there is no difference according to the self-reported emotion label for all collected structured data was rejected. The significance of the difference for each driver differed, suggesting the need for personalization of driver emotion recognition. We compared the state-of-the-art face detectors using the collected front images and presented the most suitable face detector and performance evaluation metric for driver face detection. Finally, we conducted a personalized driver emotion recognition study using the collected images and biophysiological and CAN data. The evaluation results of single-modal and multimodal using the above data suggested that multimodal data and personalization are necessary for driver emotion recognition.

Although several case studies were conducted by collecting a large-scale dataset using the proposed system design, enabling safe data collection in real-world driving, the dataset was collected by few drivers over a long period. Because the number of drivers is insufficient to generalize the case studies, these may be treated as particular cases. Based on further collected data, we will continue to study the generalization performance of multimodal personalized driver emotion recognition.

Author Contributions: Conceptualization, S.L. (Sejoon Lim) and S.L. (Sangho Lee); methodology, G.O., E.J. and R.C.K.; software, G.O. and E.J.; validation, J.H.Y., S.L. (Sejoon Lim) and S.L. (Sangho Lee); formal analysis, G.O.; investigation, G.O., J.H.Y. and S.L. (Sejoon Lim); resources, G.O., J.H.Y. and S.H.; data curation, G.O., E.J., R.C.K. and S.H.; writing—original draft preparation, G.O.; writing—review and editing, J.H.Y. and S.L. (Sejoon Lim); visualization, G.O. and E.J.; supervision, S.L. (Sejoon Lim); project administration, S.L. (Sejoon Lim) and S.L. (Sangho Lee); and funding acquisition, J.H.Y., S.L. (Sejoon Lim) and S.L. (Sangho Lee). All authors have read and agreed to the published version of the manuscript.

Funding: This research was supported by the Hyundai Motor Group, the Knowledge Service Industry Core Technology Development Program funded by the Ministry of Trade, Industry, and Energy of Korea (No. 20003519), the Basic Science Research Program of the National Research Foundation of Korea funded by the Ministry of Science, ICT, and Future Planning (No. 2021R1A2C1005433), the BK21 Program through the National Research Foundation of Korea (NRF) funded by the Ministry of Education (No. 5199990814084), and the Korea Institute of Police Technology (KIPoT) grant funded by the Korea government (KNPA) (No. 092021C26S03000, Development of infrastructure information integration and management technologies for real time traffic safety facility operation).

Institutional Review Board Statement: The study was conducted according to the guidelines of the Declaration of Helsinki and approved by the Institutional Review Board of Kookmin University (protocol code: KMU-202104-HR-264; date of approval: 2 June 2021).

Informed Consent Statement: Informed consent was obtained from all subjects involved in the study.

Acknowledgments: The authors thank Junghwan Ryu, Taesan Kim, and Joonghoo Park for building a vehicle with a data collection system and Youngdong Kwon and Myengkyu Lee for setting representative emotions and GUI design.

Conflicts of Interest: The authors declare no conflict of interest.

Abbreviations

The following abbreviations are used in this manuscript:

CAN	Controller area network
HMI	Human–machine interaction
GUI	Graphical user interface
UX	User experience
SAM	Self-assessment manikin
RGB	Red green blue
IR	Infrared
E4	E4 wristband
EDA	Electrodermal activity
PPG	Photoplethysmography
IBI	Interbeat interval
HR	Heart rate
OBD	On-board diagnostics
IoU	Intersection over union
TP	True positive
FP	False positive
PR	Presicion–recall
AP	Average precision
FN	False negative

Appendix A

The part describes terminologies and variables used in the main text. Table A1 contains details of terminologies and variables.

Table A1. Deficition of terminologies and variables used on the main text.

Expression	Definition	Unit
R_v	Sample rate of the video data	Hz
R_a	Sample rate of the audio data	Hz
R_s	Sample rate of the self-reporting	Hz
R_c	Sample rate of the CAN data	Hz
I_r	Request time interval of HMI application	s
I_{rr}	Re-request time interval of HMI application	s
I_s	Skip time interval of HMI application	s
K	Mileage for completing the train data collection	km
H_0	Null hypothesis of the statistical hypothesis test	-
H_1	Alternative hypothesis of the statistical hypothesis test	-
S_f	Single-modal recognition model of the front image	-
S_s	Single-modal recognition model of the side image	-
S_b	Single-modal recognition model of the bio-phyological	-
S_c	Single-modal recognition model of the CAN	-
M	Multimodal recognition model	-
N	Total number of representative emotions	-
s	Second	-
bpm	Beats per minute	-
g	Gravitationnal acceleration	m/s^2
FPS	Frame per second	-

References

1. Rouast, P.V.; Adam, M.T.; Chiong, R. Deep learning for human affect recognition: Insights and new developments. *IEEE Trans. Affect. Comput.* **2019**, *12*, 524–543. [CrossRef]
2. Underwood, G.; Chapman, P.; Wright, S.; Crundall, D. Anger while driving. *Transp. Res. Part F Traffic Psychol. Behav.* **1999**, *2*, 55–68. [CrossRef]
3. Jeon, M. Don't cry while you're driving: Sad driving is as bad as angry driving. *Int. J. Hum.-Comput. Interact.* **2016**, *32*, 777–790. [CrossRef]
4. Hassib, M.; Braun, M.; Pfleging, B.; Alt, F. Detecting and influencing driver emotions using psycho-physiological sensors and ambient light. In Proceedings of the IFIP Conference on Human-Computer Interactionr, Paphos, Cyprus, 2–6 September 2019; pp. 721–742.
5. Gao, H.; Yüce, A.; Thiran, J.P. Detecting emotional stress from facial expressions for driving safety. In Proceedings of the 2014 IEEE International Conference on Image Processing (ICIP), Paris, France, 27–30 October 2014; pp. 5961–5965.
6. Oh, G.; Ryu, J.; Jeong, E.; Yang, J.H.; Hwang, S.; Lee, S.; Lim, S. Drer: Deep learning-based driver's real emotion recognizer. *Sensors* **2021**, *21*, 2166. [CrossRef]
7. Healey, J.A.; Picard, R.W. Detecting stress during real-world driving tasks using physiological sensors. *IEEE Trans. Intell. Transp. Syst.* **2005**, *6*, 156–166. [CrossRef]
8. Angkititrakul, P.; Petracca, M.; Sathyanarayana, A.; Hansen, J.H. UTDrive: Driver behavior and speech interactive systems for in-vehicle environments. In Proceedings of the 2007 IEEE Intelligent Vehicles Symposium, Istanbul, Turkey, 13–15 June 2007; pp. 566–569.
9. Singh, R.R.; Conjeti, S.; Banerjee, R. Biosignal based on-road stress monitoring for automotive drivers. In Proceedings of the 2012 National Conference on Communications (NCC), Kharagpur, India, 3–5 February 2012; pp. 1–5.
10. Jones, C.; Jonsson, I.M. Using paralinguistic cues in speech to recognise emotions in older car drivers. In *Affect and Emotion in Human-Computer Interaction*; Springer: Berlin/Heidelberg, Germany, 2008; pp. 229–240.
11. Ma, Z.; Mahmoud, M.; Robinson, P.; Dias, E.; Skrypchuk, L. Automatic detection of a driver's complex mental states. In Proceedings of the International Conference on Computational Science and Its Applications, Trieste, Italy, 3–6 July 2017; pp. 678–691.
12. Kato, T.; Kawanaka, H.; Bhuiyan, M.S.; Oguri, K. Classification of positive and negative emotion evoked by traffic jam based on electrocardiogram (ECG) and pulse wave. In Proceedings of the 2011 14th International IEEE Conference on Intelligent Transportation Systems (ITSC), Washington, DC, USA, 5–7 October 2011; pp. 1217–1222.

13. Taib, R.; Tederry, J.; Itzstein, B. Quantifying driver frustration to improve road safety. In Proceedings of the CHI'14 Extended Abstracts on Human Factors in Computing Systems, Toronto, ON, Canada, 26 April–1 May 2014; pp. 1777–1782.
14. Ihme, K.; Dömeland, C.; Freese, M.; Jipp, M. Frustration in the face of the driver: A simulator study on facial muscle activity during frustrated driving. *Interact. Stud.* **2018**, *19*, 487–498. [CrossRef]
15. Zepf, S.; Hernandez, J.; Schmitt, A.; Minker, W.; Picard, R.W. Driver emotion recognition for intelligent vehicles: A survey. *ACM Comput. Surv.* **2020**, *53*, 1–30. [CrossRef]
16. Ortega, J.D.; Kose, N.; Cañas, P.; Chao, M.A.; Unnervik, A.; Nieto, M.; Otaegui, O.; Salgado, L. Dmd: A large-scale multi-modal driver monitoring dataset for attention and alertness analysis. In Proceedings of the European Conference on Computer Vision, Glasgow, UK, 23–28 August 2020; pp. 387–405.
17. Jegham, I.; Khalifa, A.B.; Alouani, I.; Mahjoub, M.A. A novel public dataset for multimodal multiview and multispectral driver distraction analysis: 3MDAD. *Signal Process. Image Commun.* **2020**, *88*, 115960. [CrossRef]
18. Martin, M.; Roitberg, A.; Haurilet, M.; Horne, M.; Reiß, S.; Voit, M.; Stiefelhagen, R. Drive&act: A multi-modal dataset for fine-grained driver behavior recognition in autonomous vehicles. In Proceedings of the IEEE/CVF International Conference on Computer Vision, Seoul, Korea, 27 October–2 November 2019; pp. 2801–2810.
19. Deo, N.; Trivedi, M.M. Looking at the driver/rider in autonomous vehicles to predict take-over readiness. *IEEE Trans. Intell. Veh.* **2019**, *5*, 41–52. [CrossRef]
20. Song, T.; Zheng, W.; Song, P.; Cui, Z. EEG emotion recognition using dynamical graph convolutional neural networks. *IEEE Trans. Affect. Comput.* **2018**, *11*, 532–541. [CrossRef]
21. Tao, W.; Li, C.; Song, R.; Cheng, J.; Liu, Y.; Wan, F.; Chen, X. EEG-based emotion recognition via channel-wise attention and self attention. *IEEE Trans. Affect. Comput.* **2020** . [CrossRef]
22. Li, S.; Deng, W. Deep facial expression recognition: A survey. *IEEE Trans. Affect. Comput.* **2020**. 2020.2981446. [CrossRef]
23. Kawaguchi, N.; Matsubara, S.; Takeda, K.; Itakura, F. Multimedia data collection of in-car speech communication. In Proceedings of the 7th European Conference on Speech Communication and Technology, Aalborg, Denmark, 3–7 September 2001.
24. Bradley, M.M.; Lang, P.J. Measuring emotion: The self-assessment manikin and the semantic differential. *J. Behav. Ther. Exp. Psychiatry* **1994**, *25*, 49–59. [CrossRef]
25. McCarthy, C.; Pradhan, N.; Redpath, C.; Adler, A. Validation of the Empatica E4 wristband. In Proceedings of the 2016 IEEE EMBS International Student Conference (ISC), Ottawa, ON, Canada, 29–31 May 2016; pp. 1–4.
26. Ragot, M.; Martin, N.; Em, S.; Pallamin, N.; Diverrez, J.M. Emotion recognition using physiological signals: Laboratory vs. wearable sensors. In Proceedings of the International Conference on Applied Human Factors and Ergonomics, Los Angeles, CA, USA, 17–21 July 2017; pp. 15–22.
27. Shiffman, S.; Stone, A.A.; Hufford, M.R. Ecological momentary assessment. *Annu. Rev. Clin. Psychol.* **2008**, *4*, 1–32. [CrossRef]
28. Jeon, M.; Walker, B.N. What to detect? Analyzing factor structures of affect in driving contexts for an emotion detection and regulation system. In Proceedings of the 55th Annual Meeting of the Human Factors and Ergonomics Society, Human Factors and Ergonomics Society, Las Vegas, NV, USA, 19–23 September 2011; Volume 55, pp. 1889–1893.
29. Russell, J.A. A circumplex model of affect. *J. Personal. Soc. Psychol.* **1980**, *39*, 1161. [CrossRef]
30. Schauss, A.G. Tranquilizing effect of color reduces aggressive behavior and potential violence. *J. Orthomol. Psychiatry* **1979**, *8*, 218–221.
31. Kruskal, W.H.; Wallis, W.A. Use of ranks in one-criterion variance analysis. *J. Am. Stat. Assoc.* **1952**, *47*, 583–621. [CrossRef]
32. Ostertagova, E.; Ostertag, O.; Kováč, J. Methodology and application of the Kruskal-Wallis test. In *Applied Mechanics and Materials*; Trans Tech Publications Ltd.: Bäch, Switzerland, 2014; Volume 611, pp. 115–120.
33. Kruskal, W.H. Historical notes on the Wilcoxon unpaired two-sample test. *J. Am. Stat. Assoc.* **1957**, *52*, 356–360. [CrossRef]
34. Hart, A. Mann-Whitney test is not just a test of medians: Differences in spread can be important. *Bmj* **2001**, *323*, 391–393. [CrossRef]
35. Jain, V.; Learned-Miller, E. *Fddb: A Benchmark for Face Detection in Unconstrained Settings*; Technical Report UMCS-2010-009; University of Massachusetts: Amherst, MA, USA, 2010 .
36. Yang, S.; Luo, P.; Loy, C.C.; Tang, X. Wider face: A face detection benchmark. In Proceedings of the IEEE Conference on Computer Vision and Pattern Recognition, Las Vegas, NV, USA, 27–30 June 2016; pp. 5525–5533.
37. Viola, P.; Jones, M. Rapid object detection using a boosted cascade of simple features. In Proceedings of the 2001 IEEE Computer Society Conference on Computer Vision and Pattern Recognition, CVPR 2001, Kauai, HI, USA, 8–14 December 2001; Volume 1; p. I.
38. Dalal, N.; Triggs, B. Histograms of oriented gradients for human detection. In Proceedings of the 2005 IEEE Computer Society Conference on Computer Vision and Pattern Recognition (CVPR'05), San Diego, CA, USA, 20–25 June 2005; Volume 1, pp. 886–893.
39. Viola, P.; Jones, M.J. Robust real-time face detection. *Int. J. Comput. Vis.* **2004**, *57*, 137–154. [CrossRef]
40. King, D.E. Max-margin object detection. *arXiv* **2015**, arXiv:1502.00046.
41. Oh, G.; Jeong, E.; Lim, S. Causal affect prediction model using a past facial image sequence. In Proceedings of the IEEE/CVF International Conference on Computer Vision, Montreal, BC, Canada, 11–17 October 2021; pp. 3550–3556.
42. Fang, H.S.; Xie, S.; Tai, Y.W.; Lu, C. RMPE: Regional Multi-person Pose Estimation. In Proceedings of the IEEE International Conference on Computer Vision, Venice, Italy, 22–29 October 2017.

Article

Identification of Video Game Addiction Using Heart-Rate Variability Parameters

Jung-Yong Kim [1], Hea-Sol Kim [1], Dong-Joon Kim [2,*], Sung-Kyun Im [2] and Mi-Sook Kim [3]

1. Department of HCI, Hanyang University ERICA, Ansan-si 15588, Gyeonggi-do, Korea; jungkim@hanyang.ac.kr (J.-Y.K.); kimheasol@hanyang.ac.kr (H.-S.K.)
2. Department of Industrial and Management Engineering, Hanyang University ERICA, Ansan-si 15588, Gyeonggi-do, Korea; skalon87@gmail.com
3. Department of Clothing and Textiles, Kyung Hee University, Seoul 02447, Korea; mskim@khu.ac.kr
* Correspondence: whatsdream@naver.com

Abstract: The purpose of this study is to determine heart rate variability (HRV) parameters that can quantitatively characterize game addiction by using electrocardiograms (ECGs). 23 subjects were classified into two groups prior to the experiment, 11 game-addicted subjects, and 12 non-addicted subjects, using questionnaires (CIUS and IAT). Various HRV parameters were tested to identify the addicted subject. The subjects played the *League of Legends* game for 30–40 min. The experimenter measured ECG during the game at various window sizes and specific events. Moreover, correlation and factor analyses were used to find the most effective parameters. A logistic regression equation was formed to calculate the accuracy in diagnosing addicted and non-addicted subjects. The most accurate set of parameters was found to be pNNI20, RMSSD, and LF in the 30 s after the "being killed" event. The logistic regression analysis provided an accuracy of 69.3% to 70.3%. AUC values in this study ranged from 0.654 to 0.677. This study can be noted as an exploratory step in the quantification of game addiction based on the stress response that could be used as an objective diagnostic method in the future.

Keywords: HRV parameter; game addiction; *League of Legends*; stress response; sensitivity; specificity; logistic regression

1. Introduction

The game industry is growing, with a market size of more than US $123.4 billion worldwide. South Korea is ranked fifth in the world, with 6.7% of the world market share [1], and accounts for 55.8% of Korea's content industry exports in 2018 [2]. Ryu and Lee [3] stated that such booming of the game industry has a positive influence on society, including stress management, the realization of the ideal self, and physical ability improvement. In particular, in the current COVID-19 environment, online games are recognized as a complementary means of social distancing [4,5]. However, Internet game players are not protected from becoming addicted to gaming. This addiction problem could adversely affect personal life as well as family and society, and has become a serious public health issue. Byun and Lee [6] found that Internet addiction is closely related to the increased frequency and duration of Internet use, and leads to anxiety, fear, depression, and obsessive-compulsive disorder, with adolescents being vulnerable target users. Koepp et al. [7] observed that dopamine is secreted from the brains of addicted adolescents with a similar pattern to that of drug addiction.

Adverse effects on adolescents have been studied by many authors [8–10]. In particular, it is notable that the most influential factor causing Internet addiction is stress due to excessive competition, and that adolescents exposed to excessive stress sources were readily immersed in the Internet [11]. Adolescents often experience alienation or loneliness when they are addicted to Internet games [12]. They relieve the stress related to daily life and loneliness by using internet games, which were easily accessible [13]. The higher the level of stress, the more they tended to fall into game addiction [14]. According to a study by Lee [15], game addiction prevents adolescents from coping with stress sources properly, causing various psychological problems and stress responses. Likewise, the literature indicates that Internet game addiction and mental stress are closely related.

In recent years, heart rate variability (HRV) has been used in many studies to evaluate stress levels [16–19]. Since stress affects the autonomic nervous system (ANS), HRV controlled by the ANS is often referenced as a stress indicator. A number of studies on HRV parameters have been conducted in this regard. Taelman et al. [20] and Vuksanović and Gal [21] observed that the mean of the NN interval, which is often expressed as the RR interval, and the standard deviation of all NN intervals (SDNN) decreased significantly under mental stress. Taelman et al. [20] and Tharion et al. [22] showed that pNN50 (percentage of successive RR intervals greater than 50 ms) is significantly decreased under stress. Papousek et al. [23] and Traina et al. [24] reported an increase in the low-frequency power range (LF), a decrease in the high-frequency power range (HF), and a significant increase in the LF/HF ratio when subjects experience stress. Park et al. [25] tested the newly developed measuring system to examine electrocardiograms (ECGs) and found a consistent increase in HR and SDNN as the level of addiction increased. At the same time, the LF and LF/HF parameters showed an obvious increasing trend at a high level of addiction.

On the other hand, Hafeez et al. [26] used EEGs to classify game addicts and non-addicts using cluster analysis and pattern discrimination. They introduced a statistical method to quantify the addiction phenomenon, and Hafeez et al. [27] and Kim et al. [28] identified the theta and theta/alpha parameters of the right occipital region as the discriminating variables between addicts and non-addicts. Likewise, the attempt to quantify the particular characteristics of addiction is an ongoing topic for researchers. If such a numerically quantifiable approach can be successful and assist physicians in identifying an addicted patient, they will be able to treat the patient more efficiently and objectively. Therefore, in this study, the authors are challenged to search for a quantifiable indicator of addiction in ECG response by investigating various HRV parameters. The purpose of this study is to extract quantitative HRV parameters that characterize the particular stress response of game addicts. To achieve this research goal, an exhaustive approach was performed by testing all the candidate parameters collected using window sizes of 30, 60, 90, and 120 s.

2. Methods

2.1. Subjects

A total of 23 male students participated in the experiment. The mean age was 23 years (±3 years). Eleven participants were addicted, and 12 of them were non-addicted. They were categorized using the Compulsive Internet Use Scale (CIUS) by Meerkerk et al. [29], and the Internet Addiction Test (IAT) by Young and De Abreu [30]. Based on CIUS, subjects with 2.5 or higher were categorized as addicted, and those with scores less than 1.5 were categorized as non-addicted [31]. An IAT score of 50 or higher has been used to classify the game-addicted by many researchers [10,32–35]. In this study, a subject was categorized as a game addict only when the subject met both the IAT and CIUS standards. For non-addicted subjects, an IAT score of 40 or lower was required. 14 addicted subjects and 14 non-addicted subjects were selected. 3 addicted subjects and 2 non-addicted subjects were discarded due to a technical error in the measurement system. Controlling the compounding effect of gender in this study, only male participants were tested in this study. Alcohol consumption was prohibited for 24 h before the start of the experiment, and smoking and coffee consumption were prohibited for 1 h before the start of the experiment. A fee was paid to the participants. The experiment was conducted in accordance with the regulations under consideration by the Institutional Review Board of Hanyang University in the Republic of Korea (IRB approval number: HYU-2019-08-004-1).

2.2. Apparatus

The questionnaires used to categorize subjects into two groups prior to the experiment were the CIUS by Meerkerk et al. [29] and the IAT by Young and De Abreu [30,36].

League of Legends by Riot Games Inc. (Los Angeles, CA, USA) was chosen for the experiment. This game was one of the most frequently played games among internet game players [37], and the frequent battles in the game made players experience a simulated life and death situation associated with probable stress reactions.

For data collection, an auxiliary channel of QEEG-64FX by LAXTHA Inc. (Daejeon, Republic of Korea) was used for ECG measurements (Figure 1). A data collection program called Telescan was used. The data sampling rate was set to 500 Hz. The experiment was conducted in a room equipped with a computer, a table, and a chair, where other external stimuli were restrained.

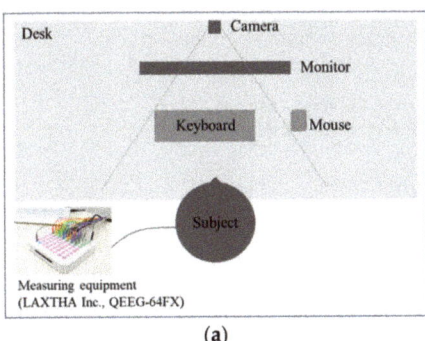

Figure 1. ECG measurement equipment. (a) Top view of experimental set-up, (b) The experimental scene.

2.3. Experimental Design

The experiment was designed to test HRV parameters to determine whether they could differentiate subjects into two groups: addicted and non-addicted. A between-subjects design was used in this study. The independent variables were the addiction status of the group, and the dependent variables were 14 parameters, including 7 time-domain variables and 7 frequency-domain variables. The time-domain parameters are NN interval average (RR interval average), SDNN, SDSD, pNNI50, pNNI20, RMSSD, and heart rate average (Table 1). The frequency-domain parameters are LF, HF, LF/HF ratio, LFnu, HFnu, total power, and VLF (Table 2). This study observed specific events during gameplay, including a "killed event", when a player's character was killed by an opponent, and a "killing event", when the player killed an opponent. The data collection window sizes for these events were 30 s, 60 s, 90 s, and 120 s, respectively, to consider the possible delay of the response.

2.4. Procedure

Positive electrode was placed in the V1 location (between the right rib 3 and 4), and the negative electrode was placed in the left infraclavicular fossa according to the standard limb guidance method [39]. The experimental procedure was briefly explained to the subject, and the ECG sensors were attached and tested to ensure that stable signals were obtained for 1 min while the subjects were relaxing. A "normal game", which is a practice game that does not affect the player's score, was played for familiarization; a "ranked game", which is a competing game affecting the player's score, was played for 30–40 min. For players' immersion in the game, the ranked game was played based on the individual skill level. Subjects played a "normal game" once and a "ranked game" twice, while the ECG was obtained. Subjects were not informed about the addiction test score; thus, they did not know whether they were categorized in the addicted group or not. The detailed experimental procedure is shown in Figure 2.

Table 1. Time-domain variables for heart rate variability [38].

Variable	Description	Equation
Mean NNI	Mean NN intervals	$\frac{1}{N}\sum_{i=1}^{N} RR_i$
SDNN	Standard deviation of all NN intervals	$\sqrt{\frac{1}{N}\sum_{i=1}^{N}(RR_i - \overline{RR})}$
SDSD	Standard deviation of differences between adjacent NN intervals	$\sqrt{\frac{1}{N-1}\sum_{i=1}^{N-1}(|RR_i - RR_{i+1}| - \overline{RRdif})^2}$
pNNI50	pNN50 count divided by the total number of all NN intervals (%)	$\frac{\sum_{i=1}^{M}\{|RR_{i+1}-RR_i| > 50 \text{ ms}\}}{N} \times 100$
pNNI20	pNN20 count divided by the total number of all NN intervals (%)	$\frac{\sum_{i=1}^{M}\{|RR_{i+1}-RR_i| > 200 \text{ ms}\}}{N} \times 100$
RMSSD	The square root of the mean of the sum of the squares of differences between adjacent NN intervals	$\sqrt{\frac{1}{N-1}\sum_{i=1}^{N-1}(RR_{i+1} - RR_i)^2}$
Mean HR	Mean heart rate	$\frac{1}{N}\sum_{i=1}^{N} HR_i$

Table 2. Frequency-domain variables of heart rate variability [38].

Variable	Description	Frequency Range
LF	Power in low-frequency range	0.04–0.15 Hz
HF	Power in high-frequency range	0.15–0.4 Hz
LF/HF ratio	Sympathovagal balance	
LFnu	LF power in normalized units: (LF/(total power − VLF)) × 100	
HFnu	HF power in normalized units: (HF/(total power − VLF)) × 100	
Total Power	The variance of NN intervals over the temporal segment	Approximately ≤ 0.4 Hz
VLF	Power in very low-frequency range	≤0.04 Hz

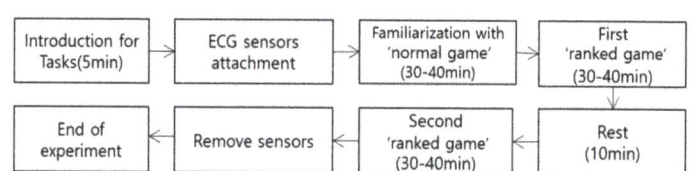

Figure 2. Experimental process.

2.5. Data Analysis

The data were analyzed in batches using Python, and time series analysis and frequency analysis were performed at the same time. The parameters used for time series analysis were extracted by using the Christov ECG R-peak segmentation algorithm. The extracted parameters were NN interval average, SDNN, RMSSD, pNNI50, pNNI20, SDSD, and heart rate average. The signal was also extracted and transformed into frequency parameters using the fast Fourier transform. Welch's periodogram was applied to estimate the spectral properties of the HRV signals, using a Hanning window. VLF (power in very-low-frequency ranges, 0.0033–0.04 Hz), LF (power in low-frequency ranges, 0.04–0.15 Hz), HF (Power in high-frequency ranges, 0.15–0.4 Hz), and total power (Power in all the frequency ranges, ≤0.4) were obtained by the sum of the power in the relevant frequency range of the spectrum. Based on these power values, the values of LF/HF ratio, LFnu, and HFnu were calculated.

Normality was tested by using Kolmogorov–Smirnov test for individual data set. The dataset with a low normality value was graphically examined to ensure an adequate level of normality. During the process, illegal outliers were treated. The *t*-test was performed ($p < 0.1$) to find the parameters and window size that statistically differentiate two groups: the addicted and non-addicted. The statistical analysis was an exhaustive process used to identify the set of most effective parameters and the window size. A correlation analysis was also performed to determine the redundancy of parameters, and a factor analysis was performed to choose the main parameters representing the characteristics of each group. Finally, a logistic regression analysis was conducted to test the sensitivity and specificity of the statistical model in identifying addicted or non-addicted subjects based on the current experimental data. The analysis process is illustrated in Figure 3. Statistical analysis was performed using SPSS Statistics 24.

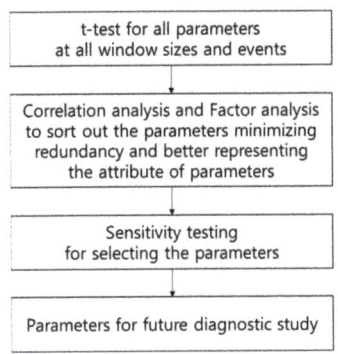

Figure 3. Data analysis process.

3. Results

An elimination process was used to sort out the best combination of parameters out of 14 parameters from 4 window sizes through statistical analyses.

3.1. The t-Test Results between Groups by Window Size

There were no significant differences in average parameter values between the addicted and non-addicted groups for the entire window sizes during the experiment ($p > 0.1$).

3.2. The t-Test Results between Groups after Specific Event

There was no significant difference of HRV parameters between groups for window sizes of 30 s, 60 s, 90 s, and 120 s after "killing events" ($p > 0.1$). However, as shown in Tables 3 and 4, the HRV parameters measured for window sizes of 30 s and 60 s after "killed events" showed a significant difference in some parameters between the two groups. In particular, pNNI20 and LF showed a significant difference ($p < 0.05$), and a marginally significant difference was observed for SDSD, RMSSD, and total power ($p < 0.1$).

3.3. Correlation Analysis and Factor Analysis with HRV Parameters

A correlation analysis was performed to examine the redundancy of the parameters in differentiating between the two groups. LF and pNNI20 with significant *p*-values ($p < 0.05$) in the *t*-test indicated a low correlation coefficient (0.264). Both could be used to improve statistical power in differentiating the two groups. On the other hand, SDSD and RMSSD showed a correlation coefficient of 1.000, and the total power and LF indicated a coefficient of 0.958. Thus, only one parameter was used to build the statistical model. Therefore, the correlation analysis suggested that the combination of the [pNNI20, LF, SDSD] or [pNNI20, LF, RMSSD] parameter set could be the best combination of parameters with the least redundancy.

Factor analysis was also performed to examine whether the selected parameters covered various factors of the data (Figure 4). The parameters with high eigenvalues for Factor 1 were RMSSD, SDSD, pNNI_50, and pNNI_20, and the parameters with high eigenvalues for Factor 2 were LF, total power, and SDNN. That is, the [pNNI20, LF, SDSD] or [pNNI20, LF, RMSSD] parameter set from the correlation analysis (Table 5) were found to have the highest eigenvalues for both Factor 1 and Factor 2 (Table 6). Therefore, the final combination of parameters for statistical modeling was [pNNI20, LF, RMSSD] or [pNNI20, LF, SDSD]. In logistic regression modeling, [pNNI20, LF, RMSSD] was arbitrarily selected to test the model performance in this study because both RMSSD and SDSD were highly correlated with each other (r = 1.000).

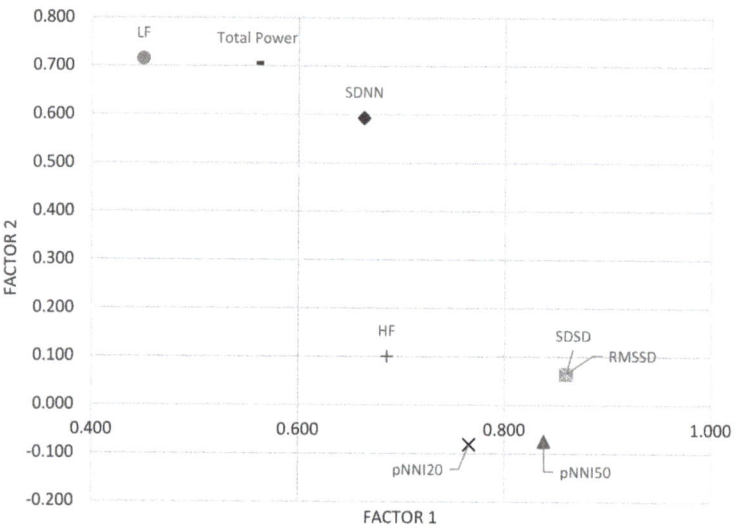

Figure 4. Factor analysis results.

Table 3. The *t*-test results for data from 30 s window size after "killed event"; mean (±standard deviation).

Parameter	Addicted Group	Non-Addicted Group	*p*-Value
MeanNNI	697.72 (±81.42)	672.16 (±82.23)	0.134
SDNN	45.31 (±14.52)	41.2 (±14.53)	0.174
SDSD	28 (±14.05)	23.35 (±12.77)	* 0.093
pNNI50	7.15 (±9.41)	5.21 (±10.9)	0.368
pNNI20	37.53 (±16.67)	29.15 (±17)	** 0.018
RMSSD	28.1 (±14.04)	23.45 (±12.78)	* 0.093
MeanHR	87.5 (±9.86)	90.91 (±10.72)	0.116
LF	1009.35 (±798.74)	679.67 (±596.46)	** 0.020
HF	326.26 (±312.39)	230.59 (±312.38)	0.141
LF/HF ratio	5.16 (±6.8)	6.45 (±10.06)	0.489
LFnu	71.93 (±19.13)	72.41 (±19.2)	0.904
HFnu	28.07 (±19.13)	27.59 (±19.2)	0.904
Total Power	1640.71 (±1084.45)	1260.54 (±970.6)	* 0.072

** $p < 0.05$, * $p < 0.1$.

Table 4. The *t*-test results for data from 60 s window size after "killed event"; mean (±standard deviation).

Parameter	Addicted Group	Non-Addicted Group	*p*-Value
MeanNNI	714.36 (± 76.27)	690.67 (± 80.38)	0.154
SDNN	53.97 (± 23.08)	50.21 (± 18.52)	0.380
SDSD	29.24 (± 13.76)	28.96 (± 21.42)	0.943
pNNI50	7.84 (± 9.18)	7.14 (± 12.55)	0.768
pNNI20	37.88 (± 16.86)	32.56 (± 17.04)	0.137
RMSSD	29.29 (± 13.78)	28.98 (± 21.42)	0.939
MeanHR	85.52 (± 9.35)	88.8 (± 10.15)	0.115
LF	973.06 (± 630.72)	753.93 (± 536.36)	* 0.071
HF	303.23 (± 268.49)	315.82 (± 459.75)	0.880
LF/HF ratio	4.21 (± 2.75)	4.51 (± 4.3)	0.708
LFnu	77.15 (± 8.37)	74.02 (± 14.85)	0.244
HFnu	22.85 (± 8.37)	25.98 (± 14.85)	0.244
Total Power	1781.33 (± 1129.5)	1672.64 (± 1251.71)	0.668
VLF	505.03 (± 441.69)	602.89 (± 539.92)	0.356

* $p < 0.1$.

Table 5. Correlation coefficients among heart rate variability parameters.

	SDNN	SDSD	pNNI50	pNNI20	RMSSD	LF	HF	TOTAL POWER
SDNN								
SDSD	0.684 **							
pNNI50	0.477 **	0.639 **						
pNNI20	0.405 **	0.480 **	0.797 **					
RMSSD	0.686 **	1.000 **	0.639 **	0.480 **				
LF	0.614 **	0.337 **	0.265 **	0.264 **	0.337 **			
HF	0.483 **	0.608 **	0.402 **	0.267 **	0.607 **	0.545 **		
Total Power	0.731 **	0.460 **	0.320 **	0.275 **	0.461 **	0.958 **	0.680 **	

** Correlation is significant at the 0.01 level (both sides).

Table 6. The eigenvalues from factor analysis.

	Factor 1	Factor 2
SDNN	0.664	0.593
SDSD	0.860	0.064
pNNI50	0.838	−0.075
pNNI20	0.766	−0.081
RMSSD	0.860	0.066
LF	0.450	0.715
HF	0.686	0.101
Total Power	0.559	0.705

3.4. Logistic Regression Models

Logistic regression models were developed using the selected parameters. A total of 15 mathematical equations were designed to test the maximum sensitivity and specificity of the parameters using natural logarithms and squares. In terms of identifying the addicted group, the sensitivity was computed, and ranged from 0.324 to 0.400; the specificity ranged from 0.828 to 0.922. The overall accuracy ranged from 67.7% to 70.3%. The model with the highest specificity of 0.922 was constructed using pNNI20, ln(RMSSD), and LF. The model with the highest sensitivity of 0.400 was obtained using ln(pNNI20), $(RMSSD)^2$, and ln(LF). The model with the highest overall accuracy of 70.3% was obtained using pNNI20, ln(RMSSD), and LF. The second-highest overall accuracy model (69.7%) was obtained using ln(pNNI20), ln(RMSSD), and $(LF)^2$. The results are summarized in Table 7.

Table 7. Summary results of four logistic regression models with the highest accuracy.

Model No.	Parameter	Model Equation	Sensitivity	Specificity	Accuracy (%)
Model 1	pNNI20 RMSSD LF	$1/(1 + \exp(-(-1.705 + 0.048 \times pNNI20 - 0.035 \times RMSSD + 0.0005 \times LF)))$	0.324	0.906	69.3
Model 2	ln(pNNI20) (RMSSD)2 ln(LF)	$1/(1 + \exp(-(-9.911 + 2.309 \times \ln(pNNI20) + 0.0003 \times (RMSSD)^2 + 0.272 \times \ln(LF))))$	0.400	0.828	67.7
Model 3	pNNI20 ln(RMSSD) LF	$1/(1 + \exp(-(-1.232 + 0.054 \times pNNI20 - 1.292 \times \ln(RMSSD) + 0.0006 \times LF)))$	0.324	0.922	70.3
Model 4	ln(pNNI20) ln(RMSSD) (LF)2	$1/(1 + \exp(-(-5.5417 + 2.505 \times \ln(pNNI20) - 1.271 \times \ln(RMSSD) + 0.0000002 \times LF^2)))$	0.891	0.343	69.7

3.5. Characteristics of Distributions Affecting the Sensitivity and Specificity

The true positive rate (sensitivity) was less than 0.4 in the above analysis, which is not good enough to provide a diagnosis of addiction for medical treatment. Such a relatively low sensitivity could be a part of the outcome based on the logistic regression model to maximize the total accuracy. To see the characteristics of the probability distribution of the data, Figures 5–7 are shown under the assumption of a normal distribution. As shown, there is a substantial overlap between distributions that could make either sensitivity or specificity low. From the observations, the criterion beta used for decision-making seemed to be biased to a conservative standard rather than a liberal one, considering that the specificity was much higher than the sensitivity. For example, for Model 1, with a maximum accuracy of 72.3%, the cut-off point associated beta value was set to 0.523, and the sensitivity and specificity were computed as 0.324 and 0.953, respectively. If a different cutoff value was then used, such as 0.372 in Model 2, the sensitivity and specificity can be computed as 0.656 and 0.703, respectively, with 67.3% accuracy.

3.6. Area under the Curve (AUC) Values

Figure 8 shows the ROC curves of the four models. The AUC value of 0.677 was for Model 1, 0.655 for Model 2, 0.673 for Model 3, and 0.654 for Model 4. According to Hosmer and Lemeshow's study [40], models having an AUC value of 0.5 or less have no discriminating power. A model can be considered acceptable only if the AUC value is between 0.7 and 0.8, and a model has excellent discriminating power if the AUC value is between 0.8 and 0.9. Thus, the AUC value of the current logistic regression model is close to the acceptable level, but further refinement is required for the model to be acceptable.

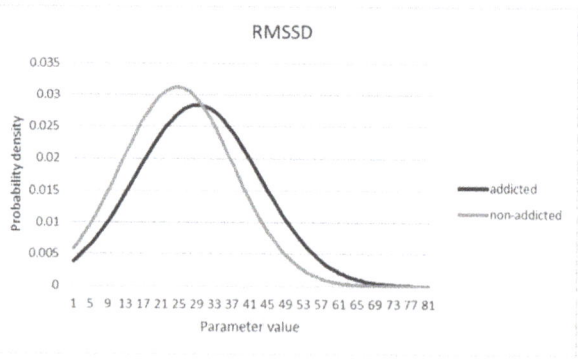

Figure 5. Probability density distribution of RMSSD parameter.

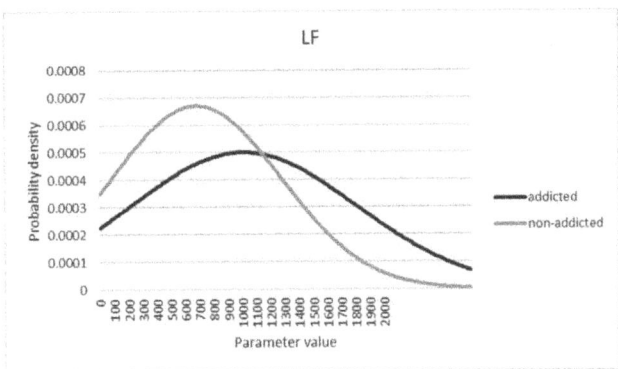

Figure 6. Probability density distribution of LF parameter.

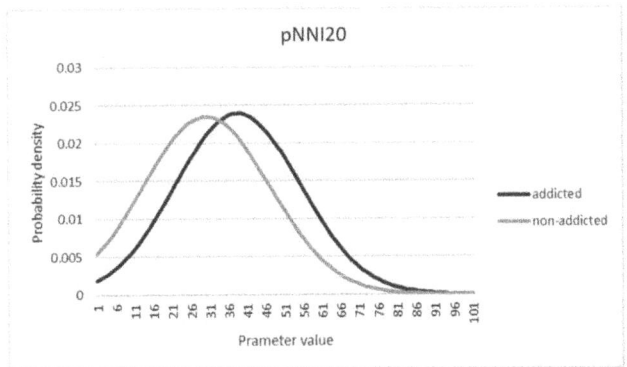

Figure 7. Probability density distribution of pNNI20 parameter.

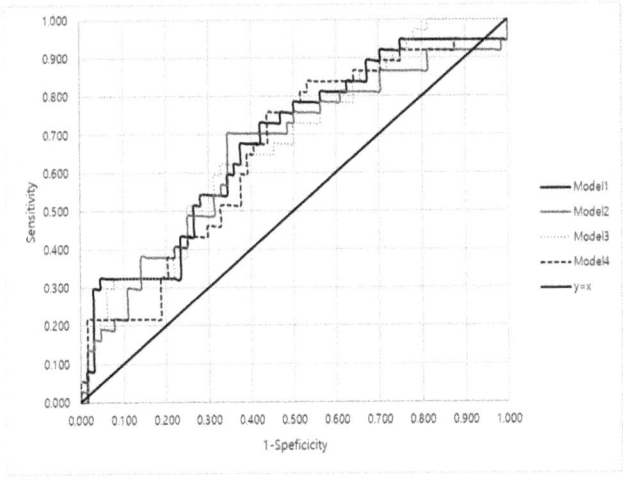

Figure 8. Receiver operating characteristic (ROC) curves of four representative models.

4. Discussion

The study showed that "being killed" in a virtual situation generated a greater signal response among the addicted subjects than non-addicted subjects. Klimt et al. [41] mentioned that a shift in self-perception would occur while enjoying the game and identifying oneself with the game character or when playing games experiencing flow or psychological mastery. Turkay and Kinzer [42] stated that the customization process of avatars by players could greatly influence players to identify themselves as game characters. Therefore, it is reasonable to think that such an affective attachment with an avatar could psychologically influence the players, and this phenomenon could be even more severe among addicted subjects than non-addicted ones.

Regarding the model building, three different statistical methods were used to select the parameters to build the best logistic regression model. Through the t-test, the pNNI20 and LF parameters were selected because they showed the most significant results ($p < 0.05$) in differentiating the two groups 30 s after the "being killed" event. This means that both time-domain and frequency-domain parameters could be effective in statistically discriminating between the two groups. The total power parameter showed a significant p-value (<0.072); however, it was not selected for the final logistic model because it was highly correlated with the LF parameter ($r = 0.958$) to avoid redundancy. In addition, the RMSSD (or SDSD) parameter was used for the logistic regression model because it showed the highest eigenvalue (0.86) of Factor 1 in the factor analysis. The LF parameter with a significant p-value in the t-test also showed the highest eigenvalue (0.715) for Factor 2, which was used for the final logistic regression model.

The final parameters selected in this study were found to be associated with the stress response based on previous studies. Bernardi et al. [43] evaluated HRV parameters under the mentally stressful situation of a subject performing arithmetic while speaking or reading, and they observed the increased power of LF when subjects were hurrying to perform the calculation task. Huang et al. [44] found that RMSSD and the combination of various variables had a positive correlation with mental fatigue induced by mental stress. According to a study by Jang et al. [45], RMSSD was also found to have a marginal correlation with tension ($r = 0.268$, $p = 0.039$), depression ($r = 0.356$, $p = 0.005$), fatigue ($r = 0.259$, $p = 0.041$), and frustration ($r = 0.304$, $p = 0.018$). Lee et al. [46] observed changes in HRV during physical and mental stress in patients with depression, and they reported a significant increase in RMSSD during the stress period compared with the rest period. Mallinani et al. [47] explained that increased sympathetic activity could be functionally characterized by an increase in the LF component in terms of LF–HF balance. Kim et al. [48] reviewed the function of HRV parameters and concluded that low parasympathetic activity was frequently related to a decrease in HF and an increase in LF.

To investigate the efficacy of the regression model in diagnosing game addiction patients, the AUC values were calculated and compared with the reference values. The computed AUC value in this study ranged from 0.654 to 0.677, which is known to have insufficient accuracy for field applications. This indicated that the increased stress response of the addicted during a "killed event" was statistically meaningful, but it might not fully reflect the symptom of addiction that the players were experiencing. Regarding the sensitivity and specificity score, the sensitivity was computed and ranged from 0.324 to 0.400, and the specificity ranged from 0.828 to 0.922 based on the logistic regression model with the default cut-off point used as a decision criterion. However, the values could change when different cut-off points were used. For now, the AUC value was less than 0.7, which could expect only less-than-accurate decision-making. Therefore, it is necessary to test the model performance under various experimental conditions. At any rate, it is important to understand the nature of HRV parameters among addicted game players, who have been very responsive to stressful stimuli, which was worthwhile to investigate further for quantification of addictive symptoms during game playing.

5. Conclusions

In this study, the difference in HRV parameters between the addicted and non-addicted group was measured during game playing, and it was found that pNNI20, RMSSD, and LF reflected the difference in stress response sensitively for a window size of 30 s after a "being killed" event. To identify the difference between the game-addicted and non-addicted subjects, the AUC score was computed and found to be less than accurate. The quantification of the psychophysiological response of the addictive game was a challenging task, as was shown in this study, but it is worth pursuing the prevention and rehabilitation of addicted patients in the future. For further study, various types and greater numbers of subjects need to be tested for better representation of the addiction symptoms. Additional mathematical exploration using artificial intelligence techniques could be another option for analyzing bio-information with a high level of variability and probable irregularity. It would also be intriguing to examine and compare the HRV parameters to other psychophysiological signals to identify the unknown patterns of game addiction.

Author Contributions: J.-Y.K., H.-S.K., D.-J.K., S.-K.I. and M.-S.K. drafted parts of the manuscript and reviewed and edited the full manuscript. All authors have read and agreed to the published version of the manuscript.

Funding: This research was funded by the Ministry of Education and the Ministry of Science and ICT, grant numbers No.NRF-2018R1D1A1B07050786 and No. 2020-0-01343.

Institutional Review Board Statement: This experiment was conducted in accordance with the regulations under consideration by the Institutional Review Board of Hanyang University in the Republic of Korea (IRB approval number: HYU-2019-08-004-1).

Informed Consent Statement: Informed consent was obtained from all subjects involved in the study.

Data Availability Statement: Not applicable.

Acknowledgments: This research was partly supported by a National Research Foundation of Korea (NRF) grant funded by the Ministry of Education (No.NRF-2018R1D1A1B07050786, Development of the algorithm to identify the EEG pattern of game addicts) and the Institute of Information & Communications Technology Planning & Evaluation (IITP) grant funded by the Korean government (MSIT) (No. 2020-0-01343, Artificial Intelligence Convergence Research Center (Hanyang University ERICA)).

Conflicts of Interest: The authors declare no conflict of interest.

References

1. Korea Creative Content Agency. *Accounts Settlement and 2018 Issue Analysis*; Ministry of Culture, Sports and Tourism of Republic of Korea: Naju-si, Korea, 2018.
2. Korea Creative Content Agency. *2018 Overseas Content Market Analysis*; Ministry of Culture, Sports and Tourism of Republic of Korea: Naju-si, Korea, 2019.
3. Ryu, S.H.; Lee, W.B. Social function of game as a leisure activity. *J. Digit. Converg.* **2012**, *10*, 245–251.
4. King, D.L.; Delfabbro, P.H.; Billieux, J.; Potenza, M.N. Problematic online gaming and the COVID-19 pandemic. *J. Behav. Addict.* **2020**, *9*, 184–186. [CrossRef]
5. Abel, T.; McQueen, D. The COVID-19 pandemic calls for spatial distancing and social closeness: Not for social distancing! *Int. J. Public Health* **2020**, *65*, 231. [CrossRef]
6. Byun, Y.S.; Lee, H.S. Impact of internet addiction on mental health in adolescents. *J. Korean Acad. Community Health Nurs.* **2007**, *18*, 460–468.
7. Koepp, M.J.; Gunn, R.N.; Lawrence, A.D.; Cunningham, V.J.; Dagher, A.; Jones, T.; Brooks, D.J.; Bench, C.J.; Grasby, P. Evidence for striatal dopamine release during a video game. *Nature* **1998**, *393*, 266–268. [CrossRef] [PubMed]
8. Kwon, S.J.; Kim, K.H.; Lee, H.S. Computer Game Addiction and Physical Health of Korea Children: Mediating Effects of Anxiety. *Surv. Res.* **2005**, *6*, 33–50.
9. Lee, Y.K.; Chae, K.M. Relations of computer game addiction and social relationship, adjustment of adolescent. *Korean J. Clin. Psychol.* **2006**, *25*, 711–726.
10. Kim, K.W. The study on the internet addiction influencing factor and coping strategies for juvenile. *J. Korea Soc. Comput. Inf.* **2009**, *14*, 157–165.
11. Gharib, K.; Homayouni, A.; Yanesari, M.K. P02-90-High levels of stress and addiction to internet. *Eur. Psychiatry* **2010**, *25*, 1. [CrossRef]

12. Hwang, H.; Ryu, S.J. The relationship between internet addiction propensity and psychosocial characteristics of christian college students: Focusing on self-control, self-esteem, self-efficacy, interpersonal efficacy, and loneliness. *Korean Assoc. Christ. Couns. Psychol.* **2008**, *16*, 321–348.
13. Park, B.S.; Park, S. Multiple Mediating Effects of Family, Friend and Teacher Relationship on the Relation between Stress and Internet Game Addiction of Adolescents. *Health Soc. Welf. Rev.* **2016**, *36*, 61–88.
14. Lee, S.J.; Kim, G.A.; Hong, C.H. The effects of internet use motivation and stress coping on adolescent's pathological internet use: Focused on gender difference. *Korean J. Woman Psychol.* **2011**, *16*, 265–284.
15. Lee, H.G. Social and psychological variables predicting violence game addiction of adolescents in the internet. *Korean J. Dev. Psychol.* **2002**, *14*, 55–79.
16. Ahn, J.W.; Ku, Y.; Kim, H.C. A novel wearable EEG and ECG recording system for stress assessment. *Sensors* **2019**, *19*, 1991. [CrossRef]
17. Correia, B.; Dias, N.; Costa, P.; Pêgo, J.M. Validation of a wireless bluetooth photoplethysmography sensor used on the earlobe for monitoring heart rate variability features during a stress-inducing mental task in healthy individuals. *Sensors* **2020**, *20*, 3905. [CrossRef]
18. Cho, D.; Ham, J.; Oh, J.; Park, J.; Kim, S.; Lee, N.-K.; Lee, B. Detection of stress levels from biosignals measured in virtual reality environments using a kernel-based extreme learning machine. *Sensors* **2017**, *17*, 2435. [CrossRef]
19. Salai, M.; Vassányi, I.; Kósa, I. Stress detection using low cost heart rate sensors. *J. Healthc. Eng.* **2016**, *2016*, 5136705. [CrossRef] [PubMed]
20. Taelman, J.; Vandeput, S.; Vlemincx, E.; Spaepen, A.; Van Huffel, S. Instantaneous changes in heart rate regulation due to mental load in simulated office work. *Eur. J. Appl. Physiol.* **2011**, *111*, 1497–1505. [CrossRef]
21. Vuksanović, V.; Gal, V. Heart rate variability in mental stress aloud. *Med. Eng. Phys.* **2007**, *29*, 344–349. [CrossRef] [PubMed]
22. Tharion, E.; Parthasarathy, S.; Neelakantan, N. Short-term heart rate variability measures in students during examinations. *Natl. Med. J. India* **2009**, *22*, 63–66. [PubMed]
23. Papousek, I.; Nauschnegg, K.; Paechter, M.; Lackner, H.K.; Goswami, N.; Schulter, G. Trait and state positive affect and cardiovascular recovery from experimental academic stress. *Biol. Psychol.* **2010**, *83*, 108–115. [CrossRef] [PubMed]
24. Traina, M.; Cataldo, A.; Galullo, F.; Russo, G. Effects of anxiety due to mental stress on heart rate variability in healthy subjects. *Minerva Psichiatr.* **2011**, *227*, 31.
25. Park, M.C.; Jung, H.C.; Kim, T.S. Design of a stress measurement system for state recognition of game addicts. *J. Korea Soc. Comput. Inf.* **2017**, *22*, 87–93.
26. Hafeez, M.; Dawood, I.M.; Kim, J.Y.; Kim, D.J. Study of Game Addiction Pattern by Using Spectral Analysis of EEG. In Proceedings of the Extended Abstracts of HCI Korea, Jeongseon-gun, Korea, 27–29 January 2016; pp. 62–64.
27. Hafeez, M.; Kim, D.J.; Im, S.K.; Kim, J.Y. The Cross Correlation and Power Spectrum Analysis of EEG Attributes between mobile Game Addicts and Non-Addicts. In Proceedings of the HCI Korea, Jeongseon-gun, Korea, 31 January–2 February 2018; pp. 997–999.
28. Kim, D.J.; Kim, H.; Im, S.; Oh, M.S.; Kim, J.; Kim, J.Y. Analysis of EEG Parameters Characteristics for High-Risk Users of Online Games. In Proceedings of the 2019 Spring Conference of ESK, Seogwipo-si, Korea, 15–18 May 2019; p. 237.
29. Meerkerk, G.J.; Van Den Eijnden, R.J.; Vermulst, A.A.; Garretsen, H.F. The compulsive internet use scale (CIUS): Some psychometric properties. *Cyberpsychol. Behav.* **2009**, *12*, 1–6. [CrossRef]
30. Young, K.S.; De Abreu, C.N. *Internet Addiction: A Handbook and Guide to Evaluation and Treatment*; John Wiley & Sons: Hoboken, NJ, USA, 2010.
31. Littel, M.; Van den Berg, I.; Luijten, M.; van Rooij, A.J.; Keemink, L.; Franken, I.H. Error processing and response inhibition in excessive computer game players: An event-related potential study. *Addict. Biol.* **2012**, *17*, 934–947. [CrossRef] [PubMed]
32. Peng, X.; Cui, F.; Wang, T.; Jiao, C. Unconscious processing of facial expressions in individuals with Internet Gaming Disorder. *Front. Psychol.* **2017**, *8*, 1059. [CrossRef]
33. Kim, E.J.; Lee, S.Y.; Oh, S.G. The validation of Korean adolescent internet addiction scale (K-AIAS). *Korean J. Clin. Psychol.* **2003**, *22*, 125–139.
34. Samaha, A.A.; Fawaz, M.; El Yahfoufi, N.; Gebbawi, M.; Abdallah, H.; Baydoun, S.A.; Ghaddar, A.; Eid, A.H. Assessing the psychometric properties of the internet addiction test (IAT) among Lebanese college students. *Front. Public Health* **2018**, *6*, 365. [CrossRef]
35. Nam, Y.O. A study on the psychosocial variables of the youth's addiction to internet and cyber sex and their problematic behavior. *Korean J. Soc. Welf.* **2002**, *50*, 173–207.
36. Kim, H.; Ha, J.; Chang, W.-D.; Park, W.; Kim, L.; Im, C.-H. Detection of craving for gaming in adolescents with internet gaming disorder using multimodal biosignals. *Sensors* **2018**, *18*, 102.
37. Most Popular Core PC Games. Available online: https://newzoo.com/insights/rankings/top-20-core-pc-games (accessed on 10 April 2021).
38. Camm, A.J.; Malik, M.; Bigger, J.T.; Breithardt, G.; Cerutti, S.; Cohen, R.; Coumel, P.; Fallen, E.; Kennedy, H.; Kleiger, R. Heart rate variability: Standards of measurement, physiological interpretation and clinical use. Task Force of the European Society of Cardiology and the North American Society of Pacing and Electrophysiology. *Ann. Noninvasive Electrocardiol.* **1996**, *1*, 151–181.
39. Drew, B.J.; Califf, R.M.; Funk, M.; Kaufman, E.S.; Krucoff, M.W.; Laks, M.M.; Macfarlane, P.W.; Sommargren, C.; Swiryn, S.; Van Hare, G.F. Practice standards for electrocardiographic monitoring in hospital settings: An American Heart Association scientific

statement from the Councils on Cardiovascular Nursing, Clinical Cardiology, and Cardiovascular Disease in the Young: Endorsed by the International Society of Computerized Electrocardiology and the American Association of Critical-Care Nurses. *Circulation* **2004**, *110*, 2721–2746.
40. Hosmer, D.W., Jr.; Lemeshow, S. *Applied Logistic Regression*, 2nd ed.; John Wiley & Sons: New York, NY, USA, 2000; pp. 160–164.
41. Klimmt, C.; Hefner, D.; Vorderer, P. The video game experience as "true" identification: A theory of enjoyable alterations of players' self-perception. *Commun. Theory* **2009**, *19*, 351–373.
42. Turkay, S.; Kinzer, C.K. The effects of avatar-based customization on player identification. In *Gamification: Concepts, Methodologies, Tools, and Applications*; IGI Global: Hershey, PA, USA, 2015; pp. 247–272.
43. Bernardi, L.; Wdowczyk-Szulc, J.; Valenti, C.; Castoldi, S.; Passino, C.; Spadacini, G.; Sleight, P. Effects of controlled breathing, mental activity and mental stress with or without verbalization on heart rate variability. *J. Am. Coll. Cardiol.* **2000**, *35*, 1462–1469. [CrossRef]
44. Huang, S.; Li, J.; Zhang, P.; Zhang, W. Detection of mental fatigue state with wearable ECG devices. *Int. J. Med. Inform.* **2018**, *119*, 39–46. [CrossRef]
45. Jang, E.H.; Kim, A.Y.; Yu, H.Y. Relationships of psychological factors to stress and heart rate variability as stress responses induced by cognitive stressors. *Sci. Emot. Sensib.* **2018**, *21*, 71–82. [CrossRef]
46. Lee, J.H.; Yu, J.; Ryu, S.H.; Ha, J.H.; Jeon, H.J.; Park, D.H. Change of heart rate variability in depressive disorder after physical or psychological stress. *Sleep Med. Psychophysiol.* **2018**, *25*, 15–20. [CrossRef]
47. Malliani, A.; Pagani, M.; Lombardi, F.; Cerutti, S. Cardiovascular neural regulation explored in the frequency domain. *Circulation* **1991**, *84*, 482–492. [CrossRef] [PubMed]
48. Kim, H.G.; Cheon, E.J.; Bai, D.S.; Lee, Y.H.; Koo, B.H. Stress and heart rate variability: A meta-analysis and review of the literature. *Psychiatry Investig.* **2018**, *15*, 235. [CrossRef] [PubMed]

Article

Individual's Social Perception of Virtual Avatars Embodied with Their Habitual Facial Expressions and Facial Appearance

Sung Park [1,*], Si Pyoung Kim [2] and Mincheol Whang [3]

1. School of Design, Savannah College of Art and Design, Savannah, GA 31401, USA
2. Department of Emotion Engineering, Sangmyung University, Jongno-gu, Seoul 03016, Korea; 201731114@sangmyung.kr
3. Department of Human Centered Artificial Intelligence, Sangmyung University, Jongno-gu, Seoul 03016, Korea; whang@smu.ac.kr
* Correspondence: spark@scad.edu

Abstract: With the prevalence of virtual avatars and the recent emergence of metaverse technology, there has been an increase in users who express their identity through an avatar. The research community focused on improving the realistic expressions and non-verbal communication channels of virtual characters to create a more customized experience. However, there is a lack in the understanding of how avatars can embody a user's signature expressions (i.e., user's habitual facial expressions and facial appearance) that would provide an individualized experience. Our study focused on identifying elements that may affect the user's social perception (similarity, familiarity, attraction, liking, and involvement) of customized virtual avatars engineered considering the user's facial characteristics. We evaluated the participant's subjective appraisal of avatars that embodied the participant's habitual facial expressions or facial appearance. Results indicated that participants felt that the avatar that embodied their habitual expressions was more similar to them than the avatar that did not. Furthermore, participants felt that the avatar that embodied their appearance was more familiar than the avatar that did not. Designers should be mindful about how people perceive individuated virtual avatars in order to accurately represent the user's identity and help users relate to their avatar.

Keywords: virtual avatar; virtual human; virtual character; embodied conversational agent; social interaction; empathy

1. Introduction

Humans communicate with others via verbal and non-verbal communication. Through dyadic social interaction, people elicit the other's intention and emotion [1]. Facial expressions represent non-verbal communication channels [2]. The face is the most recognizable region and has unique characteristics that represent an individual [3]. Humans are born with an innate capability to sense and perceive the most important person (i.e., mother) at the early stage of life. Infants are known to discriminate facial features starting at two months after birth [4], and they also prefer facial features over other shapes and forms [5]. Hiding one's face implies the concealment of one's identity. For example, covering a face with a mask may be considered negative social behavior [6].

The rapid advancement of VR (Virtual Reality) technology facilitates the introduction of expressive services tailored to the metaverse. Virtual experiences using HMD (Head-Mounted Display) are now prevalent in households due to video games. In addition, the AR (Augmented Reality) industry is growing through mobile platforms with the availability of engaging entertainment services. Naturally, virtual avatars, a conduit that connects the virtual world to the user, have gained much attention. Many users are interested in projecting or extending their identities through avatars in the internet's social landscape.

There are various ways to express oneself through a virtual avatar. The most direct way is to apply one's physical characteristics to an avatar that embodies the user's facial appearance or proportions [7]. Studies are also considering the application of a user's habitual expressions based on facial muscle movement [8]. A virtual avatar with the user's unique signature may elicit social responses such as perceived similarity and familiarity.

1.1. Habitual Facial Expressions and Facial Appearance

The human face consists of 20 facial muscles. Humans communicate through an interplay of these muscles, which produce expressions. Facial expressions enable social communication, which abides by shared rules [9]. They are a powerful source of visual information that embodies the individual's emotions, behavioral predisposition, and intention [10]. Humans can infer the interaction partner's psychological state through facial expressions and identify their traits [11]. In psychology, an individual's traits are, by definition, their habitual pattern of thoughts or affect.

Facial expressions are individual behavioral habits that consist of patterned muscle movement. Such patterns include unique muscle characteristics (e.g., the intensity of the movement of each facial muscle). As a result of these individual differences, people can reliably discriminate themselves from others [8].

On the other hand, facial appearance provides a person's unique identity from the physical features, specifically face and head. Although the perception of appearance relies on many environmental factors (e.g., head pose, lighting conditions), there are descriptive characteristics of a particular individual, such as the location of the eye, nose, and mouth. In our study, we used such facial landmarks to identify critical regions of the face by defining their coordinates (x,y) on the facial image.

Visual perception plays an integral part in facial recognition, which also applies to recognizing oneself. The easiest way to look at oneself is through a mirror. Being able to recognize one's own face is one of the critical prerequisites of self-consciousness and self-identity. Only humans and a few animals may recognize themselves through a mirror [12]. For humans, this ability develops at the age of two. This ability correlates with empathic and altruistic behavior.

Humans feel a sense of closeness to familiar entities. They also feel more intimate with objects that they are repeatedly exposed to, even without interacting with these (i.e., mere exposure) [13]. An object to which a person is familiarized through repetitive exposure may elicit positive responses [14,15]. For example, stimuli such as names [16] or photos [17] may elicit positive responses after repeated exposure. This phenomenon may also be observed with facial perception. When participants viewed a specific face repetitively, they described it as more familiar, similar, and attractive than those who did not [18].

Humans belong to social circles of varying size. Individuals have a higher chance of getting exposed to a member in the same group than to a member in a different group. When exposed to identical situations, people in the same group tend to exhibit similar responses. The more members express different responses, the lesser the probability of sustaining the group [19].

Exhibiting a similar response to an identical stimulus is related to empathy. In a dyadic interaction, an empathic response is manifested by mimicking the other's facial expressions or gestures [20]. Sustaining a similar expression or empathic response for a long time results in the repeated utilization of the respective muscles responsible for empathic expressions. Repetitive use of certain muscles affects bone structure and as a result, leads to an appearance that is similar to that of the significant other [21].

Furthermore, perceived similarity is known to entail a positive face-to-face interaction. People are predisposed to think that in dyadic socialization, a part of their partner's attitude, values, and beliefs is similar to theirs [22,23]. People tend to like and trust people who have a similar physical appearance more than those who do not [24].

1.2. Virtual Avatar

The term avatar is derived from a Sanskrit word and connotes the incarnation of a deity. In modern society, the user's mental model of an avatar is that it is an alter ego of the user that can interact with other virtual avatars in a virtual world [25]. Recently, the need for a virtual avatar has not only come from games, movies, advertisements, and remote collaboration but has extended to medical practice and crime investigation. Research, design, and development explore the avatar model and how it can imitate users in real time. Realistic animation is possible by depicting the movement based on bone and muscle structure, considering the real-world laws of physics.

In general, the more similar the illustration of a virtual avatar is to the user, the more immersive their experience [26,27]. Nevertheless, a very realistic but imperfect depiction of a user may lead to negative feelings [28]. Virtual characteristics that reach a certain point of human likeness tend to elicit a feeling of eeriness.

Much research has been conducted on the interaction channels of virtual avatars. There has been much attention on non-verbal expressions such as the gaze, the facial expression, and gestures of an avatar. For example, minute movements of the pupil add a sense of immersion and social presence. Studies found that participants perceived a higher level of social presence when communicating via richer media than through a text-based medium [29–31].

In a virtual environment, users may use their virtual avatar to represent themselves. Users tend to prefer an avatar that embodies their unique and exclusive characteristics that differentiate them from the others. Some people prefer an avatar that is similar to themselves, while others prefer their avatar to be an idealized version of themselves. Users who adopted such avatars reported higher satisfaction and attachment [32]. Users are more motivated to use avatars that have a facial appearance similar to theirs than those that do not [24].

However, the majority of avatar illustrations and expressions do not consider the individual's facial characteristics. Applying individualized facial habits or appearances does not require sophisticated technology and is viable with the current computer systems available to the mass. However, software that can animate such virtual avatars needs to be developed with investment and resources.

Another reason why individuated avatars are not prevalent involves the users. Many users do not recognize their own facial habits and would have trouble customizing the facial characteristics by themselves. It would be necessary for the application to capture and analyze the user's facial movements and suggest a personalized avatar for approval before use. The users may feel that this is a hassle, not to mention that there is resistance from users against taking a video of their own face. Most importantly, research lacks an understanding of common elements applicable to individuated virtual avatars. Specifically, we do not clearly understand the social effects of personalized virtual avatars with individualized features. Would people prefer avatars with their appearance or habitual expressions? Would people perceive a similarity between the avatar and themselves? Would people be able to relate to the avatar and use it for their profile in a social networking service?

1.3. Research Goal

Humans have universally recognizable expressions. Ekman found a universal relationship between facial muscle movements and specific emotions (e.g., happiness, sadness, anger, fear, surprise, disgust, interest) [33]. Despite the universality, individual differences exist in the *intensity* of each muscle movement. Researchers also found that the asymmetrical measures of facial regions identify stable individual differences [34].

A facial habit results from a habitual personal pattern that exhibits a unique individual signature. Facial recognition based on these individual differences in expression analyzes the movement pattern of facial muscles to discriminate individuals [8].

Another factor to consider is the individual's appearance. The perception of a form is necessary to identify an object [35]. The holistic form is a pivotal component required to distinguish an individual [36].

In summary, our research aims to evaluate the perceived social effect of a virtual avatar using two markers: (1) habitual facial expressions captured through the *intensity* of muscle movement and (2) facial appearance identified using facial *landmarks*. The research hypotheses are summarized accordingly in Table 1. We added the third hypothesis because both facial habit and facial appearance involve the facial muscle, and therefore, an interaction may occur. Thus, we intend to analyze whether facial habits (independent variable) have a different effect on the social constructs (dependent variables) depending on facial appearance (independent variable).

Table 1. Research hypotheses.

	Research Hypotheses
H1	A virtual avatar that displays the participant's habitual expressions will elicit the following perceived social constructs more than a virtual avatar that does not: Perceived similarity Perceived familiarity Perceived attraction Perceived liking Perceived involvement
H2	A virtual avatar that has a similar facial appearance to the participant will elicit the following perceived social constructs more than a virtual avatar that does not: Perceived similarity Perceived familiarity Perceived attraction Perceived liking Perceived involvement
H3	There is an interaction between the participant's habitual expressions and facial appearance.

In short, the study aims to evaluate people's social perception of an avatar that embodies the unique and individual characteristics of the user. We planned to investigate the interaction of the two independent variables (facial appearance, facial habit) and their respective main effects.

2. Methods

2.1. Participants

Forty-five university students were recruited as participants. The participants' average age was 23.78 years (SD = 2.88), with 20 males and 25 females. We recommended that the participants get sufficient sleep the day before the experiment. We selected participants with a corrective vision of 0.7 or above to ensure the participants' reliable recognition of visual stimuli. All participants were briefed on the purpose and procedure of the experiment and signed a consent form. Participants were given participation fees as compensation.

2.2. Materials

2.2.1. Video Stimulus

The current study used a video stimulus to elicit participants' facial responses to produce data to create an individuated avatar. We used video materials known to evoke emotions, which were empirically verified by an experiment conducted in and provided by Stanford University (n = 411, [37]).

For each emotional state (positive and negative), we selected two candidate stimuli from Stanford's materials [37]. We conducted a manipulation check on all candidate materials. With regard to the positive stimuli, participants perceived the two video stimuli as positive. The results did not show a significant difference from those of the Stanford study.

However, there was no significant change in the facial expression of participants when the negative stimuli were exposed. In a follow-up questionnaire, participants reported having a negative emotional state but did not display a negative facial expression. Since the current experiment requires valid participant data on emotional expression to be applied to a virtual avatar, we decided not to include stimuli evoking a negative emotional state.

2.2.2. Video Analysis

We used Open Face, which is open-source software that enables face recognition with deep neural networks [38]. We used AU (Action Units) as the basic unit for appraisal from the Facial Action Code System (FACS) [39]. Figure 1 depicts the process. We first normalized the facial region from the participants' videos. The video was organized as a sequence of images of fixed size (200 × 200 pixels). From this image sequence, we elicited the intensity of AU movement and the 68 facial landmarks (see Figure 2). The landmarks extract the coordinates (x,y) of key facial regions (e.g., the eye, eyebrows, nose, lips, and chin). The movement and intensity of AU were identified from the AU vector data in HOG (Histograms of Oriented Gradients) [40]. We elicited the individual's habitual expression data from the AU movement intensity. We elicited the individual's facial appearance from the landmark data.

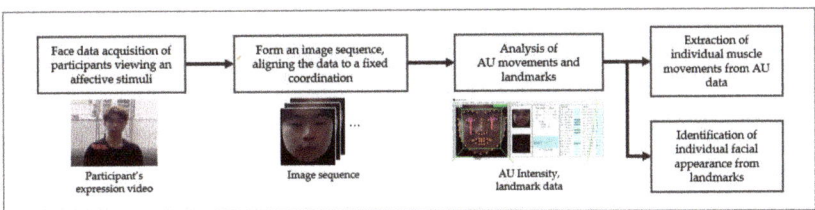

Figure 1. The analytical process of identifying individual muscle movements and facial appearance.

Figure 2. The 68 facial landmarks used to identify the participant's facial appearance.

2.2.3. Virtual Avatar

We designed two baseline avatars, male and female, to embody the participant's expressive habits and facial appearances (see Figure 3). For the female model, we modified a public model available from an open source [41]. To visualize the muscle movement, we produced AU-based blend shapes using the animation software Maya (Autodeck). We used blend shapes that morphed the lower face of the virtual avatar for a more natural look. Table 2 shows the relationship between blend shapes and facial muscles.

Figure 3. The baseline virtual avatar models in the study.

Table 2. The blend shape type based on the virtual avatar's AU and facial appearance.

Blend Shape	Description	Muscular Basis
AU1	Inner brow raiser	Frontalis, Pars medialis
AU2	Outer brow raiser	Frontalis, Pars lateralis
AU4	Brow lowerer	Depressor glabellae, Depressor supercilli, Corrugator supercilli
AU5	Upper lid raiser	Levator palpebrae superioris
AU6	Cheek raiser	Orbicularis oculi, Pars orbitalis
AU7	Lid tightener	Orbicularis oculi, Pars palpebralis
AU9	Nose wrinkler	Levator labii superioris alaeque nasi
AU10	Upper lip raiser	Levator labii superioris, Caput infraorbitalis
AU12	Lip corner puller	Zygomaticus major
AU14	Dimpler	Buccinator
AU15	Lip corner depressor	Depressor anguli oris (Triangularis)
AU17	Chin raiser	Mentalis
AU20	Lip stretcher	Risorius
AU23	Lip tightener	Orbicularis oris
AU25	Lips part	Depressor labii, Relaxation of mentalis (AU17), Orbicularis oris
AU26	Jaw drop	Masseter, Temporal and Internal pterygoid relaxed
AU28	Lip suck	Orbicularis oris
AU45	Blink	Relaxation of levator palpebrae and Contraction of orbicularis oculi, Pars palpebralis.
Shape1	Expansion of the lower jaw bone	Mandible ramus extension
Shape2	Contraction of the lower jaw bone	Mandible ramus compression
Shape3	Expansion of the lower jaw	Chin extension
Shape4	Contraction of the lower jaw	Chin compression

We used the Unity 3D engine to render and animate the virtual avatar [42]. Figure 4 depicts the two versions of the avatar with the participant's facial signature (facial appearances, habitual expression) applied. How participants viewed such variations and what was measured will be explained in Section 2.3 (Experiment Procedure).

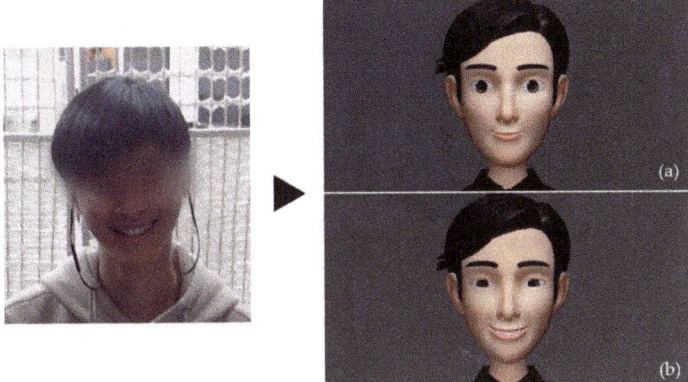

Figure 4. An example of the baseline virtual human morphed based on the participant's (**a**) facial appearances and (**b**) habitual expression.

2.2.4. Subjective Appraisal of Social Constructs

The current study investigated participants' perceptions (similarity, familiarity, attraction, liking, and involvement) of virtual avatars. All constructs involve the subjective appraisal by participants rather than an objective quantitative measurement. Table 3 depicts their operational definition. Each construct was measured on a 7-point Likert scale. For example, the seven items of similarity were *slightly*, *somewhat*, and *extremely* toward both ends (dissimilar and similar) with *neutral* in the middle.

Table 3. The operational definition of the social constructs of interest.

Social Construct	Operational Definition
Similarity	The degree to which the participant *believes* the virtual avatar's appearance is similar to themselves.
Familiarity	The degree to which the participant is *familiar* with the virtual avatar's appearance.
Attraction	The degree to which the participant is *attracted to* the virtual avatar.
Liking	The degree to which the participant *likes or dislikes* the virtual avatar.
Involvement	The degree to which the participant *relates to* or *empathizes with* the virtual avatar.

Similarity connotes the degree to which the user sees themselves as similar with the avatar. Some research includes attitudinal similarity (e.g., personality, attitude, belief system) in the definition [18,43]. However, in this study, we limited the definition to only include the physical likeliness to the participant and formulated the survey question accordingly. We purposely designed the study to eliminate interaction with the virtual avatar to investigate the effect of its mere presence without any convoluted variables that may arise from interactions. Since there is no interaction with the virtual avatar, it is extremely difficult to validly assess attitudinal similarity.

It is important to emphasize that we investigated *perceived* similarity as opposed to actual similarity. Researchers have made a clear distinction between the two constructs [44]. *Actual* similarity is measurable and quantifiable using standardized personality assessment. As the paper will discuss later, the relationship between similarity and attraction is critical.

Some research studies suggest that only perceived similarity is a prerequisite to eliciting attraction [45–47]; other research emphasizes the importance of actual similarity [48]. In this study, mainly for consistency with other perceived constructs, we investigated the perceived similarity.

Perceived familiarity was measured to assess the degree to which participants were familiar with the virtual avatar that had the participant's facial characteristics applied. In interpersonal and social science literature, this construct connotes "being knowledgeable" or acquainted with a person [18,49] or a concept [50,51]. That is, a priori knowledge is necessary to measure perceived familiarity. For example, in psychology, after an interaction (e.g., phone call, discussion) with a person, the participant felt subjective familiarity with the person similar to what they would feel with a close friend [49]. Other studies measured familiarity using objective quantitative measures, such as the amount of exposure to a person's photo and not just focusing on perception [18].

Some studies use the terms perceived familiarity and resemblance (perceived similarity) interchangeably [49]; however, we measured the two constructs (perceived similarity and perceived familiarity) independently. The literature suggests that the two constructs correlate and have a causal relationship, with attraction as a mediating variable [18]. In our study, we minimized interaction with the virtual humans (e.g., conversation) to test the mere exposure effect.

Since the pioneering work of Byrne [52] (for a review of attraction as a research paradigm, see [53]), researchers have investigated interpersonal attraction in relationships [54]. Researchers widely accept Newcomb's definition of attraction as the most comprehensive one, and it is defined as follows: "Attraction refers to any direction orientation (on the part of one person toward another) which may be described in terms of sign and intensity" (Page 6) [55].

Studies on attraction generally investigate the relationship between the independent variables (e.g., attitudinal similarity, physical attractiveness) and the attraction response as a dependent variable. It is critical to note that attraction is distinguished from attractiveness, i.e., characteristics (e.g., attractive personality, good looks) that attract others [56]. In our study, we obtained the participant's perceived attraction (dependent variable) to the virtual avatar, which varied according to different facial features (independent variable). The intensity of attraction depends on many factors such as their relationship (e.g., parent–child, wife–husband) and the duration of interaction (e.g., long-term, first acquaintance) [57].

Perceived liking, as a construct, is defined as the degree to which the participant likes or dislikes the other person in a dyad. A causal pattern consists between the perception of being liked and liking the other [58]. Compared to attraction, perceived liking has a corresponding place on a like–dislike spectrum, whereas attraction is located on an attraction–repulsion spectrum [59].

In psychology, involvement connotes *approach* predispositions (e.g., empathy, sympathy, challenge) as opposed to distance, which refers to *avoidance* predispositions (e.g., antipathy, irritation, boredom) [24]. The two constructs are unipolar. Involvement refers to the degree to which the participants relate to and empathize with the virtual avatar. Since empathy is mainly dependent on the task and context [60,61]), we provided the context that the virtual agent would be used in a profile for a social networking service.

2.3. Procedure

Figure 5 outlines the experiment procedure. The experiment was conducted twice, with an interval of one week between the two sessions (i.e., Session #1 and Session #2).

In the first experiment, the participants were briefed about the purpose of the experiment and the procedures. Then, participants viewed the two affective stimuli from the display in a relaxed position (see Figure 6). Participants were guided not to force any expression but display the natural expression felt from the viewing. The web camera on display recorded a video of the participant's facial responses for 90 s. Then, the participants left the experiment after a brief explanation of the second experiment session.

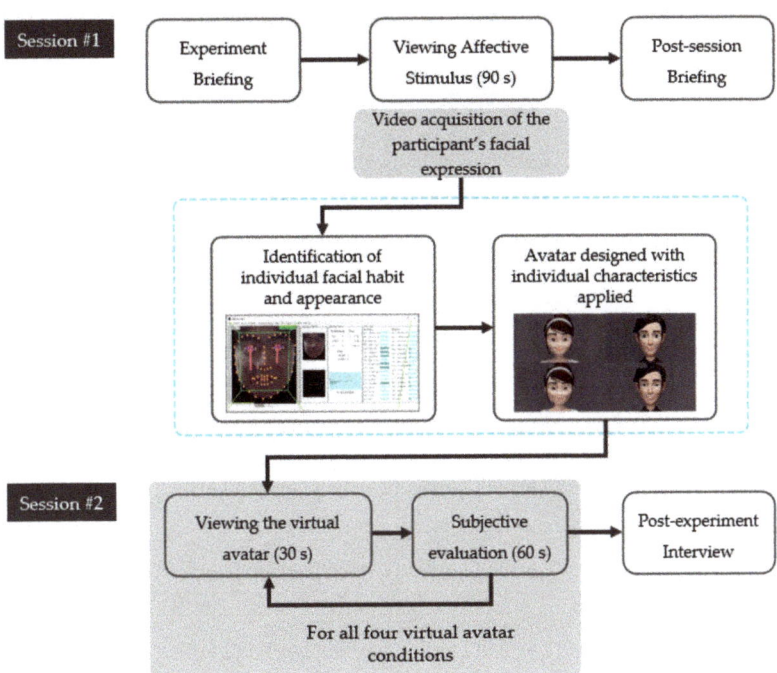

Figure 5. The experiment consists of two sessions, with one week in between for each participant.

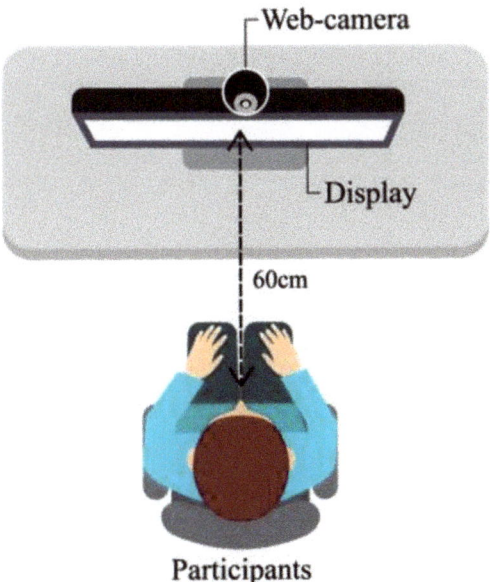

Figure 6. The experiment environment.

In between the two sessions, we produced the following four virtual avatars for the second experiment session based on the data acquired from the participants:

(1) An avatar with both the habitual facial expression and appearance applied;
(2) An avatar with only the facial appearance applied;
(3) An avatar with only the habitual facial expression applied;
(4) Baseline avatar with none of the individual data applied.

For an avatar without any habitual facial expression applied (2 and 4), AU movement based on the literature was applied instead [39]. For an avatar without any facial appearance applied (3 and 4), the original baseline appearance of the avatar was used (Figure 3).

Then, the participants viewed these virtual avatar stimuli. The study used a 2 × 2 within-subject design. There were two levels of habitual expression (applied or not) and facial appearance (applied or not), respectively.

Every participant viewed all four virtual avatar types. The order of the virtual avatar was randomized using a Latin square to counter the potential learning and fatigue effect. After viewing the avatar for 30 s, the participant responded to a subjective questionnaire.

Interaction with the virtual human was limited to mere exposure as opposed to an interactive one (e.g., conversation). The strength of the subjective response was contingent on the nature of the task [62] and may have elicited a confounding effect, which would be difficult to identify.

2.4. Statistical Analysis

To understand the effects of the two independent variables (habitual facial expression, facial appearance), we conducted a two-way ANOVA on the participant's subjective evaluation of the four avatars.

Data from participants who did not exhibit any facial expressions during the experiment were excluded during the acquisition process. The exclusion criteria are outlined as follows. First, we divided the non-expression interval and the expression interval. The latter was defined based on the average expression data. The intensities of AU 6 (Cheek raiser) and AU 12 (Lip corner puller) during the expression interval were compared to those of the non-expression interval. If the intensity during the expression interval was less than the non-expression interval or non-existent, we excluded the participant's data. The Latin square factors were tested to examine whether the order affected the dependent variable. The Latin square order did not affect data, so all results were collapsed over these variables.

3. Results

3.1. Similarity

The results of analysis of subjective perception involving similarity are as follows. Figure 7 depicts participants' responses to the different avatars that varied according to two factors (facial habit and facial appearance). The Y-axis indicates the average of subjective Likert ratings. The results showed no significant interaction between Facial Habit × Facial Appearance, $F(1, 163) = 2.517, p > 0.11$. Of particular importance, the results showed that Facial Habit had a significant main effect, $F(1, 81) = 5.182, p < 0.05$. On the other hand, Facial Appearance had no significant main effect, $F(1, 81) = 0.576, p > 0.44$.

3.2. Familiarity

The results of analysis of subjective perception involving familiarity are as follows. Figure 8 depicts participants' responses to the different avatars that varied according to two factors (facial habit and facial appearance). The Y-axis indicates the average of subjective Likert ratings. The results showed no significant interaction between Facial Habit × Facial Appearance, $F(1, 163) = 0.004, p > 0.94$. Of particular importance, the results showed that Facial Appearance had a significant main effect, $F(1, 81) = 4.182, p < 0.05$, whereas Facial Habit had no significant effect, $F(1, 81) = 0.966, p > 0.32$.

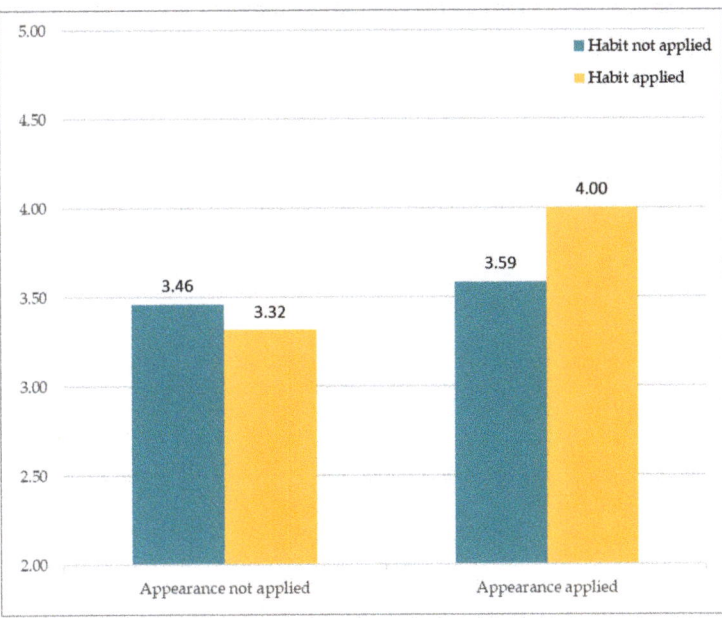

Figure 7. Subjective appraisal of perceived similarity.

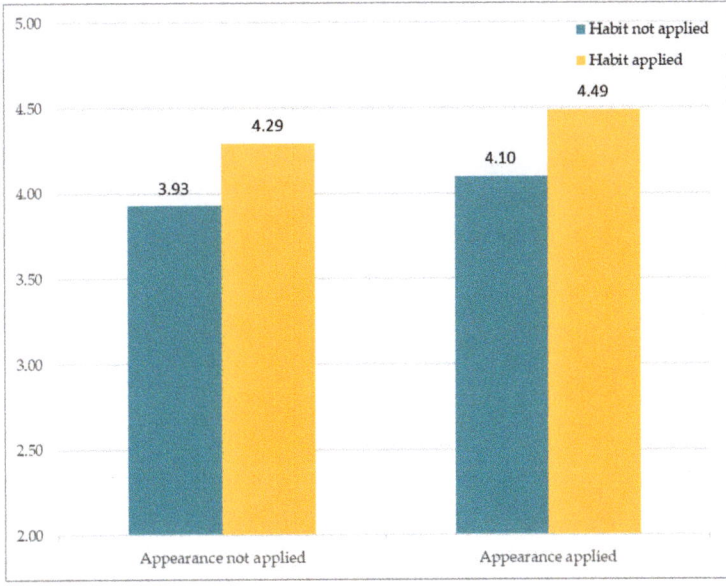

Figure 8. Subjective appraisal of perceived similarity.

3.3. Attraction

The results of the analysis of subjective perception involving attraction are as follows. Figure 9 depicts participants' responses to the different avatars that varied according to two factors (Facial Habit and Facial Appearance). The Y-axis indicates the average of subjective Likert ratings. The results showed no significant interaction between Facial Habit

× Facial Appearance, F(1, 163) = 2.3, $p > 0.13$. Both Facial Appearance, F(1, 81) = 0.047, $p > 0.82$, and Facial Habit, F(1, 81) = 0.631, $p > 0.42$, had no significant main effect.

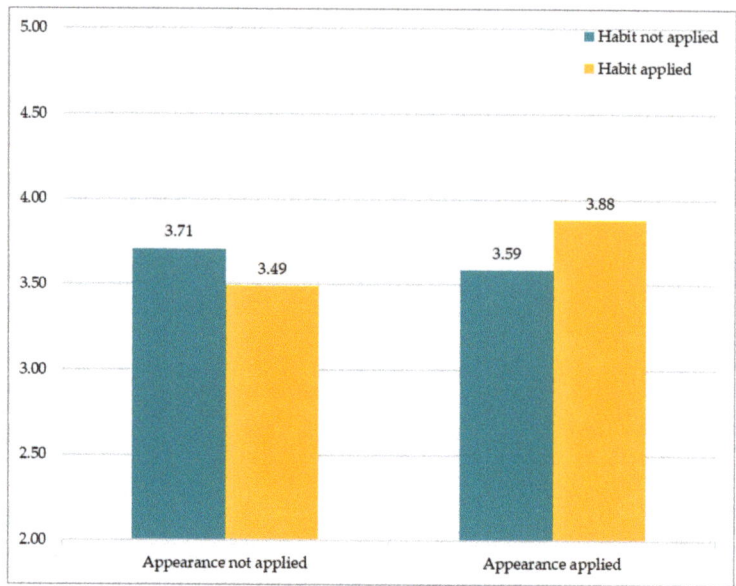

Figure 9. Subjective appraisal of perceived attraction.

3.4. Liking

The results of analysis of subjective perception involving liking are as follows. Figure 10 depicts participants' responses to the different avatars that varied according to two factors (Facial Habit and Facial Appearance). The Y-axis indicates the average of subjective Likert ratings. There was no significant interaction between Facial Habit × Facial Appearance, F(1, 163) = 1.165, $p > 0.28$. Both Facial Appearance, F(1, 81) = 0.004, $p > 0.94$, and Facial Habit, F(1, 81) = 2.133, $p > 0.14$, had no significant main effect.

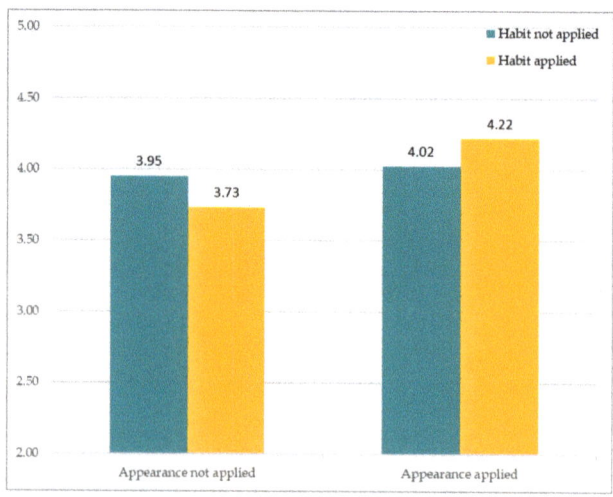

Figure 10. Subjective appraisal of perceived liking.

3.5. Involvement

The results of analysis of subjective perception related to involvement are as follows. Figure 11 depicts participants' responses to the different avatars that varied according to two factors (Facial Habit and Facial Appearance). The Y-axis indicates the average of subjective Likert ratings. The results showed no significant interaction between Facial Habit × Facial Appearance, $F(1, 163) = 0.221$, $p > 0.63$. Both Facial Appearance, $F(1, 81) = 0.055$, $p > 0.81$, and Facial Habit, $F(1, 81) = 0.221$, $p > 0.63$, had no significant main effect.

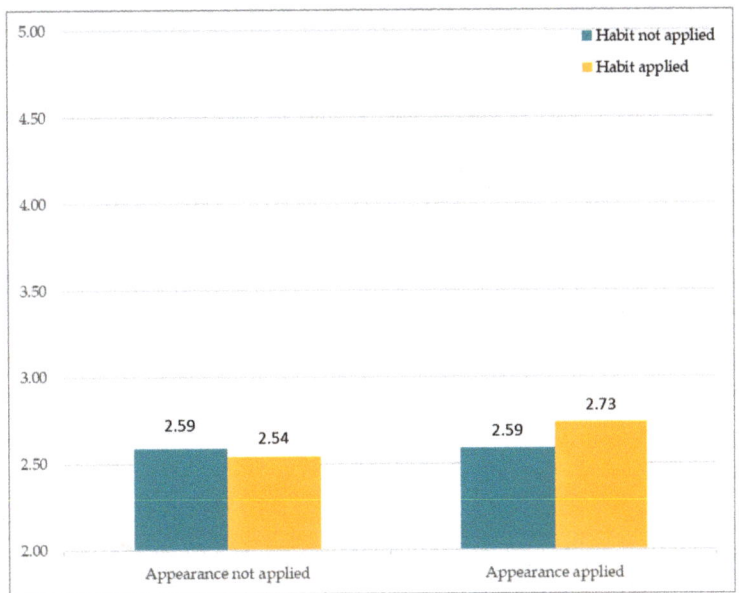

Figure 11. Subjective appraisal of perceived involvement.

3.6. The Correlations between Social Perceptions

We conducted a bivariate correlation analysis to understand the relationship among participant's social perceptions of the virtual avatars (see Table 4). The results show a significant correlation in all pairs of the analysis. The correlation between perceived attraction and liking was the highest ($r = 0.695$, $p < 0.01$) (see Figure 12). The implications of the correlation results will be discussed, integrating results from other analyses.

Table 4. The Pearson correlation coefficients between perceived social constructs ($n = 164$, p *** < 0.01).

	Similarity	Familiarity	Attraction	Liking	Involvement
Similarity		0.425 ***	0.304 ***	0.432 ***	0.376 ***
Familiarity			0.597 ***	0.564 ***	0.500 ***
Attraction				0.659 ***	0.588 ***
Liking					0.499 ***
Involvement					

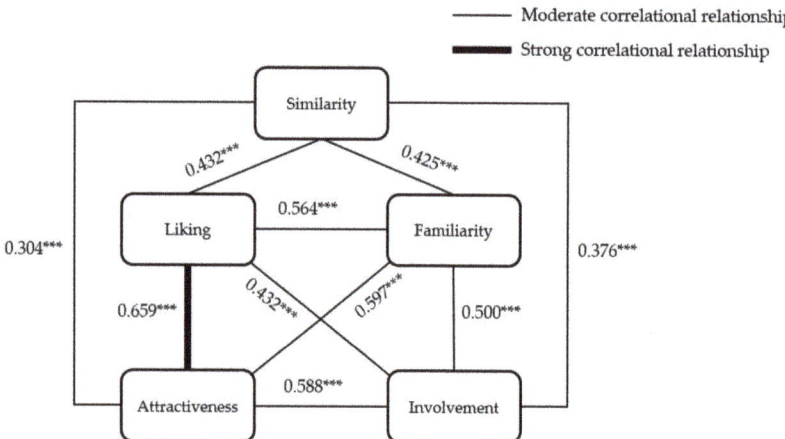

Figure 12. The correlational relationship between social constructs. *** $p < 0.001$

3.7. Data Categorization

Thus far, we identified that facial habit had a main effect on similarity, while facial appearance had a main effect on familiarity. However, these variables had no effects on attraction. As discussed in the operational definitions, attraction is based on a person's liking for the other, and perceived liking in the initial stage of interaction may lead to feelings of attraction [58]. Our results also show that among the constructs, perceived attraction and liking have the highest correlation (r = 0.695, $p < 0.01$).

However, attraction is a much larger and multifaceted construct [63]. Based on the pioneering work by Byrne [64], both perceived similarity and liking lead to attraction, and many researchers have attempted to understand the exact interplay and different weights of the two on attraction [44]. Therefore, we conducted a two-way ANOVA on the sum of perceived liking and similarity (i.e., data categorization) of the four avatar conditions (see Figure 13). The Y-axis indicates the addition of the Likert ratings of perceived liking and similarity.

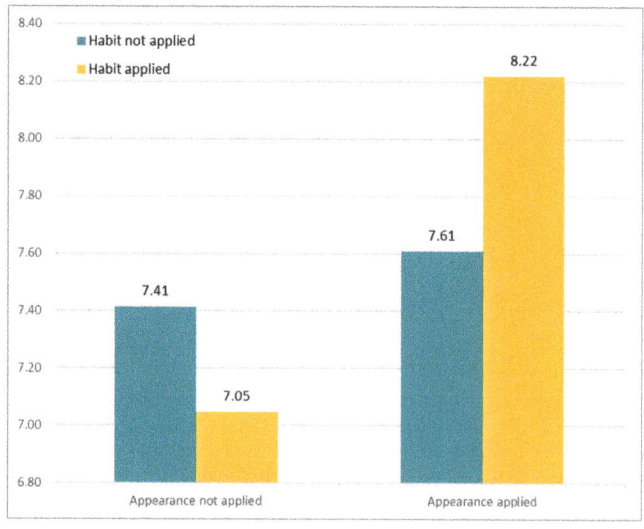

Figure 13. Subjective appraisal of the sum of similarity and liking.

The results showed that Facial Habit had a significant main effect, $F(1, 81) = 4.836$, $p < 0.05$, whereas Facial Appearance had no significant main effect, $F(1, 81) = 0.610, p > 0.69$. Furthermore, there was no significant interaction between Facial Habit × Facial Appearance, $F(1, 163) = 2.467, p > 0.12$.

The research investigated the participant's social perception (similarity, familiarity, attraction, liking, and involvement) of virtual avatars engineered with the participant's unique facial signature (facial appearance, facial habit). In summary, the participants perceived significant similarity to an avatar with habitual expression applied compared to the avatar that did not ($p < 0.05$). In addition, habitual expressions also significantly affected the sum of perceived similarity and perceived liking ($p < 0.05$). The participants perceived familiarity with the avatar with facial appearance applied compared to the avatar that did not ($p < 0.05$).

4. Discussion and Conclusions

To our knowledge, this is the first research to reveal that participants can perceive similarity to a virtual human that had their characteristic facial movements (i.e., habitual pattern), which has significant implications for the design of virtual agents. The virtual human community had long researched the effects of virtual agent realism. The consensus is that behavioral realism is more critical than visual realism in eliciting believability [27]. The suspension of disbelief refers to the deliberate avoidance of critical thinking, whereas a reality check involves deciding what is possible or not in the real world [65,66]. Thus, behavioral realism is more socially engaging and believable than visual realism [27].

In the context of this study, the effect of perceived similarity of a virtual agent to oneself is consistent with research findings on believability. Specifically, while participants did not perceive similarity in virtual avatars to which their facial appearance were applied (i.e., visual realism), they perceived similarity in virtual avatars to which their facial habits were applied (i.e., behavioral realism). This implies that designs may go beyond anthropomorphic design. For example, future research may conduct studies using animal-inspired avatars with facial features (e.g., eyes) and see if participants can perceive similarity to these avatars when their facial movements are applied.

There is much empirical evidence that similarity, as a social construct, elicits attraction [44], and this relationship is regarded as "one of the most robust relationships in all of the behavioral sciences (p. 281)" [67]. Researchers found a positive linear relationship between similarity and attraction (i.e., the law of attraction) [68]. However, the various virtual avatars had no significant effect on attraction. This may be due to interaction being limited to one-time mere exposure. We purposely limited interaction to exclude variables (e.g., perception of personality) that may influence the perceived measures, which may have been brought on by prolonged interaction. Perceived similarity is influenced not only by physical appearance [69] but also attitude [70] and personality [71]. Future studies may add a persona to the virtual avatar to test the complexities of perceived similarity.

The study's limitation in understanding the effects of an individuated avatar on attraction is apparent. Since perceived attraction is a multifaceted construct, it typically requires more interaction, building up from initial liking [58]. Future studies may investigate the degree of attraction as a function of time or when participants interact with the individuated virtual avatar. The perceived relationship also influences attraction; thus, future studies need to address the relationship between the avatar (e.g., companion, butler, assistant) and the participant carefully.

Nevertheless, through data categorization, we found that habitual expressions had a main effect on the sum of perceived similarity and perceived liking ($p < 0.05$). Since the interplay between perceived similarity and liking leads to attraction [64], these results suggest that an individuated avatar may elicit attraction with prolonged interaction.

Additionally, the individualized virtual avatars had no significant effect on perceived involvement. Although we provided the context that the virtual agent would be used as part of a profile for a social networking service, we also acknowledge that many users do not use profiles similar to their appearance. Future studies should cluster the participants based on who use or intend to use avatars with a similar appearance as an alter ego and assess their perceived responses accordingly.

The perceived familiarity with a virtual avatar to which the participant's facial appearance was applied may be due to the participant's repetitive exposure to their reflections in mirrors or still photos of themselves. Repetitive exposure elicits familiarity [13]. On the other hand, people may not be familiar with their habitual expressions during various emotional states.

Finally, the study is limited in that the virtual avatars were designed based on only positive emotional expressions. Future research on individualized virtual avatars should also include negative or complex emotions.

Author Contributions: S.P.: methodology, validation, formal analysis, investigation, writing, review, editing; S.P.K.: conceptualization, methodology, software, validation, formal analysis, investigation, resources, data curation, writing, visualization, project administration; M.W.: conceptualization, methodology, writing, review, supervision, funding acquisition. All authors have read and agreed to the published version of the manuscript.

Funding: This work was supported by the National Research Foundation of Korea (NRF) grant funded by the Korean government (MSIT) (NRF-2020R1A2B5B02002770).

Institutional Review Board Statement: The study was conducted according to the guidelines of the Declaration of Helsinki and approved by the Institutional Review Board of Sangmyung University (protocol code BE2017-20, approved at 22 September 2017).

Informed Consent Statement: Informed consent was obtained from all subjects involved in the study. Written informed consent has been obtained from the subjects to publish this paper.

Conflicts of Interest: The authors declare no conflict of interest.

References

1. Patterson, M.L. Invited article: A parallel process model of nonverbal communication. *J. Nonverbal. Behav.* **1995**, *19*, 3–29. [CrossRef]
2. Ekman, P.; Friesen, W.V. *Emotions Revealed: Recognizing Faces and Feelings to Improve Communication and Emotional*; Calavia Balduz, J.M., López-Palop de Piquer, B., Laita de Roda, P., Eds.; Holt Paper-Back: New York, NY, USA, 2007.
3. Sproull, L.; Subramani, M.; Kiesler, S.; Walker, J.H.; Waters, K. When the interface is a face. *Hum.-Comput. Interact.* **1996**, *11*, 97–124. [CrossRef]
4. Morton, J.; Johnson, M.H. CONSPEC and CONLERN: A two-process theory of infant face recognition. *Psychol. Rev.* **1991**, *98*, 164. [CrossRef]
5. Bond, E.K. Perception of form by the human infant. *Psychol. Bull.* **1972**, *77*, 225. [CrossRef]
6. Diener, E.; Fraser, S.C.; Beaman, A.L.; Kelem, R.T. Effects of deindividuation variables on stealing among Halloween trick-or-treaters. *J. Pers. Soc. Psychol.* **1976**, *33*, 178. [CrossRef]
7. Rhodes, G. Looking at faces: First-order and second-order features as determinants of facial appearance. *Perception* **1988**, *17*, 43–63. [CrossRef] [PubMed]
8. Cohn, J.F.; Schmidt, K.; Gross, R.; Ekman, P. Individual differences in facial expression: Stability over time, relation to self-reported emotion, and ability to inform person identification. In Proceedings of the 4th IEEE International Conference on Multimodal Interfaces, Pittsburgh, PA, USA, 16 October 2002; p. 491.
9. Ekman, P. Facial expression and emotion. *Am. Psychol.* **1993**, *48*, 384. [CrossRef]
10. Patterson, M.L. *Nonverbal Behavior: A Functional Perspective*; Springer: New York, NY, USA, 2012.
11. Fridlund, A.J. *Human Facial Expression: An Evolutionary View*; Academic Press: Cambridge, MA, USA, 2014.
12. Gallup, G.G., Jr.; Anderson, J.R.; Shillito, D.J. The mirror test. *Cogn. Anim. Empir. Theor. Perspect. Anim. Cogn.* **2002**, 325–333. Available online: https://courses.washington.edu/ccab/Gallup%20on%20mirror%20test.pdf (accessed on 3 September 2021).
13. Zajonc, R.B. Attitudinal effects of mere exposure. *J. Pers. Soc. Psychol.* **1968**, *9*, 1. [CrossRef]
14. Saegert, S.; Swap, W.; Zajonc, R.B. Exposure, context, and interpersonal attraction. *J. Pers. Soc. Psychol.* **1973**, *25*, 234. [CrossRef]
15. Swap, W.C. Interpersonal attraction and repeated exposure to rewarders and punishers. *Personal. Soc. Psychol. Bull.* **1977**, *3*, 248–251. [CrossRef]
16. Harrison, A.A.; Tutone, R.M.; McFadgen, D.G. Effects of frequency of exposure of changing and unchanging stimulus pairs on affective ratings. *J. Pers. Soc. Psychol.* **1971**, *20*, 102. [CrossRef]
17. Hamm, N.H.; Baum, M.R.; Nikels, K.W. Effects of race and exposure on judgments of interpersonal favorability. *J. Exp. Soc. Psychol.* **1975**, *11*, 14–24. [CrossRef]

18. Moreland, R.L.; Zajonc, R.B. Exposure effects in person perception: Familiarity, similarity, and attraction. *J. Exp. Soc. Psychol.* **1982**, *18*, 395–415. [CrossRef]
19. Parkinson, B.; Fischer, A.H.; Manstead, A.S.R. *Emotion in Social Relations: Cultural, Group, and Interpersonal Processes*; Psychology Press: London, UK, 2005.
20. Hoffman, M.L. Toward a theory of empathic arousal and development. In *The Development of Affect*; Springer: Berlin/Heidelberg, Germany, 1978; pp. 227–256.
21. Zajonc, R.B.; Adelmann, P.K.; Murphy, S.T.; Niedenthal, P.M. Convergence in the physical appearance of spouses. *Motiv. Emot.* **1987**, *11*, 335–346. [CrossRef]
22. Benjafield, J.; Adams-Webber, J. Assimilative projection and construct balance in the repertory grid. *Br. J. Psychol.* **1975**, *66*, 169–173. [CrossRef]
23. Ross, L.; Greene, D.; House, P. The 'false consensus effect': An egocentric bias in social perception and attribution processes. *J. Exp. Soc. Psychol.* **1977**, *13*, 279–301. [CrossRef]
24. Van Vugt, H.C.; Bailenson, J.N.; Hoorn, J.F.; Konijn, E.A. Effects of facial similarity on user responses to embodied agents. *ACM Trans. Comput. Interact.* **2008**, *17*, 1–27. [CrossRef]
25. Bailenson, J.N.; Yee, N. Digital chameleons: Automatic assimilation of nonverbal gestures in immersive virtual environments. *Psychol. Sci.* **2005**, *16*, 814–819. [CrossRef]
26. Garau, M.; Slater, M.; Vinayagamoorthy, V.; Brogni, A.; Steed, A.; Sasse, M.A. The impact of avatar realism and eye gaze control on perceived quality of communication in a shared immersive virtual environment. In Proceedings of the SIGCHI Conference on Human Factors in Computing Systems, Fort Lauderdale, FL, USA, 5–10 April 2003; pp. 529–536.
27. Bailenson, J.N.; Yee, N.; Merget, D.; Schroeder, R. The effect of behavioral realism and form realism of real-time avatar faces on verbal disclosure, nonverbal disclosure, emotion recognition, and copresence in dyadic interaction. *Presence Teleoperators Virtual Environ.* **2006**, *15*, 359–372. [CrossRef]
28. Mori, M. The uncanny valley: The original essay by masahiro mori. *IEEE Robot.* **2017**. Available online: https://www.semanticscholar.org/paper/The-Uncanny-Valley%3A-The-Original-Essay-by-Masahiro-Mori-MacDorman/243242898b3148b32a31df5f884d4c4f01ea4e61 (accessed on 3 September 2021).
29. Bente, G.; Rüggenberg, S.; Krämer, N.C.; Eschenburg, F. Avatar-mediated networking: Increasing social presence and interpersonal trust in net-based collaborations. *Hum. Commun. Res.* **2008**, *34*, 287–318. [CrossRef]
30. Appel, J.; von der Pütten, A.; Krämer, N.C.; Gratch, J. Does humanity matter? Analyzing the importance of social cues and perceived agency of a computer system for the emergence of social reactions during human-computer interaction. *Adv. Hum.-Comput. Interact.* **2012**, *2012*, 324694. [CrossRef]
31. Kim, H.; Suh, K.-S.; Lee, U.-K. Effects of collaborative online shopping on shopping experience through social and relational perspectives. *Inf. Manag.* **2013**, *50*, 169–180. [CrossRef]
32. Ducheneaut, N.; Wen, M.-H.; Yee, N.; Wadley, G. Body and mind: A study of avatar personalization in three virtual worlds. In Proceedings of the SIGCHI Conference on Human Factors in Computing Systems, Boston, MA, USA, 4–9 April 2009; pp. 1151–1160.
33. Ekman, P. Universal facial expressions of emotions. *Calif. Ment. Health Res. Dig.* **1970**, *8*, 151–158.
34. Liu, Y.; Schmidt, K.L.; Cohn, J.F.; Mitra, S. Facial asymmetry quantification for expression invariant human identification. *Comput. Vis. Image Underst.* **2003**, *91*, 138–159. [CrossRef]
35. Livingstone, M.; Hubel, D. Segregation of form, color, movement, and depth: Anatomy, physiology, and perception. *Science* **1988**, *240*, 740–749. [CrossRef]
36. Bruce, V.; Young, A. Understanding face recognition. *Br. J. Psychol.* **1986**, *77*, 305–327. [CrossRef]
37. Samson, A.C.; Kreibig, S.D.; Soderstrom, B.; Wade, A.A.; Gross, J.J. Eliciting positive, negative and mixed emotional states: A film library for affective scientists. *Cogn. Emot.* **2016**, *30*, 827–856. [CrossRef]
38. Schroff, F.; Kalenichenko, D.; Philbin, J. Facenet: A unified embedding for face recognition and clustering. In Proceedings of the IEEE Conference on Computer Vision and Pattern Recognition, Boston, MA, USA, 7–12 June 2015; pp. 815–823.
39. Ekman, P.; Friesen, W.V. *Manual of the Facial Action Coding System (FACS)*; Consulting Psychologists Press: Palo Alto, CA, USA, 1978.
40. Baltrušaitis, T.; Mahmoud, M.; Robinson, P. Cross-dataset learning and person-specific normalisation for automatic action unit detection. In Proceedings of the 2015 11th IEEE International Conference and Workshops on Automatic Face and Gesture Recognition (FG), Ljubljana, Slovenia, 4–8 May 2015; Volume 6, pp. 1–6.
41. Alvarez, J.; Garcia, M. 2015. Available online: https://www.meryproject.com/ (accessed on 20 August 2021).
42. Unity Real-Time Development Platform. Available online: https://unity.com/ (accessed on 20 August 2021).
43. Folkes, V.S. Forming relationships and the matching hypothesis. *Personal. Soc. Psychol. Bull.* **1982**, *8*, 631–636. [CrossRef]
44. Montoya, R.M.; Horton, R.S.; Kirchner, J. Is actual similarity necessary for attraction? A meta-analysis of actual and perceived similarity. *J. Soc. Pers. Relat.* **2008**, *25*, 889–922. [CrossRef]
45. Condon, J.W.; Crano, W.D. Inferred evaluation and the relation between attitude similarity and interpersonal attraction. *J. Pers. Soc. Psychol.* **1988**, *54*, 789. [CrossRef]
46. Hoyle, R.H. Interpersonal attraction in the absence of explicit attitudinal information. *Soc. Cogn.* **1993**, *11*, 309–320. [CrossRef]

47. Ptacek, J.T.; Dodge, K.L. Coping strategies and relationship satisfaction in couples. *Personal. Soc. Psychol. Bull.* **1995**, *21*, 76–84. [CrossRef]
48. Byrne, D.; Gouaux, C.; Griffitt, W.; Lamberth, J.; Murakawa, N.; Prasad, M.; Prasad, A.; Ramirez, M., III. The ubiquitous relationship: Attitude similarity and attraction: A cross-cultural study. *Hum. Relat.* **1971**, *24*, 201–207. [CrossRef]
49. White, G.L.; Shapiro, D. Don't I know you? Antecedents and social consequences of perceived familiarity. *J. Exp. Soc. Psychol.* **1987**, *23*, 75–92. [CrossRef]
50. Ladwig, P.; Dalrymple, K.E.; Brossard, D.; Scheufele, D.A.; Corley, E.A. Perceived familiarity or factual knowledge? Comparing operationalizations of scientific understanding. *Sci. Public Policy* **2012**, *39*, 761–774. [CrossRef]
51. Noble, A.M.; Klauer, S.G.; Doerzaph, Z.R.; Manser, M.P. Driver training for automated vehicle technology–Knowledge, behaviors, and perceived familiarity. In Proceedings of the Human Factors and Ergonomics Society Annual Meeting, Seattle, WA, USA, 28 October–1 November 2019; Volume 63, pp. 2110–2114.
52. Byrne, D. Attitudes and attraction. In *Advances in Experimental Social Psychology*; Elsevier: Amsterdam, The Netherlands, 1969; Volume 4, pp. 35–89.
53. Byrne, D. An overview (and underview) of research and theory within the attraction paradigm. *J. Soc. Pers. Relat.* **1997**, *14*, 417–431. [CrossRef]
54. Berscheid, E.; Hatfield, E. *Interpersonal Attraction*; Addison-Wesley Reading: Boston, MA, USA, 1969; Volume 69.
55. Newcomb, T.M. *The Acquaintance Process as a Prototype of Human Interaction*; Holt, Rinehart & Winston: New York, NY, USA, 1961.
56. Aron, A.; Lewandowski, G. Psychology of interpersonal attraction. *Int. Encycl. Soc. Behav. Sci.* **2001**, 7860–7862.
57. Huston, T.L. *Foundations of Interpersonal Attraction*; Elsevier: Amsterdam, The Netherlands, 2013.
58. Backman, C.W.; Secord, P.F. The effect of perceived liking on interpersonal attraction. *Hum. Relat.* **1959**, *12*, 379–384. [CrossRef]
59. Slane, S.; Leak, G. Effects of self-perceived nonverbal immediacy behaviors on interpersonal attraction. *J. Psychol.* **1978**, *98*, 241–248. [CrossRef]
60. Davis, M.H. *Empathy: A Social Psychological Approach*; Routledge: London, UK, 2018.
61. De Vignemont, F.; Singer, T. The empathic brain: How, when and why? *Trends Cogn. Sci.* **2006**, *10*, 435–441. [CrossRef] [PubMed]
62. Park, S.; Catrambone, R. Social Responses to Virtual Humans: The Effect of Human-Like Characteristics. *Appl. Sci.* **2021**, *11*, 7214. [CrossRef]
63. Duck, S.E.; Perlman, D.E. *Understanding Personal Relationships: An Interdisciplinary Approach*; Sage Publications, Inc.: Thousand Oaks, CA, USA, 1985.
64. Byrne, D.; Griffitt, W. Similarity versus liking: A clarification. *Psychon. Sci.* **1966**, *6*, 295–296. [CrossRef]
65. Johnston, O.; Thomas, F. *The Illusion of Life: Disney Animation*; Disney Editions: New York, NY, USA, 1981.
66. Bates, J. The role of emotion in believable agents. *Commun. ACM* **1994**, *37*, 122–125. [CrossRef]
67. Berger, C.R. Proactive and retroactive attribution processes in interpersonal communications. *Hum. Commun. Res.* **1975**, *2*, 33–50. [CrossRef]
68. Byrne, D.; Rhamey, R. Magnitude of positive and negative reinforcements as a determinant of attraction. *J. Pers. Soc. Psychol.* **1965**, *2*, 884. [CrossRef] [PubMed]
69. Peterson, J.L.; Miller, C. Physical attractiveness and marriage adjustment in older American couples. *J. Psychol.* **1980**, *105*, 247–252. [CrossRef]
70. Yeong Tan, D.T.; Singh, R. Attitudes and attraction: A developmental study of the similarity-attraction and dissimilarity-repulsion hypotheses. *Personal. Soc. Psychol. Bull.* **1995**, *21*, 975–986. [CrossRef]
71. Banikiotes, P.G.; Neimeyer, G.J. Construct importance and rating similarity as determinants of interpersonal attraction. *Br. J. Soc. Psychol.* **1981**, *20*, 259–263. [CrossRef]

MDPI
St. Alban-Anlage 66
4052 Basel
Switzerland
Tel. +41 61 683 77 34
Fax +41 61 302 89 18
www.mdpi.com

Sensors Editorial Office
E-mail: sensors@mdpi.com
www.mdpi.com/journal/sensors